Study Guide

Solutions to Selected Problems
General, Organic, and Biological Chemistry

SEVENTH EDITION

H. Stephen Stoker
Weber State University

Prepared by

Danny V. White

Joanne A. White

Australia • Brazil • Mexico • Singapore • United Kingdom • United States

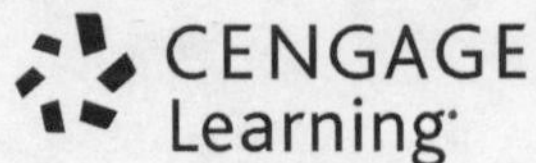

WCN: 01-100-101

ISBN: 978-1-305-08108-6

Cengage Learning
20 Channel Center Street
Boston, MA 02210
USA

Cengage Learning is a leading provider of customized learning solutions with office locations around the globe, including Singapore, the United Kingdom, Australia, Mexico, Brazil, and Japan. Locate your local office at: **www.cengage.com/global**.

Cengage Learning products are represented in Canada by Nelson Education, Ltd.

To learn more about Cengage Learning Solutions, visit **www.cengage.com**.

Purchase any of our products at your local college store or at our preferred online store **www.cengagebrain.com**.

Printed in the United States of America
3 4 5 6 7 22 21 20 19 18

Contents

Preface

If the study of chemistry is new to you, you are about to gain a new perspective on the material world. You will never again look at the objects and substances around you in quite the same way. Knowledge of the invisible structure and organization of matter, the "how and why" of chemical change, will help to demystify many occurrences in the world around you. Chemistry is not an isolated academic study. We use it throughout our lives to appreciate and understand the world and to make responsible choices in that world.

The purpose of this study guide is to help you in your study of the textbook, *General, Organic, and Biological Chemistry*, by H. Stephen Stoker, by providing summaries of the text and additional practice exercises. As you use this Study Guide, we suggest that you follow the steps below.

1. Read the overview for the chapter to get a general idea of the facts and concepts in each chapter.

2. Read the section summaries and work the practice exercises as you come to them. Write out the answers even if you are sure you understand the concepts. This will help you to check your understanding of the material. Refer to the answers at the end of the chapter as soon as you have answered each practice exercise. By checking your answers, you will know whether to review or continue to the next section.

3. When you have finished answering the practice exercises, take the self-test at the end of the chapter. Check your answers with the answer key at the end of the chapter. If there are any questions that you answer incorrectly or do not understand, refer to the chapter section numbers in the answer key and review that material. The Solutions section of this book contains answers to selected problems from the textbook.

Chemistry is a discipline of patterns and rules. Once your mind has begun to understand and accept these patterns, the time you have spent on repetition and review will be well rewarded by a deeper total picture of the world around you. As teachers, we have enjoyed preparing this study guide and hope that it will assist you in your study of chemistry.

Danny and Joanne White

Basic Concepts About Matter — Chapter 1

Chapter Overview

Why is the study of chemistry important to you? Chemistry produces many substances of practical importance to us all: building materials, foods, medicines. For anyone entering one of the life sciences, such as the health sciences, agriculture, or forestry, an understanding of chemistry leads to an understanding of the many life processes.

In this chapter you will be studying some of the fundamental ideas and the language of chemistry. You will characterize three states of matter, differentiate between physical properties and chemical properties, and identify two different types of mixtures. You will describe elements and compounds and practice using symbols and formulas.

Practice Exercises

1.1 **Matter** (Sec. 1.1) is anything that has **mass** (Sec. 1.1) and occupies space. Matter exists in three **physical states** (Sec. 1.2): solid, liquid, and gas. Complete the following table indicating the properties of each of these states of matter.

State	Definite shape?	Definite volume?
solid	yes	
liquid		
gas		

1.2 The **physical properties** (Sec. 1.3) of a substance can be observed without changing the identity of the substance. **Chemical properties** (Sec. 1.3) are observed when a substance changes or resists changing to another substance. Complete the following table:

Property	Physical	Chemical	Insufficient information
A liquid boils at 100°C.			
A solid forms a gas when heated.			
A metallic solid exposed to air forms a white solid.			
Butane is flammable in air.			
Chrlorine is a greenish-yellow gas.			

1.3 A **physical change** (Sec. 1.4) is a change in shape or form but not in composition. A **chemical change** (Sec. 1.4) produces a new substance; that is, the composition is changed.

Classify the following processes as physical or chemical changes by writing the correct word in the second column:

Process	Physical or chemical change
An ice cube melts, producing water.	
A wood block burns, producing ashes and gases.	
Salad oil freezes, producing a solid.	
Sugar dissolves in hot tea.	
A wood block is split into smaller pieces.	
Butter becomes rancid.	

1.4 Matter can be classified as a **pure substance** (Sec. 1.5) or a **mixture** (Sec. 1.5). Mixtures of substances may be either **homogeneous** (Sec. 1.5), one phase, uniform throughout, or **heterogeneous** (Sec. 1.5), visibly different phases (parts). Indicate whether each of the following mixtures is homogeneous or heterogeneous, and give a reason for your choice:

Mixture	Homogeneous or heterogeneous?	Why?
apple juice (water, sugar, fruit juice)		
cornflakes and milk		
fruit salad (sliced bananas, grapes, oranges)		
brass (copper and zinc)		

1.5 Complete the following diagram organizing some terms from this chapter:

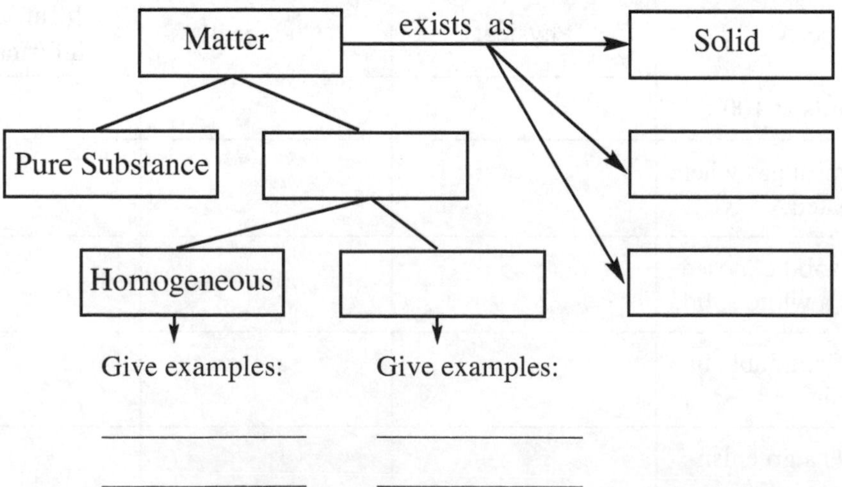

1.6 **Elements** (Sec. 1.6) are **pure substances** (Sec. 1.5) that cannot be broken down into simpler pure substances. **Compounds** (Sec. 1.6) can be broken down into two or more simpler pure substances by chemical means. Complete the following table:

	Substance is an element	Substance is a compound	Insufficient information to make a classification
Substance A reacts violently with water.			
Substance B can be broken into simpler substances by chemical processes.			
Cooling substance C at 350°C turns it from a liquid into a solid.			
Substance D cannot decompose into simpler substances by chemical processes.			

1.7 In the following table, write the **chemical symbol** (Sec. 1.8) or name for each element:

Name	Chemical Symbol
calcium	
copper	
argon	
nickel	
magnesium	

Chemical Symbol	Name
C	
Ne	
Zr	
Pb	
Fe	

1.8 An **atom** (Sec. 1.9) is the smallest particle of an element that retains the identity of that element. A **molecule** (Sec. 1.9) is a tightly bound group of two or more atoms that functions as a unit. The **chemical formula** (Sec. 1.10) for a molecule is made up of the symbol for each element in the molecule and a subscript indicating the number of atoms of the element.

In the table below, indicate whether each of the units is an atom or a molecule, and classify each substance as an element or a compound:

Unit	Atom or molecule?	Element or compound?
N_2		
Zn		
HCN		
Au		
CH_4		

1.9 **Homoatomic molecules** (Sec. 1.9) are made up of atoms of one element; **heteroatomic molecules** (Sec. 1.9) contain atoms of two or more elements.

Indicate whether each of the molecules in the table below is homoatomic or heteroatomic, classify the molecule according to the number of atoms it contains (diatomic, triatomic, etc.), and tell how many atoms of each element are in each molecule.

Molecule	Homoatomic or heteroatomic?	Type of molecule	Number of atoms of each element
NH_3			
O_2			
CO_2			
HNO_3			
HF			

1.10 Write a chemical formula for each of the following compounds based on the information given:

a. A molecule of limonene contains 10 atoms of carbon and 16 atoms of hydrogen.

b. A molecule of nitric acid is pentatomic and contains the elements hydrogen, nitrogen, and oxygen. Each molecule of nitric acid contains only one atom of hydrogen and one of nitrogen.

1.11 a. How many atoms of each type are in one unit of $(NH_4)_2CO_3$?

b. What is the total number of atoms in one unit of $(NH_4)_2CO_3$?

Self-Test

True-false: Indicate whether the following statements are true or false. If the statement is false, give the word or phrase that may be substituted for the underlined portion to make the statement true.

1. Matter is anything that has volume and occupies space.
2. Gases have no definite shape or volume.
3. Liquids take the shape of their container and completely occupy the volume of the container.
4. A mixture of oil and water is an example of a homogeneous mixture.
5. The evaporation of water from salt water is an example of a chemical change.
6. Elements cannot be broken down into simpler pure substances by chemical means.
7. Sugar dissolving in water is an example of a chemical change.

8. A chemical property describes the ability of a substance to undergo change to form a new substance or to resist such change.
9. A mixture is a chemical combination of two or more pure substances.
10. Synthetic (laboratory-produced) elements are all radioactive.
11. The most common (abundant) element in the universe is oxygen.
12. Two-letter chemical symbols are always the first two letters of the element's name.
13. Molecules that are made up of two or more kinds of atoms are homoatomic.
14. The simplest kind of molecule that can exist is a diatomic molecule.
15. A salad dressing made up of vinegar, oil, and dissolved salt is a mixture containing two phases.

Multiple choice:

16. One of the three states of matter is the solid state. A solid has:
    a. definite volume but no definite shape
    b. no definite volume and no definite shape
    c. definite volume and shape
    d. no definite volume but definite shape
    e. none of these combinations of characteristics

17. An example of a homogeneous mixture is:
    a. sand and water b. salt and water c. wood and water
    d. oil and water e. none of these

18. An example of a physical change is:
    a. iron rusting b. sugar dissolving in coffee
    c. gasoline burning in a car engine d. coal burning
    e. none of these

19. An example of a chemical change is:
    a. iron rusting b. coal burning
    c. gasoline burning in a car engine d. a, b, and c are all correct
    e. none of these

20. The chemical symbol for the element iron is:
    a. FE b. Fe c. F d. Ir e. none of these

21. The name for the element Ne is:
    a. neodymium b. neon c. neptunium
    d. nitrogen e. none of these

22. $MgCO_3$ is a compound that is composed of which elements?
    a. magnesium, chlorine, iron b. magnesium, carbon, neon
    c. manganese, carbon, oxygen d. magnesium, carbon, oxygen
    e. none of these

23. The total number of atoms in one molecule of CH_4O is:
    a. 3 b. 4 c. 5 d. 6 e. none of these

24. On the basis of its chemical formula, which of the following substances is an element?
   a. NH_3 b. Cl_2 c. CO_2 d. CO e. none of these

25. Which of the following is the formula for a chemical compound?
   a. Fe b. Fm c. O_2 d. HI e. none of these

26. Which of the following formulas indicates a triatomic molecule?
   a. HCN b. HCl c. HF d. HNO_3 e. none of these

Answers to Practice Exercises

1.1

State	Definite shape?	Definite volume?
solid	yes	yes
liquid	no	yes
gas	no	no

1.2

Property	Physical	Chemical	Insufficient information
A liquid boils at 100 °C.	X (No new substance is formed.)		
A solid forms a gas when heated.			X (Gas may or may not be a new substance.)
A metallic solid exposed to air forms a white solid.		X (A new substance is formed.)	
Butane is flammable in air.		X (Burning produces new substances.)	
Chlorine is a greenish-yellow gas.	X (Describes chlorine; no new substance formed.)		

1.3

Process	Physical or chemical change
An ice cube melts, producing water.	physical
A wood block burns, producing ashes and gases.	chemical
Salad oil freezes, producing a solid.	physical
Sugar dissolves in hot tea.	physical
A wood block is split into smaller pieces.	physical
Butter becomes rancid.	chemical

1.4

Mixture	Homogeneous or heterogeneous?	Why?
apple juice (water, sugar, fruit juice)	homogeneous	uniform throughout
cornflakes and milk	heterogeneous	visibly different parts
fruit salad (sliced bananas, grapes, oranges)	heterogeneous	visibly different parts
Brass (copper and zinc)	homogeneous	uniform throughout

1.5

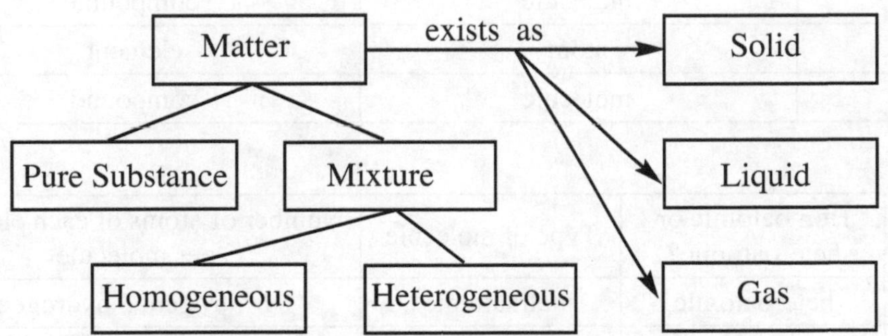

Give examples of homogeneous mixtures:

air

maple syrup

Give examples of heterogeneous mixtures:

smoke

concrete

(These are a few of many possible examples.)

1.6

	Substance is an element	Substance is a compound	Insufficient information to make a classification
Substance A reacts violently with water.			X
Substance B can be broken into simpler substances by chemical processes.		X	
Cooling substance C at 350°C turns it from a liquid into a solid.			X
Substance D cannot decompose into simpler substances by chemical processes.	X		

1.7

Name	Chemical Symbol
calcium	Ca
copper	Cu
argon	Ar
nickel	Ni
magnesium	Mg

Chemical Symbol	Name
C	carbon
Ne	neon
Zr	zirconium
Pb	lead
Fe	iron

1.8

Unit	Atom or molecule?	Element or compound?
N_2	molecule	element
Zn	atom	element
HCN	molecule	compound
Au	atom	element
CH_4	molecule	compound

1.9

Molecule	Homoatomic or heteroatomic?	Type of molecule	Number of atoms of each element per molecule
NH_3	heteroatomic	tetraatomic	1 nitrogen, 3 hydrogen
O_2	homoatomic	diatomic	2 oxygen
CO_2	heteroatomic	triatomic	1 carbon, 2 oxygen
HNO_3	heteroatomic	pentatomic	1 hydrogen, 1 nitrogen, 3 oxygen
HF	heteroatomic	diatomic	1 hydrogen, 1 fluorine

1.10 a. $C_{10}H_{16}$ b. HNO_3

1.11 a. 2 nitrogen atoms, 8 hydrogen atoms, 1 carbon atom, 3 oxygen atoms
b. The total is 14 atoms in one unit of $(NH_4)_2CO_3$.

Answers to Self-Test

The numbers in parentheses refer to sections in your textbook.
1. F; mass (1.1) **2**. T (1.2) **3**. F; gases (1.2) **4**. F; heterogeneous (1.5) **5**. F; physical change (1.4)
6. T (1.6) **7**. F; physical change (1.4) **8**. T (1.3) **9**. F; physical combination (1.5) **10**. T (1.7)
11. F; hydrogen (1.7) **12**. F; often (1.8) **13**. F; heteroatomic (1.9) **14**. T (1.9)
15. T (1.5) **16**. c (1.2) **17**. b (1.5) **18**. b (1.4) **19**. d (1.4) **20**. b (1.8) **21**. b (1.8) **22**. d (1.8)
23. d (1.10) **24**. b (1.10) **25**. d (1.10) **26**. a (1.9)

Measurements in Chemistry — Chapter 2

Chapter Overview

In chemistry the preferred system of measurement is the modern metric (SI) system. In this chapter you will study some of the common units used in measuring length, volume, mass, temperature, and heat in the metric system. You will solve problems involving measurement using the method of dimensional analysis, in which units associated with numbers are used as a guide in setting up the calculations.

You will learn to use scientific notation to express large and small numbers efficiently and will practice using the number of significant figures that corresponds to the accuracy of the measurements being made. You will also use equations to calculate density, to calculate heat loss or gain involved in a temperature change, and to convert temperature from one temperature scale to another.

Practice Exercises

2.1 In the metric system the basic units of length, mass, and volume are the **meter** (m), the **gram** (g), and the **liter** (L) (Sec. 2.2). Other units are derived from these basic units by using prefixes. For example: 1 kilogram = 1000 grams

Complete the following table using information from Table 2.1 of your textbook.

Basic Unit	Prefix	Derived Unit	Value of derived unit in terms of the basic unit
meter	kilo-		
liter	deci-		
gram	milli-		
meter	centi-		
liter	milli-		

2.2 **Exact numbers** (Sec. 2.3) occur in definitions, in counting, and in simple fractions. **Inexact numbers** (Sec. 2.3) are obtained from measurements. Classify the numbers in the following statements as exact or inexact by marking the correct column.

Number	Exact	Inexact
A bag of sugar weighs 5 pounds.		
The temperature was 104°F in the shade.		
There were 107 people in the airplane.		
An octagon has eight sides.		
One meter equals 100 centimeters.		
1/3 is a simple fraction.		
The swimming pool was 25 meters long.		

2.3 **Significant figures** (Sec. 2.4) are the digits in any **measurement** (Sec. 2.1) that are known with certainty plus one digit that is uncertain or estimated. These guidelines will help you in determining the number of significant figures:

1. All nonzero digits are significant. (23.4 m has three significant figures)
2. Zeros in front of nonzero digits (leading zeros) are not significant. (0.00025 has two significant figures)
3. Zeros between nonzero digits (confined zeros) are significant. (2.005 has four significant figures)
4. Zeros at the end of a number (trailing zeros) are significant if a decimal point is present (1.60 has three significant figures) but are not significant if there is no decimal point (500 has one significant figure, 500. has three significant figures).

The position of the last significant figure specifies the measurement's uncertainty. For example, 3.91 has three significant figures. It has an uncertainty in the hundredths place; the magnitude of the uncertainty is ±0.01.

Using the guidelines above, complete the following table.

Number	Number of significant figures	Last significant digit	Position of uncertainty	Magnitude of uncertainty
1570				
45932				
103.045				
0.0340				
298442.0				

2.4 In **rounding off** (Sec. 2.5) a number to a certain number of significant figures:

1. Look at the first digit to be deleted.
2. If that digit is 4 or less, drop that digit and all those to the right of it. If that digit is 5 or more, increase the last significant digit by one.

Example: Round 7.3589 to three significant figures. Because 7.35 has the correct number of significant figures and 8 is greater than 5, increase 7.35 to 7.36.

In the following table, round each number to the given number of significant figures.

Number	Rounded to 3 significant figures	Rounded to 2 significant figures
2763		
0.003628		
65.20		
2.989		
2.22278		
2.2278		

2.5 When multiplying and dividing measurements, the number of significant figures in the answer is the same as the number of significant figures in the measurement that contains the fewest significant figures.

Example: 4.2 m x 3.12 m = 13.104 m^2 = 13 m^2
(4.2 has fewer significant figures than 3.12 does, so the answer has two significant figures.)

Exact numbers (such as three people or twelve eggs in a dozen) do not limit the number of significant figures.

Carry out each mathematical operation in the following table, and round the answer to the correct number of significant figures. Assume there are no exact numbers in any calculation.

Problem	Answer before rounding off	Rounded to correct number of significant figures
160 x 0.32		
482 x 0.00358		
72 ÷ 1.37		
0.0485 ÷ 88.342		

2.6 When measurements are added and subtracted, the answer can have no more digits to the right of the decimal point than the measurement with the least number of decimal places.

Example: 3.58 m + 7.2 m = 10.78 m = 10.8 m

Carry out the mathematical operations indicated below, and round the answer to the correct number of significant figures. Assume there are no exact numbers in any calculation.

Problem	Answer before rounding off	Rounded to correct number of significant figures
153 + 4521		
483 + 0.223		
1.097 + 0.34		
744 – 36		
8093 – 0.566		
0.345 – 0.0221		

2.7 **Scientific notation** (Sec. 2.6) is a convenient way of expressing very large or very small numbers in a compact form. In this notation numbers are written in the form $A \times 10^n$ (a coefficient A multiplied by an exponential term 10^n where n is a whole number.) The coefficient A is a number with a single nonzero digit to the left of the decimal point. To convert a number from decimal notation to scientific notation use the following steps:

1. Write the original number.

2. Move the decimal point to a position just to the right of the first nonzero digit.
3. Count the number of places the decimal point was moved. This number (n) will be the exponent of 10.
4. If the decimal point is moved to the left, the exponent will be positive; if the decimal point is moved to the right, the exponent will be negative.

Examples:

1. $2300 = 2.3 \times 10^3$ (decimal moved 3 places to the left)
2. $43{,}010{,}000 = 4.301 \times 10^7$ (decimal moved 7 places to the left)
3. $0.0072 = 7.2 \times 10^{-3}$ (decimal moved 3 places to the right)

Note that only significant figures become a part of the coefficient (A).

Complete the following tables:

Decimal number	Scientific notation
4378	
783	
8400.0	
0.00362	
0.093200	

Scientific notation	Decimal number
6.389×10^6	
3.34×10^1	
4.55×10^{-3}	
9.08×10^{-5}	
2.0200×10^{-2}	

2.8 When numbers in exponential form (as in scientific notation) are added or subtracted, the exponents of 10 must be the same. Use the correct number of significant figures.

Example:

$(1.53 \text{ x } 10^{-3}) + (7.2 \text{ x } 10^{-4})$ (can be added in this form only with a calculator)

$(1.53 \text{ x } 10^{-3}) + (0.72 \text{ x } 10^{-3}) = 2.25 \text{ x } 10^{-3}$

Carry out the following addition and subtraction problems, and express each answer in scientific notation with the correct number of significant figures.

1. $(4.54 \times 10^4) + (1.804 \times 10^4) =$ ______________________
2. $(8.522 \times 10^{-3}) + (1.3 \times 10^{-4}) =$ ______________________
3. $(5.631 \times 10^4) - (1.52 \times 10^4) =$ ______________________
4. $(2.94 \times 10^{-4}) - (5.866 \times 10^{-5}) =$ ______________________

2.9 When multiplying numbers in exponential form, multiply the coefficients and add the exponents of 10. For example:

1. (two positive exponents) $(1.2 \times 10^3) \times (4.7 \times 10^5) = 5.6 \times 10^8$
2. (negative and positive exponents) $(2.3 \times 10^4) \times (1.8 \times 10^{-8}) = 4.1 \times 10^{-4}$
3. $(5.1 \times 10^{-2}) \times (7.2 \times 10^5) = 37 \times 10^3 = 3.7 \times 10^4$

When dividing numbers in exponential form, divide the coefficients and subtract the exponents of 10.

Examples:

1. $(4.8 \times 10^8) \div (3.4 \times 10^3) = 1.4 \times 10^5$
2. $(6.5 \times 10^4) \div (3.7 \times 10^{-3}) = 1.8 \times 10^7$

Carry out the mathematical operations below. Express answers to the correct number of significant figures:

1. $(4.155 \times 10^3) \times (1.50 \times 10^6) =$ ____________________
2. $(7.36 \times 10^2) \times (4.711 \times 10^3) =$ ____________________
3. $(1.7 \times 10^{-3}) \times (3.363 \times 10^{-5}) =$ ____________________
4. $(4.09 \times 10^6) \div (2.9001 \times 10^3) =$ ____________________
5. $(3.1413 \times 10^4) \div (7.83 \times 10^5) =$ ____________________
6. $(5.204 \times 10^{-6}) \div (4.1 \times 10^3) =$ ____________________
7. $(6.61 \times 10^{-6}) \div (7.278 \times 10^{-3}) =$ ____________________

2.10 A **conversion factor** (2.7) is a ratio that shows how two units of measurement are related to each other. Conversion factors always come in pairs. For example, 1 m = 1000 mm; the two conversion factors for this relationship are:

$$\frac{1\ \text{m}}{10^3\ \text{mm}} \quad \text{or} \quad \frac{10^3\ \text{mm}}{1\ \text{m}}$$

Conversion factors may be defined (within the same system of units) or measured (between different systems of units).

Write the two conversion factors for each of the following relationships:

Relationship	Conversion factors	Defined or measured?
10^3 mL = 1 L		
1 kg = 10^3 g		
1 in = 2.54 cm		

2.11 **Dimensional analysis** (2.8) is a method of problem solving in which a given quantity is multiplied by one or more conversion factors to give an answer with the desired units. If the correct conversion factors are chosen, all of the units will cancel except for the units desired for the answer.

Example: How many millimeters (mm) are in 2.41 meters (m)?

$$2.41\ \text{m} \times \frac{10^3\ \text{mm}}{1\ \text{m}} = (2.41 \times 10^3)\left(\frac{\text{m} \times \text{mm}}{\text{m}}\right) = 2410\ \text{mm}$$

Use the method of dimensional analysis to solve the problems below. Choose one or more conversion factors that will give the answer in the correct units. Use the conversion factors in Table 2.2 of your textbook for conversion between the metric and English systems.

Problem	Conversion factor setup	Answer with units
3.89 km = ? cm	$3.89\ \text{km} \times \dfrac{10^3\ \text{m}}{1\ \text{km}} \times \dfrac{10^2\ \text{cm}}{1\ \text{m}}$	
7.89×10^4 cm = ? m		
0.987 mm = ? km		
4.05×10^{-4} cm = ? km		
5.5×10^5 L = ? mL		
3.6 m = ? in.		
8.2 mL = ? qt		
57 fl oz = ? mL		

2.12 **Density** (Sec. 2.9) is the ratio of the **mass** (Sec. 2.2) of an object to the volume occupied by that object.

$$\text{Density} = \frac{\text{mass}}{\text{volume}}$$

In the space below, use the formula for density to calculate the density of an object that has a mass of 123 g and a volume of 17 cm^3.

2.13 Density can be used as a conversion factor (dimensional analysis) to solve problems involving mass and volume.

Example: What mass will a cube of aluminum have if its volume is 32 cm^3? Aluminum has a density of 2.7 g/cm^3.

We choose the conversion factor that will give the answer (mass) in grams, and set up a relationship involving density.

$$32\ \text{cm}^3 \times \frac{2.7\ \text{g}}{1.0\ \text{cm}^3} = 86\ \text{g}$$

Using density as a conversion factor, solve the following problems. First write the relationship with conversion factors and units. Then solve the equation.

1. Ethanol has a density of 0.789 g/mL. What is the mass of 458 mL of ethanol?

2. Copper has a density of 8.92 g/cm^3. What is the volume of 8.97 kg of copper?

3. What is the mass of a cube of gold (density = 19.3 g/ cm^3) that is 1.00 m on a side?

2.14 The three different temperature scales commonly used in chemistry are Celsius and Kelvin, both part of the metric system, and Fahrenheit, part of the English system. Conversion from one scale to another can be made using the following equations.

$$K = {}^{\circ}C + 273 \quad \text{or} \quad {}^{\circ}C = K - 273$$

$${}^{\circ}F = 9/5({}^{\circ}C) + 32 \quad \text{or} \quad {}^{\circ}C = 5/9({}^{\circ}F - 32)$$

Use the equations above to convert the boiling points of the following compounds to the indicated temperature scales.

Boiling point	Substituted Equation	Temperature
ethyl acetate (nail polish remover) 77.0°C	°F = 9/5(77.0) + 32	°F
toluene (additive in gasoline) 111°C		°F
isopropyl alcohol (rubbing alcohol) 180°F		°C
naphthalene (moth balls) 424°F		°C
propane (fuel for camping stoves) –42°C		K
methane (natural gas) –260°F		K

Self-Test

True-false: Indicate whether the following statements are true or false. If the statement is false, give the word or phrase that may be substituted for the underlined portion to make the statement true.

1. The basic metric unit of volume is the milliliter.
2. The basic unit of mass in the metric system is the kilogram.
3. In scientific notation, a coefficient between 1 and 10 is multiplied by a power of 10.
4. When numbers in exponential form are multiplied or divided, the exponents of 10 must be the same.
5. In rounding numbers, the number of digits is determined by the number of significant figures in the measurement.
6. The density of an object is the ratio of its weight to its volume.
7. The uncertainty of the measurement 4.902 m is in the hundredths place.
8. In the metric system, the prefix micro- means one-thousandth (0.001, 10^{-3}).
9. The number 0.002010 has four significant figures.
10. On the Kelvin temperature scale, all temperature readings are positive.
11. A measure of the total quantity of matter in an object is that object's weight.
12. In determining significant figures, leading zeros are never significant.

Multiple choice:

13. The correct way of expressing 4174 in scientific notation is:

    a. 4.174×10^2 b. 4.174×10^3 c. 4.174×10^4
    d. 4.2×10^3 e. none of these

14. The number 0.005140 should be written in scientific notation as:

    a. 5.140×10^{-2} b. 514×10^{-4} c. 5.14×10^{-3}
    d. 5.140×10^{-3} e. none of these

15. The sum of 5472 plus 1946 would be written in scientific notation as:

    a. 7418 b. 7.418×10^3 c. 7.4×10^3
    d. 7.42×10^3 e. none of these

16. The product of 8311 times 0.01452 would be written in scientific notation as:

    a. 120.7 b. 1.21×10^2 c. 1.207×10^2
    d. 1.2×10^2 e. none of these

17. When 1.487 is rounded to three significant figures, the correct answer is:

    a. 1.480 b. 1.490 c. 1.48 d. 1.49 e. none of these

18. In converting grams to milligrams, the known quantity of grams should be multiplied by which of these conversion factors?

    a. 1 g/1000 mg b. 1000 mg/1 g c. 1 g/1000000 µg
    d. 1000000 µg/1 g e. none of these

19. How many milliliters are in 25.2 kilograms of a liquid that has a density of 0.833 g/mL?

a. 3.03×10^4 mL b. 21.0 mL c. 2.10×10^4 mL
d. 30.3 mL e. none of these

20. A temperature of 112 °C would be equal to what temperature on the Kelvin scale?

a. 110 K b. 112 K c. 273 K
d. 322 K e. 385 K

21. Which of the following measurements has three significant figures?

a. 1.050 b. 2301 mL c. 0.0702 g
d. 16.20 m e. none of these

22. A temperature of 22°C would have which of these values on the Fahrenheit scale?

a. –6.6°F b. 54°F c. 97°F d. 72°F e. none of these

23. A nanogram is

a. 1000 g b. 10^{-3} g c. 10^{-6} g d. 10^{-9} g c. 10^{-12} g

24. What unit would be most appropriate for measuring the mass of a car?

a. meter b. centimeter c. gram d. kilogram e. none of these

Answers to Practice Exercises

2.1

Basic Unit	Prefix	Derived Unit	Value of derived unit in terms of the basic unit
meter	kilo-	kilometer	1 km = 1000 m
liter	deci-	deciliter	1 dL = 0.1 L
gram	milli-	milligram	1 mg = 0.001 g
meter	centi-	centimeter	1 cm = 0.01 m
liter	milli-	milliliter	1 mL = 0.001 L

2.2

Number	Exact	Inexact
A bag of sugar weighs 5 pounds.		X
The temperature was 104°F in the shade.		X
There were 107 people in the airplane.	X	
An octagon has eight sides.	X	
One meter equals 100 centimeters.	X	
1/3 is a simple fraction.	X	
The swimming pool was 25 meters long.		X

2.3

Number	Number of significant figures	Last significant digit	Position of uncertainty	Magnitude of uncertainty
1570	3	7	tens	±10
45932	5	2	ones	±1
103.045	6	5	thousandths	±0.001
0.0340	3	0	ten thousandths	±0.0001
298442.0	7	0	tenths	±0.1

2.4

Number	Rounded to 3 significant figures	Rounded to 2 significant figures
2763	2760	2800
0.003628	0.00363	0.0036
65.20	65.2	65
2.989	2.99	3.0
2.22278	2.22	2.2
2.2278	2.23	2.2

2.5

Problem	Answer before rounding off	Rounded to correct number of significant figures
160 × 0.32	51.2	51
482 × 0.00358	1.72556	1.73
72 ÷ 1.37	52.554744	53
0.0485 ÷ 88.342	0.000549	0.000549

2.6

Problem	Answer before rounding off	Rounded to correct number of significant figures
153 + 4521	4674	4674
483 + 0.223	483.223	483
1.097 + 0.34	1.437	1.44
744 – 36	708	708
8093 – 0.566	8092.434	8092
0.345 – 0.0221	0.3229	0.323

2.7

Decimal number	Scientific notation
4378	4.378×10^3
783	7.83×10^2
8400.0	8.4000×10^3
0.00362	3.62×10^{-3}
0.093200	9.3200×10^{-2}

Scientific notation	Decimal number
6.389×10^6	6,389,000
3.34×10^1	33.4
4.55×10^{-3}	0.00455
9.08×10^{-5}	0.0000908
2.0200×10^{-2}	0.020200

2.8
1. $(4.54 \times 10^4) + (1.804 \times 10^4) = 6.34 \times 10^4$
2. $(8.522 \times 10^{-3}) + (1.3 \times 10^{-4}) = 8.7 \times 10^{-3}$
3. $(5.631 \times 10^4) - (1.52 \times 10^4) = 4.11 \times 10^4$
4. $(2.94 \times 10^{-4}) - (5.866 \times 10^{-5}) = 2.35 \times 10^{-4}$

2.9
1. $(4.155 \times 10^3) \times (1.50 \times 10^6) = 6.23 \times 10^9$
2. $(7.36 \times 10^2) \times (4.711 \times 10^3) = 34.7 \times 10^5 = 3.47 \times 10^6$
3. $(1.7 \times 10^{-3}) \times (3.363 \times 10^{-5}) = 5.7 \times 10^{-8}$
4. $(4.09 \times 10^6) \div (2.9001 \times 10^3) = 1.41 \times 10^3$
5. $(3.1413 \times 10^4) \div (7.83 \times 10^5) = 0.401 \times 10^{-1} = 4.01 \times 10^{-2}$
6. $(5.204 \times 10^{-6}) \div (4.1 \times 10^3) = 1.3 \times 10^{-9}$
7. $(6.61 \times 10^{-6}) \div (7.278 \times 10^{-3}) = 9.08 \times 10^{-4}$

2.10

Relationship	Conversion factors	Defined or measured?
1000 mL = 1 L	$\frac{1\ \text{L}}{10^3\ \text{mL}}$ or $\frac{10^3\ \text{mL}}{1\ \text{L}}$	Defined
1 kg = 1000 g	$\frac{1\ \text{kg}}{10^3\ \text{g}}$ or $\frac{10^3\ \text{g}}{1\ \text{kg}}$	Defined
1 in = 2.54 cm	$\frac{1\ \text{in}}{2.54\ \text{cm}}$ or $\frac{2.54\ \text{cm}}{1\ \text{in}}$	Measured

2.11

Problem	Relationship with units	Answer with units
3.89 km = ? cm	$3.89\ \text{km} \times \frac{10^3\ \text{m}}{1\ \text{km}} \times \frac{10^2\ \text{cm}}{1\ \text{m}}$	3.89×10^5 cm
7.89×10^4 cm = ? m	$7.89 \times 10^4\ \text{cm} \times \frac{1\ \text{m}}{10^2\ \text{cm}}$	7.89×10^2 m
0.987 mm = ? km	$0.987\ \text{mm} \times \frac{1\ \text{m}}{10^3\ \text{mm}} \times \frac{1\ \text{km}}{10^3\ \text{m}}$	9.87×10^{-7} km
4.05×10^{-4} cm = ? km	$4.05 \times 10^{-4}\ \text{cm} \times \frac{1\ \text{m}}{10^2\ \text{cm}} \times \frac{1\ \text{km}}{10^3\ \text{m}}$	4.05×10^{-9} km
5.5×10^5 L = ? mL	$5.5 \times 10^5\ \text{L} \times \frac{10^3\ \text{mL}}{1\ \text{L}}$	5.5×10^8 mL
3.6 m = ? in.	$3.6\ \text{m} \times \frac{39.4\ \text{in}}{1.00\ \text{m}}$	1.4×10^2 in.
8.2 mL = ? qt	$8.2\ \text{ml} \times \frac{1\ \text{L}}{10^3\ \text{mL}} \times \frac{1.00\ \text{qt}}{0.946\ \text{L}}$	8.7×10^{-3} qt
57 fl oz = ? mL	$57\ \text{fl oz} \times \frac{1.00\ \text{mL}}{0.0338\ \text{fl oz}}$	1.7×10^3 mL

2.12 $\text{Density} = \dfrac{\text{mass}}{\text{volume}} = \dfrac{123\text{ g}}{17\text{ cm}^3} = 7.2\text{ g/cm}^3$

2.13 1. Use density as a conversion factor:

$$\text{mass} = \frac{0.789\text{ g}}{1\text{ mL}} \times 458\text{ mL} = 361\text{ g}$$

2. Use two conversion factors to solve this problem.

$$\text{volume} = \frac{1.00\text{ cm}^3}{8.92\text{ g}} \times \frac{10^3\text{ g}}{1\text{ kg}} \times 8.97\text{ kg} = 1.01 \times 10^3\text{ cm}^3$$

3. Use two conversion factors.

$$\text{mass} = \left(1.00\text{ m} \times \frac{10^2\text{ cm}}{1\text{ m}}\right)^3 \times \frac{19.3\text{ g}}{1.00\text{ cm}^3} = 1.00 \times 10^6\text{ cm}^3 \times \frac{19.3\text{ g}}{1.00\text{ cm}^3} = 1.93\text{ g} \times 10^7\text{ g}$$

2.14

Boiling point	Substituted equation	Temperature
ethyl acetate (nail polish remover) 77.0°C	°F = 9/5(77.0) + 32	171°F
toluene (additive in gasoline) 111°C	°F = 9/5(111) + 32	232°F
isopropyl alcohol (rubbing alcohol) 180°F	°C = 5/9(180 – 32)	82°C
naphthalene (moth balls) 424°F	°C = 5/9(424 – 32)	218°C
propane (fuel for camping stoves) –42°C	K = (–42) + 273	231 K
methane (natural gas) –260°F	°C = 5/9(–260–32) = – 162°C K = (–162) + 273	111 K

Answers to Self-Test

The numbers in parentheses refer to sections in your textbook.
1. F; liter (2.2) **2**. F; gram (2.2) **3**. T (2.6) **4**. F; added or subtracted (2.6) **5**. T (2.4)
6. F; its mass to its volume (2.9) **7**. F; thousandths (2.4) **8**. F; milli- (2.2) **9**. T (2.4) **10**. T (2.10)
11. F; mass (2.2) **12**. T (2.4) **13**. b (2.6) **14**. d (2.6) **15**. b (2.6) **16**. c (2.5 and 2.6) **17**. d (2.4)
18. b (2.7) **19**. a (2.9) **20**. e (2.10) **21**. c (2.4) **22**. d (2.10) **23**. d (2.2) **24**. d (2.2)

Atomic Structure and the Periodic Table — Chapter 3

Chapter Overview

All matter is made of basic building blocks called atoms, and these atoms have a substructure. In this chapter you will describe the three basic subatomic particles in terms of mass, charge, and location and calculate the number of each of the three types of particles in an atom using the atomic number and the mass number of that atom. You will learn to describe an isotope from its symbol, and you will calculate the average atomic mass of an element. Using the electron configuration of an element and the principle of the distinguishing electron, you will be able to classify the elements into groups with similar properties.

As you study the structure of atoms, you will develop an understanding of how atoms bond together to form the many substances that make up our world. By studying the periodic law, you will begin to predict the properties of elements according to their positions in the periodic table.

Practice Exercises

3.1 Atoms are made up of even smaller particles called **subatomic particles** (Sec. 3.1). The three types of subatomic particles found in atoms are **electrons, protons,** and **neutrons** (Sec. 3.1). Complete the table below summarizing properties of these subatomic particles and their location within atoms.

Type of particle	Relative mass	Electric charge	Location
electron			
proton			
neutron			

3.2 The **atomic number** (Sec. 3.2) of an **element** (Sec. 3.2) is the number of protons in the **nucleus** (Sec. 3.1) of an atom of that element. Because the net electrical charge on an atom is zero, the number of protons in an atom is equal to the number of electrons. Thus, the atomic number also gives the number of electrons in a neutral atom. The **mass number** (Sec. 3.2) of an atom is the total number of **nucleons** (protons plus neutrons) (Sec. 3.1) in the nucleus. Therefore, the number of neutrons in an atom can be found by subtraction:

Mass number – atomic number = number of neutrons

Use the relationships above and the **periodic table** (Sec. 3.4) to complete the following table:

Atomic number	Mass number	Number of protons	Number of neutrons	Number of electrons	Symbol of element
6			6		
	39	19			
			77		Xe
	64			29	
		35	45		

3.3 **Isotopes** (Sec. 3.3) are atoms of an element that have the same number of protons (*Z*) but different numbers of neutrons and, therefore, different mass numbers (*A*). Isotopes are usually represented by **complete chemical symbol notation** (Sec. 3.3).

$^{A}_{Z}X$ The superscript is the mass number, or *A*.
The subscript is the atomic number, or *Z*.

For example, the isotope carbon-14 is: $^{14}_{6}C$

Complete the following table:

Isotope	*A*	*Z*	Protons	Neutrons	Electrons	Nucleons
$^{40}_{20}Ca$						
$^{40}_{18}Ar$						
$^{23}_{11}Na$						
$^{37}_{17}Cl$						
$^{35}_{17}Cl$						

3.4 The **atomic mass** (Sec. 3.3) of an element is an average mass of the mixture of isotopes that reflects the relative abundance of the isotopes as they occur in nature. The atomic mass can be calculated by multiplying the relative mass of each isotope by its fractional abundance and then totaling the products. For example:

Magnesium is composed of 78.7% $^{24}_{12}Mg$, 10.1% $^{25}_{12}Mg$ and 11.2% $^{26}_{12}Mg$. To find the atomic mass for magnesium, multiply each isotope's percent abundance by its mass, and add these products together:

0.787 x 23.99 amu = 18.9 amu
0.101 x 24.99 amu = 2.52 amu
0.112 x 25.98 amu = 2.91 amu
24.3 amu

An element has two common isotopes: 80.4% of the atoms have a mass of 11.01 amu and 19.6% of the atoms have a mass of 10.01 amu. In the space below, set up the equations and calculate the atomic mass for this element. Identify the element.

Atomic mass =__________ Element: ______________

3.5 According to the **periodic law** (Sec. 3.4), when elements are arranged in order of increasing atomic number, elements with similar chemical properties occur at periodic intervals. The periodic table represents this statement graphically: elements with similar properties are found in the same **group** (Sec. 3.4) or vertical column. The horizontal rows are known as **periods** (Sec. 3.4). A steplike line in the periodic table separates the **metals** (Sec. 3.5) on the left from the **nonmetals** (Sec. 3.5) on the right. Metals typically have the physical properties of luster, thermal conductivity, electrical conductivity and malleability, while nonmetals typically lack these properties.

Refer to your periodic table (Fig. 3.3) for information to complete the table below:

Element	Group	Period	Metal	Nonmetal
Be	IIA	2	X	
Na				
N				
Br				
O				
Sn				
K				

Which two elements in the table above would you expect to have similar chemical properties and why?

3.6 Four groups of elements in the periodic table have common names. Complete the table below to check your knowledge of these groups.

Group	Common Name	Physical and Chemical Properties
IA		
IIA		
	halogens	
		unreactive gases, undergo few chemical reactions

3.7 The main energy levels of the electrons in an atom are the **electron shells** (Sec. 3.6). The electron shells are divided into **electron subshells** (Sec. 3.6), which are in turn divided into **electron orbitals** (Sec. 3.6). An electron shell is indicated by a number (1, 2, 3 …), a subshell by a letter (*s*, *p*, *d*, or *f*). Using information from Section 3.6 of your textbook, determine the maximum number of electrons that can be contained in each of the following:

2s subshell	2 electrons
3s orbital	
1st electron shell	
2p orbital	

5d orbital	
5p subshell	
4d subshell	
2nd electron shell	

3.8 The **electron configuration** (Sec. 3.7) of an atom specifies the number of electrons in each electron subshell of the atom. In the electron configuration, a superscript on the subshell designation (*s*, *p*, *d*, or *f*) indicates the number of electrons contained in that subshell.

Example: The electron configuration for $_6C$ is $1s^2 2s^2 2p^2$. This shows that the atom has 2 electrons in the 1*s* subshell, 2 in the 2*s* subshell, and 2 in the 2*p* subshell.

Write the electron configurations for the elements below. Use Figure 3.10 in your textbook to determine the order in which the electron orbitals are filled.

Element	Electron configuration
neon	
chlorine	
iron	

3.9 An **orbital diagram** (Sec. 3.7) is a notation that shows how many electrons an atom has in each of its occupied electron orbitals. Each arrow in the diagram indicates an electron. Electron spin is denoted by the direction of the arrow.

Example: $_6C$ 1*s* [↓↑] 2*s* [↓↑] 2*p* [↑][↑][ ]

If there is only one electron in an orbital, it is referred to as an **unpaired electron** (Sec. 3.7). In the example, you can see that carbon has two unpaired electrons.

Note that in filling an orbital diagram, electrons are added to the lowest possible energy level first. In a subshell containing more than one orbital (*p*, *d*, or *f*), one electron is added to each orbital in the subshell (unpaired electrons). The next electrons are added to each half-filled orbital, filling each orbital (paired electrons) before moving on to the next subshell.

Draw arrows (indicating electrons) in the orbital diagram for each element below:

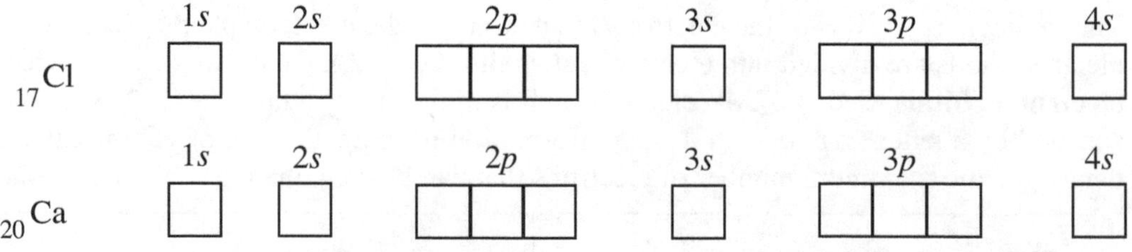

Do either of the orbital diagrams you completed above contain any unpaired electrons?

____

3.10 We can classify an element by determining the subshell of its **distinguishing electron** (Sec. 3.8), the last electron added to the electron configuration when the subshells are filled in order of increasing energy. If the distinguishing electron is added to an *s* or a *p* subshell, the element is a **representative element** (Sec. 3.9), and if the distinguishing electron completes a *p* subshell, the element is a **noble gas** (Sec. 3.9). If the distinguishing electron is added to a *d* subshell, the element is a **transition element** (Sec. 3.9); if it is added to an *f* subshell, the element is an **inner transition element** (Sec. 3.9).

Using Figures 3.12 and 3.13 in your textbook, determine the electron subshell of the distinguishing electron (d.e.) for each element below, and classify the element as a representative element, a noble gas, a transition element, or an inner transition element.

Element	Subshell of d.e.	Classification
Mg	3*s*	representative
Ti		
Ar		
S		

Element	Subshell of d.e.	Classification
Pb		
Xe		
U		
Zr		

Self-Test

True-false: Indicate whether the following statements are true or false. If the statement is false, give the word or phrase that may be substituted for the underlined portion to make the statement true.

1. Of the three basic subatomic particles found in atoms, the <u>nucleus</u> is the smallest.
2. Most of the mass of an atom is located in the <u>nucleus</u>.
3. The nucleus contains <u>electrons and protons</u>.
4. The three naturally occurring isotopes of hydrogen have <u>similar</u> physical properties.
5. The atomic number of an element is the number of <u>protons and neutrons</u>.
6. The mass number is the total number of <u>protons and electrons</u> in the atom.
7. Electron orbitals have different shapes: *s*-orbitals are <u>spherical</u>.
8. The nuclei of different isotopes of a specific element contain different numbers of <u>neutrons</u>.
9. The four dominant or "building block" elements of the human body are all <u>nonmetals</u>.
10. The periodic law states that when elements are arranged in order of <u>increasing atomic number</u>, elements with similar properties occur at periodic intervals.
11. In the modern periodic table, the horizontal rows are called <u>groups</u>.
12. <u>Metals</u> are substances that have a high luster and are malleable.
13. Metals are on the <u>left</u> side of the periodic table.
14. Nonmetals are <u>good</u> conductors of electricity.
15. In atoms of the <u>noble gases</u>, the outermost *s* and *p* subshells of electrons are filled.
16. A transition element is characterized by a distinguishing electron in a <u>*d*-orbital</u>.
17. The two elements that have the atomic numbers 4 and 16 would have <u>different</u> chemical properties.
18. The elements of the Periodic Table are arranged in order of increasing number of <u>neutrons</u>.

19. The sum of the number of neutrons and the number of protons in an atom is called the <u>mass number</u>.

Multiple choice:

20. The nucleus of an atom contains these basic particles:
    a. electrons and protons b. neutrons and electrons c. protons and neutrons
    d. only neutrons e. none of these

21. Isotopes of a specific element vary in the following manner:
    a. Electron numbers are different. b. Neutron numbers are different.
    c. Proton numbers are different. d. Neutron and proton numbers are different.
    e. none of these

22. The element $^{48}_{22}Ti$ has the following electron configuration:
    a. $1s^22s^22p^63s^23p^63d^{10}4s^2$ b. $1s^22s^22p^63s^23p^63d^4$
    c. $1s^22s^22p^63s^23p^64s^23d^2$ d. $1s^22s^22p^63s^23p^64s^23d^4$
    e. none of these

23. In the isotope $^{81}_{35}Br$ how many neutrons are in the nucleus?
    a. 35 b. 46 c. 81 d. 116 e. none of these

24. How many nucleons are there in an atom of cesium, $^{133}_{55}Cs$?
    a. 55 b. 78 c. 133 d. 188 e. none of these

25. The element that has the electron configuration $1s^22s^22p^63s^23p^64s^23d^{10}4p^6$ is:
    a. $_{10}Ne$ b. $_{54}Xe$ c. $_{18}Ar$ d. $_{36}Kr$ e. none of these

26. In the periodic table, the elements on the far left side are classified as:
    a. metals b. noble gases c. transition metals
    d. nonmetals e. none of these

27. In the periodic table, the elements called noble gases are in:
    a. Group IA b. Group IIA c. Group VIIA
    d. Group VA e. none of these

28. The distinguishing electron for $_{19}K$ would be found in what subshell?
    a. 3*s* b. 2*p* c. 3*d* d. 4*s* e. none of these

29. How many unpaired electrons are there in an atom of nitrogen?
    a. 1 b. 2 c. 3 d. 4 e. none of these

Answers to Practice Exercise

3.1

Type of particle	Relative mass	Electric charge	Location
electron	1	–1	outside the nucleus
proton	1837	+1	inside the nucleus
neutron	1839	0	inside the nucleus

3.2

Atomic number	Mass number	Number of protons	Number of neutrons	Number of electrons	Symbol of element
6	12	6	6	6	C
19	39	19	20	19	K
54	131	54	77	54	Xe
29	64	29	35	29	Cu
35	80	35	45	35	Br

3.3

Isotope	*A*	*Z*	Protons	Neutrons	Electrons	Nucleons
${}^{40}_{20}Ca$	40	20	20	20	20	40
${}^{40}_{18}Ar$	40	18	18	22	18	40
${}^{23}_{11}Na$	23	11	11	12	11	23
${}^{37}_{17}Cl$	37	17	17	20	17	37
${}^{35}_{17}Cl$	35	17	17	18	17	35

3.4 0.804 x 11.01 amu = 8.85 amu
0.196 x 10.01 amu = <u>1.96 amu</u>
10.81 amu

Atomic mass = <u>10.81 amu</u> Element: <u>boron</u>

3.5

Element	Group	Period	Metal	Nonmetal
Be	IIA	2	X	
Na	IA	3	X	
N	VA	2		X
Br	VIIA	4		X
O	VIA	2		X
Sn	IVA	5	X	
K	IA	4	X	

We would expect Na and K to have similar chemical properties; they are in the same group of the periodic table.

3.6

Group	Common Name	Physical and Chemical Properties
IA	akali metals	soft, shiny metals, react readily with water
IIA	alkaline earth metals	soft, shiny metals, react moderately with water
VIIA	halogens	reactive, colored elements, F_2 and Cl_2 are gases at room temperature
VIIIA	noble gases	unreactive gases, undergo few chemical reactions

3.7

2s subshell	2 electrons
3s orbital	2 electrons
1st electron shell	2 electrons
2p orbital	2 electrons

5d orbital	2 electrons
5p subshell	6 electrons
4d subshell	10 electrons
2nd electron shell	8 electrons

3.8

Element	Electron configuration
neon	$1s^22s^22p^6$
chlorine	$1s^22s^22p^63s^23p^5$
iron	$1s^22s^22p^63s^23p^64s^23d^6$

3.9

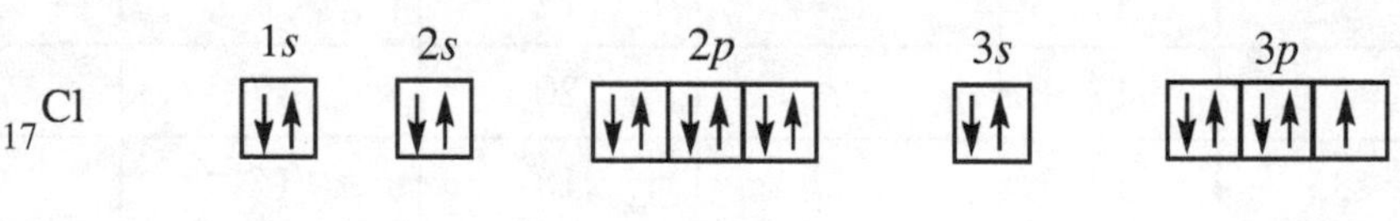

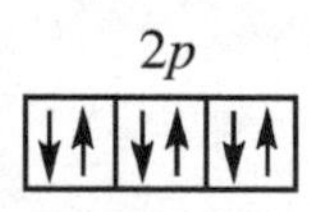

Yes. $_{17}$Cl contains one unpaired 3*p* electron.

3.10

Element	Subshell of d.e.	Classification
Mg	3*s*	representative
Ti	3*d*	transition
Ar	3*p*	noble gas
S	3*p*	representative

Element	Subshell of d.e.	Classification
Pb	6*p*	representative
Xe	5*p*	noble gas
U	5*f*	inner transition
Zr	4*d*	transition

Answers to Self-Test

The numbers in parentheses refer to sections in your textbook.
1. F; electron (3.1) **2**. T (3.1) **3**. F; protons and neutrons (3.1) **4**. F; different (3.3) **5**. F; protons (3.2) **6**. F; protons and neutrons (3.2) **7**. T (3.6) **8**. T (3.3) **9**. T (3.6) **10**. T (3.4) **11**. F; periods (3.4) **12**. T (3.5) **13**. T (3.5) **14**. F; poor (3.5) **15**. T (3.9) **16**. T (3.9) **17**. T (3.4) **18**. F; protons (3.4) **19**. T (3.2) **20**. c (3.1) **21**. b (3.3) **22**. c (3.7) **23**. b (3.3) **24**. c (3.3) **25**. d (3.7) **26**. a (3.5) **27**. e; Group VIIIA (3.9) **28**. d (3.8) **29**. c (3.7)

Chemical Bonding: The Ionic Bond Model — Chapter 4

Chapter Overview

Atoms are rarely found singly or separately. They are almost always found in close association with other atoms, in groups held together by chemical bonds. In this chapter you will see how electrons from one or more atoms are transferred to another atom or atoms to form ionic bonds.

The electron configuration of the atoms of an element determines the chemical properties of that element. You will identify the valence electrons of an atom using the electron configuration of the atom and the element's group number in the periodic table. You will draw the Lewis symbols for atoms and use these symbols to show electron transfer in ionic bond formation. You will predict the chemical formulas for ionic compounds and name these compounds.

Practice Exercises

4.1 **Chemical bonds** (Sec. 4.1) are the binding forces that hold atoms together in more complex units. The two general types of chemical bonds are **ionic bonds** (Sec. 4.1), which involve the transfer of electrons from one or more atoms to another atom or atoms, and **covalent bonds** (Sec. 4.1), which involve the sharing of electrons between atoms. Ionic substances and molecular (covalently-bonded) substances have definite differences in physical properties. Use information in Sec. 4.1 of your textbook to complete the following table:

	Ionic substances	Molecular substances
Formation of chemical bonds by:	electron transfer	
Melting point:		
Electrical conduction:		
Physical state at room temperature:		

4.2 The **valence electrons** (Sec. 4.2) of an atom are the electrons in the outermost electron shell. Write the electron configuration and give the number of valence electrons and the Group number in the periodic table for atoms of each of the following elements:

Element	Electron configuration	Number of valence electrons	Group number
lithium			
beryllium			
boron			
phosphorus			
sulfur			

4.8 To name a compound containing a metal with a variable ionic charge, use a Roman numeral after the metal name to indicate the charge on the metal ion. Give the chemical formulas and the names of the ionic compounds prepared by combining the following ions:

	Cl^-	O^{2-}
K^+		
Pb^{2+}		
Fe^{3+}		
Sn^{4+}		

4.9 Single atoms can lose or gain electrons to form **monatomic ions** (Sec. 4.10). A **polyatomic ion** (Sec. 4.10) is a **covalently bonded** (Sec. 4.1) group of atoms having a charge. (For example: NO_3^-, nitrate ion, is a negative polyatomic ion; H_3O^+, hydronium ion, is a positive polyatomic ion.) In writing the chemical formula for an ionic compound, treat a polyatomic ion as a unit. If more than one of these ions is required for charge balance, enclose the ion in parentheses and put the number of ions outside the parentheses. Give the chemical formula for one formula unit of each ionic compound in the table below:

ion	Br^-	NO_3^-	CO_3^{2}	PO_4^{3-}
Na^+	NaBr			
Ca^{2+}		$Ca(NO_3)_2$		
NH^{4+}				
Al^{+3}				

Self-Test

True-false: Indicate whether the following statements are true or false. If the statement is false, give the word or phrase that may be substituted for the underlined portion to make the statement true.

1. Lewis symbols show the number of inner electrons of an atom.
2. Valence electrons determine the chemical properties of an element.
3. A negative ion is formed when an element loses an electron.
4. Metals tend to gain electrons to attain the configuration of a noble gas.
5. Bromine would accept an electron to attain the configuration of the noble gas krypton.
6. An ionic bond results from the sharing of one or more pairs of electrons between atoms.
7. The maximum number of valence electrons for any element is four.
8. The most stable electron configuration is that of the noble gases.
9. A binary ionic compound is formed from a metal that can donate electrons and a nonmetal that can accept electrons.

10. In naming binary ionic compounds, the full name of the metallic element is given first.
11. In binary ionic compounds, the fixed-charge metals are generally found in Groups VIIA and VIIIA.
12. A polyatomic ion is a group of atoms that is held together by ionic bonds and has acquired a charge.
13. When rock salt (NaCl), an ionic solid, dissolves in water, Na and Cl atoms are produced.
14. An atom of silicon has three valence electrons.
15. In the formation of the ionic compound $MgBr_2$, electrons are lost by bromine atoms.

Multiple choice:

16. The binary ionic compound RbI would be called:

a rubidium(I) iodide b. rubidium iodate c. rubidium iodine
d. rubidium iodide e. none of these

17. The chemical formula for the binary ionic compound silver sulfide is:

a. SiS b. AgS c. Ag_2S d. AgS_2 e. none of these

18. In the electron configuration for sulfur, $1s^22s^22p^63s^23p^4$, what electron shell number determines the valence electrons?

a. 1 b. 2 c. 3 d. 2 and 3 e. none of these

19. Which of these ions would be isoelectronic with Ca^{2+}?

a. K^+ b. Ba^{2+} c. Br^-
d. Al^{3+} e. none of these

20. The electron configuration of a noble gas is:

a. $1s^22s^2$ b. $1s^22s^22p^4$ c. $1s^22s^22p^63s^23p^2$
d. $1s^22s^22p^63s^23p^6$ e. none of these

21. The electron configuration of the ion S^{2-} is:

a. $1s^22s^22p^6$ b. $1s^22s^22p^63s^23p^4$ c. $1s^22s^22p^63s^23p^6$
d. $1s^22s^22p^63s^23p^64s^2$ e. none of these

22. In the compound Na_3N, the total number of electrons accepted by the nitrogen atom is:

a. 1 b. 2 c. 3 d. 4 e. none of these

23. In the ionic compound calcium phosphate, how many polyatomic ions (phosphate ions) are in 1 formula unit?

a. 1 b. 2 c. 3 d. 4 e. none of these

24. The Lewis symbol for a Group VA element would have dots representing the following number of valence electrons:

a. 2 b. 3 c. 4 d. 5 e. none of these

25. Which of the following pairs of elements would form a binary ionic compound?

a. sulfur and oxygen b. bromine and chlorine c. magnesium and bromine
d. oxygen and hydrogen e. none of these

26. At room temperature, an ionic compound would be in which physical state?

    a. gas b. liquid c. solid
    d. gas and liquid e. none of these

27. The smallest whole-number repeating ratio of ions in an ionic compound is called the

    a. molecular formula b. formula unit c. binary compound
    d. ionic formula e. polyatomic structure

28. What is the formula for the ionic compound that forms between ammonium ions and sulfide ions?

    a. NH_4S_2 b. $(NH_4)S_2$ c. NH_4S
    d. $(NH_4)_2S_2$ e. none of these

29. In the formation of CaI_2, how many electrons are transferred?

    a. 4 b. 2 c. 0 d. 1 e. none of these

Answers to Practice Exercises

4.1

	Ionic substances	Molecular substances
Formation of chemical bonds	electron transfer	electron sharing
Melting point	high	low
Electrical conduction	good	poor
Physical state	solids	gases, liquids, low-melting solids

4.2

Element	Electron configuration	Number of valence electrons	Group number
lithium	$1s^22s^1$	1	IA
beryllium	$1s^22s^2$	2	IIA
boron	$1s^22s^22p^1$	3	IIIA
phosphorus	$1s^22s^22p^63s^23p^3$	5	VA
sulfur	$1s^22s^22p^63s^23p^4$	6	VIA

4.3

Group	Valence electrons	Lewis symbol
Group IA	1	X·
Group IVA	4	·Ẋ· (with one dot below)
Group VIIA	7	:Ẋ: (with two dots below)

Element	Valence electrons	Lewis symbol
sulfur	6	:Ṡ: (with one dot below)
bromine	7	·Br: (with two dots above and two below)
magnesium	2	·Mg·

4.4

Element	Group number	Electrons lost/gained	Ion formed	Noble gas
Na	IA	1 lost	Na^{+}	Ne
Br	VIIA	1 gained	Br^{-}	Kr
S	VIA	2 gained	S^{2-}	Ar
Ca	IIA	2 lost	Ca^{2+}	Ar
N	VA	3 gained	N^{3-}	Ne

4.5

Element	Electrons lost	Symbol for ion
tin	2	Sn^{2+}
tin	4	Sn^{4+}
cobalt	2	Co^{2+}

Element	Electrons lost	Symbol for ion
cobalt	3	Co^{3+}
iron	2	Fe^{2+}
iron	3	Fe^{3+}

4.6

Lewis symbol	Lewis symbol	Formation of ionic compound (Lewis structures)
K•	•Br:	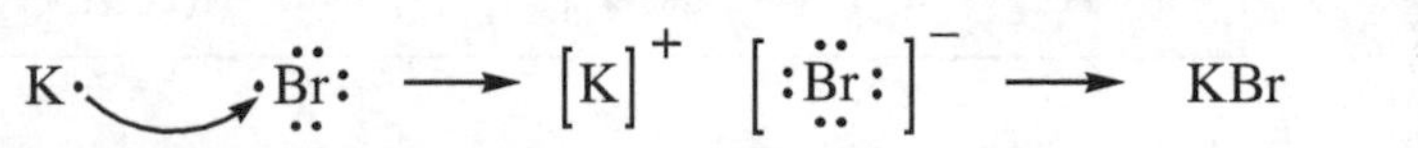 K• + •Br: → $[K]^{+}$ $[:Br:]^{-}$ → KBr
•Ca•	•I:	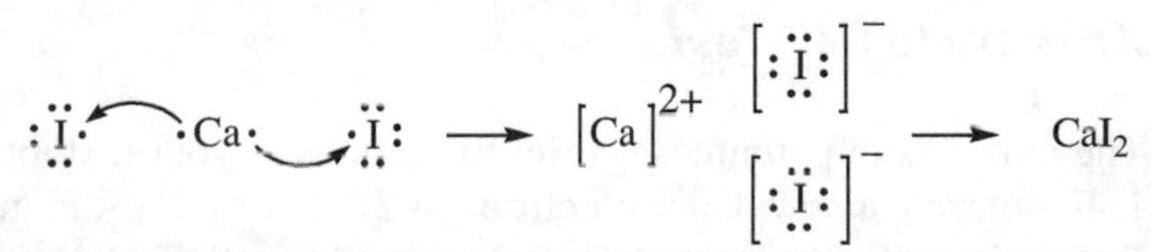 :I• + •Ca• + •I: → $[Ca]^{2+}$ $[:I:]^{-}$ $[:I:]^{-}$ → CaI_2
Sr:	•S:	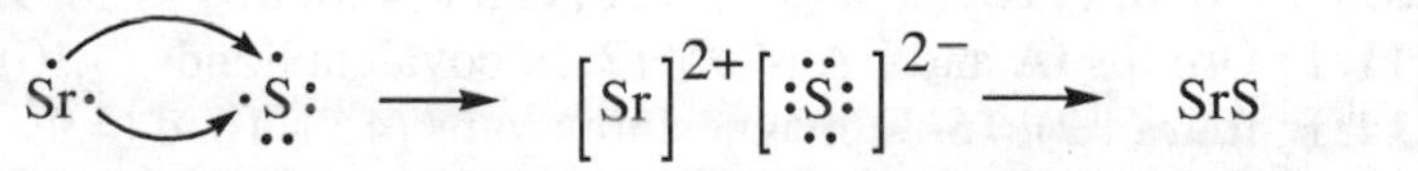 Sr: + •S: → $[Sr]^{2+}$ $[:S:]^{2-}$ → SrS

4.7

Elements	Ions formed	Formula unit	Name of ionic compound
potassium and chlorine	K^{+}, Cl^{-}	KCl	potassium chloride
beryllium and iodine	Be^{2+}, I^{-}	BeI_2	beryllium iodide
sodium and sulfur	Na^{+}, S^{2-}	Na_2S	sodium sulfide
strontium and oxygen	Sr^{2+}, O^{2-}	SrO	strontium oxide
aluminum and fluorine	Al^{3+}, F^{-}	AlF_3	aluminum fluoride
cesium and bromine	Cs^{+}, Br^{-}	CsBr	cesium bromide
calcium and oxygen	Ca^{2+}, O^{2-}	CaO	calcium oxide
aluminum and sulfur	Al^{3+}, S^{2-}	Al_2S_3	aluminum sulfide

4.8

	Cl^-	O^{2-}
K^+	KCl potassium chloride	K_2O potassium oxide
Pb^{2+}	$PbCl_2$ lead(II) chloride	PbO lead(II) oxide
Fe^{3+}	$FeCl_3$ iron(III) chloride	Fe_2O_3 iron(III) oxide
Sn^{4+}	$SnCl_4$ tin(IV) chloride	SnO_2 tin(IV) oxide

4.9

ion	Br^-	NO_3^-	CO_3^{2}	PO_4^{3-}
Na^+	$NaBr$	$NaNO_3$	Na_2CO_3	Na_3PO_4
Ca^{2+}	$CaBr_2$	$Ca(NO_3)_2$	$CaCO_3$	$Ca_3(PO_4)_2$
NH_4^+	NH_4Br	NH_4NO_3	$(NH_4)_2CO_3$	$(NH_4)_3PO_4$
Al^{+3}	$AlBr_3$	$Al(NO_3)_3$	$Al_2(CO_3)_3$	$AlPO_4$

Answers to Self-Test

The numbers in parentheses refer to sections in your textbook.
1. F; outermost or valence electrons (4.2) **2**. T (4.2) **3**. F; gains (4.4) **4**. F; lose (4.5) **5**. T (4.5) **6**. F; covalent (4.1) **7**. F; eight (4.5) **8**. T (4.3) **9**. T (4.9) **10**. T (4.9) **11**. F; Groups IA and IIA (4.9) **12**. F; covalent bonds (4.10) **13**. F; Na^+ and Cl^- ions (4.8) **14**. F; four (4.2) **15**. F; magnesium atoms (4.7) **16**. d (4.9) **17**. c (4.9) **18**. c (4.2) **19**. a (4.5) **20**. d (4.3) **21**. c (4.5) **22**. c (4.7) **23**. b (4.11) **24**. d (4.2) **25**. c (4.5, 4.9) **26**. c (4.1) **27**. b (4.8) **28**. e; $(NH_4)_2S$ (4.11) **29**. b (4.6)

Chemical Bonding: The Covalent Bond Model

Chapter 5

Chapter Overview

As we saw in Chapter 4, ionic bonds are formed by the transfer of electrons. Covalent bonds are the result of the sharing of electrons between atoms. Covalent bonds join nonmetallic atoms together to form molecules.

In this chapter you will use Lewis structures to indicate the various types of covalent bonds in molecules. You will study the concept of electronegativity differences between atoms and how this determines whether a bond is ionic or covalent, polar or nonpolar. You will use VSEPR theory to predict the three-dimensional shape of molecules and determine molecular polarity.

Practice Exercises

5.1 Remember that the valence electrons of an atom (the electrons in the atom's outermost shell) are the electrons involved in bond formation. Draw Lewis symbols showing valence electrons for the following elements:

·Mg·				
Mg	C	P	Br	Ar

5.2 A **covalent bond** (Sec. 5.1) is a chemical bond resulting from two nuclei attracting the same shared electrons, the **bonding electrons** (Sec. 5.2). **Nonbonding electrons** (Sec. 5.2) are pairs of electrons that are not shared. Molecules tend to be stable when each atom in the molecule shares in an octet of electrons. (For hydrogen, an "octet" is only two electrons.)

Draw the Lewis structures for the molecules below. Circle the bonding electrons in each **single covalent bond** (Sec. 5.3) in the molecule:

H(:)Br: (with lone pairs above and below Br)				
HBr	F_2	BrI	H_2O	CH_4

5.3 Shared electron pairs can also be represented by dashes or bond lines. Redraw the Lewis structures for the molecules above using dashes to represent the covalent bonds and pairs of dots to represent the nonbonding electrons.

HBr	F_2	BrI	H_2O	CH_4

Calculate the electronegativity difference between each pair of atoms, and indicate whether the bond formed between them will be ionic, nonpolar covalent, or polar covalent. (Electronegativities are given in Figure 5.11 of your textbook.) Indicate the direction of bond polarity using both arrow and delta notations.

Pair of atoms	Electronegativity difference	Type of bond	Arrow notation	Delta notation
Na and F				
Br and Br				
S and O				
P and Br				
Mg and Br				
N and Cl				

5.10 **Molecular polarity** (Sec. 5.11) is a measure of the total electron distribution over a molecule, rather than over just one bond. A molecule whose bonds are polar may be a **polar molecule** or a **nonpolar molecule** (Sec. 5.11), depending on its molecular geometry. Individual bond polarities may cancel one another in a highly symmetrical molecule, resulting in a nonpolar molecule.

Classify the molecules in Practice Exercise 5.8 as polar molecules or nonpolar molecules.

Molecular formula	Molecular geometry	Number of polar bonds	Polar or nonpolar molecule?

5.11 Go back to Practice Exercise 5.2. Below each of the Lewis structures that you drew in this exercise, write whether the molecule is polar or nonpolar. Next to each polar molecule, indicate for each bond (using δ^- or δ^+) the approximate positions of the partial negative and partial positive charges (the polarity of the bonds).

5.12 **Binary molecular compounds** (Sec. 5.12) are made up of two nonmetallic elements. In naming binary molecular compounds, name the element of lower electronegativity first, followed by the stem of the more electronegative nonmetal and the suffix *-ide*. Include prefixes to indicate the number of atoms of each nonmetal. See Table 5.1 in your textbook for numerical prefixes. Name the following molecular compounds:

a. $SiCl_4$ ____________________

b. CO_2 ____________________

c. PI_3 ____________________

d. N_2O_4 ____________________

Self-Test

True-false: Indicate whether the following statements are true or false. If the statement is false, give the word or phrase that may be substituted for the underlined portion to make the statement true.

1. Carbon dioxide is a polar molecule.
2. Covalent bond formation between nonmetal atoms involves electron transfer.
3. Nonbonding electrons are pairs of valence electrons that are not shared between atoms having a covalent bond.
4. A nitrogen molecule, N_2, would have a double covalent bond between the two nitrogen atoms.
5. Carbon can form multiple covalent bonds with other nonmetallic elements.
6. A coordinate covalent bond is a covalent bond formed when both electrons of a shared pair are donated by one atom.
7. According to VSEPR, the electron groups in the valence shell arrange themselves to maximize the repulsion between the electron groups.
8. According to VSEPR, a water molecule, H_2O, would have a linear arrangement of the valence electron groups.
9. According to VSEPR theory convention, a double bond counts as one electron group.
10. Electronegativity is a measure of the relative repulsion that an atom has for the shared electrons in a bond.
11. Electronegativity values increase from left to right across periods in the periodic table.
12. The bond between fluorine and bromine would be a nonpolar covalent bond.
13. Nonbonding electron pairs are important in determining the shape of a molecule.
14. The number of electrons represented by a bond line is 2.
15. Iodine is the most electronegative element in the periodic table.

Multiple choice:

16. Which of the following pairs of atoms would form a covalent bond?

a. sulfur and oxygen b. potassium and iodine c. magnesium and bromine
d. calcium and fluorine e. none of these

17. Which of the following pairs of atoms would form a nonpolar covalent bond?

a. nitrogen and oxygen b. fluorine and fluorine c. calcium and iodine
d. potassium and bromine e. none of these

5.9

Pair of atoms	Electronegativity difference	Type of bond	Arrow notation	Delta notation
Na and F	3.1	ionic	$\overset{+\!\!\longrightarrow}{\text{Na—F}}$	δ^+ Na—F δ^-
Br and Br	0.0	nonpolar covalent	none	none
S and O	1.0	polar covalent	$\overset{+\!\!\longrightarrow}{\text{S—O}}$	δ^+ S—O δ^-
P and Br	0.7	polar covalent	$\overset{+\!\!\longrightarrow}{\text{P—Br}}$	δ^+ P—Br δ^-
Mg and Br	1.6	ionic	$\overset{+\!\!\longrightarrow}{\text{Mg—Br}}$	δ^+ Mg—Br δ^-
N and Cl	0.0	nonpolar covalent	none	none

5.10

Molecular formula	Molecular geometry	Number of polar bonds	Polar or nonpolar molecule?
CBr_4	tetrahedral	4	nonpolar
CH_2O	trigonal planar	3	polar
CS_2	linear	2	nonpolar
H_2S	angular	2	polar
NCl_3	trigonal pyramidal	3	polar

5.12

δ^+ H:Br: δ^-	:F:F:	δ^- :Br:I: δ^+	δ^+ H:O: (H below, δ^+; δ^- above)	H:C:H with H above and below
polar	nonpolar	polar	polar	nonpolar

5.13 a. $SiCl_4$ silicon tetrachloride

b. CO_2 carbon dioxide

c. PI_3 phosphorus triiodide

d. N_2O_4 dinitrogen tetroxide

Answers to Self-Test

The numbers in parentheses refer to sections in your textbook.
1. F; nonpolar (5.11) **2**. F; sharing (5.1) **3**. T (5.2) **4**. F; triple (5.3) **5**. T (5.4) **6**. T (5.5)
7. F; minimize (5.8) **8**. F; an angular (5.8) **9**. T (5.8) **10**. F; attraction (5.9) **11**. T (5.9)
12. F; a polar covalent (5.10) **13**. T (5.8) **14**. T (5.2) **15**. F; fluorine (5.9) **16**. a (5.9) **17**. b (5.10)
18. d (5.2) **19**. a (5.10) **20**. d (5.2) **21**. a (5.4) **22**. d (5.8) **23**. d (5.8) **24**. e (5.9) **25**. c (5.12)

Chemical Calculations: Formula Masses, Moles, and Chemical Equations — Chapter 6

Chapter Overview

Calculation of the ratios and masses of the substances involved in chemical reactions is very important in many chemical processes. Central to these calculations is the concept of the mole, a convenient counting unit for atoms and molecules.

In this chapter you will learn to determine the formula mass of a substance and the number of moles of a substance. You will practice writing and balancing chemical equations, and you will learn to use these equations in determining the amounts of substances that react and are produced in chemical reactions.

Practice Exercises

6.1 The **formula mass** (Sec. 6.1) of a compound is the sum of the atomic masses, expressed in atomic mass units (amu), of the atoms in the chemical formula of the substance. Calculate the formula mass, rounded to the hundredths place, for each of the compounds below. Atomic masses are given on the inside front cover of your textbook.

Formula	Calculations with atomic masses	Formula mass
KBr		
$CaCl_2$		
Na_2CO_3		
$(NH_4)_3PO_4$		

6.2 The **mole** (Sec. 6.2 and 6.3) is a useful unit for counting numbers of atoms and molecules. The number of particles in a mole is 6.02×10^{23}, which is known as **Avogadro's number** (Sec. 6.2). Use the definition of Avogadro's number as a conversion factor in the calculations below:

Moles	Quantity with conversion factor	Number of atoms
1.00 mole He atoms		
2.60 moles Na atoms		
0.316 mole Ar atoms		

6.3 The mass of 1 mole, the **molar mass** (Sec. 6.3), of any substance is its formula mass (amu) expressed in grams. Answer the questions below to check your understanding of molar mass.

1. What is the mass in amu of one carbon atom? ______________________________
2. What is the molar mass of carbon? ______________________________
3. How many atoms of carbon are in 1 molar mass of carbon? ________________
4. What is the mass of three moles of carbon atoms? ______________________
5. How many carbon atoms are in 36.03 g of carbon? ______________________

6.4 Use dimensional analysis and a conversion factor derived from the definition of molar mass (either moles/gram or grams/mole) to find either the mass or the number of moles for the quantities below:

Given quantity	Relationship with conversion factor	Calculated
1.00 mole H_2O		g H_2O
2.53 moles H_2O		g H_2O
0.519 mole H_2O		g H_2O
1.00 g NaBr		mole NaBr
417 g NaBr		mole NaBr
0.322 g NaBr		mole NaBr

6.5 The subscripts in a chemical formula show the number of atoms of each element per formula unit. They also show the number of moles of atoms of each element in one mole of the substance (atoms/molecule = moles of atoms/moles of molecules). Determine the moles of carbon atoms in the following problems:

Moles of compound	Carbon atoms/molecule	Moles of carbon atoms
1.00 mole $C_{10}H_{16}$ (limonene)		
13.5 moles C_2H_6O (ethanol)		
0.705 mole C_5H_{12} (pentane)		

6.6 In solving problems involving chemical formulas, the number of grams of one substance cannot be compared directly to the number of grams of another substance; however, moles can be related to moles quite easily by looking at subscripts in chemical formulas. Use the map in Fig. 6.9 of your textbook to set up conversion factors, and use dimensional analysis to solve the problems below:

a. How many moles of KCl would be found in 0.125 g of KCl?

b. How many molecules of KCl would be found in 0.125 g of KCl?

c. Calculate the number of grams of Cl in 12.5 g of KCl.

d. Calculate the number of fluorine atoms in 1.77 g of AlF_3.

e. Calculate the number of grams of F in 1.77 g of AlF_3.

6.7 The description of a chemical reaction can be expressed efficiently with the chemical formulas and symbols of a **chemical equation** (Sec. 6.6). The substances that react (the reactants) are placed on the left, and those that are produced (products) are on the right. The arrow in the chemical equation is read as "to produce." Plus signs on the left side mean "reacts with," and plus signs on the right are read as "and."

Write chemical equations for the following chemical reactions:

a. Hydrogen chloride reacts with sodium hydroxide to produce sodium chloride and water.

b. Silver nitrate and potassium bromide react with one another to produce silver bromide and potassium nitrate.

6.8 To be useful in chemical calculations, chemical equations must be **balanced** (Sec. 6.6); that is, the number of atoms of each element must be the same on both sides of the chemical equation. A suggested method for balancing chemical equations is in Section 6.6 of your textbook. Remember: use the **equation coefficients** (Sec. 6.6) to balance chemical equations, but do not change the subscripts within the chemical formulas.

Balance the following chemical equations:

a. $H_2 + Cl_2 \rightarrow HCl$

b. $AgNO_3 + H_2S \rightarrow Ag_2S + HNO_3$

c. $P + O_2 \rightarrow P_2O_3$

d. $HCl + Ba(OH)_2 \rightarrow BaCl_2 + H_2O$

e. $H_2O_2 \rightarrow O_2 + H_2O$

6.9 Balanced chemical equations tell us how much product we can expect from a given amount of reactant, or how much reactant to use for a specific amount of product. The chemical equation gives the ratio of the numbers of atoms and molecules involved in the chemical reaction; it also gives the ratio of the numbers of moles of the substances involved.

Since the ratios of moles of reactants and products are known from the coefficients of the balanced equation, these mole ratios can be used as conversion factors. Consider the balanced chemical equation:

$$4Na + O_2 \rightarrow 2Na_2O$$

a. How many moles of Na_2O can be produced from 4.00 moles of Na?

b. How many moles of Na are required to produce 1.00 mole of Na_2O?

Use the mole diagram in Figure 6.11 of your textbook to determine the sequence of steps and the conversion factors to use in solving the following problems. Show the setup with conversion factors, and solve for the unknown quantity in each of the problems.

c. How many moles of Na_2O could be produced from 2.18 g of sodium?

d. How many grams of Na_2O could be produced from 5.15 g of Na?

6.10 Balance the chemical equation: $Al + Cl_2 \rightarrow AlCl_3$

a. How many moles of chlorine gas will react with 0.160 mole of aluminum?

b. The **theoretical yield** (Sec. 6.9) for a chemical reaction is the maximum amount of a product that can be obtained from given amounts of reactants. Calculate the theoretical yield, in grams, of aluminum chloride that could be produced from 5.27 moles of aluminum. (Assume that enough Cl_2 is present.)

c. What is the theoretical yield, in grams, of aluminum chloride that could be produced from 14.0 g of chlorine gas?

d. The **actual yield** (Sec. 6.9) is the amount of product obtained experimentally from a chemical reaction. It must be measured rather than calculated.

The **percent yield** (Sec. 6.9) is the ratio of the actual (experimental) yield to the theoretical (calculated) yield, multiplied by 100 (to give percent).

$$\text{percent yield} = \frac{\text{actual yield}}{\text{theoretical yield}} \times 100$$

If the actual yield of $AlCl_3$ in part c is 12.1 g, what is the percent yield?

Self-Test

True-false: Indicate whether the following statements are true or false. If the statement is false, give the word or phrase that may be substituted for the underlined portion to make the statement true.

1. The mass of 1 mole of helium atoms is <u>the same as</u> the mass of 1 mole of gold atoms.
2. In a balanced chemical equation, the total number of atoms on the reactant side is <u>equal to</u> the total number of atoms on the product side.
3. Formula masses are calculated on the <u>$^{16}_{8}O$ relative-mass</u> scale.
4. In balancing a chemical equation, do not change the <u>coefficients</u> within the formulas.
5. Atomic mass and formula mass are both expressed in <u>amu</u>.

6. In a chemical equation, the reactants are the materials to the right of the arrow.
7. The number of atoms in a mole of H_2O is equal to 6.02×10^{23}.
8. One mole of glucose ($C_6H_{12}O_6$) contains 6 moles of carbon atoms.
9. In a chemical equation, the products are the materials that are consumed.
10. A mole of nitrogen gas (N_2) contains 6.02×10^{23} nitrogen atoms.
11. One mole of Cl_2 molecules equals Avogadro's number of chlorine atoms.
12. The mass of one molar mass of carbon is 12.01 amu.
13. In a balanced chemical equation, the sum of the coefficients on each side of the equation must be equal to one another.
14. The actual yield of a chemical reaction is the maximum amount of product that can be obtained from given amounts of reactants.

Multiple choice:

15. How many atoms are contained in 6.8 moles of calcium?

    a. 4.1×10^{24} b. 1.6×10^{26} c. 6.2×10^{22} d. 3.3×10^{23} e. none of these

16. What is the formula mass for iron(III) carbonate?

    a. 115.86 amu b. 287.57 amu c. 291.73 amu d. 171.71 amu e. none of these

17. What is the weight in grams of 4.72 moles of $NaHCO_3$?

    a. 84.1 b. 283 c. 264 d. 396 e. none of these

18. A mole of butane contains 4 moles of carbon atoms and 10 moles of hydrogen atoms. Its formula is:

    a. C_6H_6 b. H_6C_{10} c. C_4H_{10} d. H_4C_6 e. none of these

19. What is the total number of moles of all atoms in 6.55 moles of $(NH_4)_2CO_3$?

    a. 52.4 b. 91.7 c. 111 d. 157 e. none of these

20. The conversion factor used in changing grams of O_2 to moles of O_2 is:

    a. 16.00 g/1 mole b. 1 mole/16.00 g c. 32.00 mole/1 g
    d. 1 mole/32.00 g e. none of these

21. When oxygen gas and hydrogen gas combine to form water, which of the following is true? (Hint: Write the balanced chemical equation.)

    a. 2 moles of O_2 produce 1 mole of H_2O
    b. 2 moles of H_2 react with 1 mole of H_2O
    c. 1 mole of O_2 produces 1 mole of H_2O
    d. 2 moles of H_2 produce 2 moles of H_2O
    e. none of these

22. Using the chemical equation you wrote in Question 21, find the mass of water in grams that would be produced by the complete reaction of 4.00 g of oxygen gas.

    a. 4.50 g b. 36.0 g c. 7.32 g d. 14.7 g e. none of these

23. The mass of 0.560 mole of methanol, CH_4O, would equal:

    a. 32.0 amu b. 32.0 g c. 17.9 amu d. 17.9 g e. none of these

6.7 a. $HCl + NaOH \rightarrow NaCl + H_2O$ b. $AgNO_3 + KBr \rightarrow AgBr + KNO_3$

6.8 a. $H_2 + Cl_2 \rightarrow 2HCl$ b. $2AgNO_3 + H_2S \rightarrow Ag_2S + 2HNO_3$

c. $4P + 3O_2 \rightarrow 2P_2O_3$ d. $2HCl + Ba(OH)_2 \rightarrow BaCl_2 + 2H_2O$

e. $2H_2O_2 \rightarrow O_2 + H_2O$

6.9 Balanced equation: $4Na + O_2 \rightarrow 2Na_2O$

a. $4.00 \text{ moles Na} \times \dfrac{2 \text{ moles } Na_2O}{4 \text{ moles Na}} = 2.00 \text{ moles } Na_2O$

b. $1.00 \text{ mole } Na_2O \times \dfrac{4 \text{ moles Na}}{2 \text{ moles } Na_2O} = 2.00 \text{ moles Na}$

c. $2.18 \text{ g Na} \times \dfrac{1 \text{ mole Na}}{22.99 \text{ g Na}} \times \dfrac{2 \text{ moles } Na_2O}{4 \text{ moles Na}} = 0.0474 \text{ mole } Na_2O$

d. $5.15 \text{ g Na} \times \dfrac{1 \text{ mole Na}}{22.99 \text{ g Na}} \times \dfrac{2 \text{ moles } Na_2O}{4 \text{ moles Na}} \times \dfrac{61.98 \text{ g } Na_2O}{1 \text{ mole } Na_2O} = 6.94 \text{ g } Na_2O$

6.10 $2Al + 3Cl_2 \rightarrow 2AlCl_3$ (molar mass of $AlCl_3$: 133.33 g/mole)

a. $0.160 \text{ mole Al} \times \dfrac{3 \text{ moles } Cl_2}{2 \text{ moles Al}} = 0.240 \text{ mole } Cl_2$

b. $5.27 \text{ moles Al} \times \dfrac{2 \text{ moles } AlCl_3}{2 \text{ moles Al}} \times \dfrac{133.33 \text{ g } AlCl_3}{1 \text{ mole } AlCl_3} = 703 \text{ g } AlCl_3$

c. $14.0 \text{ g } Cl_2 \times \dfrac{1 \text{ mole } Cl_2}{70.90 \text{ g } Cl_2} \times \dfrac{2 \text{ moles } AlCl_3}{3 \text{ moles } Cl_2} \times \dfrac{133.33 \text{ g } AlCl_3}{1 \text{ mole } AlCl_3} = 17.6 \text{ g } AlCl_3$

d. $\text{percent yield} = \dfrac{\text{actual yield}}{\text{theoretical yield}} \times 100 = \dfrac{12.1 \text{ g}}{17.6 \text{ g}} \times 100 = 68.8\%$

Answers to Self-Test

1. F; less than (6.3) **2**. T (6.6) **3**. F; $^{12}_{6}C$ relative mass (6.1) **4**. F; subscripts (6.6)
5. T (6.1) **6**. F; products (6.6) **7**. F; molecules (6.4) **8**. T (6.4) **9**. F; produced (6.6)
10. F; molecules (6.4) **11**. F; two times Avogadro's number (6.2) **12**. F; 12.01 g (6.3)
13. F; total number of atoms (6.6) **14**. F; theoretical yield (6.9) **15**. a (6.4) **16**. c (6.1)
17. d (6.4) **18**. c (6.4) **19**. b (6.4) **20**. d (6.5) **21**. d (6.6) **22**. a (6.8) **23**. d (6.3) **24**. b (6.3)
25. d (6.6) **26**. d (6.8) **27**. a (6.9) **28**. b (6.9) **29**. c (6.9)

Gases, Liquids, and Solids — Chapter 7

Chapter Overview

The physical states of matter and the behavior of matter in these states are determined by the behavior of the particles (atoms, molecules, ions) of which matter is made. The movements and interactions of these particles are described by the kinetic molecular theory of matter.

In this chapter you will study the five core concepts of the kinetic molecular theory and the ways in which these concepts explain the physical behavior of matter. You will use the gas laws to describe quantitatively various changes in the conditions of pressure, temperature, and volume of matter in the gaseous state. You will study three types of **intermolecular forces** (Sec. 7.13) that affect liquids and solids and their changes of state.

Practice Exercises

7.1 According to the **kinetic molecular theory of matter** (Sec. 7.1), the differing physical properties of the three states of matter, **solids, liquids, and gases** (Sec. 7.2), are determined by the **potential energy** (cohesive forces) and the **kinetic energy** (disruptive forces) between particles of each state (Sec. 7.1).

The properties of density, **compressibility**, and **thermal expansion** differ for the three states of matter according to which forces (cohesive or disruptive) are dominant in a given state. In the table below, identify the states of matter that have the given characteristics.

Characteristic	State(s)
Low compressibility	
Large volume increase with temperature	
High density	
Dominant type of energy is kinetic	
Cohesive forces dominant over disruptive forces	
Definite volume	
Potential energy is dominant over kinetic energy	
Small thermal expansion	
Particles in constant motion	

7.2 In describing the quantitative behavior of gases using relationships called gas laws (Sec. 7.3), four variables are involved: pressure (*P*), temperature (*T*), volume (*V*), and amount (*n*). In the table below, state the units commonly used in describing these variables:

Temperature	Pressure	Volume	Amount

7.3 Gases can be described by quantitative relationships called **gas laws** (Sec. 7.3). According to **Boyle's law** (Sec. 7.4), the volume of a fixed amount of gas is inversely proportional to the **pressure** (Sec. 7.3) of the gas if the temperature is constant. The mathematical expression of Boyle's law is: $P_1 \times V_1 = P_2 \times V_2$

Set up and complete the following problems using the mathematical expression of Boyle's law. The subscripts refer to the initial conditions (1) and the new conditions (2).

a. The pressure on 2.45 L of helium is changed from 2340 mm Hg to 3580 mm Hg at 50.5°C. What is the new volume?

b. The pressure on 12.5 L of nitrogen gas is doubled from 1.00 atm to 2.00 atm, and the temperature is held constant. What is the new volume of the nitrogen gas?

c. The volume of 8.24 L of gas at 3630 mm Hg is increased to 16.4 L. If the temperature is held constant, what is the new pressure?

7.4 According to **Charles's law** (Sec. 7.5), the volume of a fixed amount of gas at constant pressure is directly proportional to the Kelvin temperature of the gas:

$V_1/T_1 = V_2/T_2$

Set up and complete the following problems using Charles's law.

a. The temperature of 4.71 L of gas is reduced from 278°C to 122°C. If the pressure remains constant, what is the new volume of the gas?

b. The volume of 14.5 L of a gas at 345 K is increased to 20.5 L, and pressure is held constant. What is the new temperature of the gas?

7.5 Boyle's law and Charles's law can be expressed as a single equation called the **combined gas law** (Sec. 7.6):

$$\frac{P_1 \times V_1}{T_1} = \frac{P_2 \times V_2}{T_2}$$

Use the combined gas law to solve the following problems.

a. The volume of a fixed amount of gas is 5.72 L at 30°C and 1.25 atm. If the gas is heated to 50°C and compressed to a volume of 4.50 L, what will be the new pressure?

b. A fixed amount of gas at 514 K and 338 mm Hg is heated to 311°C and 507 mm Hg. What will be the final volume of the gas, if the initial volume is 14.2 L?

7.6 The relationship between the four gas variables at a specific set of conditions is described by the **ideal gas law** (Sec. 7.7): $PV = nRT$

The value of R (the ideal gas constant) is 0.0821 atm • L/mole • K.

To solve the problems below, rearrange the ideal gas law to solve for the unknown variable.

a. What is the volume of 1.49 moles of helium with a pressure of 1.21 atm at 224°C?

b. What is the temperature of neon gas, when 0.339 mole of neon gas is in a 5.72-liter tank and the pressure gauge reads 2.53 atm?

7.7 **Dalton's law of partial pressures** (Sec. 7.8) states that the total pressure exerted by a mixture of gases is the sum of the **partial pressures** (Sec. 7.8) of the individual gases:

$$P_T = P_1 + P_2 + P_3 + \cdots$$

Using Dalton's law of partial pressures, complete the following problems:

a. What is the total pressure exerted by a mixture of helium and argon? The partial pressures of helium and argon are $P_{He} = 270$ mm Hg and $P_{Ar} = 400$ mm Hg.

b. What is the partial pressure of oxygen gas in a mixture of O_2, CO_2 ($P_{CO_2} = 341$ mm Hg), and CO ($P_{CO} = 114$ mm Hg) if the total pressure of the mixture is 744 mm Hg?

7.8 Use the following table summarizing the gas laws to review your knowledge of this chapter:

Law	Quantities held constant	Variables	Equation
Boyle's law			
			$\frac{V_1}{T_1} = \frac{V_2}{T_2}$
Combined gas law			
Ideal gas law	$R = 0.0821$ atm • L/mole • K		
Dalton's law of partial pressures			

7.9 A **change of state** (Sec. 7.9) is a process in which matter changes from one state to another. If heat is absorbed during the process, the change is **endothermic** (Sec. 7.9); if heat is released during the process, the change is **exothermic** (Sec. 7.9).

a. Complete the following table with the correct term for the physical change involved.
b. Write "endothermic" or "exothermic" for each physical change.

	To solid	To liquid	To gas
From solid			
From liquid			
From gas			

7.10 **Hydrogen bonds** (Sec. 7.13) are very strong **dipole-dipole interactions** (Sec. 7.13) between molecules that occur when hydrogen is chemically bonded to fluorine, oxygen, or nitrogen. The hydrogen atom in this case is almost a "bare" nucleus and has a strong δ+ charge. It is strongly attracted to the δ– charge of a pair of electrons on a small electronegative atom (F, O, or N) of another molecule.

Complete the following table by indicating for each substance whether hydrogen bonding can occur between individual molecules or with a water molecule, and explain briefly.

Molecule	Hydrogen bonding between molecules?	Hydrogen bonding with water molecules?
HF		
HI		
NH_3		
CH_4		
CO		
CH_3CH_2OH		
CH_3–O–CH_3		

Self-Test

True-false: Indicate whether the following statements are true or false. If the statement is false, give the word or phrase that may be substituted for the underlined portion to make the statement true.

1. Boyle's law states that for a given mass of gas at constant temperature, the volume of the gas varies directly with pressure.
2. Charles's law states that for a given mass of gas at constant pressure, the volume of the gas varies directly with temperature.

3. Gases cool when they are compressed.
4. Assuming that the temperature and number of moles of gas remain constant, doubling the volume of a gas will double the pressure of the gas.
5. One atmosphere is the pressure required to support 760 mm of Hg.
6. The total pressure exerted by a mixture of gases is equal to the sum of the partial pressures of the gases.
7. The vapor pressure of a liquid decreases as temperature increases.
8. As temperature increases, the kinetic energy of molecules in a liquid decreases.
9. The energy resulting from the attractions and repulsions between charged particles in matter is a part of that matter's potential energy.
10. Liquids are very compressible because there is a lot of empty space between particles.
11. Foods cook faster in a pressure cooker because the boiling point of water is lower than it is at normal atmospheric pressure.
12. A volatile liquid is one that has a high vapor pressure.
13. A state of equilibrium may exist between a liquid and its vapor in a closed container.
14. The weakest type of intermolecular force is the London Force.
15. The temperature at which a liquid's vapor pressure is equal to atmospheric pressure is the liquid's boiling point.

Multiple choice:

16. A fixed amount of oxygen gas with a volume of 5.00 L at 1 atm and 273 K was heated to 402 K and the pressure was doubled. What was the new volume of the oxygen gas?

    a. 3.68 L b. 5.00 L c. 1.71 L d. 14.7 L e. none of these

17. The pressure of a gas at 300 K is 2.00 atm. What is its pressure in torr?

    a. 760 torr b. 380 torr c. 3040 torr d. 1520 torr e. none of these

18. Two gases, nitrogen and oxygen, in the same container, have a total pressure of 600 mm Hg. If the partial pressure of oxygen equals the partial pressure of nitrogen, what is the partial pressure of oxygen?

    a. 600 mm Hg b. 400 mm Hg c. 300 mm Hg
    d. 200 mm Hg e. none of these

19. How many moles of helium are in a 28.4 L balloon at 45°C and 1.03 atm?

    a. 0.271 mole b. 0.0453 mole c. 6.98 moles
    d. 1.12 moles e. none of these

20. The strongest intermolecular forces between water molecules are:

    a. ionic bonds b. covalent bonds c. hydrogen bonds
    d. London forces e. none of these

21. Hydrogen bonding would *not* occur between two molecules of which of these compounds?

    a. HF b. CH_4 c. CH_3NH_2
    d. H_2O e. CH_3OH

22. London forces would be the strongest attractive forces between two molecules of which of these substances?
    a. HF b. BrCl c. F_2
    d. H_2O e. none of these

23. Compared to liquids that have no hydrogen bonding, liquids that have significant hydrogen bonding have a:
    a. higher vapor pressure b. higher boiling point
    c. lower condensation temperature d. greater tendency to evaporate
    e. none of these

24. The pressure on 526 mL of gas is increased from 755 mm Hg to 974 mm Hg. If the temperature remains constant, what is the new volume?
    a. 408 mL b. 633 mL c. 215 mL
    d. 387 mL e. none of these

25. Which of the following changes is endothermic?
    a. condensation b. freezing c. sublimation
    d. deposition e. none of these

26. What will increase the pressure of a gas in a closed container?
    a. decreasing the temperature of the gas
    b. adding more gas to the container
    c. increasing the volume of the container
    d. replacing the gas with the same number of moles of a different gas
    e. both b and c

Answers to Practice Exercises

7.1

Characteristic	State(s)
Low compressibility	solid and liquid
Large volume increase with temperature	gas
High density	solid and liquid
Dominant type of energy is kinetic	gas
Cohesive forces dominant over disruptive forces	solid
Definite volume	solid and liquid
Potential energy is dominant over kinetic energy	solid
Small thermal expansion	solid and liquid
Particles in constant motion	liquid and gas

7.2

Temperature	Pressure	Volume	Amount
Kelvin (K)	mm of Hg (torr) atmospheres (atm)	liters (L) milliliters (mL)	moles

7.3 a. $V_2 = \frac{P_1 \times V_1}{P_2} = \frac{2340 \text{ mm Hg} \times 2.45 \text{ L}}{3580 \text{ mm Hg}} = 1.60 \text{ L}$

b. $V_2 = \frac{P_1 \times V_1}{P_2} = \frac{1.00 \text{ atm} \times 12.5 \text{ L}}{2.00 \text{ atm}} = 6.25 \text{ L}$

c. $P_2 = \frac{P_1 \times V_1}{V_2} = \frac{3630 \text{ mm Hg} \times 8.24 \text{ L}}{16.4 \text{ L}} = 1820 \text{ mm Hg}$

7.4 a. $V_2 = \frac{V_1 \times T_2}{T_1} = \frac{4.71 \text{ L} \times 395 \text{ K}}{551 \text{ K}} = 3.38 \text{ L}$

b. $T_2 = \frac{V_2 \times T_1}{V_1} = \frac{20.5 \text{ L} \times 345 \text{ K}}{14.5 \text{ L}} = 488 \text{ K}$

7.5 a. $P_2 = \frac{P_1 \times V_1 \times T_2}{T_1 \times V_2} = \frac{1.25 \text{ atm} \times 5.72 \text{ L} \times 323 \text{ K}}{4.50 \text{ L} \times 303 \text{ K}} = 1.69 \text{ atm}$

b. $V_2 = \frac{P_1 \times V_1 \times T_2}{P_2 \times T_1} = \frac{338 \text{ mm Hg} \times 14.2 \text{ L} \times 584 \text{ K}}{507 \text{ mm Hg} \times 514 \text{ K}} = 10.8 \text{ L}$

7.6 a. $V = \frac{n \times R \times T}{P} = \frac{1.49 \text{ moles} \times 0.0821 \text{ atm L/mole K} \times 497 \text{ K}}{1.21 \text{ atm}} = 50.2 \text{ L}$

b. $T = \frac{P \times V}{n \times R} = \frac{2.53 \text{ atm} \times 5.72 \text{ L}}{0.339 \text{ mole} \times 0.0821 \text{ atm L/mole K}} = 5.20 \times 10^2 \text{ K}$

7.7 a. $P_{\text{total}} = P_{\text{He}} + P_{\text{Ar}} = 270 \text{ mm Hg} + 400 \text{ mm Hg} = 670 \text{ mm Hg}$

b. $P_{\text{total}} = P_{O_2} + P_{CO_2} + P_{CO}$

$P_{O_2} = P_{\text{total}} - P_{CO_2} - P_{CO} = 744 \text{ mm Hg} - 341 \text{ mm Hg} - 114 \text{ mm Hg} = 289 \text{ mm Hg}$

7.8

Law	Quantities held constant	Variables	Equation
Boyle's law	T, n	P, V	$P_1V_1 = P_2V_2$
Charles's law	P, n	V, T	$\frac{V_1}{T_1} = \frac{V_2}{T_2}$
Combined gas law	n	P, V, T	$\frac{P_1 \times V_1}{T_1} = \frac{P_2 \times V_2}{T_2}$
Ideal gas law	$R = 0.0821$ atm • L/mole • K	P, V, T, n	$PV = nRT$
Dalton's law of partial pressures	V, T, n	P_i, P_T	$P_T = P_1 + P_2 + P_3 + \cdots$

7.9

	To solid	To liquid	To gas
From solid		melting; endothermic	sublimation; endothermic
From liquid	freezing; exothermic		evaporation; endothermic
From gas	deposition; exothermic	condensation; exothermic	

7.10

Molecule	Hydrogen bonding between molecules?	Hydrogen bonding with water molecules?
HF	Yes. H is bonded to F.	Yes. H is bonded to O (H_2O) and F (HF); both molecules are very polar.
HI	No. H is not bonded to F, O, or N.	No. The HI dipole is not strong enough to attract the polar H_2O molecule.
NH_3	Yes. H is bonded to N.	Yes. H is bonded to O (H_2O) and N (NH_3); both molecules are very polar.
CH_4	No. H is not bonded to F, O, or N.	No. CH_4 has no nonbonding electron pairs and is a nonpolar molecule.
CO	No. H is not bonded to F, O, or N.	Yes. The H (H_2O) is attracted to nonbonding electron pair of O (CO).
CH_3CH_2OH	Yes. H is bonded to O.	Yes. H is bonded to O in both molecules.
CH_3–O–CH_3	No. H is not bonded to F, O, or N.	Yes. The H (H_2O) is attracted to nonbonding electron pair of O (CH_3–O–CH_3).

Answers to Self-Test

The numbers in parentheses refer to sections in your textbook.
1. F; varies inversely (7.4) **2**. T (7.5) **3**. F; become hotter (7.5) **4**. F; halve (7.4)
5. T (7.3) **6**. T (7.8) **7**. F; increases (7.11) **8**. F; increases (7.1) **9**. T (7.1)
10. F; gases (7.2) **11**. F; higher (7.12) **12**. T (7.11) **13**. T (7.11) **14**. T (7.13) **15**. T (7.12)
16. a (7.6) **17**. d (7.3) **18**. c (7.8) **19**. d (7.7) **20**. c (7.13) **21**. b (7.13) **22**. c (7.13) **23**. b (7.13)
24. a (7.4) **25**. c (7.9) **26**. b (7.8)

Solutions

Chapter 8

Chapter Overview

Many chemical reactions take place in solutions, particularly in water solutions. The properties of water make it a vital part of all living systems.

In this chapter you will define terms associated with solutions, study how solutions form, and calculate the concentrations of solutions using various units. You will study osmotic pressure and the factors that control the important process of osmosis.

Practice Exercises

8.1 A **solution** (Sec. 8.1) is a homogeneous mixture consisting of a **solvent** (Sec. 8.1) and one or more **solutes** (Sec. 8.1). The solvent is the substance present in the greatest amount.

In the table below, identify the solute and the solvent in each of the solutions:

Solution	Solute	Solvent
10.0 g of potassium chloride in 70.0 g of water		
80.0 g of ethyl alcohol in 50.0 g of water		
40.0 g of potassium iodide in 55.0 g of water		
30.0 mL of ethyl alcohol in 40.0 mL of methyl alcohol		

8.2 The **solubility** (Sec. 8.2) of a substance is the amount of the substance that will dissolve in a given amount of solvent. Solubility depends on several factors: the temperature of the solution, the pressure on the solution, the nature of the solvent. The rate at which a substance dissolves to form a solution depends on how fast the particles come in contact with the solvent.

In the table below, indicate whether each of the changes in conditions would increase or decrease the solubility or the rate of solution of a given solid substance in water.

Change of conditions	Solubility in water	Rate of solution in water
raising the temperature		
crushing or grinding the solid		
adding more of the solid		
agitating the solid/solvent mixture		

8.3 The solubility of a substance can be predicted to some extent by the generalization that substances of like polarity tend to be more soluble in each other than substances that differ in polarity: "like dissolves like." However, the solubility of ionic compounds is more complex. Table 8.2 in your textbook gives solubility guidelines for ionic compounds in water.

Predict the solubility of each of the substances below in the two solvents, water and benzene.

Substance	Water (polar)	Benzene (nonpolar)
$CaCO_3$ (ionic solid)		
$NaNO_3$ (ionic solid)		
K_2SO_4 (ionic solid)		
petroleum jelly (nonpolar solid)		
butane (nonpolar liquid)		
acetone (polar liquid)		

8.4 The **concentration** (Sec. 8.5) of a solution is the amount of **solute** (Sec. 8.1) present in a specified amount of solution. One way of expressing concentration is **percent by mass** (Sec. 8.5), the mass of solute divided by the mass of solution multiplied by 100.

$$\text{percent by mass} = \%(\text{m/m}) = \frac{\text{mass of solute}}{\text{mass of solution}} \times 100$$

a. What is the percent by mass, %(m/m), concentration of NaCl in a solution prepared by dissolving 14.8 g of NaCl in 122 g of water?

b. How many grams of NaCl should be added to 225 g of water to prepare a 7.52%(m/m) NaCl solution? Hint: Remember that mass of water (solvent) does not equal mass of solution; %(m/m) cannot be used directly as a conversion factor. First, find the mass of solvent from the %(m/m). Then use the ratio of mass of solute to mass of solvent as a conversion factor.

8.5 **Percent by volume** (Sec. 8.5) is a percentage unit used when the solute and the solvent are both liquids or both gases.

$$\text{Percent by volume} = \%(\text{v/v}) = \frac{\text{volume of solute}}{\text{volume of solution}} \times 100$$

Calculate the percent by volume for the following solution:
25.0 mL of ethyl alcohol is added to enough water to make 155 mL of solution.

8.6 Another commonly used concentration unit is **mass-volume percent** (Sec. 8.5), the mass of solute divided by the volume of solution:

$$\text{Mass-volume percent} = \%(\text{m/v}) = \frac{\text{mass of solute (g)}}{\text{volume of solution (mL)}} \times 100$$

a. How many grams of KCl must be added to 250.0 mL of water to prepare a 9.82%(m/v) solution? Hint: Since volume of water (mL) is equal to volume of solution (mL), %(m/v) can be used as a conversion factor.

b. Calculate the mass-volume percent of the following solution:
10.5 g of sugar added to enough water to make a solution having a volume of 164 mL.

8.7 The **molarity** (Sec. 8.6) of a solution is a ratio giving the number of moles of solute per liter of solution:

$$\text{Molarity (M)} = \frac{\text{moles of solute}}{\text{liters of solution}}$$

a. A solution with a volume of 425 mL is prepared by dissolving 2.64 moles of $CaCl_2$ in water. What is the molarity?

b. If 7.21 g of KCl is dissolved in enough water to prepare 0.333 L of solution, what is the molarity?

8.8 The definition of molarity can be used as a conversion factor to relate liters of solution to moles of solute. Use dimensional analysis to solve for the correct variable in the following problems.

a. How many grams of $CaCl_2$ were used to prepare 0.250 L of a 0.143 M $CaCl_2$ solution?

b. How many liters of solution would be needed to produce 21.5 g of $CaCl_2$ from a 0.842 M $CaCl_2$ solution?

8.9 **Dilution** (Sec. 8.7) is the process in which more solvent is added to a solution in order to lower its concentration. The simple relationship used for dilution is as follows: the concentration of the stock solution times the volume of the stock solution is equal to the concentration of the **diluted solution** (Sec. 8.2) times the volume of the diluted solution.

$$C_s \times V_s = C_d \times V_d$$

a. If 255 mL of water is added to 325 mL of a 0.477 M solution, what is the molarity of the new solution?

b. How many milliliters of water would have to be added to 250.0 mL of a 2.33 M solution to prepare a 0.551 M solution?

8.10 **Colligative properties** (Sec. 8.9) of solutions are those physical properties affected by the concentration of solute particles regardless of whether the particles are molecules or ions. Boiling point elevation, freezing point depresseion, and vapor pressure lowering are examples of colligative properties.

a. Which of the following solutes would raise the boiling point of 1.00 kg of water more – one mole of KBr or one mole of $CaBr_2$? Why?

b. Which of the following solutes would lower the freezing point of 1.00 kg of water more – two moles of glucose (a molecular solute) or one mole of $CaBr_2$? Why?

8.11 a. One mole of solute particles (ions or molecules) lowers the freezing point of 1.00 kilogram of water by 1.86°C. Calculate the freezing point of a solution containing 1.00 mole of $CaBr_2$ in 1.00 kg of water.

b. One mole of solute particles (ions or molecules) raises the boiling point of 1.00 kilogram of water by 0.51°C. Calculate the boiling point elevation when 0.300 mole of $CaBr_2$ is dissolved in 500 g of water.

8.12 **Osmosis** (Sec. 8.10) is the movement of water across a **semipermeable membrane** (Sec. 8.10) from a more dilute solution to a more **concentrated solution** (Sec. 8.2). **Osmotic pressure** (Sec. 8.10) is the amount of pressure necessary to stop this net flow of water.

Osmotic pressure depends on the number of particles of solute in solution. **Osmolarity** (Sec. 8.10) is a measure of the concentration of particles. It is the product of the molarity of the solution and the number of particles (i) produced when the solute dissociates:

$$\text{Osmolarity} = \text{molarity} \times i$$

Complete the following table on osmolarity. (Salts dissociate into ions; glucose does not.)

Molarity of solution	Osmolarity of solution
3 M KCl	
2 M $CaBr_2$	
2 M in KCl and 1 M in glucose	
3 M in $CaBr_2$ and 2 M in glucose	
2 M in $CaBr_2$ and 1 M in KBr	

8.13 Water flows across the semipermeable membrane of a cell from a solution of lower solute concentration to one of higher solute concentration. A solution outside the cell is classified with reference to the solution within the cell: a **hypotonic** solution has a lower concentration than the concentration of the solution within the cell, a **hypertonic** solution has a higher concentration of solute, and an **isotonic** solution (the prefix *iso* means equal) has the same solute concentration (Sec. 8.10).

a. In the table below, indicate which way water will flow across the cell membranes of red blood cells under the given conditions.

b. Write the numbers for the conditions below that would cause hemolysis to occur. _____ Under which conditions would crenation occur? ______________

Conditions	Water flows into cells	Water flows out of cells
1. cells immersed in concentrated NaCl solution		
2. solution around cells is hypotonic		
3. cells immersed in an isotonic solution		
4. hypertonic solution surrounds cells		
5. cells immersed in pure water		
6. cells immersed in physiological saline solution 0.9%(m/v)		

8.14 Many new terms describing solutions have been introduced in this chapter. To check your understanding, match each of the following terms with the best definition.

	1. saturated solution	a. a substance other than water is the solvent
	2. concentrated solution	b. has a higher osmotic pressure than that within cells
	3. supersaturated solution	c. contains the maximum amount of solute that will dissolve in the solvent
	4. hypotonic solution	d. osmotic pressure is equal to that within cells
	5. isotonic solution	e. contains a large amount of solute relative to the maximum amount that will dissolve in the solvent
	6. aqueous solution	f. the solvent is water
	7. dilute solution	g. contains a small amount of solute relative to the amount that could dissolve in the solvent
	8. non-aqueous solution	h. an unstable solution containing more than the maximum amount of solute that will normally dissolve
	9. hypertonic solution	i. has a lower osmotic pressure than that within cells

Self-Test

True-false: Indicate whether the following statements are true or false. If the statement is false, give the word or phrase that may be substituted for the underlined portion to make the statement true.

1. A solution that contains the maximum amount of solute that can be dissolved in the solvent is called a concentrated solution.

2. Colligative properties are properties that depend on the amount of solute dissolved in a given mass of solution.
3. Osmotic semipermeable membranes permit only dissolved salt to flow through.
4. Hypertonic solutions contain a smaller number of solute molecules than the intracellular fluid.
5. In a salt-water solution, the solute is water.
6. If the solubility of a substance is 120 g/100 mL of water, a solution containing 65 g of the substance dissolved in 50 mL of water would be unsaturated.
7. Undissolved solute is in equilibrium with dissolved solute in a saturated solution.
8. An unsaturated solution is always a dilute solution.
9. Carbon dioxide is more soluble in water when pressure increases.
10. A polar gas, such as NO_2, is insoluble in water.
11. Percent by mass (%m/m) is mass of solute divided by mass of solvent, x 100.
12. Red blood cells in a hypotonic solution may undergo hemolysis.
13. The same number of moles of NaCl are in 225 mL of a 1.55 M NaCl solution as are in 450 mL of a 1.55 M NaCl solution.
14. A suspension is a mixture containing small dispersed particles which do not settle out under the influence of gravity.

Multiple choice:

15. Which of the following compounds would *not* dissolve in water?

a. NaCl b. CCl_4 c. $CaCl_2$
d. HCl e. all would dissolve

16. A solution containing 10.0 g of NaCl in 0.500 L of solution would have what molarity?

a. 0.342 M b. 0.200 M c. 0.500 M
d. 0.174 M e. none of these

17. How many milliliters of 4.57 M potassium bromide solution would be needed to prepare 1.00 L of 2.08 M potassium bromide solution?

a. 155 mL b. 325 mL c. 455 mL d. 695 mL e. none of these

18. How much solute is present in 215 mL of a 0.500 M solution of HCl in water?

a. 1.12 moles b. 0.0566 mole c. 0.752 mole
d. 0.108 mole e. none of these

19. If 53.0 mL of a 3.00 M NaCl solution is diluted to give a solution whose molarity is 0.150 M, what is the volume of the new solution?

a. 0.520 L b. 1.06 L c. 835 mL d. 626 mL e. none of these

20. The osmolarity of a solution that is 2 M $CaCl_2$ and 2 M glucose is:

a. 4 M b. 6 M c. 8 M d. 10 M e. none of these

21. Which of the following solutions would be isotonic with 0.1 M NaCl?

a. 0.5 M $CaCl_2$ b. 0.2 M glucose c. 0.1 M sucrose
d. 0.05 M $Ca(NO_3)_2$ e. none of these

22. Water flows out of red blood cells placed in which of the following solutions?
    a. hypotonic b. isotonic c. hypertonic
    d. both a and c e. none of these

23. What mass-volume percent %(m/v) would result from dissolving 5.00 g of NaCl in enough water to form 50.0 mL of saline solution?
    a. 5.00%(m/v) b. 9.09%(m/v) c. 10.0%(m/v)
    d. 20.0%(m/v) e. none of these

24. Dissolving 7.5 g of NaCl in 50.3 g of water would yield a solution that is what percent by mass, %(m/m)?
    a. 7.50%(m/m) b. 14.9%(m/m) c. 13.0%(m/m)
    d. 74.6%(m/m) e. none of these

25. A 0.030 (m/v) NaCl solution would have a concentration of:
    a. 30 mg/dL b. 300 mg/dL c. 0.030 mg/dL d. 0.30 mg dL e. none of these

Answers to Practice Exercises

8.1

Solution	Solute	Solvent
10.0 g of potassium chloride in 70.0 g of water	10.0 g potassium chloride	70.0 g of water
80.0 g of ethyl alcohol in 50.0 g of water	50.0 g of water	80.0 g of ethyl alcohol
40.0 g of potassium iodide in 55.0 g of water	40.0 g of potassium iodide	55.0 g of water
30.0 mL of ethyl alcohol in 40.0 mL of methyl alcohol	30.0 mL of ethyl alcohol	40.0 mL of methyl alcohol

8.2

Change of conditions	Solubility in water	Rate of solution in water
raising the temperature	depends on solid (may increase or decrease)	increases
crushing or grinding the solid	no effect	increases
adding more of the solid	no effect	increases
agitating the solid/solvent mixture	no effect	increases

8.3

Substance	Water (polar)	Benzene (nonpolar)
$CaCO_3$ (ionic solid)	insoluble	insoluble
$NaNO_3$ (ionic solid)	soluble	insoluble
K_2SO_4 (ionic solid)	soluble	insoluble
petroleum jelly (nonpolar solid)	insoluble	soluble
butane (nonpolar liquid)	insoluble	soluble
acetone (polar liquid)	soluble	soluble

8.4 a. $\%(m/m) = \dfrac{\text{mass of solute}}{\text{mass of solution}} \times 100 = \dfrac{14.8\text{ g NaCl}}{14.8\text{ g NaCl} + 122\text{ g }H_2O} \times 100 = 10.8\%(m/m)$

b. 100 g solution – 7.52 g NaCl = 92.48 g H_2O (Remember: g solution = g NaCl + g H_2O)

$$225\text{ g }H_2O \times \frac{7.52\text{ g NaCl}}{92.48\text{ g }H_2O} = 18.3\text{ g NaCl}$$

8.5 $\%(v/v) = \dfrac{\text{volume of solute}}{\text{volume of solution}} \times 100 = \dfrac{25.0\text{ mL of solute}}{155\text{ mL of solution}} \times 100 = 16.1\%$

8.6 a. $\%(m/v) = 9.82\% = \dfrac{9.82\text{ g KCl}}{100\text{ mL solution}} \times 100$

mass of KCl = %(m/v) × volume of solution

$$\text{mass of KCl} = \frac{9.82\text{ g KCl}}{100\text{ mL solution}} \times 250\text{ mL of solution} = 24.6\text{ g KCl}$$

b. $\%(m/v) = \dfrac{\text{mass of solute (g)}}{\text{volume of solution (mL)}} \times 100 = \dfrac{10.5\text{ g solute}}{164\text{ mL solution}} \times 100 = 6.40\%$

8.7 a. $425\text{ mL} \times \dfrac{1\text{ L}}{1000\text{ mL}} = 0.425\text{ L}$

$$\text{Molarity (M)} = \frac{\text{moles of solute}}{\text{liters of solution}} = \frac{2.64\text{ moles }CaCl_2}{0.425\text{ L}} = 6.21\text{ M}$$

b. $7.21\text{ g KCl} \times \dfrac{1.00\text{ mole KCl}}{74.6\text{ g KCl}} = 9.66 \times 10^{-2}\text{ moles KCl}$

$$M = \frac{\text{moles of solute}}{\text{liters of solution}} = \frac{9.66 \times 10^{-2}\text{ moles KCl}}{0.333\text{ L}} = 0.290\text{ M}$$

8.8 a. $\dfrac{0.143\text{ mole }CaCl_2}{1\text{ L}} \times \dfrac{111\text{ g }CaCl_2}{1\text{ mole }CaCl_2} \times 0.250\text{ L} = 3.97\text{ g }CaCl_2$

b. $\text{liters of solution} = \text{moles} \times \dfrac{1}{M} = \text{moles} \times \dfrac{\text{liters of solution}}{\text{moles}}$

$$\text{liters} = 21.5\text{ g }CaCl_2 \times \frac{1\text{ mole }CaCl_2}{111\text{ g }CaCl_2} \times \frac{1\text{ L}}{0.842\text{ mole }CaCl_2} = 0.230\text{ L}$$

8.9 a. $(C_s \times V_s = C_d \times V_d)$

$$C_d = \frac{C_s \times V_s}{V_d} = \frac{0.477\text{ M} \times 325\text{ mL}}{325\text{ mL} + 255\text{ mL}} = 0.267\text{ M}$$

b. $$V_d = \frac{C_s \times V_s}{C_d} = \frac{2.33\text{ M} \times 250.0\text{ mL}}{0.551\text{ M}} = 1060\text{ mL}$$

$V_d - V_s = 1060\text{ mL} - 250.0\text{ mL} = 810\text{ mL}$ water added

8.10 a. One mole of $CaBr_2$ will raise the boiling point more than one mole of KBr because one mole of $CaBr_2$ will produce three moles of particles (ions) and one mole of KBr will produce two moles of ions.

b. One mole of $CaBr_2$ will lower the freezing point more than two moles of glucose. Two moles of dissolved glucose will produce two moles of particles (molecules), but one mole of dissolved $CaBr_2$ will produce three moles of particles (ions).

8.11 a. $$1.00\text{ mole } CaBr_2 \times \frac{3\text{ moles particles}}{1\text{ mole } CaBr_2} \times \frac{1.86^0\text{C}}{1\text{ mole particles}} = 5.58^0\text{C}$$

Therefore, the freezing point of the solution is 0.00 °C – 5.58 °C = –5.58 °C.

b. $$\frac{0.300\text{ moles } CaBr_2}{500\text{ g } H_2O} \times \frac{1000\text{ g } H_2O}{1\text{ kg } H_2O} = \frac{0.600\text{ moles } CaBr_2}{1\text{ kg } H_2O}$$

The boiling point elevation is:

$$0.600\text{ mole } CaBr_2 \times \frac{3\text{ moles particles}}{1\text{ mole } CaBr_2} \times \frac{0.51^0\text{C}}{1\text{ mole particles}} = 0.92^0\text{C}$$

8.12

Molarity of solution	Osmolarity of solution
3 M KCl	6 osmol
2 M $CaBr_2$	6 osmol
2 M KCl and 1 M glucose	5 osmol
3 M $CaBr_2$ and 2 M glucose	11 osmol
2 M $CaBr_2$ and 1 M KBr	8 osmol

8.13

Conditions	Water flows into cells	Water flows out of cells
1. cells immersed in concentrated NaCl solution		X
2. solution around cells is hypotonic	X	
3. cells immersed in an isotonic solution	no flow	in or out
4. hypertonic solution surrounds cells		X
5. cells immersed in pure water	X	
6. cells immersed in phys. saline solution (0.9 % m/v)	no flow	in or out

b. Write the number for the conditions under which hemolysis would occur. #2, #5
Under which conditions would crenation occur? #1, #4

8.14

c.	1. saturated solution	a. a substance other than water is the solvent
e.	2. concentrated solution	b. has a higher osmotic pressure than that within cells
h.	3. supersaturated solution	c. contains the maximum amount of solute that will dissolve in the solvent
i.	4. hypotonic solution	d. osmotic pressure is equal to that within cells
d.	5. isotonic solution	e. contains a large amount of solute relative to the maximum amount that will dissolve in the solvent
f.	6. aqueous solution	f. the solvent is water
g.	7. dilute solution	g. contains a small amount of solute relative to the amount that could dissolve in the solvent
a.	8. non-aqueous solution	h. an unstable solution containing more than the maximum amount of solute that will normally dissolve
b.	9. hypertonic solution	i. has a lower osmotic pressure than that within cells

Answers to Self-Test

The numbers in parentheses refer to sections in your textbook.
1. F; saturated (8.2) **2**. F; number of particles (8.9) **3**. F; ions and small molecules (8.10) **4**. F; larger (8.10) **5**. F; solvent (8.1) **6**. F; supersaturated (8.2) **7**. T (8.2) **8**. F; sometimes (8.2) **9**. T (8.2) **10**. F; soluble (8.4) **11**. F; solution (8.5) **12**. T (8.10) **13**. F; fewer moles (8.6) **14**. F; colloidal dispersion (8.8) **15**. b (8.4) **16**. a (8.6) **17**. c (8.6) **18**. d (8.6) **19**. b (8.6) **20**. c (8.10) **21**. b (8.10) **22**. c (8.10) **23**. c (8.5) **24**. c (8.5) **25**. a (8.5)

Chemical Reactions — Chapter 9

Chapter Overview

Chemical reactions are processes in which new substances are formed. The concepts of collision theory explain how and under what conditions chemical reactions take place.

In this chapter you will learn to recognize five basic types of chemical reactions. You will identify oxidizing agents and reducing agents in redox reactions. You will study factors that affect the rate of a chemical reaction. Not all chemical reactions go to completion; you will learn to calculate the concentrations of reactants and products in an equilibrium state.

Practice Exercises

9.1 In a **chemical reaction** (Sec. 9.1) at least one new substance is produced as the result of chemical change. Most chemical reactions can be classified in five categories. Classify the following chemical reactions as **combination, decomposition, displacement, exchange,** or **combustion reactions** (Sec. 9.1).

Reaction	Classification
a. $2NaNO_3 \rightarrow 2NaNO_2 + O_2$	
b. $H_2 + Cl_2 \rightarrow 2HCl$	
c. $2C_2H_6 + 7O_2 \rightarrow 4CO_2 + 6H_2O$	
d. $AgNO_3 + KBr \rightarrow AgBr + KNO_3$	
e. $Cu + 2AgNO_3 \rightarrow 2Ag + Cu(NO_3)_2$	

9.2 The **oxidation number** (Sec. 9.2) of an atom represents the charge that the atom would have if all the electrons in each of its bonds were transferred to the more electronegative atom of the two atoms in the bond.

Assign oxidation numbers for each type of atom in each of the following substances, using the rules for determining oxidation numbers found in Section 9.2 of your textbook.

Substance	Oxidation number
Fe	
Ne	
Br_2	
KBr	
MgO	

Substance	Oxidation number
NO_2	
NO_2^-	
PO_4^{3-}	
SO_4^{2-}	
NH_4^+	

9.3 In **oxidation-reduction (redox) reactions** (Sec. 9.2) electrons are transferred from one reactant to another reactant. Electrons are lost by the substance being **oxidized** (Sec. 9.3), so its oxidation number is increased. A substance being **reduced** (Sec. 9.3) gains electrons, and its oxidation number decreases.

In the following chemical equations, assign oxidation numbers to each atom in the reactants and products. Looking at oxidation number changes, classify the reaction as a redox or a nonredox reaction.

Reaction	Redox or nonredox
a. $KOH + HBr \rightarrow KBr + HOH$	
b. $2NaNO_3 \rightarrow 2NaNO_2 + O_2$	
c. $Cu + 2AgNO_3 \rightarrow 2Ag + Cu(NO_3)_2$	

9.4 In redox reactions an **oxidizing agent** (Sec. 9.3) accepts electrons and is reduced. A **reducing agent** (Sec. 9.3) loses electrons and is oxidized. For the chemical equations below, first assign the oxidation numbers, and then identify the oxidizing and reducing agents and the substances oxidized and reduced.

Equation	Substance oxidized	Substance reduced	Oxidizing agent	Reducing agent
a. $4Na(s) + O_2(g) \rightarrow 2Na_2O(s)$				
b. $Ca(s) + S(s) \rightarrow CaS(s)$				
c. $Mg(ClO_3)_2 \rightarrow MgCl_2 + 3O_2$				

9.5 According to **collision theory** (Sec. 9.4), a chemical reaction takes place when two reactant particles collide with a certain minimum amount of energy, called **activation energy** (Sec. 9.4), and the proper orientation. In an energy diagram, the activation energy is the energy difference between the energy of the reactants and the top of the energy "hill." Some of this energy is regained during the reaction; in an **exothermic reaction** (Sec. 9.5) energy is given off in product formation, but in an **endothermic reaction** (Sec. 9.5) energy is absorbed, so that the products are at a higher energy level than the reactants.

Sketch two energy diagrams below and label these parts on each diagram: a. average energy of reactants, b. average energy of products, c. energy absorbed or given off during the reaction, and d. activation energy.

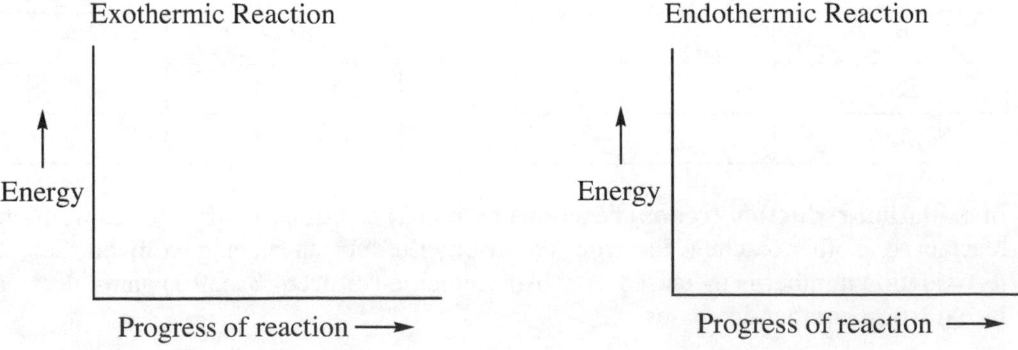

9.6 The **rate of a chemical reaction** (Sec. 9.6) is the rate at which reactants are consumed or products are formed in a given time period. Various factors affect the rate of a chemical reaction: the physical nature of reactants, reactant concentrations, reaction temperature, the presence of a **catalyst** (Sec. 9.6).

Indicate whether the listed conditions would increase or decrease the rate of the following chemical reaction:

$$A(solid) + B \rightarrow C + D + heat$$

Change in condition	Change in rate	Explanation
Decreasing the concentration of reactants		
Decreasing the temperature of the reaction		
Introduction of an effective catalyst for this reaction		
Increasing the surface area of the solid reactant by dividing the solid into smaller particles		

9.7 A **reversible reaction** (Sec. 9.7) is a chemical reaction in which two reactions (the forward reaction and the reverse reaction) occur simultaneously. When these two opposing chemical reactions occur at the same rate, the system is said to be at **chemical equilibrium** (Sec. 9.7).

$$wA + xB \rightleftharpoons yC + zD$$

Because the rates of the forward and reverse chemical reactions are the same, the concentrations of the reactants and the products remain constant. An **equilibrium constant** (Sec. 9.8) that describes numerically the extent of the reaction can be obtained by writing an equilibrium constant expression and evaluating it numerically.

$$K_{eq} = \frac{[C]^y[D]^z}{[A]^w[B]^x}$$

Write the equilibrium constant expression for each of the chemical equations below. Rules for writing these expressions are found in Section 9.8 of your textbook.

a. $2A(s) + 3B_2(g) \rightleftharpoons 2AB_3(g)$

b. $CH_4(g) + 2O_2(g) \rightleftharpoons CO_2(g) + 2H_2O(g)$

9.8 If the concentrations of reactants and products are known for a given chemical reaction at equilibrium, the equilibrium constant can be evaluated. Write the equilibrium constant expression for the following equation and substitute the given molarities to calculate a numerical value for K_{eq}.

$2HI(g) \rightleftharpoons H_2(g) + I_2(g)$ In a 1.0 L container, there are 2.3 moles HI, 0.45 mole H_2 and 0.24 mole I_2

9.9 **Le Châtelier's principle** (Sec. 9.9) considers the effects of outside forces on systems at chemical equilibrium. According to this principle, a stress applied to the system can favor the chemical reaction that will reduce the stress – either the forward reaction, in which case more product is formed, or the reverse reaction, in which case more reactants form. Some of the stresses that can cause this readjustment of the chemical equilibrium are concentration changes, temperature changes, and pressure changes.

a. Indicate whether each change in conditions below would shift the chemical equilibrium of the following chemical reaction to the left or to the right:

$CH_4(g) + 2O_2(g) \rightleftharpoons CO_2(g) + 2H_2O(g) + \text{heat}$

Change in conditions	Equilibrium Change	Explanation
increasing the concentration of O_2		
increasing the temperature of the reaction		
introduction of an effective catalyst for this reaction		
increasing the pressure exerted on the reaction		

b. Indicate whether each change in conditions below would shift the equilibrium of the following chemical reaction to the left or to the right:

$N_2(g) + 2O_2(g) + \text{heat} \rightleftharpoons 2NO_2(g)$

Change in conditions	Equilibrium Change	Explanation
increasing the concentration of O_2		
increasing the temperature of the reaction		
introduction of an effective catalyst for this reaction		
increasing the pressure exerted on the reaction		

Self-Test

True-false: Indicate whether the following statements are true or false. If the statement is false, give the word or phrase that may be substituted for the underlined portion to make the statement true.

1. The reaction $2CuO \rightarrow 2Cu + O_2$ is an example of a displacement reaction.
2. The oxidation number of a metal in its elemental state is always positive.
3. The oxidation number of oxygen in most compounds is -2.
4. A substance that is oxidized loses electrons.
5. A reducing agent gains electrons.
6. Adding heat to an endothermic reaction helps the reaction to go toward the products side.
7. The addition of a catalyst will not change the equilibrium position of a chemical reaction.
8. The rate of a reaction is not affected by the addition of a catalyst.
9. Increasing the concentration of products in a chemical equilibrium reaction shifts the equilibrium toward the product side of the reaction.
10. In a chemical equilibrium constant expression, the concentrations of the reactants are found in the numerator.
11. A large chemical equilibrium constant indicates that the equilibrium position is to the right side of the equation.
12. In chemical equilibrium constant expressions, concentrations of pure solids and pure liquids remain constant.
13. When human body temperature falls so low that the chemical reactions of the body cannot take place at a normal reaction rate, the condition is called hyperthermia.
14. Energy is released in an exothermic reaction.

Multiple choice:

15. The equation $X + YZ \rightarrow Y + XZ$ is a general equation for which type of reaction?

    a. combination b. displacement c. exchange
    d. decomposition e. combustion

16. The oxidation number of chromium in the compound $K_2Cr_2O_7$ is:

    a. +6 b. −7 c. +5 d. −3 e. none of these

17. In the reaction $Zn + Cu(NO_3)_2 \rightarrow Zn(NO_3)_2 + Cu$, the oxidizing agent is:

    a. Zn b. $Cu(NO_3)_2$ c. $Zn(NO_3)_2$ d. Cu e. none of these

18. Which of these factors does *not* affect the rate of a chemical reaction?

    a. the frequency of the collisions b. the energy of the collisions
    c. the orientation of the collisions d. the product of the collisions
    e. all of these affect rate

19. For a chemical reaction at equilibrium, the concentration of product:

    a. increases rapidly b. increases slowly c. remains the same
    d. decreases slowly e. none of these

Answer Questions 20 through 22 using the general chemical equilibrium equation:

$$A + B \rightleftharpoons C + D + \text{heat}$$

20. What is the chemical equilibrium constant for this reaction, if the following concentrations are measured at equilibrium:
   [A] = 0.20 M; [B] = 1.5 M; [C] = 5.2 M; [D] = 3.7 M?

   a. 64   b. 0.016   c. 5.7   d. 0.25   e. none of these

21. The rate of the forward chemical reaction could be increased by:

   a. decreasing the concentration of B
   b. increasing the concentration of A
   c. increasing the concentration of C
   d. both b and c
   e. none of these

22. If more A is added to the reaction mixture at equilibrium:

   a. the amount of C will increase
   b. the amount of B will increase
   c. the amount of B will decrease
   d. both a and c
   e. none of these

23. In the reaction $2Mg + O_2 \rightarrow 2MgO$, the magnesium is:

   a. reduced and is the oxidizing agent
   b. reduced and is the reducing agent
   c. oxidized and is the oxidizing agent
   d. oxidized and is the reducing agent
   e. none of these

24. The minimum total collision energy that is required for a reaction to take place is called:

   a. the oxidation energy
   b. the activation energy
   c. the reactant energy
   d. the product energy
   e. none of these

Answers to Practice Exercises

9.1

Reaction	Classification
a. $2NaNO_3 \rightarrow 2NaNO_2 + O_2$	decomposition
b. $H_2 + Cl_2 \rightarrow 2HCl$	combination
c. $2C_2H_6 + 7O_2 \rightarrow 4CO_2 + 6H_2O$	combustion
d. $AgNO_3 + KBr \rightarrow AgBr + KNO_3$	exchange
e. $Cu + 2AgNO_3 \rightarrow 2Ag + Cu(NO_3)_2$	displacement

9.2

Substance	Oxidation number
Fe	0
Ne	0
Br_2	0
KBr	K (+1), Br (–1)
MgO	Mg (+2), O (–2)

Substance	Oxidation number
NO_2	N (+4), O (–2)
NO_2^-	N (+3), O (–2)
PO_4^{3-}	P (+5), O (–2)
SO_4^{2-}	S (+6), O (–2)
NH_4^+	N (–3), H (+1)

9.3

Oxidation numbers for all atoms	Redox or nonredox
a. KOH + HBr → KBr + H_2O +1,–2,+1 +1,–1 +1,–1 +1,–2	nonredox, no oxidation numbers change
b. $2NaNO_3$ → $2NaNO_2$ + O_2 +1,+5,–2 +1,+3,–2 0	redox, N (+5 → +3) O (–2 → 0)
c. Cu + $2AgNO_3$ → 2Ag + $Cu(NO_3)_2$ 0 +1,+5,–2 0 +2,+5,–2	redox, Cu (0 → +2) Ag (+1 → 0)

9.4

Equation	Substance oxidized	Substance reduced	Oxidizing agent	Reducing agent
a. 4Na(s) + O_2(g) → $2Na_2O$(s) 0 0 +1,–2	sodium	oxygen	oxygen	sodium
b. Ca(s) + S(s) → CaS(s) 0 0 +2,–2	calcium	sulfur	sulfur	calcium
c. $Mg(ClO_3)_2$ → $MgCl_2$ + $3O_2$ +2,+5,–2 +2,–1 0	oxygen	chlorine	ClO^-_3	ClO^-_3

9.5 a. average energy of reactants, b. average energy of products, c. energy absorbed or given off during the reaction, and d. activation energy.

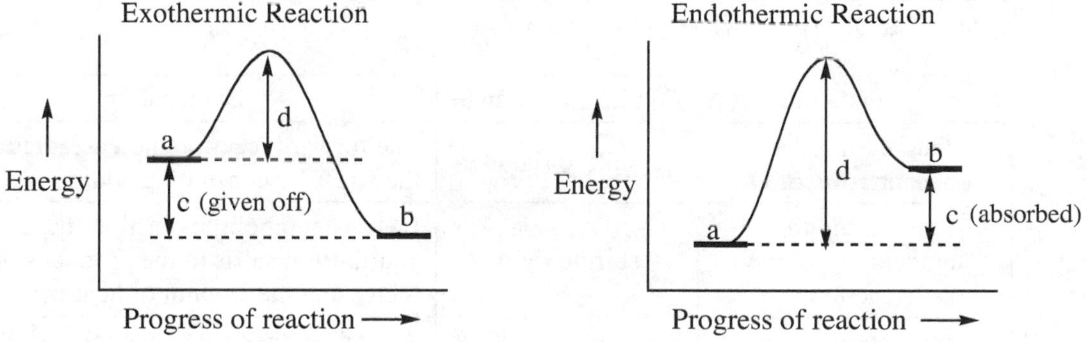

9.6

Change in conditions	Change in rate; explanation
Decreasing the concentration of reactants	Decreases reaction rate; fewer molecules collide, so fewer molecules react.
Decreasing the temperature of the reaction	Decreases reaction rate; lower kinetic energy, lower collision energy, so fewer collisions are effective.
Introduction of an effective catalyst for this reaction	Increases reaction rate; catalysts provide alternative reaction pathways that have lower energies of activation, so more collisions are effective.
Increasing the surface area of the solid reactant by dividing the solid into smaller particles	Increases reaction rate; larger surface area of solid provides more chances for collision.

9.7 a. $K_{eq} = \frac{\text{Products}}{\text{Reactants}} = \frac{[AB_3]^2}{[B_2]^3}$ b. $K_{eq} = \frac{\text{Products}}{\text{Reactants}} = \frac{[CO_2][H_2O]^2}{[CH_4][O_2]^2}$

9.8 $K_{eq} = \frac{[H_2][I_2]}{[HI]^2} = \frac{(0.45) \text{ x } (0.24)}{(2.3)^2} = 0.020$

9.9 a. $CH_4(g) + 2O_2(g) \rightleftharpoons CO_2(g) + 2H_2O(g) + \text{heat}$

Change in conditions	Equilibrium change	Explanation
increasing the concentration of O_2	shift to right	The forward reaction increases, reducing the stress by consuming additional O_2.
increasing the temperature of the reaction	shift to left	The equilibrium shifts to decrease the amount of heat produced.
introduction of an effective catalyst for this reaction	no change	A catalyst does not change the position of the equilibrium; it lowers the energy of activation.
increasing the pressure exerted on the reaction	no change	Because the moles of gas on each side of the equation are the same, pressure change would have no effect.

b. $N_2(g) + 2O_2(g) + \text{heat} \rightleftharpoons 2NO_2(g)$

Change in conditions	Equilibrium change	Explanation
increasing the concentration of O_2	shift to right	The forward reaction increases, reducing the stress by consuming additional O_2.
increasing the temperature of the reaction	shift to right	For an endothermic reaction the equilibrium shifts to the product side, increasing the amount of heat consumed.
introduction of an effective catalyst for this reaction	no change	A catalyst does not change the position of the equilibrium; it lowers the energy of activation.
increasing the pressure exerted on the reaction	shift to right	An increase in the forward reaction would relieve pressure because there are fewer moles of gas on the right.

Answers to Self-Test

The numbers in parentheses refer to sections in your textbook.
1. F; decomposition (9.1) **2**. F; zero (9.2) **3**. T (9.2) **4**. T (9.3) **5**. F; loses (9.3) **6**. T (9.5) **7**. T (9.6) **8**. F; increased (9.6) **9**. F; reactant side (9.9) **10**. F; denominator (9.8) **11**. T (9.8) **12**. T (9.8) **13**. F; hypothermia (9.6) **14**. T (9.5) **15**. b (9.1) **16**. a (9.2) **17**. b (9.3) **18**. d (9.4) **19**. c (9.7) **20**. a (9.8) **21**. b (9.6) **22**. d (9.9) **23**. d (9.3) **24**. b (9.4)

Acids, Bases, and Salts — Chapter 10

Chapter Overview

Acids, bases, and salts play a central role in much of the chemistry that affects our daily lives. Learning the terms and concepts associated with these compounds will give you a greater understanding of the chemistry of the human body, and of the ways in which chemicals are manufactured.

In this chapter you will learn to identify acids and bases according to the Arrhenius and Brønsted-Lowry definitions, write equations for acid and base dissociations in water, and calculate pH, a measure of acidity. You will write equations for the hydrolysis of the salt of a weak acid or a weak base, and you will study the actions of buffers.

Practice Exercises

10.1 According to the Arrhenius acid-base theory, the **dissociation** (Sec. 10.1) of an **Arrhenius acid** (Sec.10.1) in water produces hydrogen ions (H^+), and the dissociation of an **Arrhenius base** in water produces hydroxide ions (OH^-). Arrhenius acids and bases have certain properties that help us identify them. In the table below, specify whether each property is that of an Arrhenius acid or of an Arrhenius base.

Property	Acid or Base?
has a sour taste	
turns blue litmus red	
has a slippery feel	

Property	Acid or base?
turns red litmus blue	
has a bitter taste	
produces OH^- in water	

10.2 According to the **Brønsted-Lowry** (Sec. 10.2) definitions of an acid and a base, an acid is a proton (H^+ ion) donor and a base is a proton (H^+ ion) acceptor; the base must have a nonbonding pair of electrons to accept the proton from the acid.

The **conjugate base** (Sec. 10.2) of an acid is the species that remains when an acid loses a proton. The **conjugate acid** (Sec. 10.2) of a base is the species formed when a base accepts a proton. The presence of a Brønsted-Lowry acid implies the presence of a Brønsted-Lowry base; the two are complementary.

$$\underset{\text{Acid}}{HA} + \underset{\text{Base}}{B} \rightleftharpoons \underset{\text{Conjugate base}}{A^-} + \underset{\text{Conjugate acid}}{HB^+}$$

Give the formula of the conjugate acid or conjugate base for the following substances:

Base	Conjugate acid
NH_3	
BrO_3^-	
HCO_3^-	

Acid	Conjugate base
HCO_3^-	
$HClO_2$	
HNO_3	

10.3 a. Write the equation for the acid-base reaction that occurs when HCl gas is added to water.
b. Label the acids and the bases. Identify the two conjugate acid-base pairs.

10.4 A molecule or an ion that can act as either a Brønsted-Lowry acid or a Brønsted-Lowry base is called an **amphiprotic substance** (Sec. 10.2). The structural requirements for an amphiprotic substance are the presence of a hydrogen atom and of a lone pair of electrons.

a. Write two equations showing the amphiprotic behavior of bicarbonate ion (HCO_3^-) in aqueous solution. Label the acids and the bases in each equation.

b. Water is the most common amphiprotic substance. Write two equations showing the amphiprotic behavior of water. Label the acids and bases in each equation.

10.5 In an acid-base reaction, a **monoprotic acid** (Sec. 10.3) donates one proton (H^+ ion) per molecule to a base, but a **diprotic acid** can donate two protons (H^+ ions) per molecule, and a **triprotic acid** can donate three protons (H^+ ions) per molecule. Complete the following two reactions involving the diprotic acid H_2CO_3, and label the acids and bases. Use one mole of OH^- per mole of acid in each equation.

a. $H_2CO_3(aq) + OH^-(aq) \rightleftharpoons$

b. $HCO_3^-(aq) + OH^-(aq) \rightleftharpoons$

10.6 A **strong acid** (Sec. 10.4) in aqueous solution transfers 100% of its protons to water. Table 10.1 in your textbook is a list of some strong acids. A **weak acid** (Sec. 10.4) in aqueous solution transfers only a small percentage of its protons to water. An **acid ionization constant** (Sec. 10.5), K_a, is the equilibrium constant for the dissociation of a weak acid in water. K_a is a measure of the **strength** (Sec. 10.4) of a weak acid. The smaller the K_a, the weaker the acid.

$$HA(aq) + H_2O(l) \rightleftharpoons H_3O^+(aq) + A^-(aq) \qquad K_a = \frac{[H_3O^+][A^-]}{[HA]}$$

a. Write the ionization equation and the acid ionization constant expression (K_a) for the ionization of nitrous acid, HNO_2, in water.

b. Strong bases ionize completely in aqueous solutions to give metal ions and hydroxide ions (Table 10.2 in your textbook). Weak bases are partially ionized in aqueous solution. The **base ionization constant** (Sec. 10.5), K_b, gives a measure of the strength of a weak base. Write the ionization equation and the ionization constant expression (K_b) for the ionization of ethylamine, $C_2H_5NH_2$, in water. (The nitrogen atom accepts a proton.)

10.7 If the concentration of an acid and its percent ionization are known, the acid ionization constant (K_a) can be calculated. Use the general equation in Problem 10.6 (above).

A 0.0150 M solution of an acid, HA, is 22% ionized at equilibrium. First, determine the individual ion concentrations, and then calculate the K_a for this acid.

10.8 **Salts** (Sec. 10.6) are compounds made up of positive metal or polyatomic ions, and negative nonmetal or polyatomic (except hydroxide) ions. Identify each of the following compounds as an acid, a base, or a salt.

Compound	Acid, base, or salt?
HCl	
NaCl	
H_2SO_4	

Compound	Acid, base, or salt?
NaOH	
$CaBr_2$	
$Ba(OH)_2$	

10.9 Soluble salts dissolved in water are completely dissociated into ions in solution. Write balanced equations for the **dissociation** (Sec. 10.1) of the following soluble ionic compounds in water.

a. KI

b. Na_3PO_4

c. CaI_2

d. Na_2CO_3

10.10 **Neutralization** (Sec. 10.7) is the reaction between an acid and a hydroxide base to form a salt and water. Complete the following neutralization equations by adding the missing products or reactants. Balance the equations, keeping in mind that H^+ and OH^- ions react in a one-to-one ratio to form water. Under each reactant molecule, write "acid" or "base", and under each product molecule, write "salt" or "water".

a. HCl + NaOH $\rightarrow$

b. $\rightarrow$ $CaCl_2$ + H_2O

c. $\rightarrow$ $Sr_3(PO_4)_2$ + H_2O

10.11 In pure water an extremely small number of water molecules transfer protons to form the ions H_3O^+ and OH^-.

$$H_2O + H_2O \rightleftharpoons H_3O^+ + OH^-$$

Equal concentrations of H_3O^+ and OH^- are produced by this self-ionization, and each is equal to 1×10^{-7} M. This value can be used to calculate the **ion product constant** (Sec. 10.8).

Ion product constant for water = $[H_3O^+] \times [OH^-] = (1 \times 10^{-7})(1 \times 10^{-7}) = 1 \times 10^{-14}$

The ion product constant relationship is true for water solutions as well as pure water, so it can be used to calculate the concentration of either H_3O^+ or OH^- if the concentration of the other ion is known.

Using the ion product constant for water, determine the following concentrations. Under each answer tell whether the solution is acidic or basic.

Given concentration	Substituted equation	Answer
$[H_3O^+] = 2.4 \times 10^{-6}$ M		$[OH^-] =$
$[OH^-] = 3.2 \times 10^{-8}$ M		$[H_3O^+] =$

10.12 Because hydronium ion concentrations (measure of solution acidity) in aqueous solutions have a very large range of values, a more practical way to represent acidity is by using the **pH scale** (Sec. 10.9):

$$pH = -\log[H_3O^+]$$

For an acidic solution, $pH < 7$; for a basic solution, $pH > 7$; for a neutral solution, $pH = 7$.

Using your calculator, complete the following pH relationships:

$[H_3O^+]$	pH	Acidic, basic, or neutral
1.0×10^{-4}		
1.0×10^{-9}		
2.8×10^{-3}		
7.9×10^{-8}		

10.13 Using the definition of pH and the ion product constant ($[H_3O^+] \times [OH^-] = 1 \times 10^{-14}$), complete the table below with the missing pH values and/or concentrations of ions. Tell whether the given solution is acidic, basic, or neutral.

pH	$[H_3O^+]$	$[OH^-]$	Acidic, basic, or neutral?
a. 5.00			
b.	1.0×10^{-2}		
c.		6.3×10^{-11}	
d. 10.40			

10.14 Acid strength can also be expressed in terms of pK_a:

$$pK_a = -\log K_a$$

a. Determine the pK_a of an acid whose ionization constant is $K_a = 1.02 \times 10^{-7}$.

b. If Acid A has a pK_a of 8.69 and Acid B has a pK_a of 11.62, which is the stronger acid?

10.15 **Salt Hydrolysis** (Sec. 10.11) is the reaction of a salt with water to produce hydronium ion or hydroxide ion or both. Salts formed from weak acids or weak bases react with water (hydrolyze) to form their "parent" weak acids or weak bases.

a. The salt $NaHCO_3$ dissociates in water: $NaHCO_3 \rightarrow Na^+ + HCO^-_3$
Na^+ does not hydrolyze, since NaOH is a strong base. Write the hydrolysis equation for HCO^-_3, the negative ion of the weak acid H_2CO_3.

Is the aqueous solution of $NaHCO_3$ acidic, basic, or neutral? ____________________

b. The salt NH_4Cl dissociates in water: $NH_4Cl \rightarrow NH_4^+ + Cl^-$
Cl^- does not hydrolyze, because HCl is a strong acid. Write the hydrolysis equation for NH_4^+, the positive ion of the weak base NH_3.

Is the aqueous solution of NH_4Cl acidic, basic, or neutral? ____________________

10.16 A **buffer** (Sec. 10.12) is a solution that resists a change in pH when small amounts of acid or base are added. Buffers consist of one of the following combinations in aqueous solution: 1) a weak acid and the salt of its conjugate base or 2) a weak base and the salt of its conjugate acid. These are known as **conjugate acid-base pairs** (Sec. 10.2).

Predict whether each of the following pairs of substances could function as a buffer in an aqueous solution. Explain your answer.

Pair of substances	Explanation
KOH, KCl	
HI, NaI	
NH_3, NH_4I	
H_3PO_4, NaH_2PO_4	

10.17 Buffers contain a substance that reacts with and removes added base and a substance that reacts with and removes added acid. Write two equations to show the buffering action in each of the following aqueous solutions.

a. NH_3/NH_4I:

1) with a small amount of added base (OH^-)

2) with a small amount of added acid (H_3O^+)

b. H_3PO_4/NaH_2PO_4:

1) with a small amount of added base (OH^-)

2) with a small amount of added acid (H_3O^+)

10.18 The pH of a buffered solution may be found by using the Henderson-Hasselbalch equation:

$$pH = pK_a + \log\frac{[A^-]}{[HA]}$$

Calculate the pH of each of these buffer solutions:

Given values	Substituted equation	Answer
$[HA] = 0.34$ M, $[A^-] = 0.51$ M, $pK_a = 5.48$		pH =
$[HA] = 0.27$ M $[A^-] = 0.55$ M, $K_a = 8.4 \times 10^{-5}$		pH =

10.19 An **electrolyte** (Sec. 10.14) is a substance that forms ions in aqueous solution and thus conducts electricity. A **strong electrolyte** (Sec. 10.14) is a substance that dissociates completely into ions in aqueous solution. Salts, strong acids, and strong bases are strong electrolytes. A **weak electrolyte** (Sec. 10.14) is only partially ionized in aqueous solution. Weak acids and weak bases are weak electrolytes.

Classify each of the substances below as a weak electrolyte or a strong electrolyte.

Formula	Weak electrolyte	Strong electrolyte
H_2SO_4		
NH_3		
NH_4Cl		
MgI_2		

10.20 The concentration of an ion in solution is often specified using equivalents as a unit. One **equivalent** (Eq) (Sec. 10.15) equals the molar amount of an ion needed to supply one mole of positive or negative charge. For ions with a +1 or -1 charge, 1 mole of ion = 1 Eq of ion. For ions with +2 or -2 charge, 1 mole of ion = 2 Eq of ion, and for ions of +3 or -3 charge, 1 mole of ion = 3 Eq of ion.

When ions are present in low concentrations, it may be more convenient to use milliequivalents (mEq): 1 mEq = 10^{-3} Eq or 1 Eq = 1000 mEq

Example: Using dimensional analysis, determine the number of mEq/L of Cl^- in a one-liter solution containing 1 mole Cl^-.

$$\frac{1 \text{ mole } Cl^-}{L} \times \frac{1 \text{ Eq } Cl^-}{1 \text{ mole } Cl^-} \times \frac{10^3 \text{ mEq } Cl^-}{1 \text{ Eq } Cl^-} = \frac{10^3 \text{ mEq } Cl^-}{L}$$

Practice converting moles/liter of solution to Eq/L and mEq/L. Show your calculations:

Moles/liter	Eq/L	mEq/L
1 mole Na^+/L		
0.2 mole NH_4^+/L		
0.2 mole Ca^{2+}/L		
0.3 mole PO_4^{3-}/L		

10.21 Milliequivalents are often used to express the concentration of ions in body fluids because the concentration of ions is low. Use dimensional analysis to answer these questions.

a. How many moles of bicarbonate ion, HCO_3^-, are in 1.00 L of blood plasma if the HCO_3^- concentration is 24 mEq per liter of blood?

b. How many mEq of HPO_4^{2-} are in 2 mL of blood if the HPO_4^{2-} concentration is 0.004 moles/L of blood?

10.22 A charge balance must exist among the ions present in an electrolytic solution. This means that the Eq/L of positive charges and the Eq/L of negative charges must be equal. If 0.020 mole of Na_2SO_4 and 0.020 moles of NaCl are both dissolved in the same 1.0 L of water (no other electrolytes are present), determine the following concentrations. (Hint: Write the equations for the for the dissociation of the salts in water.)

Ion	Moles/L	Eq/L
Na^+		
SO_4^{2-}		
Cl^-		

Is charge balance maintained in this solution? ______________________________

10.23 **Acid-base titration** (Sec. 10.16) is a procedure used to determine the concentration of an acid or base solution. A measured volume of an acid (or a base) of known concentration is exactly reacted with a measured volume of a base (or an acid) of unknown concentration. The unknown concentration can be calculated using dimensional analysis.

a. Determine the molarity of an unknown HCl solution if 21.9 mL of 0.338 M NaOH was needed to neutralize 41.6 mL of the HCl solution. 1) Write the balanced neutralization reaction equation. 2) Use dimensional analysis to calculate the moles of HCl neutralized. 3) Use the definition of molarity to calculate the molarity of the HCl solution.

b. What is the molarity of an unknown sulfuric acid solution if 33.2 mL of 0.225 M NaOH is needed to neutralize 13.8 mL of the H_2SO_4 solution? Follow the steps used in part a.

Self-Test

True-false: Indicate whether the following statements are true or false. If the statement is false, give the word or phrase that may be substituted for the underlined portion to make the statement true.

1. The value of the ion product constant of water is always $\underline{1 \times 10^{-10}}$ at room temperature.
2. Arrhenius defined a base as a substance that, in water, produces hydroxide ions.
3. According the Brønsted-Lowry theory, NH_3 is an acid.

4. Aqueous solutions of acids have a hydronium ion concentration <u>less than</u> 1×10^{-7} moles per liter.
5. A polyprotic acid can transfer <u>two protons</u> per molecule during an acid-base reaction.
6. The pH of an acid is the negative logarithm of the <u>hydronium ion concentration</u>.
7. A neutralization reaction produces a salt and <u>a base</u>.
8. Hydrolysis of the salt of a <u>weak acid</u> and a strong base produces a solution that is basic (alkaline).
9. A solution whose hydronium ion concentration is 1.0×10^{-4} has a pH of <u>1.4</u>.
10. A buffer is a weak acid plus the salt of its conjugate <u>base</u>.
11. An amphiprotic substance can function as an acid or a <u>salt</u>.
12. The hydrolysis of HCO_3^- is the primary cause for blood's slightly <u>acidic</u> pH.
13. The main system that stabilizes blood pH is the <u>H_2CO_3/HCO_3^- buffer system</u>.
14. Anxiety and hyperventilation may cause a shift in the blood buffer system resulting in <u>acidosis</u>.

Multiple choice:

15. The pH of a solution in which $[H_3O^+] = 1.0 \times 10^{-5}$ is:
    a. –5.00 b. –1.50 c. 5.00 d. 1.50 e. none of these
16. The pH of a solution in which $[OH^-] = 1.0 \text{ x } 10^{-8}$ is:
    a. 6.00 b. 8.00 c. –8.00 d. –1.80 e. none of these
17. If the pH of a solution is 4.80, the hydronium ion concentration is:
    a. 4.0×10^{-8} M b. 8.3×10^{-4} M c. 3.2×10^{-12} M
    d. 1.6×10^{-5} M e. none of these
18. If the pH of a solution is 9.20, the hydroxide ion concentration is:
    a. 1.6×10^{-5} M b. 2.1×10^{-9} M c. 9.3×10^{-2}
    d. 7.5×10^{-10} e. none of these
19. How many milliliters of 0.512 M HCl would be required to neutralize 35.8 mL of 1.50 M KOH?
    a. 11.3 mL b. 35.8 mL c. 105 mL
    d. 183 mL e. none of these
20. The conjugate base of the acid HNO_2 is:
    a. H_2O b. H_3O^+ c. NO_2^- d. OH^- e. none of these
21. A 0.400 M solution of a monoprotic acid is 9.0% ionized. The value of K_a for this acid is:
    a. 3.6×10^{-3} b. 9.8×10^{-2} c. 2.1×10^{-4}
    d. 8.9×10^{-4} e. none of these
22. The salt K_2CO_3 is produced by the reaction of:
    a. a weak acid with a weak base b. a weak acid with a strong base
    c. a strong acid with a weak base d. a strong acid with a strong base
    e. none of these

23. Which of the following is a weak electrolyte?

a. K_3PO_4 b. HNO_3 c. $Ba(OH)_2$ d. H_2CO_3 e. none of these

24. A small amount of hydroxide ion added to the buffer H_2CO_3/HCO_3^- will produce more:

a. H_3O^+ b. HCO_3^- c. H_2CO_3 d. H_2 e. none of these

25. Arrange 0.5 M aqueous solutions of HBr, HNO_2, KBr, KNO_2 in order of increasing pH:

a. HBr, HNO_2, KBr, KNO_2 b. KNO_2, HNO_2, KBr, HBr
c. HNO_2, KBr, HBr, KNO_2 d. KBr, HBr, KNO_2, HNO_2,
e. none of these

Answers to Practice Exercises

10.1

Property	Acid or Base?
has a sour taste	acid
turns blue litmus red	acid
has a slippery feel	base

Property	Acid or base?
turns red litmus blue	base
has a bitter taste	base
produces OH^- in water	base

10.2

Base	Conjugate acid
NH_3	NH_4^+
BrO_3^-	$HBrO_3$
HCO_3^-	H_2CO_3

Acid	Conjugate base
HCO_3^-	CO_3^{2-}
$HClO_2$	ClO_2^-
HNO_3	NO_3^-

10.3

$$\underset{\text{acid}}{HCl} + \underset{\text{base}}{H_2O} \rightleftharpoons \underset{\text{conjugate base}}{Cl^-} + \underset{\text{conjugate acid}}{H_3O^+}$$

The two conjugate acid/base pairs are: HCl/Cl^- andH_3O^+/H_2O

10.4 a.

$$\underset{\text{acid}}{HCO_3^-} + \underset{\text{base}}{H_2O} \rightleftharpoons \underset{\text{conjugate base}}{CO_3^{2-}} + \underset{\text{conjugate acid}}{H_3O^+}$$

$$\underset{\text{base}}{HCO_3^-} + \underset{\text{acid}}{H_2O} \rightleftharpoons \underset{\text{conjugate acid}}{H_2CO_3} + \underset{\text{conjugate base}}{OH^-}$$

b.

$$\underset{\text{acid}}{H_2O} + \underset{\text{base}}{OH^-} \rightleftharpoons \underset{\text{conjugate base}}{OH^-} + \underset{\text{conjugate acid}}{H_2O}$$

$$\underset{\text{base}}{H_2O} + \underset{\text{acid}}{H_3O^+} \rightleftharpoons \underset{\text{conjugate acid}}{H_3O^+} + \underset{\text{conjugate base}}{H_2O}$$

10.5 a.

$$\underset{\text{acid}}{H_2CO_3(aq)} + \underset{\text{base}}{OH^-(aq)} \rightleftharpoons \underset{\text{conjugate base}}{HCO_3^-(aq)} + \underset{\text{conjugate acid}}{H_2O\,(l)}$$

b.

$$\underset{\text{acid}}{HCO_3^-(aq)} + \underset{\text{base}}{OH^-(aq)} \rightleftharpoons \underset{\text{conjugate base}}{CO_3^{2-}(aq)} + \underset{\text{conjugate acid}}{H_2O\,(l)}$$

10.6 a. $HNO_2 + H_2O \rightleftharpoons NO_2^- + H_3O^+$ $K_a = \dfrac{[H_3O^+][NO_2^-]}{[HNO_2]}$

b. $C_2H_5NH_2 + H_2O \rightleftharpoons C_2H_5NH_3^+ + OH^-$ $K_b = \dfrac{[C_2H_5NH_3^+][OH^-]}{[C_2H_5NH_2]}$

10.7 $HA + H_2O \rightleftharpoons H_3O^+ + A^-$

$$K_a = \frac{[H_3O^+][A^-]}{[HA]} = \frac{[0.0150 \times 0.22][0.0150 \times 0.22]}{[(0.0150)-(0.0150 \times 0.22)]} = \frac{[0.0033][0.0033]}{[0.0117]} = 9.3 \times 10^{-4}$$

10.8

Compound	Acid, base, or salt?
HCl	acid
NaCl	salt
H_2SO_4	acid

Compound	Acid, base, or salt?
NaOH	base
$CaBr_2$	salt
$Ba(OH)_2$	base

10.9 a. $KI \rightarrow K^+(aq) + I^-(aq)$ b. $Na_3PO_4 \rightarrow 3Na^+(aq) + PO_4^{3-}(aq)$

c. $CaI_2 \rightarrow Ca^{2+}(aq) + 2I^-(aq)$ d. $Na_2CO_3 \rightarrow 2Na^+(aq) + CO_3^{2-}(aq)$

10.10 a. $HCl + NaOH \rightarrow NaCl + H_2O$
acid base salt water

b. $2HCl + Ca(OH)_2 \rightarrow CaCl_2 + 2H_2O$
acid base salt water

c. $2H_3PO_4 + 3Sr(OH)_2 \rightarrow Sr_3(PO_4)_2 + 6H_2O$
acid base salt water

10.11

Given concentration	Substituted equation	Answer
$[H_3O^+] = 2.4 \times 10^{-6}$	$[OH^-] = \dfrac{1.00 \times 10^{-14}}{[2.4 \times 10^{-6}]} =$	$[OH^-] = 4.2 \times 10^{-9}$ acidic
$[OH^-] = 3.2 \times 10^{-8}$	$[H_3O^+] = \dfrac{1.00 \times 10^{-14}}{[3.2 \times 10^{-8}]} =$	$[H_3O^+] = 3.1 \times 10^{-7}$ acidic

10.12

$[H_3O^+]$	pH	Acidic, basic, or neutral
1.0×10^{-4}	4.00	acidic
1.0×10^{-9}	9.00	basic
2.8×10^{-3}	2.55	acidic
7.9×10^{-8}	7.10	basic

10.13

pH	$[H_3O^+]$	$[OH^-]$	Acidic, basic, or neutral
5.00	1.0×10^{-5}	1.0×10^{-9}	acidic
2.00	1.0×10^{-2}	1.0×10^{-12}	acidic
3.80	1.6×10^{-4}	6.3×10^{-11}	acidic
10.40	4.0×10^{-11}	2.5×10^{-4}	basic

10.14 a. $pK_a = 6.991$

b. Acid A ($pK_a = 8.69$) is the stronger acid.

10.15 a. $HCO_3^- + H_2O \rightleftharpoons H_2CO_3 + OH^-$ basic solution (OH^- produced)

b. $NH_4^+ + H_2O \rightleftharpoons H_3O^+ + NH_3$ acidic solution (H_3O^+ produced)

10.16

Pair of substances	Explanation
KOH, KCl	No. Strong base and strong acid-strong base salt
HI, NaI	No. Strong acid and strong acid-strong base salt
NH_3, NH_4I	Yes. Weak base and salt of its conjugate acid
H_3PO_4, NaH_2PO_4	Yes. Weak acid and salt of its conjugate base

10.17 a. 1) small amount of added base (OH^-): $NH_4^+ + OH^- \rightarrow NH_3 + H_2O$
2) small amount of added acid (H_3O^+): $NH_3 + H_3O^+ \rightarrow NH_4^+ + H_2O$

b. 1) small amount of added base (OH^-): $H_3PO_4 + OH^- \rightarrow H_2PO_4^- + H_2O$
2) small amount of added acid (H_3O^+): $H_2PO_4^- + H_3O^+ \rightarrow H_3PO_4 + H_2O$

10.18

Given values	Substituted equation	Answer
[HA] = 0.34 M, $[A^-]$ = 0.51 M, pK_a = 5.48	$pH = 5.48 + \log\left(\frac{0.51}{0.34}\right) = 5.66$	pH = 5.66
[HA] = 0.27 M, $[A^-]$ = 0.55 M, $K_a = 8.4 \text{ x } 10^{-5}$	$pH = 4.08 + \log\left(\frac{0.55}{0.27}\right) = 4.39$	pH = 4.39

10.19

Formula	Weak electrolyte	Strong electrolyte
H_2SO_4		X
NH_3	X	
NH_4Cl		X
MgI_2		X

10.20

Eq per liter of solution	mEq per liter of solution
$\frac{1 \text{ mole } Na^+}{L} \times \frac{1 \text{ Eq } Na^+}{1 \text{ mole } Na^+} = \frac{1 \text{ Eq } Na^+}{L}$	$\frac{1 \text{ Eq } Na^+}{L} \times \frac{10^3 \text{ mEq } Na^+}{1 \text{ Eq } Na^+} = \frac{1 \times 10^3 \text{ mEq } Na^+}{L}$
$\frac{0.2 \text{ mole } NH_4^+}{L} \times \frac{1 \text{ Eq } NH_4^+}{1 \text{ mole } NH_4^+} = \frac{0.2 \text{ Eq } NH_4^+}{L}$	$\frac{0.2 \text{ Eq } NH_4^+}{L} \times \frac{10^3 \text{ mEq } NH_4^+}{1 \text{ Eq } NH_4^+} = \frac{2 \times 10^2 \text{ mEq } NH_4^+}{L}$
$\frac{0.2 \text{ mole } Ca^{2+}}{L} \times \frac{2 \text{ Eq } Ca^{2+}}{1 \text{ mole } Ca^{2+}} = \frac{0.4 \text{ Eq } Ca^{2+}}{L}$	$\frac{0.4 \text{ Eq } Ca^{2+}}{L} \times \frac{10^3 \text{ mEq } Ca^{2+}}{1 \text{ Eq } Ca^{2+}} = \frac{4 \times 10^2 \text{ mEq } Ca^{2+}}{L}$
$\frac{0.3 \text{ mole } PO_4^{3-}}{L} \times \frac{3 \text{ Eq } PO_4^{3-}}{1 \text{ mole } PO_4^{3-}} = \frac{0.9 \text{ Eq } PO_4^{3-}}{L}$	$\frac{0.9 \text{ Eq } PO_4^{3-}}{L} \times \frac{10^3 \text{ mEq } PO_4^{3-}}{1 \text{ Eq } PO_4^{3-}} = \frac{9 \text{ x } 10^2 \text{ mEq } PO_4^{3-}}{L}$

10.21 a. $1.00 \text{ L blood} \times \frac{24 \text{ mEq } HCO_3^-}{1 \text{ L blood}} \times \frac{10^{-3} \text{ Eq } HCO_3^-}{1 \text{ mEq } HCO_3^-} \times \frac{1 \text{ mole } HCO_3^-}{1 \text{ Eq } HCO_3^-} = 0.024 \text{ mole } HCO_3^-$

b. $2 \text{ mL blood} \times \frac{0.004 \text{ moles } HPO_4^{2-}}{1 \text{ L blood}} \times \frac{1 \text{ L blood}}{1000 \text{ mL blood}} \times \frac{2 \text{ Eq } HPO_4^{2-}}{1 \text{ mole } HPO_4^{2-}} \times \frac{1 \text{ mEq } HPO_4^{2-}}{10^{-3} \text{ Eq } HPO_4^{2-}}$

$= 0.016 \text{ mEq } HPO_4^{2-}$

10.22

Ion	Moles/L	Eq/L
Na^+	0.060	0.060
SO_4^{2-}	0.020	0.040
Cl^-	0.020	0.020

Yes, the charge balance is maintained; positive and negative ions each have a total of 0.060 Eq/L.

10.23 a. 1) $HCl + NaOH \rightarrow NaCl + H_2O$

2) $21.9 \text{ mL NaOH x } \frac{0.338 \text{ mole NaOH}}{1000 \text{ mL NaOH}} \text{ x } \frac{1 \text{ mole HCl}}{1 \text{ mole NaOH}} = 0.00740 \text{ moles HCl}$

3) $\frac{0.00740 \text{ moles HCl}}{41.6 \text{ mL HCl}} \text{ x } \frac{1000 \text{ mL HCl}}{1.00 \text{ L HCl}} = 0.178 \text{ M HCl}$

b. 1) $H_2SO_4 + 2NaOH \rightarrow Na_2SO_4 + 2H_2O$

2) $33.2 \text{ mL NaOH} \times \frac{0.225 \text{ mole NaOH}}{1000 \text{ mL NaOH}} \times \frac{1 \text{ mole } H_2SO_4}{2 \text{ moles NaOH}} = 0.00374 \text{ moles } H_2SO_4$

3) $\frac{0.00374 \text{ moles } H_2SO_4}{13.8 \text{ mL } H_2SO_4} \times \frac{1000 \text{ mL } H_2SO_4}{1.00 \text{ L } H_2SO_4} = 0.271 \text{ M } H_2SO_4$

Answers to Self-Test

The numbers in parentheses refer to sections in your textbook.
1. F; 1.00 x 10^{-14} (10.8) **2**. T (10.1) **3**. F; base (10.2) **4**. F; more than (10.8) **5**. F; two or more (10.3) **6**. T (10.9) **7**. F; water (10.7) **8**. T (10.11) **9**. F; 4.00 (10.9) **10**. T (10.12) **11**. F; base (10.2) **12**. F; basic (10.11) **13**. T (10.12) **14**. F; alkalosis (10.12) **15**. c (10.9) **16**. a (10.9) **17**. d (10.9) **18**. a (10.9) **19**. c (10.16) **20**. c (10.2) **21**. a (10.5) **22**. b (10.7) **23**. d (10.14) **24**. b (10.12) **25**. a (10.11)

Nuclear Chemistry — Chapter 11

Chapter Overview

Chemical reactions involve the exchange or sharing of the electrons of atoms as chemical bonds are broken or formed. In **nuclear reactions** (Sec. 11.1) an atom's nucleus changes, absorbing or emitting particles or rays.

In this chapter you will compare three kinds of nuclear radiation and write equations for radioactive decay. You will study the rate of radioactive decay, defining the concept of the half-life of a radionuclide and using it in calculations. You will compare nuclear fission and nuclear fusion, study the effects of ionizing radiation on the human body, and learn about some of the ways in which radiation and radionuclides are used in medicine.

Practice Exercises

11.1 The **radioactive decay** (Sec. 11.3) of naturally radioactive substances results in the emission of three types of radiation: **alpha particles, beta particles,** and **gamma rays** (Sec. 11.2). They differ in mass and charge. Complete the following table.

Type of radiation	Mass number	Charge	Symbol
alpha particle			
beta particle			
gamma ray			

11.2 The emission of an alpha particle ($^{4}_{2}\alpha$) from a nucleus results in the formation of a **nuclide** (Sec. 11.1) of a different element. The **daughter nuclide** (Sec. 11.3) has an atomic number (subscript *Z*) that is two less and a mass number (superscript *A*) that is four less than the **parent nuclide** (Sec. 11.3).

In a **balanced nuclear equation** (Sec. 11.3), the sums of the mass numbers (*A*) and the sums of the atomic numbers (*Z*) on both sides of the equation are equal. Write the balanced nuclear equation for alpha particle decay of the following nuclides. Use complete symbols ($^{A}_{Z}X$):

a. $^{149}_{65}Tb \rightarrow$

b. $^{231}_{91}Pa \rightarrow$

11.3 In beta particle decay, a neutron in the nucleus is transformed into a proton and a beta particle. Write a balanced nuclear equation for the formation of a beta particle from a neutron. Use complete symbols:

11.4 Beta particle decay results in the formation of a nuclide of a different element. The mass number of the daughter nuclide is the same as that of the parent nuclide; the atomic number of the daughter nuclide is one greater than that of the parent nuclide. Write balanced nuclear equations for the beta particle decay of the following nuclides. Use complete symbols.

a. $^{31}_{14}Si \rightarrow$

b. $^{59}_{26}Fe \rightarrow$

11.5 Not all **radioactive nuclides** (Sec. 11.1) decay at the same rate; the more unstable the nucleus, the faster it decays. The **half-life** (Sec. 11.4) of a substance (the amount of time for half of a given quantity of nuclide to decay) is a measure of the nuclide's stability.

The amount of radioactive material remaining after radioactive decay can be calculated from the following formula, where n = number of half-lives.

$$\begin{pmatrix}\text{Amount of radionuclide}\\ \text{undecayed after } n \text{ half-lives}\end{pmatrix} = \begin{pmatrix}\text{original amount}\\ \text{of radionuclide}\end{pmatrix} \times \left(\frac{1}{2^n}\right)$$

In the following problems, an original sample of 0.43 mg of plutonium-239 is used (half-life = 24,400 years).

a. How much plutonium-239 would remain after three half-lives?

b. How much plutonium-239 would remain after 122,000 years?

11.6 The number of half-lives that have elapsed since the original measurement can be determined by the fraction of the original nuclide that remains.

A sample of iron-59 after 135 days has 1/8 of the iron-59 of the original sample. What is the half-life of iron-59?

11.7 In a **transmutation reaction** (Sec. 11.5) a nuclide of one element changes into a nuclide of another element. Transmutation can sometimes be brought about by bombardment of nuclei with small particles traveling at very high speeds.

Balance these **bombardment reaction** (Sec. 11.5) equations by supplying the missing parts:

a. ${}^{10}_{5}\text{B} + \_\_\_\_ \rightarrow {}^{7}_{3}\text{Li} + {}^{4}_{2}\alpha$

b. ${}^{7}_{3}\text{Li} + {}^{1}_{1}\text{p} \rightarrow \_\_\_\_ + {}^{4}_{2}\alpha$

c. ${}^{10}_{5}\text{B} + {}^{4}_{2}\alpha \rightarrow {}^{13}_{7}\text{N} + \_\_\_\_$

d. $\_\_\_\_ + {}^{14}_{7}\text{N} \rightarrow {}^{247}_{99}\text{Es} + 5\ {}^{1}_{0}\text{n}$

11.8 Radionuclides with high atomic numbers decay through a series of steps to reach a stable nuclide of lower atomic number. A few of these steps in a **radioactive decay series** (Sec. 11.6) are shown below. Balance the equations by filling in the missing parts:

step 1: ${}^{210}_{82}\text{Pb} \rightarrow \_\_\_\_ + \_\_\_\_$ (β-emission)

step 2: $\_\_\_\_ \rightarrow {}^{210}_{84}\text{Po} + \_\_\_\_$ (β-emission)

step 3: ${}^{210}_{84}\text{Po} \rightarrow \_\_\_\_ + \_\_\_\_$ (α-emission)

11.9 Radiation can affect an atom or a molecule chemically in two ways: formation of an **ion pair** (an electron and a positive ion, Sec. 11.8) or formation of a **free radical** (an atom, molecule, or ion with an unpaired electron, Sec. 11.8).

a. **Ionizing radiation** (Sec. 11.8) has enough energy to remove an electron from an atom or a molecule, and **nonionizing radiation** (Sec. 11.8) does not. Give two examples of each:

ionizing radiation ____

nonionizing radiation____

b. Explain why free radicals are dangerous to living cells.

11.10 The three types of naturally occurring radioactive emissions differ in their ability to penetrate matter (Sec. 11.9) and, therefore, in their biological effects. Complete the following table summarizing properties of the three types of radiation:

Type	Speed	Penetration	Biological damage
alpha particles			
beta particles			
gamma radiation			

11.11 In **nuclear medicine** (Sec. 11.11), radionuclides are used for both diagnostic purposes (determination or detection of disease) and therapeutic purposes (treatment of disease).

a. Most radionuclides used in diagnosis are γ-emitters. Complete the table below showing some radionuclides commonly used in diagnosis.

Determination or detection of:	Nuclide commonly used:
1.	1. iodine-123
2. blood flow in impaired heart muscle	2.
3.	3. gallium-67
4. detection of circulatory problems	4.
5. determination of bile duct obstruction	5.

b. Radionuclides that are commonly used in radiation therapy to destroy abnormal cells in the human body are listed in Table 11.5 in your textbook. Tell what type of radiation is emitted by each of the three commonly-used nuclides named below:

yttrium-90 _____________; cobalt-60 ___________; radium-226 ____________

11.12 Two types of nuclear reactions used as sources of energy are **nuclear fission** (a large nucleus splits into two medium-sized nuclei with the release of free neutrons and a large amount of energy, Sec. 11.12) and **nuclear fusion** (two small nuclei collide with one another to produce a larger nucleus and a large amount of energy, Sec. 11.12).

a. The equation below shows one of the possible fission processes of uranium-235. Balance the equation by putting in the number of neutrons that are given off in the reaction.

$$^{235}_{92}\text{U} + {}^{1}_{0}n \rightarrow {}^{139}_{56}\text{Ba} + {}^{94}_{36}\text{Kr} + \_\_\_ {}^{1}_{0}n$$

b. Complete the following nuclear fusion reaction, which is the basis of the hydrogen bomb and which is also being studied very actively as a future energy source.

$$^{3}_{1}\text{H} + {}^{2}_{1}\text{H} \rightarrow$$

Self-Test

True-false: Indicate whether the following statements are true or false. If the statement is false, give the word or phrase that may be substituted for the underlined portion to make the statement true.

1. Elements retain their identity during <u>nuclear reactions</u>.
2. <u>Alpha particles</u> are the same type of radiation as X rays.
3. When an atom loses an alpha particle, its atomic number is <u>decreased</u> by 2.
4. Beta particles are more penetrating than <u>alpha particles</u>.
5. The <u>proton-to-neutron</u> ratio is approximately 1.5 for the heaviest stable elements.
6. <u>Nuclear fission</u> is the reaction that provides the sun's energy.
7. The combination of two small nuclei to produce a larger nucleus is called <u>nuclear fusion</u>.
8. Ionizing radiation knocks <u>protons</u> off some atoms so that they become ions.
9. The energy involved in a nuclear reaction is <u>smaller</u> than the energy involved in a chemical reaction.
10. One <u>diagnostic application</u> of radionuclides is the use of ionizing radiation to kill cancer cells.
11. In a Geiger counter, radiation is detected by the <u>ionization</u> of argon gas in a metal tube.
12. A free radical is an atom or molecule with <u>a positive charge</u> whose formation can be caused by radiation.
13. Different isotopes of an element have practically identical <u>chemical properties</u>.
14. Indoor exposure to airborne radon-222 is harmful to humans because of the <u>*β*-particle</u> decay of radon-222, which produces polonium-218.
15. A transmutation process is a nuclear reaction in which a nuclide of one element is changed into a nuclide of <u>the same</u> element.
16. Technetium-99m is useful in diagnostic medicine because it emits <u>beta particles</u>.
17. After three half-lives, the amount of an 8.0 g sample of a radionuclide that has decayed is <u>1.0 g.</u>

Multiple choice:

18. The nuclide produced by the emission of an alpha particle from the platinum nuclide ${}^{186}_{78}\text{Pt}$ is:

    a. ${}^{188}_{80}\text{Hg}$ b. ${}^{182}_{76}\text{Os}$ c. ${}^{184}_{75}\text{Re}$ d. ${}^{186}_{79}\text{Au}$ e. none of these

19. The nuclide produced by beta emission from the barium radionuclide ${}^{142}_{56}\text{Ba}$ is:

    a. ${}^{142}_{57}\text{La}$ b. ${}^{143}_{55}\text{Cs}$ c. ${}^{138}_{54}\text{Xe}$ d. ${}^{146}_{58}\text{Ce}$ e. none of these

20. The half-life for tritium ${}^{3}_{1}\text{H}$ is 12.26 years. After 36.78 years, how much of an original 0.40-g sample of tritium will remain?

    a. 0.35 g b. 0.20 g c. 0.10 g d. 0.050 g e. none of these

21. Four radionuclides have the half-lives listed below. Which of the four has the most stable nucleus?

    a. 22 days b. 4000 years c. 16 minutes
    d. 56 seconds e. all are very unstable

22. An alpha particle is made up of:
    a. two protons and four neutrons
    b. two protons and two neutrons
    c. two neutrons and two electrons
    d. two protons and two electrons
    e. none of these

23. Radiation is harmful to the human body because:
    a. it causes the formation of free radicals
    b. it causes molecules to fragment
    c. it produces ions in the body
    d. a and b only
    e. a, b, and c

24. Fusion reactions are not generally used to provide energy on Earth because:
    a. they give off too little energy
    b. a very high temperature is required to start the reactions
    c. the reactions proceed only at very low pressures
    d. the starting materials (reactants) are too expensive
    e. none of these

25. Radiation therapy includes the following use of radiation:
    a. radiographs of bone tissue
    b. radiation of cancer cells
    c. dental X rays
    d. use of radioactive tracer
    e. all of the above

Answers to Practice Exercises

11.1

Type of radiation	Mass number	Charge	Symbol
alpha particle	4	+2	${}^{4}_{2}\alpha$
beta particle	0	–1	${}^{0}_{-1}\beta$
gamma ray	0	0	${}^{0}_{0}\gamma$

11.2 a. ${}^{149}_{65}\text{Tb} \rightarrow {}^{145}_{63}\text{Eu} + {}^{4}_{2}\alpha$ b. ${}^{231}_{91}\text{Pa} \rightarrow {}^{227}_{89}\text{Ac} + {}^{4}_{2}\alpha$

11.3 ${}^{1}_{0}\text{n} \rightarrow {}^{1}_{1}\text{p} + {}^{0}_{-1}\beta$

11.4 a. ${}^{31}_{14}\text{Si} \rightarrow {}^{31}_{15}\text{P} + {}^{0}_{-1}\beta$ b. ${}^{59}_{26}\text{Fe} \rightarrow {}^{59}_{27}\text{Co} + {}^{0}_{-1}\beta$

11.5 a. $$\left(\begin{array}{l}\text{Amount of radionuclide}\\ \text{undecayed after 3 half - lives}\end{array}\right) = (0.43\text{ mg}) \text{ x} \left(\frac{1}{2^3}\right) = (0.43\text{ mg}) \text{ x} \left(\frac{1}{8}\right) = 0.054\text{ mg}$$

b. $$\text{Number of half - lives} = \frac{122{,}000\text{ years}}{24{,}400\text{ years / half - life}} = 5\text{ half - lives}$$

$$\left(\begin{array}{l}\text{Amount of radionuclide}\\ \text{undecayed after 5 half - lives}\end{array}\right) = (0.43\text{ mg}) \text{ x} \left(\frac{1}{2^5}\right) = (0.43\text{ mg}) \text{ x} \left(\frac{1}{32}\right) = 0.013\text{ mg}$$

11.6 $$\left(\frac{1}{2}\right) \text{x} \left(\frac{1}{2}\right) \text{x} \left(\frac{1}{2}\right) = \frac{1}{8} = \frac{1}{2^3} = \frac{1}{2^n};\ n = 3 = \text{number of half-lives}$$

$$\frac{135\text{ days}}{3\text{ half - lives}} = 45\text{ days} = 1\text{ half-life of iron-59}$$

11.7 a. ${}^{10}_{5}B + {}^{1}_{0}n \rightarrow {}^{7}_{3}Li + {}^{4}_{2}\alpha$ b. ${}^{7}_{3}Li + {}^{1}_{1}p \rightarrow {}^{4}_{2}He + {}^{4}_{2}\alpha$

c. ${}^{10}_{5}B + {}^{4}_{2}\alpha \rightarrow {}^{13}_{7}N + {}^{1}_{0}n$ d. ${}^{238}_{92}U + {}^{14}_{7}N \rightarrow {}^{247}_{99}Es + 5\,{}^{1}_{0}n$

11.8 step 1: ${}^{210}_{82}Pb \rightarrow {}^{210}_{83}Bi + {}^{0}_{-1}\beta$

step 2: ${}^{210}_{83}Bi \rightarrow {}^{210}_{84}Po + {}^{0}_{-1}\beta$

step 3: ${}^{210}_{84}Po \rightarrow {}^{206}_{82}Pb + {}^{4}_{2}\alpha$

11.9 a. ionizing radiation – cosmic rays, X rays, ultraviolet light

nonionizing radiation – radio waves, microwaves, infrared light, visible light

b. A free radical is a very reactive species that can produce undesirable chemical reactions inside a living cell. Free radicals can also react with other molecules to produce new free radicals in a chain reaction.

11.10

Type	Speed	Penetration	Biological damage
alpha particles	slow	very little, stopped by skin or paper	cannot penetrate skin; ingestion damages internal organs
beta particles	faster	penetrating, stopped by wood or aluminum foil	can cause severe skin burns; ingestion causes damage to internal organs
gamma rays	fastest (speed of light)	very penetrating, some stopped by lead or concrete	penetrates deeply into organs, bones, and other tissue; causes serious damage to all tissues

11.11 a.

Determination or detection of:	Nuclide commonly used:
1. thyroid activity	1. iodine-123
2. blood flow in impaired heart muscle	2. thallium-201
3. sites of infection	3. gallium-67
4. detection of circulatory problems	4. sodium-24
5. determination of bile duct obstruction	5. technetium-99m

b. yttrium-90 – beta and gamma; cobalt-60 – gamma; radium-226 – alpha and gamma

11.12 a. ${}^{235}_{92}U + {}^{1}_{0}n \rightarrow {}^{139}_{56}Ba + {}^{94}_{36}Kr + 3\,{}^{1}_{0}n$

b. ${}^{3}_{1}H + {}^{2}_{1}H \rightarrow {}^{4}_{2}He + {}^{1}_{0}n$

Answers to Self-Test

The numbers in parentheses refer to sections in your textbook.
1. F; chemical reactions (11.13) **2**. F; gamma rays (11.2) **3**. T (11.3) **4**. T (11.9)
5. F; neutron-to-proton (11.1) **6**. F; nuclear fusion (11.12) **7**. T (11.12) **8**. F; electrons (11.8)
9. F; larger (11.13) **10**. F; therapeutic application (11.11) **11**. T (11.7)
12. F; an unpaired electron (11.8) **13**. T (11.13) **14**. F; α-particle (11.10) **15**. F; another (11.5)
16. F; gamma rays (11.11) **17**. F; 7.0 g (11.4) **18**. b (11.3) **19**. a (11.3) **20**. d (11.4) **21**. b (11.4)
22. b (11.2) **23**. e (11.9) **24**. b (11.12) **25**. b (11.11)

Saturated Hydrocarbons — Chapter 12

Chapter Overview

Carbon compounds are the basis of life on Earth; all organic materials are carbon-based. The hydrocarbons, which you will study in this chapter, make up petroleum and are thus an important part of the industrial world, as fuel and in the manufacture of synthetic materials.

In this chapter you will find out how carbon forms such a vast variety of compounds. You will draw various types of structural formulas for alkanes and cycloalkanes and name them according to the IUPAC rules. You will identify and draw constitutional isomers and *cis-trans* isomers. You will write equations for the two major reactions of hydrocarbons and name halogenated hydrocarbons.

Practice Exercises

12.1 **Organic chemistry** (Sec. 12.1) is the study of **hydrocarbons** (compounds of hydrogen and carbon, Sec. 12.3) and **hydrocarbon derivatives** (compounds that contain carbon and hydrogen and one or more additional elements, Sec. 12.3). Carbon has four valence electrons and forms four covalent bonds (four shared pairs of electrons). These bonds may be: four single bonds; two single bonds and one double bond; two double bonds; or one single bond and one triple bond.

Using one line (–) for a single bond, two lines (=) for a double bond, and three lines (≡) for a triple bond, draw **structural formulas** (Sec. 12.5) for the following hydrocarbons:

CH_4	C_2H_4	C_2H_2

12.2 Hydrocarbons may be **saturated** (containing only single carbon-to-carbon bonds, Sec. 12.3) or **unsaturated** (containing at least one carbon-to-carbon double or triple bond, Sec. 12.3) and cyclic (carbon atoms arranged in a ring) or acyclic (structural arrangement is not cyclic). Acyclic saturated hydrocarbons are called **alkanes** (Sec. 12.4).

Methane, ethane, propane, butane are alkanes containing one, two, three, and four carbon atoms, respectively. Complete the following table. (Keep in mind that each carbon atom in an alkane forms four single covalent bonds.)

Alkane	Molecular formula	Total number of atoms	Number of C–H bonds	Number of C–C bonds
methane				
ethane				
propane				
butane				

What is the general molecular formula for alkanes? ______________________

What is the geometrical arrangement of the atoms bonded to each carbon atom in an alkane?

____

12.3 Some common types of representations of alkanes are the **expanded structural formula** (Sec. 12.5), which shows all atoms and all bonds, the **condensed structural formula** (Sec. 12.5), which shows groupings of atoms, and the **skeletal structural formula** (Sec. 12.5), which shows carbon atoms and bonds but omits hydrogen atoms.

Complete the following table of structural representations. (All of the carbon atoms in the molecules represented below are connected in a straight chain.)

Molecular formula	Expanded structural formula	Condensed structural formula	Skeletal structural formula
C_2H_6	H H | | H–C–C–H | | H H		
C_4H_{10}			
		CH_3–CH_2–CH_2–CH_2–CH_3	
			C–C–C–C–C–C

12.4 **Isomers** (Sec. 12.6) are compounds that have the same molecular formula but differ in the way their atoms are arranged. **Constitutional isomers** (Sec. 12.6), also known as structural isomers, are isomers that differ in the order in which atoms are attached to each other within molecules.

Isomers of alkanes may be **continuous-chain** or **branched-chain** (Sec. 12.6).

Draw and name the nine constitutional isomers of heptane, C_7H_{16}. Use skeletal structural formulas. Rules for naming alkanes can be found in Section 12.8 of your textbook.

12.5 **Conformations** (Sec. 12.7) are differing orientations of a molecule made possible by rotation about a carbon-carbon single bond. Conformations are not isomers, because one conformation can change to another without breaking or forming bonds.

Using skeletal structural formulas, draw two possible conformations of the straight-chain alkane, heptane C_7H_{16}.

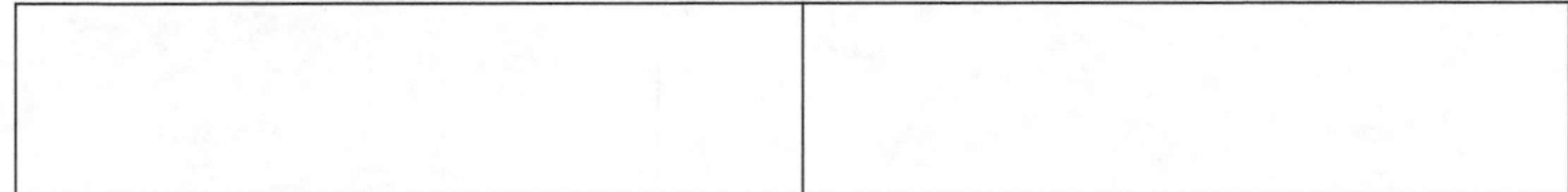

12.6 The IUPAC rules for naming organic compounds make it possible to give each compound a name that uniquely identifies it. Using that name, the structural formula of the molecule can be drawn.

Using the rules for nomenclature found in Section 12.8 in your textbook, give the IUPAC name for each of the following condensed structural formulas:

<table>
<tr><td>CH

$$\begin{array}{l}CH_3 \qquad\qquad\quad CH_3\\ \;| \qquad\qquad\qquad\;\; |\\ CH_2-CH_2-CH-CH_3\end{array}$$

a.</td><td>

$$\begin{array}{r}CH_3\;\;CH_3\\ |\qquad\;|\;\;\;\\ CH_3-CH_2-CH_2-CH-CH-CH_3\end{array}$$

b.</td></tr>
<tr><td>

$$\begin{array}{l}\qquad\qquad\qquad\quad\;\; CH_3\\ \qquad\qquad\qquad\quad\;\;\; |\\ CH_3 \qquad\qquad\qquad\; CH_2\;\;CH_3\\ \;| \qquad\qquad\qquad\qquad |\qquad\;\; |\\ CH_2-CH_2-CH_2-CH-CH-CH_3\end{array}$$

c.</td><td>

$$\begin{array}{l}\qquad\;\; CH_3\\ \qquad\;\;\; |\\ \qquad\;\; CH_2\\ \qquad\;\;\; |\\ CH_3\;\;CH_2 \qquad\qquad\; CH_3\;\;CH_3\\ \;|\qquad\;\;\; | \qquad\qquad\qquad\; |\qquad\;\; |\\ CH_2-CH-CH_2-CH-CH-CH_2-CH_3\end{array}$$

d.</td></tr>
</table>

12.7 Condensed structural formulas can be further condensed in two ways: linear (straight line) condensed structural formulas, in which parentheses are used to designate side chains on the longest carbon chain; and **line-angle structural formulas** (Sec. 12.9), in which a carbon-carbon bond is represented by a straight line segment, and a carbon atom is understood to be present at every point where two lines meet and at the ends of the lines. Line-angle structural formulas give a truer representation of the three-dimensional structure of the molecule than linear condensed structural formulas do.

Convert each condensed structural formula in Practice Exercise 12.6 above to both a linear condensed structural formula and a line-angle structural formula. Part a. is an example.

a. CH_3-CH-(CH_3)-CH_2-CH_2-CH_3	b.
c.	d.

12.8 Once you have learned the IUPAC rules for naming organic compounds, you can translate the name of an alkane into a structural formula.

Draw a linear (straight line) condensed structural formula and a line-angle structural formula for each of the following four alkanes:

a. 3-methylhexane	b. 2,2-dimethylpentane
c. 3,5-dimethylheptane	d. 2,2,4,4-tetramethylhexane

12.9 Each carbon atom in a hydrocarbon can be classified according to the number of other carbon atoms to which it forms bonds: a **primary carbon atom** (1°) is bonded to one other carbon atom, a **secondary carbon atom** (2°) to two carbon atoms, a **tertiary carbon atom** (3°) to three carbon atoms, and a **quaternary carbon atom** (4°) to four other carbon atoms (Sec. 12.10).

Using the structural formulas that you drew in Practice Exercise 12.8, determine the total number of primary, secondary, tertiary, and quaternary carbons in each compound.

Compound	Number of 1° carbon atoms	Number of 2° carbon atoms	Number of 3° carbon atoms	Number of 4° carbon atoms
3-methylhexane				
2,2-dimethylpentane				
3,5-dimethylheptane				
2,2,4,4-tetramethylhexane				

12.10 There are four common **alkyl groups** (Sec. 12.8) containing **branched chains** (Sec. 12.6, 12.11) that you should learn to name and draw.

Complete the following table by drawing the condensed structural formulas for these four important branched-chain alkyl groups:

a. isopropyl	b. *tert*-butyl	c. isobutyl	d. *sec*-butyl

12.11 In a **cycloalkane** (Sec. 12.12), the carbon atoms are attached to one another in a ringlike arrangement. The general formula for cycloalkanes is C_nH_{2n}. IUPAC naming procedures for cycloalkanes are found in Section 12.13 of your textbook. Line-angle structural formulas are often used to represent cycloalkane structures.

Give the IUPAC names for the following cycloalkanes:

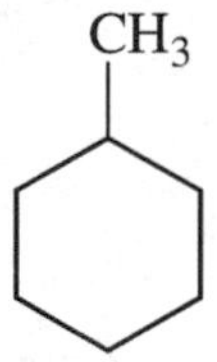

CH_3	CH_3 / $-CH_3$		
a.	b.	c.	d.

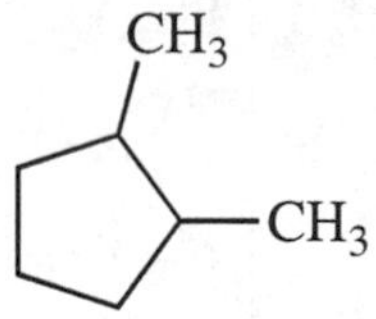

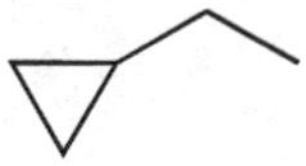

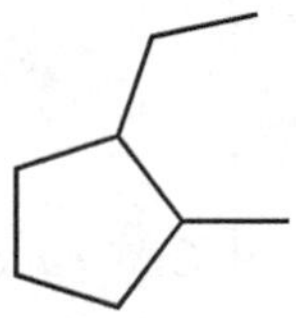

12.12 The problems in this exercise give a review of structural concepts and constitutional isomerism in alkanes and cycloalkanes. Compare the following pairs of molecules. Use line-angle notation to draw a structural formula for each molecule. Write the molecular formula for each molecule, and tell whether the molecules in each pair are isomers of one another.

a. methylcyclohexane and ethylcyclobutane

b. 2,3-dimethylpentane and 2,2,3-trimethylbutane

c. ethylcyclopropane and 2-methylbutane

12.13 **Stereoisomers** (Sec. 12.14) have the same molecular and structural formulas, but different orientations of atoms in space. The form of stereoisomerism associated with cycloalkanes is ***cis-trans* isomerism** (Sec. 12.14). *Cis-trans* isomers have the same molecular and structural formulas but different spatial arrangements of their atoms because rotation is restricted around bonds. The prefix *cis-* in the compound's name indicates that atoms or groups are on the same side of the molecule; *trans-* indicates that they are on opposite sides.

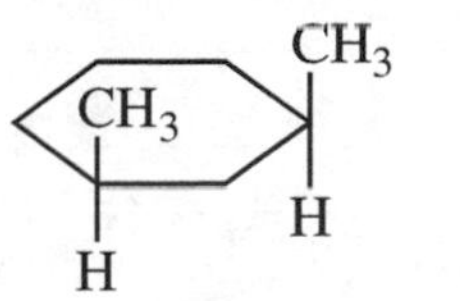

Give the IUPAC names for the *cis-trans* isomers in parts a and b.

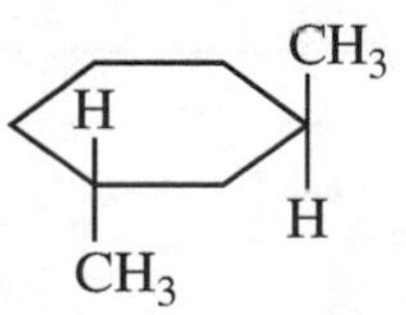

CH_3 H H CH_3	CH_3 CH_3 H H
a.	b.

12.14 In the spaces below, draw structural formulas for:

a. a cyclohexane ring with two methyl groups, one on carbon 1 and one on carbon 2.

b. a cyclohexane ring with two methyl groups on carbon-1.

c. Determine whether *cis-trans* isomerism is possible for the structures you drew in parts a. and b. (If so, draw both isomers.) Give the IUPAC name for each structure.

a.	b.
Cis-trans isomerism? ______ name:	*Cis-trans* isomerism? ______ name:

12.15 Some of the physical properties of alkanes are influenced by the nonpolarity of alkane molecules. Predict whether each of these physical properties of alkanes would be greater than or less than the property of a polar molecule of similar molecular weight.

a. Solubility of an alkane in water: ____________________

b. Boiling point of an alkane: ____________________

12.16 Alkanes undergo few reactions. One of their principal reactions is **combustion** (Sec. 12.17). Complete combustion is the chemical reaction of a hydrocarbon with oxygen to produce carbon dioxide and water. Incomplete combustion may occur when not enough oxygen is available; in this case, some carbon monoxide (CO, a poisonous gas) or elemental carbon may be formed.

Write balanced chemical equations for the complete combustion of the following alkanes:

a. pentane:

b. 2,3-dimethylhexane:

12.17 The other important chemical reaction of alkanes is **halogenation** (Sec. 12.17), a substitution reaction in which a halogen atom is substituted for a hydrogen atom. Complete the general equation below for the substitution of a single hydrogen atom in an alkane with a halogen atom, X. (The symbol R is used to represent an alkyl group in a structural formula. For example, R—H is a generic representation for an alkane.)

$$R{-}H \; + \; X_2 \xrightarrow[\text{light}]{\text{heat or}}$$

12.18 The compounds shown below are products of the halogenation of alkanes. Give the IUPAC name for each compound:

$CH_3{-}CH_2{-}CH_2{-}\underset{\displaystyle Cl}{\underset{\vert}{CH}}{-}CH_3$ a.	$CH_3{-}CH_2{-}\overset{\displaystyle Cl}{\overset{\vert}{CH}}{-}CH_2{-}Cl$ b.
$CH_3{-}CH_2{-}\overset{\displaystyle Cl}{\overset{\vert}{CH}}{-}\underset{\displaystyle CH_3}{\underset{\vert}{CH}}{-}CH_3$ c.	(cyclopentane ring with H and Br on one carbon, Br and H on adjacent carbon: H, Br, H, Br) d.

12.19 Draw and name the four constitutional isomers of dichloropropane ($C_3H_6Cl_2$), obtained by the substitution in propane of two atoms of chlorine for two atoms of hydrogen.

a.	b.
c.	d.

Self-Test

True-false: Indicate whether the following statements are true or false. If the statement is false, give the word or phrase that may be substituted for the underlined portion to make the statement true.

1. The smallest alkane is ethane.
2. Alkyl groups cannot rotate around the single bonds between the carbons in the ring of a cycloalkane.
3. Straight-chain pentane would have a higher boiling point than its constitutional isomer, 2,2-dimethylpropane.
4. Carbon atoms can form compounds containing chains and rings because carbon has six valence electrons.
5. A hydrocarbon contains only carbon, hydrogen, and oxygen atoms.
6. A saturated hydrocarbon contains at least one carbon-carbon double bond or triple bond.
7. The alkane 2-methylpentane contains one tertiary carbon atom.
8. An acyclic hydrocarbon contains carbon atoms arranged in a ring structure.
9. The alkyl group $-CH_2-CH_2-CH_3$ is named propanyl.
10. It takes a minimum of three carbon atoms to form a cyclic arrangement of carbon atoms.
11. The general formula for a cycloalkane is C_nH_{2n}.
12. Natural gas consists mainly of ethane.
13. Constitutional isomers have different physical properties.
14. The reaction of alkanes with oxygen to form carbon dioxide and water is a substitution reaction.
15. The IUPAC name for isopropyl chloride is 2-chloropropane.
16. HFCs are more reactive than CFCs.

Multiple choice:

17. Which of the following is an organic compound?

    a. $MgCO_3$ b. CH_3Cl c. CO d. NaCN e. none of these

18. The straight-chain alkane having the molecular formula C_6H_{14} is called:

    a. hexane b. heptane c. pentane
    d. nonane e. none of these

19. Using IUPAC rules for naming, if the parent compound is pentane, an ethyl group could be attached to carbon:

a. 1 b. 2 c. 3 d. 4 e. 5

20. How many constitutional isomers can butane have?

a. two b. three c. four d. five e. six

21. Cyclopropane has the following molecular formula:

a. CH_4 b. C_2H_6 c. C_3H_8 d. C_4H_{10} e. none of these

22. Which of the following compounds is an isomer of hexane?

a. methylcyclopentane b. 2-methylbutane c. 3-ethylpentane
d. 2-methylpentane e. none of these

23. Which of the following compounds forms *cis-trans* isomers?

a. 2,3-dimethylpentane b. 2,2-dimethylpentane
c. 1,1-dimethylcyclopentane d. 1,2-dimethylcyclopentane
e. none of these

24. Choose the correct name for this compound:

a. *trans*-1,3-dibromocyclohexane
b. *cis*-1,3-dibromohexane
c. *trans*-1,5-dibromocyclohexane
d. *cis*-1,2-dibromocyclohexane
e. none of these

Br H H Br

25. Which of the following compounds can have constitutional isomers?

a. CH_3Cl b. C_3H_7Cl c. C_3H_8 d. C_2H_5Cl e. none of these

26. Which of the following is a correct IUPAC name?

a. 2-methylcyclobutane b. *cis*-2,3-dimethylpentane
c. 1-methylbutane d. 3-ethylhexane
e. *cis*-1,2-methylpropane

27. Another name for chloroform is:

a. trichloromethane b. methylene chloride c. carbon tetrachloride
d. methyl chloride e. none of these

28. How many possible isomers can be written for dimethylcyclobutane?

a. two b. three c. four d. five e. six

29. Draw structures for the following compounds:

2-bromo-3-methylbutane	*trans*-1,2-dichlorocyclopropane

Answers to Practice Exercises

12.1

<table>
<tr><td>CH_4</td><td>C_2H_4</td><td>C_2H_2</td></tr>
<tr><td><pre>
  H
  |
H-C-H
  |
  H
</pre></td><td><pre>
  H H
  | |
H-C=C-H
</pre></td><td><pre>
H-C≡C-H
</pre></td></tr>
</table>

12.2

Alkane	Molecular formula	Total number of atoms	Number of C–H bonds	Number of C–C bonds
methane	CH_4	5	4	0
ethane	C_2H_6	8	6	1
propane	C_3H_8	11	8	2
butane	C_4H_{10}	14	10	3

The general molecular formula for alkanes is: C_nH_{2n+2}

The geometrical arrangement of the atoms bonded to each atom in an alkane is: tetrahedral

12.3

<table>
<tr><th>Molecular formula</th><th>Expanded structural formula</th><th>Condensed structural formula</th><th>Skeletal structural formula</th></tr>
<tr><td>C_2H_6</td><td><pre>
  H H
  | |
H-C-C-H
  | |
  H H
</pre></td><td>$CH_3—CH_3$</td><td>C—C</td></tr>
<tr><td>C_4H_{10}</td><td><pre>
  H H H H
  | | | |
H-C-C-C-C-H
  | | | |
  H H H H
</pre></td><td>$CH_3-CH_2-CH_2-CH_3$</td><td>C–C–C–C</td></tr>
<tr><td>C_5H_{12}</td><td><pre>
  H H H H H
  | | | | |
H-C-C-C-C-C-H
  | | | | |
  H H H H H
</pre></td><td>CH_3-CH_2-CH_2-CH_2-CH_3</td><td>C–C–C–C–C</td></tr>
<tr><td>C_6H_{14}</td><td><pre>
  H H H H H H
  | | | | | |
H-C-C-C-C-C-C-H
  | | | | | |
  H H H H H H
</pre></td><td>CH_3-CH_2-CH_2-CH_2-CH_2-CH_3</td><td>C–C–C–C–C–C</td></tr>
</table>

12.4

<table>
<tr><td><pre>
C-C-C-C-C
| |
C   C
</pre>heptane</td><td><pre>
C-C-C-C-C
    | |
    C C
</pre>2,3 dimethylpentane</td><td><pre>
C-C-C-C-C
    |
    C-C
</pre>3-ethylpentane</td></tr>
</table>

C–C–C–C–C–C (CH at C2 bears C) 2-methylhexane	C–C(C)(C)–C–C–C 2,2-dimethylpentane	C–C(C)–C–C(C)–C 2,4-dimethylpentane
C–C–C(C)(C)–C–C 3,3-dimethylpentane	C–C–C–C(C)–C–C 3-methylhexane	C–C(C)(C)–C(C)–C 2,2,3-trimethylbutane

12.5

C—C—C—C (C on first C; C—C on last C) heptane	C—C C—C—C C—C heptane

There are many more possible conformations of heptane; these are just two of the representations.

12.6 a. 2-methylpentane b. 2,3-dimethylhexane
c. 3-ethyl-2-methylheptane d. 6-ethyl-3,4-dimethylnonane

12.7 a. CH_3-CH_2-CH_2-CH-(CH_3)-CH_3
b. CH_3-CH-(CH_3)-CH-(CH_3)-CH_2-CH_2-CH_3
c. CH_3-CH-(CH_3)-CH-(CH_2-CH_3)-CH_2-CH_2-CH_2-CH_3
d. CH_3-CH_2-CH-(CH_3)-CH-(CH_3)-CH_2-CH-(CH_2-CH_3)-CH_2-CH_2-CH_3

a.	b.
c.	d.

12.8 a. CH_3-CH_2-CH-(CH_3)-CH_2-CH_2-CH_3
b. CH_3-C-$(CH_3)_2$-CH_2-CH_2-CH_3
c. CH_3-CH_2-CH-(CH_3)-CH_2-CH-(CH_3)-CH_2-CH_3
d. CH_3-C-$(CH_3)_2$-CH_2-C-$(CH_3)_2$-CH_2-CH_3

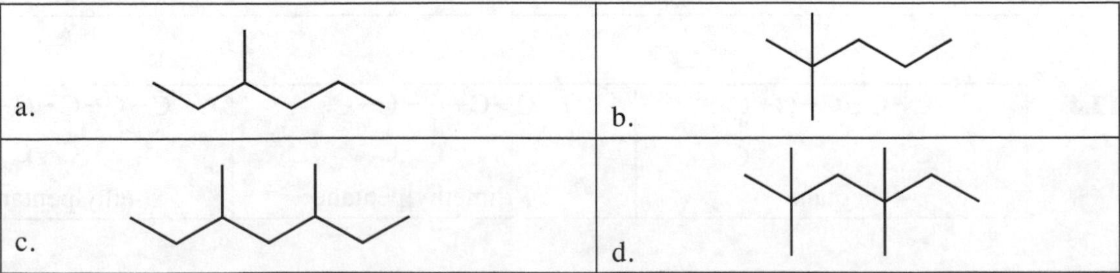

12.9

Compound	Number of 1° carbon atoms	Number of 2° carbon atoms	Number of 3° carbon atoms	Number of 4° carbon atoms
3-methylhexane	3	3	1	0
2,2-dimethylpentane	4	2	0	1
3,5-dimethylheptane	4	3	2	0
2,2,4,4-tetramethylhexane	6	2	0	2

12.10

CH_3-CH- with CH_3 on CH	CH_3-C- with CH_3 above and below C	$CH_3-CH-CH_2-$ with CH_3 on CH	CH_3-CH_2-CH- with CH_3 on CH
a. isopropyl	b. *tert*-butyl	c. isobutyl	d. *sec*-butyl

12.11 a. methylcyclohexane b. 1,2-dimethylcyclopentane
c. ethylcyclopropane d. 1-ethyl-2-methylcyclopentane

12.12

a.

methylcyclohexane C_7H_{14}

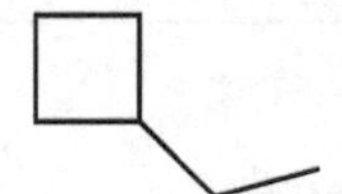

ethylcyclobutane C_6H_{12}

not isomers (different numbers of carbons and hydrogens)

b.

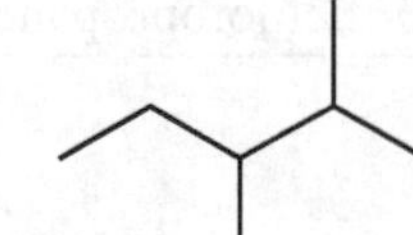

2,3-dimethylpentane C_7H_{16}

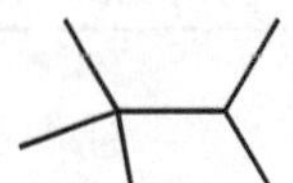

2,2,3-trimethylbutane C_7H_{16}

isomers (same molecular formula)

c.

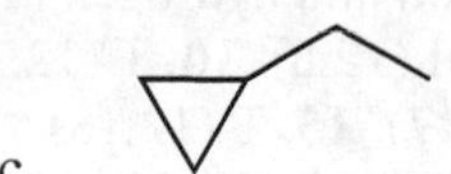

ethylcyclopropane C_5H_{10}

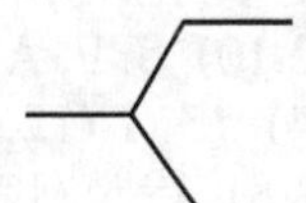

2-methylbutane C_5H_{12}

not isomers (different numbers of hydrogens)

12.13 a. *trans*-1,3-dimethylcyclohexane b. *cis*-1,3-dimethylcyclohexane

12.14

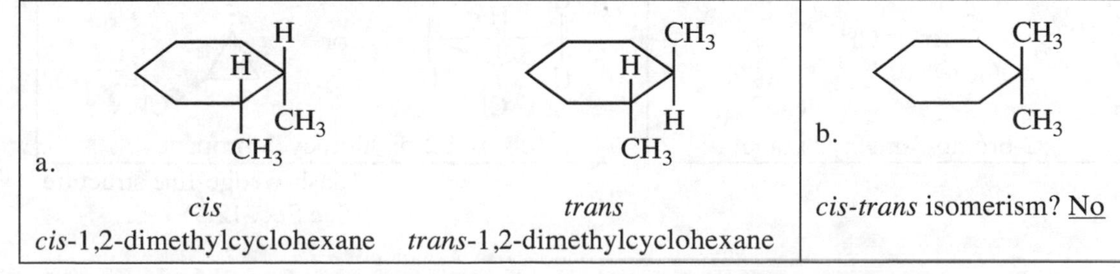

a. *cis* — *cis*-1,2-dimethylcyclohexane; *trans* — *trans*-1,2-dimethylcyclohexane

b. *cis-trans* isomerism? No

c. *Cis-trans* isomerism is possible in a., because the methyl groups can be on the same side of the molecule, or one methyl group can be above and one below. There are not *cis* and *trans* forms of the molecule in b., because both methyl groups are attached to the same carbon. The name of the molecule in b. is 1,1-dimethylcyclohexane.

12.15 a. Solubility of an alkane in water: Alkanes are nonpolar and are less soluble in water than polar molecules of comparable size. Water is polar, and "like dissolves like."

b. Boiling point of an alkane: Alkanes of low molecular weight have a lower boiling point than water because alkanes have weaker attractive forces between molecules.

12.16 a. pentane: $C_5H_{12} + 8O_2 \rightarrow 5CO_2 + 6H_2O$

b. 2,3-dimethylhexane: $2C_8H_{18} + 25O_2 \rightarrow 16CO_2 + 18H_2O$

12.17 $R{-}H + X_2 \xrightarrow[\text{light}]{\text{heat or}} R{-}X + H{-}X$

12.18 a. 2-chloropentane b. 1,2-dichlorobutane
c. 3-chloro-2-methylpentane d. *trans*-1,3-dibromocyclopentane

12.19

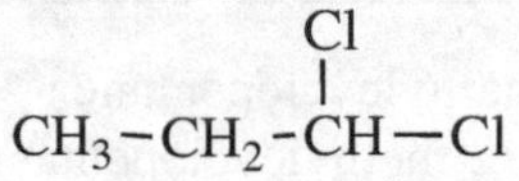 $CH_3{-}CH_2{-}CH(Cl){-}Cl$ a. 1,1-dichloropropane	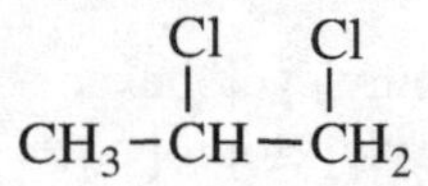 $CH_3{-}CH(Cl){-}CH_2(Cl)$ b. 1,2-dichloropropane
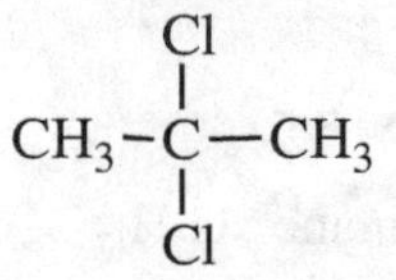 $CH_3{-}C(Cl)_2{-}CH_3$ c. 2,2-dichloropropane	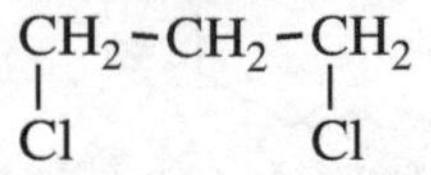 $CH_2(Cl){-}CH_2{-}CH_2(Cl)$ d. 1,3-dichloropropane

Answers to Self-Test

The numbers in parentheses refer to sections in your textbook.
1. F; methane (12.4) **2**. T (12.14) **3**. T (12.16) **4**. F; four (12.2) **5**. F; carbon and hydrogen (12.3) **6**. F; An unsaturated (12.3) **7**. T (12.10) **8**. F; A cyclic (12.12) **9**. F; propyl (12.8) **10**. T (12.12) **11**. T (12.12) **12**. F; methane (12.14) **13**. T (12.6) **14**. F; combustion (12.17) **15**. T (12.18) **16**. T (12.18) **17**. b (12.2) **18**. a (12.8) **19**. c (12.8) **20**. a (12.6) **21**. e; C_3H_6 (12.12) **22**. d (12.6) **23**. d (12.14) **24**. a (12.14) **25**. b (12.6, 12.18) **26**. d (12.8, 12.13, 12.14) **27**. a (12.12) **28**. d (12.14)
29.

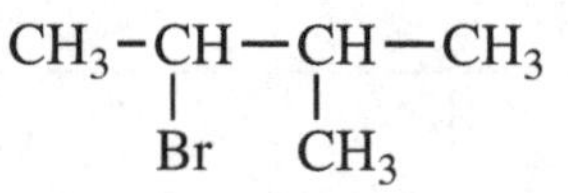 $CH_3{-}CH(Br){-}CH(CH_3){-}CH_3$ 2-bromo-3-methylbutane	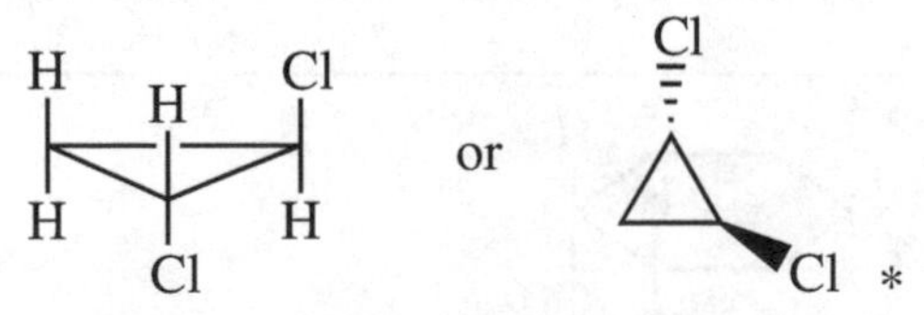 [cyclopropane ring with H, H, Cl above and H, H, Cl below] or [dash-wedge cyclopropane with Cl, Cl] * *trans*-1,2-dichlorocyclopropane

*dash-wedge-line structure
(See Sec. 12.9)

Unsaturated Hydrocarbons — Chapter 13

Chapter Overview

Saturated hydrocarbons, including the alkanes and cycloalkanes, do not contain multiple bonds; they are referred to as saturated because the carbon atoms are bonded to the maximum number of atoms to which a carbon atom can bond (four atoms).

Unsaturated hydrocarbons have one or more carbon-carbon double or triple bonds; they are unsaturated because the carbon atoms of the multiple bond are each bonded to less than four atoms. Because a multiple bond is more easily broken than a single bond, unsaturated hydrocarbons are more reactive chemically than saturated hydrocarbons.

In this chapter you will name and write structural formulas for the three types of unsaturated hydrocarbons: alkenes, alkynes, and aromatic hydrocarbons. You will learn the characteristics of these compounds and their most common reactions.

Practice Exercises

13.1 An **alkene** (Sec. 13.2) contains one or more carbon-carbon double bonds. The IUPAC names for alkenes (Sec. 13.3) are similar to those for alkanes, but the *-ene* ending replaces the *-ane* ending. The longest carbon chain must contain the double bond. The location of the double bond is indicated with a single number, that of the first carbon atom of the double bond.

Write the IUPAC names for the following alkenes and **cycloalkenes** (Sec. 13.2).

$CH_3-\underset{\displaystyle CH_3}{\underset{\vert}{C}}=CH_2$ a.	$CH_3-CH_2-CH_2-\underset{\displaystyle CH_3}{\underset{\vert}{CH}}-CH=CH_2$ b.
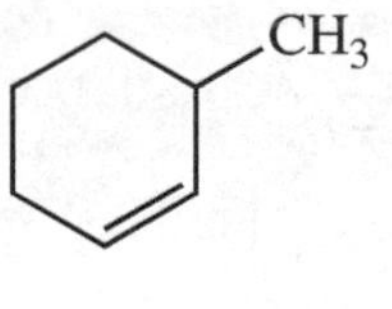 CH_3 (on cyclohexene ring) c.	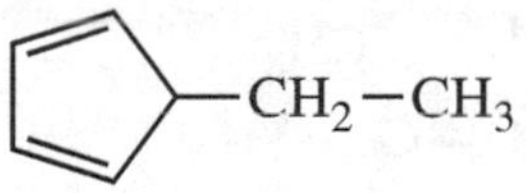 (cyclopentadiene ring)$-CH_2-CH_3$ d.

13.2 Draw structural formulas for the following alkene, diene, and cycloalkene.

a. 3-methyl-1-pentene	b. 2-methyl-1,3-pentadiene	c. 1-methylcyclopentene

13.3 Using common names for the three most frequently encountered alkenyl groups (listed in Section 13.3 in your textbook), give common names for the following structures.

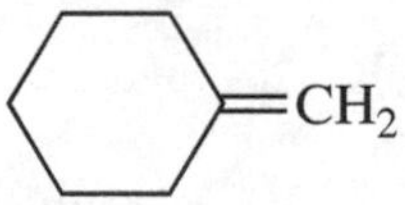 (cyclohexane ring)$=CH_2$ a.	$CH_2=CH-Br$ b.	$CH_2=CH-CH_2-I$ c.

13.4 Constitutional isomers exist for some alkenes. They may be **positional isomers** (Sec. 13.5), which have the same carbon-atom arrangement but a different location of the **functional group** (Sec. 13.1), in this case the multiple bond, or **skeletal isomers** (13.5), which differ in their carbon-atom arrangements.

a. In the space below, draw skeletal structural diagrams for the three positional isomers of 2-methylbutene, and give the IUPAC name for each compound.

b. There are two pentenes that are skeletal isomers of the three 2-methylbutenes drawn above. Draw the two pentenes and give the IUPAC name for each of them.

13.5 *Cis-trans* isomerism is possible for some alkenes (Sec. 13.6) because the double bond is rigid, preventing rotation around its axis. For *cis-trans* isomers to exist, each of the two carbon atoms of the double bond must have two different groups attached to it.
Determine whether each of these alkenes can exist as *cis-trans* isomers. If *cis-trans* isomers do exist, draw the structural formulas and give the IUPAC name for each isomer.

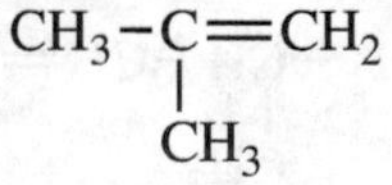 $CH_3-C(CH_3)=CH_2$ a.	$CH_3-CH=CH-CH_2-CH_3$ b.
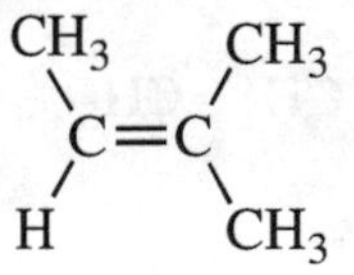 $(CH_3)(H)C=C(CH_3)(CH_3)$ (CH₃ and H on left carbon; CH₃ and CH₃ on right carbon) c.	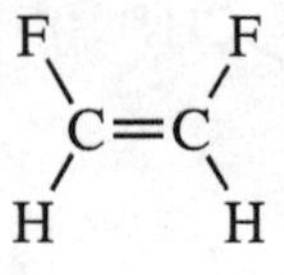 $(F)(H)C=C(F)(H)$ (F top, H bottom on each carbon) d.

e. Is *cis-trans* isomerism possible for any of the compounds you drew in Practice Exercise 13.4?________________________________

13.6 Draw structural formulas and line-angle structural formulas (Sec. 13.4) for these alkenes:

a. *trans*-3-hexene	b. *cis*-2-hexene	c. *trans*-1,2-dibromoethene

13.7 One method for the preparation of alkenes is **dehydrogenation** (Sec. 13.9), a chemical reaction in which a hydrogen atom is lost from each of two adjacent carbon atoms, and a carbon-carbon double bond is formed. Dehydrogenation reactions take place only at high temperatures and in the presence of a special catalyst.

Write a chemical equation showing the preparation of propene from propane by dehydrogenation. Use expanded structural formulas to represent reactant and products.

13.8 The most important reactions of alkenes are **addition reactions** (Sec. 13.10). **Symmetrical addition reactions** (Sec. 13.10), such as **hydrogenation** (Sec. 13.10) and **halogenation** (Sec. 13.10), involve adding identical atoms to each carbon of the double bond. In hydrogenation, a hydrogen atom is added to each carbon of the double bond by heating the alkene and H_2 in the presence of a catalyst. Halogenation involves the use of Br_2 or Cl_2 to add a halogen atom to each carbon of the double bond.

Complete the following equations. Give the structural formula for each product.

a. $CH_3-CH_2-CH=CH_2 + H_2 \xrightarrow[\text{catalyst}]{\text{Ni}}$

b. $CH_3-CH_2-CH=CH_2 + Br_2 \longrightarrow$

c. 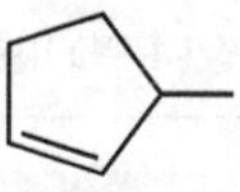 (cyclopentene ring)$-CH_3 + H_2 \xrightarrow[\text{catalyst}]{\text{Ni}}$

13.9 Addition reactions may also be **unsymmetrical** (Sec. 13.10); different atoms or groups of atoms are added to the carbons of the double bond. **Hydrohalogenation** (Sec. 13.10), the addition of a hydrogen halide (usually HCl or HBr), and **hydration** (Sec. 13.10), the addition of water in the presence of H_2SO_4 as catalyst, are important types of unsymmetrical addition.

Markovnikov's rule (Sec. 13.9) states that in an unsymmetrical addition, the hydrogen atom from the molecule being added becomes attached to the unsaturated carbon atom that already has the most hydrogen atoms.

Complete the following equations. Although two isomeric products are possible in each case, one product usually predominates. Give the structural formula for the major expected product. Use Markovnikov's rule.

a. $CH_3-CH_2-CH=CH_2 + HBr \longrightarrow$

b. $CH_3-CH_2-CH=CH_2 + H_2O \xrightarrow{H_2SO_4}$

c. 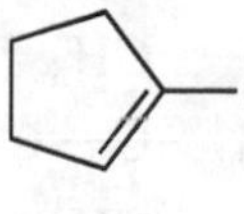 (cyclopentene ring)$-CH_3 + HBr \longrightarrow$

d. 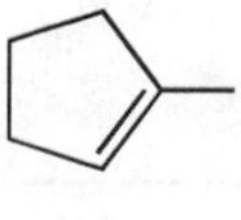 (cyclopentene ring)$-CH_3 + H_2O \xrightarrow{H_2SO_4}$

13.10 Write the name of the alkene that could be used to prepare each of the following compounds. Remember Markovnikov's rule.

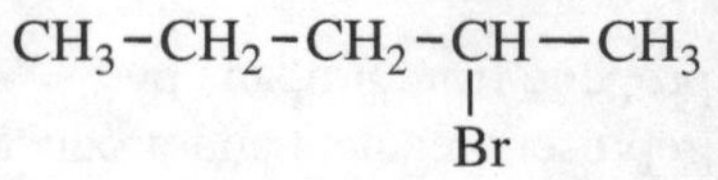 $CH_3-CH_2-CH_2-\underset{Br}{\underset{\vert}{CH}}-CH_3$ a.	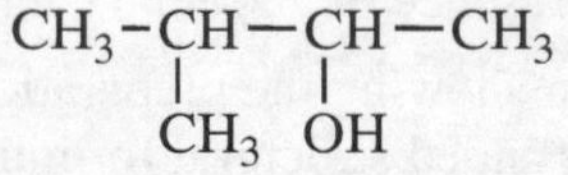 $CH_3-\underset{CH_3}{\underset{\vert}{CH}}-\underset{OH}{\underset{\vert}{CH}}-CH_3$ b.	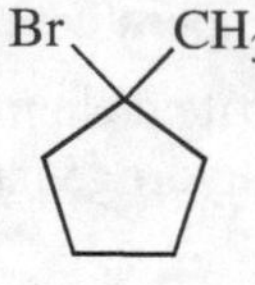 1-bromo-1-methylcyclopentane (Br and CH_3 on the same ring carbon) c.

13.11 How many molecules of hydrogen gas, H_2, would react with one molecule of each of the following compounds? Give the structure and IUPAC name for each product.

$CH_2{=}CH-CH_2-CH{=}CH_2$ a.	 cycloheptene ring structure b.

13.12 A **polymer** (Sec. 13.11) is a very large molecule composed of many identical repeating units called **monomers** (Sec. 13.11). Alkenes can form **addition polymers** (Sec. 13.11) when the alkene monomers simply add together. The double bond of the alkene is broken, and single carbon-carbon bonds form between monomers.

Draw condensed structural formulas for the monomer units from which these addition polymers were made and for the first three repeating units of the polymers.

General formula of polymer	Monomer	First three units of polymer
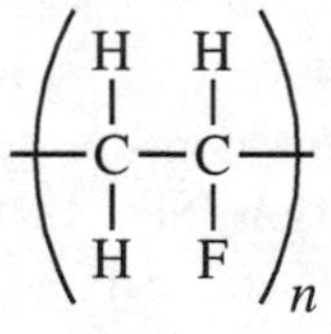 $\left(-\underset{H}{\overset{H}{C}}-\underset{F}{\overset{H}{C}}-\right)_n$		
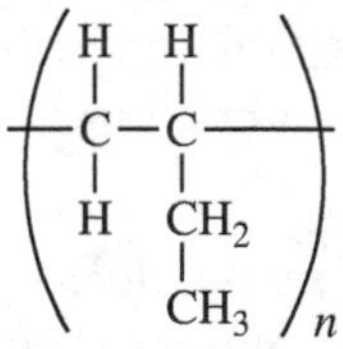 $\left(-\underset{H}{\overset{H}{C}}-\underset{CH_2CH_3}{\overset{H}{C}}-\right)_n$		

13.13 An **alkyne** (Sec. 13.12) has one or more carbon-carbon triple bonds. The rules for naming alkynes are the same as those for naming alkenes, with the ending *-yne* instead of *-ene*.

Give the IUPAC names for the following alkynes:

$CH_3-CH_2-CH_2-C{\equiv}CH$ a.	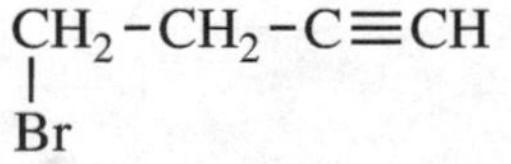 $\underset{Br}{\underset{\vert}{CH_2}}-CH_2-C{\equiv}CH$ b.	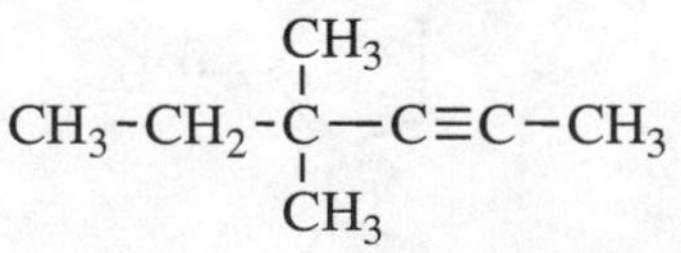 $CH_3-CH_2-\underset{CH_3}{\overset{CH_3}{C}}-C{\equiv}C-CH_3$ c.

13.14 Draw condensed structural formulas and line-angle structural formulas for these alkynes. An alkyne's triple bond has a linear structure.

a. 3-methyl-1-pentyne	b. 4-chloro-2-pentyne	c. 4,5-dimethyl-2-hexyne

13.15 Addition reactions of alkynes are similar to those of alkenes. However, two molecules of a specific reactant can add to the triple bond.

Complete the following reactions. Give the structural formula for the major expected product.

a. $CH_3{-}CH_2{-}CH_2{-}C{\equiv}CH \quad + \quad 2Cl_2 \longrightarrow$

b. $CH_3{-}CH_2{-}CH_2{-}C{\equiv}CH \quad + \quad 2HBr \longrightarrow$

c. $CH_3{-}CH_2{-}CH_2{-}C{\equiv}CH \quad + \quad 1HCl \longrightarrow$

d. $CH_3{-}CH_2{-}CH_2{-}C{\equiv}CH \quad + \quad 2H_2 \xrightarrow[\text{catalyst}]{\text{Ni}}$

13.16 **Aromatic hydrocarbons** (Sec. 13.13) are unsaturated cyclic compounds that do not readily undergo addition reactions. The carbon-carbon bonds in simple aromatic hydrocarbons, such as benzene, are all equivalent. The double and single bonds in conventional structural diagrams of these compounds are actually **delocalized bonds** (Sec. 13.13) with electrons shared among more than two atoms.

The naming of the compounds below is based on the aromatic hydrocarbon benzene. The name of a substituent on the benzene ring is used as a prefix. A benzene ring with two substituents (a disubstituted benzene) may be named using either numbered positions on the ring or one of the prefixes *ortho-* (*o*-), *meta-* (*m*-), and *para-* (*p*-) (Sec. 13.14).

Using the two different prefix systems (a. numbered and b. *ortho-*, *meta-*, *para-*), give two names for each of these compounds.

1. 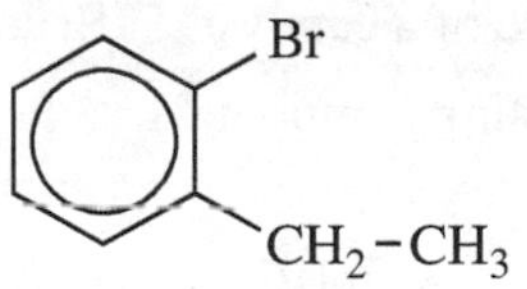 a. b.	2. 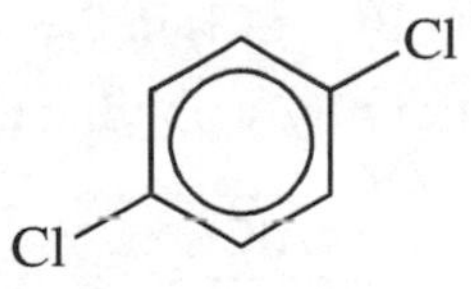 a. b.	3. 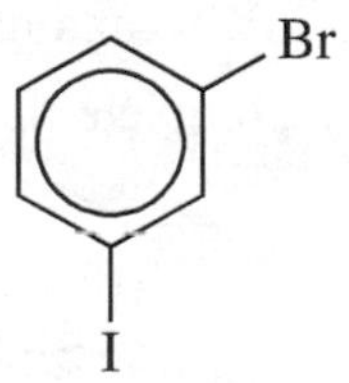 a. b.

13.17 Some substituents change the name of the parent molecule (benzene). Benzene with one methyl substituent is called toluene; with two methyl substituents, it is called xylene. Name the following compounds, using both IUPAC naming systems:

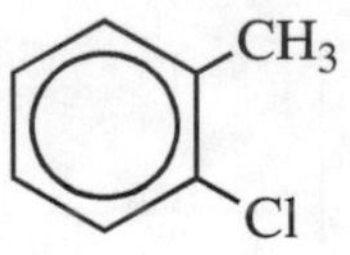 a. (benzene ring with CH_3 and Cl on adjacent carbons)	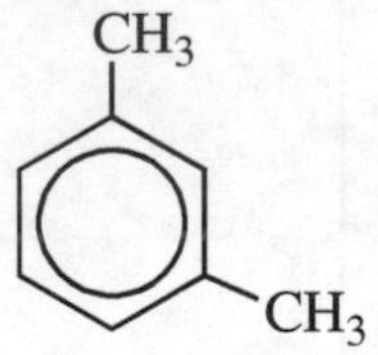 b. (benzene ring with CH_3 at top and CH_3 at meta position)	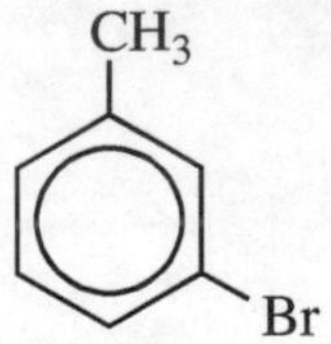 c. (benzene ring with CH_3 at top and Br at meta position)

13.18 If the group attached to the benzene ring is not easily named as a substituent, the benzene ring is treated as the attached group and called a *phenyl* group. The compound is then named as an alkane, an alkene, or an alkyne.

Write the IUPAC name for the following phenyl-substituted compounds:

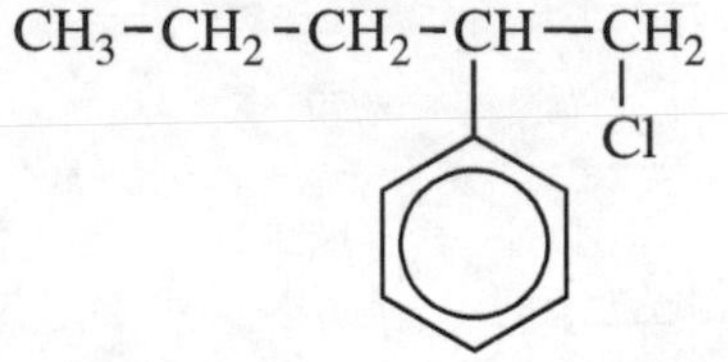 a. $CH_3-CH_2-CH_2-CH-CH_2$ (Cl on CH_2; phenyl on CH)	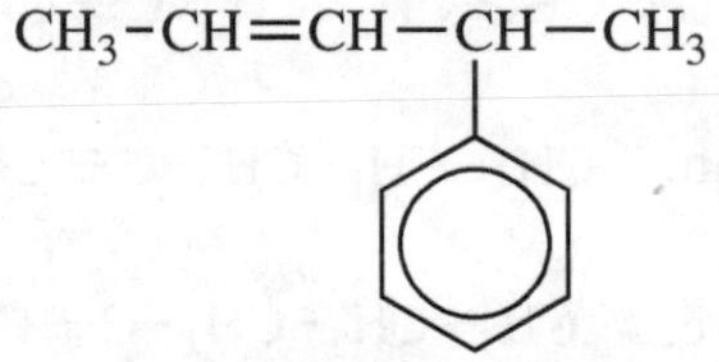 b. $CH_3-CH=CH-CH-CH_3$ (phenyl on CH)

13.19 Draw structural formulas for the following compounds:

a. 1,3-diiodobenzene	b. *p*-bromoethylbenzene	c. 3-phenyl-1-hexene

13.20 Because of the stability of the aromatic ring, aromatic hydrocarbons do not readily undergo addition reactions. The most important reactions of aromatic compounds are substitution reactions, in which an atom or group of atoms is substitituted for one of the hydrogen atoms on the benzene ring. These reactions include alkylation (using alkyl halides and the catalyst $AlCl_3$) and halogenation (using Br_2 or Cl_2 in the presence of a catalyst, $FeBr_3$ or $FeCl_3$).

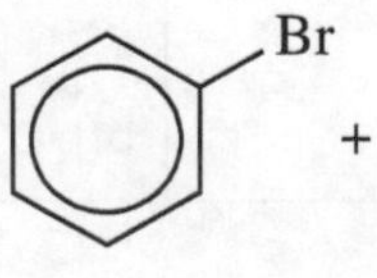

Complete the following equations by supplying the missing compounds.

a.  (benzene) + CH_3Cl $\xrightarrow{AlCl_3}$? + HCl

b. ? + ? $\xrightarrow{FeBr_3}$ (bromobenzene: benzene ring with Br) + HBr

c. [benzene] + Cl_2 $\xrightarrow{FeCl_3}$? + ?

d. ? + ? $\xrightarrow{?}$ [benzene ring]$-CH_2-CH_3$ + HCl

13.21 Use this identification exercise to review your knowledge of the structures introduced in this chapter and in Chapter 12. Using structures A through F, give the best choice for each of the terms below.

A.	B.	C.
$CH_3(H)C{=}C(H)CH_3$ (CH₃ and H on first carbon; H and CH₃ on second, trans)	$H(CH_3)C{=}C(H)CH_3$ (both H up, both CH₃ down, cis)	$CH_3-CH_2-CH_2-CH_3$
D. CH_3-[cyclohexene ring]	**E.** [benzene ring with CH_3 and adjacent Cl]	**F.** $CH_3-CH_2-C{\equiv}CH$

a. aromatic compound __________
b. isomer of compound A __________
c. alkyne __________
d. cyclic alkene __________
e. *trans*-isomer of an alkene __________
f. straight-chain alkane __________

Give IUPAC names for A through F:
g. Structure A __________
h. Structure B __________
i. Structure C __________
j. Structure D __________
k. Structure E __________
l. Structure F __________

Self-Test

True-false: Indicate whether the following statements are true or false. If the statement is false, give the word or phrase that may be substituted for the underlined portion to make the statement true.

1. The functional group of an alkyne is a carbon-carbon double bond.
2. The general molecular formula for an alkene is C_nH_{2n}.
3. When naming alkenes by the IUPAC system, select as the parent carbon chain the longest chain that contains at least one carbon atom of the double bond.
4. The name of a cycloalkene does not include the numbered location of the double bond.
5. The carbon atoms at each end of a double bond have a trigonal planar arrangement of bonds.
6. The common name for the simplest alkyne is ethylene.
7. Benzene undergoes an addition reaction with bromine in the presence of a catalyst.
8. Propene is the simplest alkene that has *cis-trans* isomerism.
9. The reverse reaction for hydrogenation is hydration.

10. *Cis-trans* isomers <u>are not</u> constitutional isomers.
11. If 2-butene is halogenated with bromine gas, the most probable product is <u>2-bromobutane</u>.
12. Hydration of an alkene produces <u>an alcohol</u>.
13. Polyethylene contains <u>many</u> double bonds.

Multiple choice:

14. According to Markovnikov's rule, addition of HBr to the double bond of 1-hexene would produce:

    a. 1-bromohexane b. 1,2-dibromohexane c. 2-bromo-1-hexene
    d. 2-bromohexane e. none of these

15. The geometry of the carbon-carbon triple bond of an alkyne is:

    a. linear b. trigonal planar c. tetrahedral
    d. angular e. none of these

16. Which of the following is a correct name according to IUPAC rules?

    a. 2-methylbenzene b. 1-chlorobenzene c. 1-bromo-2-chlorobenzene
    d. 2,4-dichlorobenzene e. none of these

17. Another name for *meta*-dichlorobenzene is:

    a. 1,2-dichlorobenzene b. 2,3-dichlorobenzene c. 1,3-dichlorobenzene
    d. 1,4-dichlorobenzene e. none of these

18. How many mono-chloro isomers of 1-butene are possible?

    a. four b. one c. five d. three e. none of these

19. Aromatic compounds have structures based on what parent molecule?

    a. benzene b. cyclopropane c. cyclohexane
    d. hexane e. none of these

20. Which of the following compounds does *not* have the formula C_5H_8?

    a. 1-methylcyclobutene b. 3-methyl-1-butyne c. 2-methyl-1-butene
    d. 1,3-pentadiene e. isoprene

21. If you wished to prepare 2-bromo-3-methylbutane by the addition of HBr to an alkene, which of these alkenes would you use?

    a. 2-methyl-2-butene b. 2-methyl-1-butene c. 3-methyl-1-butene
    d. 3-methyl-2-butene e. none of these

22. Which of the following compounds would be the most likely to undergo an addition reaction with HCl?

    a. toluene b. cyclohexene c. benzene
    d. heptane e. none of these

23. Which of the following is a fused-ring aromatic compound?

    a. terpene b. isoprene c. xylene
    d. naphthalene e. toluene

24. The antioxidant substance that acts as a Vitamin A precursor is:

    a. limonene   b. *β*-carotene   c. benzene
    d. menthol   e. retinal

Answers to Practice Exercises

13.1 a. 2-methylpropene
b. 3-methyl-1-hexene
c. 3-methylcyclohexene
d. 5-ethyl-1,3-cyclopentadiene

13.2

$CH_3-CH_2-\underset{\displaystyle CH_3}{\underset{\mid}{CH}}-CH=CH_2$	$CH_2=\underset{\displaystyle CH_3}{\underset{\mid}{C}}-CH=CH-CH_3$	(cyclopentene ring)$-CH_3$
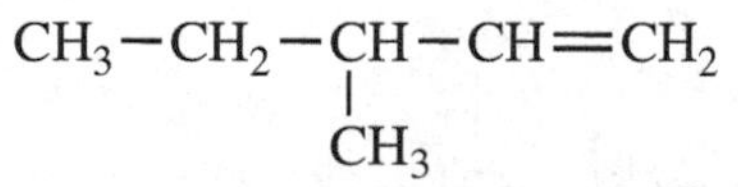 a. 3-methyl-1-pentene	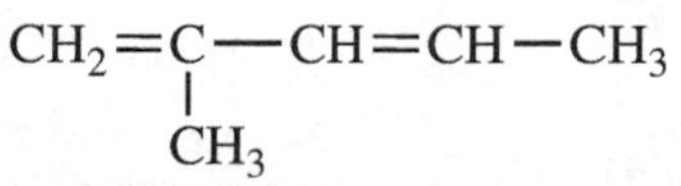 b. 2-methyl-1,3-pentadiene	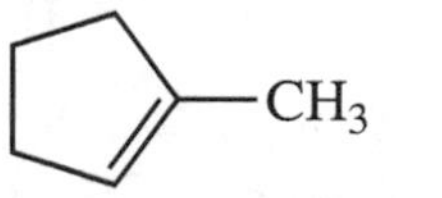 c. 1-methylcyclopentene

13.3 a. methylene cyclohexane
b. vinyl bromide
c. allyl iodide

13.4 a.

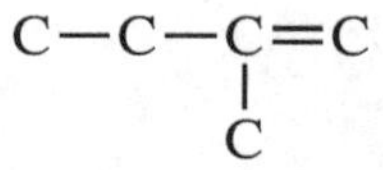

$C-C-\underset{\displaystyle C}{\underset{\mid}{C}}=C$ 2-methyl-1-butene

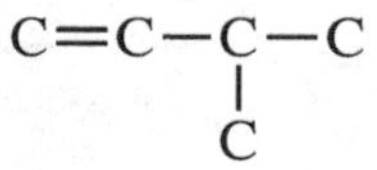

$C=C-\underset{\displaystyle C}{\underset{\mid}{C}}-C$ 3-methyl-1-butene

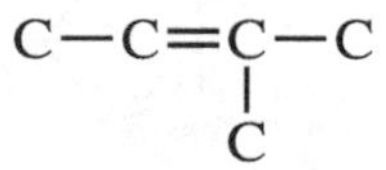

$C-C=\underset{\displaystyle C}{\underset{\mid}{C}}-C$ 2-methyl-2-butene

b.

$C=C-C-C-C$ 1-pentene

$C-C=C-C-C$ 2-pentene

13.5 a. no *cis-trans* isomers (2-methylpropene)
b. yes; *trans*-2-pentene and *cis*-2-pentene

$CH_3(H)C=C(H)CH_2-CH_3$	$CH_3(H)C=C(H)CH_2-CH_3$
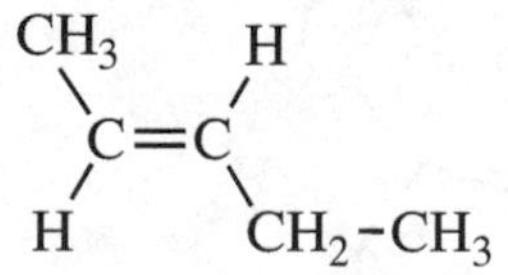 *trans*-2-pentene	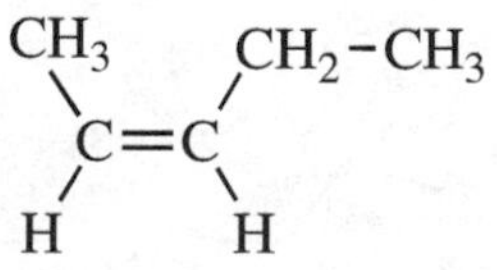 *cis*-2-pentene

c. no *cis-trans* isomers (2-methyl-2-butene)
d. yes; *cis*-1,2-difluoroethene and *trans*-1,2-difluoroethene

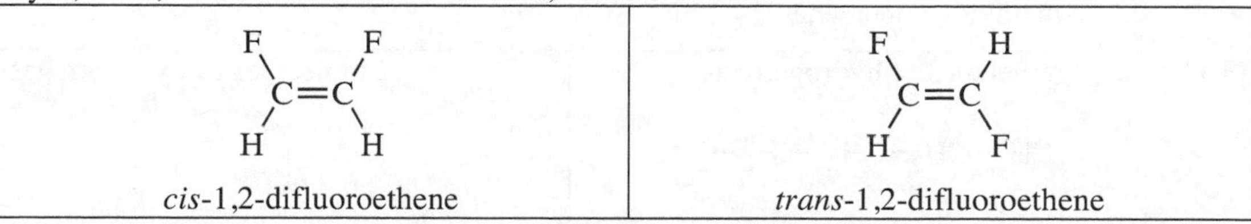

cis-1,2-difluoroethene   *trans*-1,2-difluoroethene

e. *Cis-trans* isomerism is possible for 2-pentene, but not for the other four compounds in Practice Exercise 13.4.

13.6

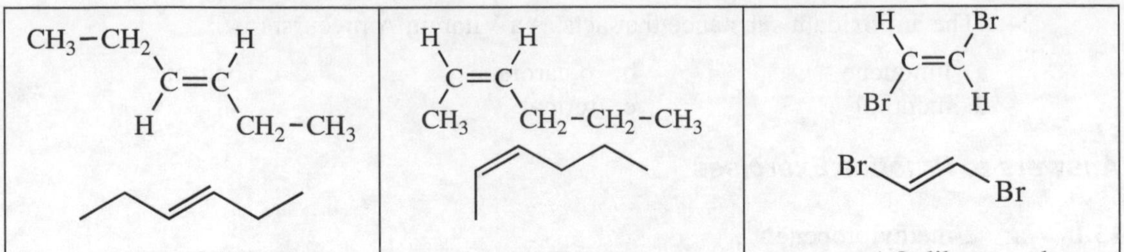

a.	b.	c.
$CH_3-CH_2(H)C=C(H)CH_2-CH_3$	$H(CH_3)C=C(H)CH_2-CH_2-CH_3$	$H(Br)C=C(Br)H$
a. *trans*-3-hexene	b. *cis*-2-hexene	c. *trans*-1,2-dibromoethene

13.7

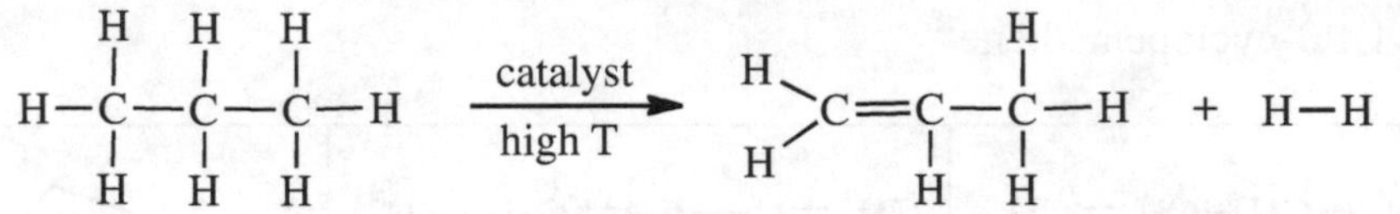

$$CH_3-CH_2-CH_3 \xrightarrow[\text{high T}]{\text{catalyst}} H_2C=CH-CH_3 + H-H$$

13.8

a. $CH_3-CH_2-CH=CH_2 + H_2 \xrightarrow[\text{catalyst}]{\text{Ni}} CH_3-CH_2-CH_2-CH_3$

b. 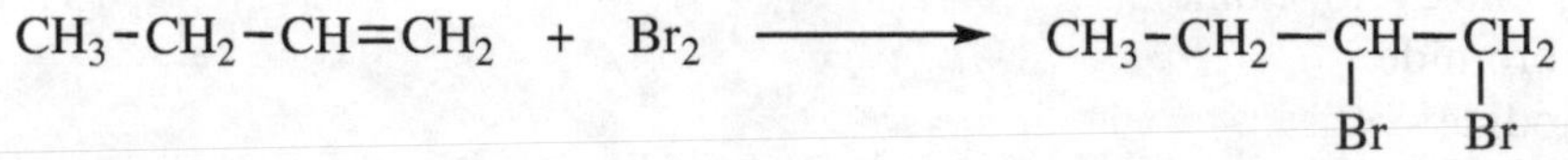 $CH_3-CH_2-CH=CH_2 + Br_2 \longrightarrow CH_3-CH_2-CH(Br)-CH_2(Br)$

c. 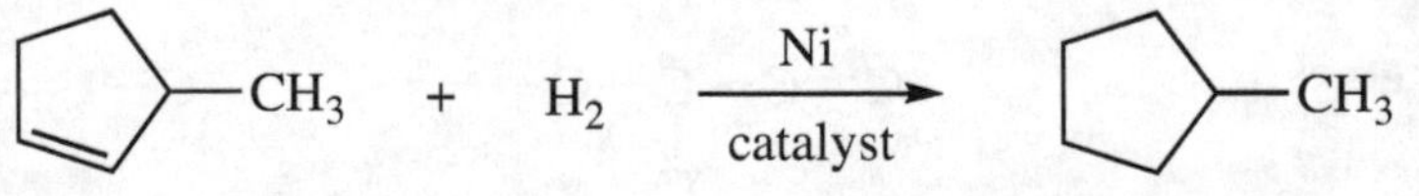 3-methylcyclopentene $-CH_3$ + H_2 $\xrightarrow[\text{catalyst}]{\text{Ni}}$ methylcyclopentane $-CH_3$

13.9

a. 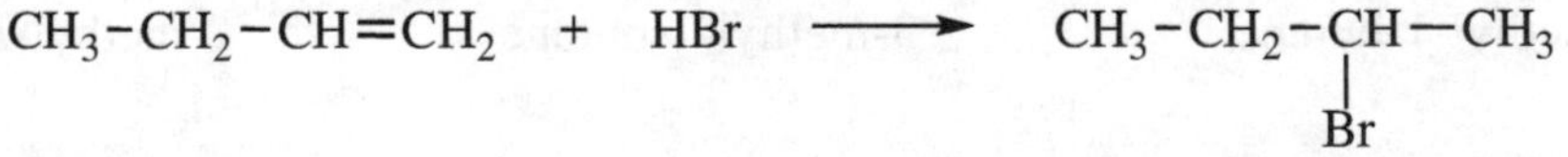 $CH_3-CH_2-CH=CH_2 + HBr \longrightarrow CH_3-CH_2-CH(Br)-CH_3$

b. 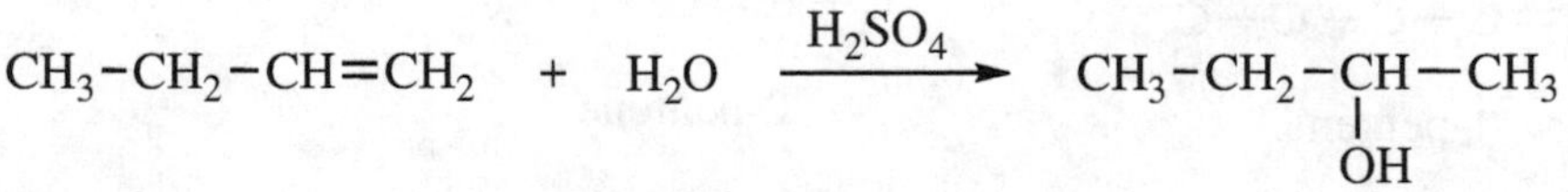 $CH_3-CH_2-CH=CH_2 + H_2O \xrightarrow{H_2SO_4} CH_3-CH_2-CH(OH)-CH_3$

c. 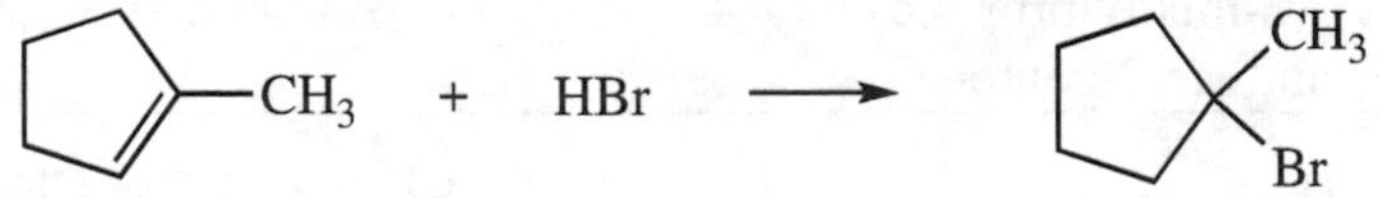 1-methylcyclopentene $-CH_3$ + HBr $\longrightarrow$ cyclopentane with CH_3 and Br on the same carbon

d. 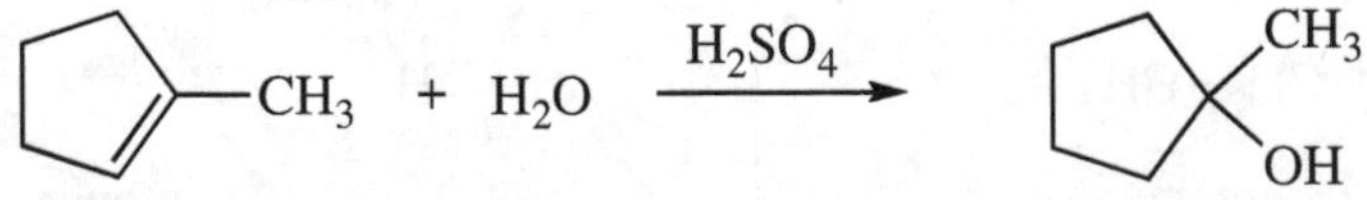 1-methylcyclopentene $-CH_3$ + H_2O $\xrightarrow{H_2SO_4}$ cyclopentane with CH_3 and OH on the same carbon

13.10 a. 1-pentene
b. 3-methyl-1-butene
c. 1-methylcyclopentene

13.11

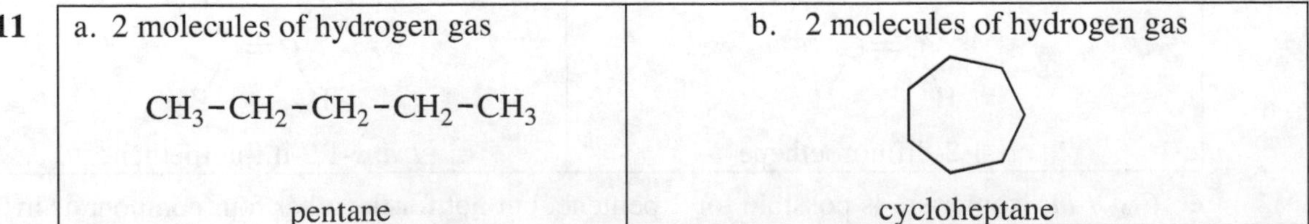

a. 2 molecules of hydrogen gas	b. 2 molecules of hydrogen gas
$CH_3-CH_2-CH_2-CH_2-CH_3$	
pentane	cycloheptane

13.12

General formula of polymer	Monomer	First three units of polymer
$\left(\begin{array}{c}\text{H}\ \ \text{H}\\ |\ \ \ |\\ -\text{C}-\text{C}-\\ |\ \ \ |\\ \text{H}\ \ \text{F}\end{array}\right)_n$	$CH_2{=}CH{-}F$	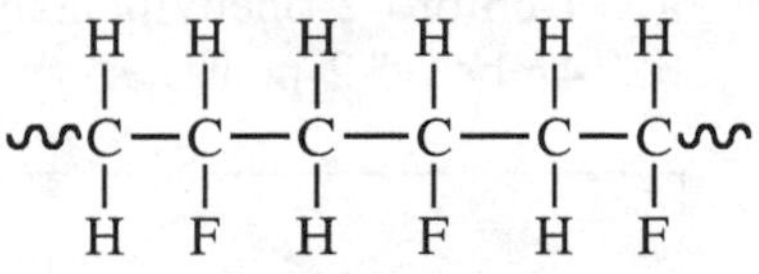 $\sim C(H)(H)-C(H)(F)-C(H)(H)-C(H)(F)-C(H)(H)-C(H)(F)\sim$
$\left(\begin{array}{c}\text{H}\ \ \text{H}\\ |\ \ \ |\\ -\text{C}-\text{C}-\\ |\ \ \ |\\ \text{H}\ \ CH_2\\ |\\ CH_3\end{array}\right)_n$	$CH_2{=}CH{-}CH_2{-}CH_3$	$\sim C(H)(H)-C(H)(CH_2CH_3)-C(H)(H)-C(H)(CH_2CH_3)-C(H)(H)-C(H)(CH_2CH_3)\sim$

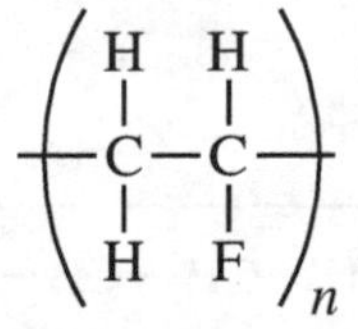

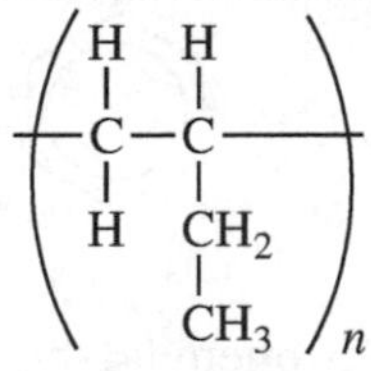

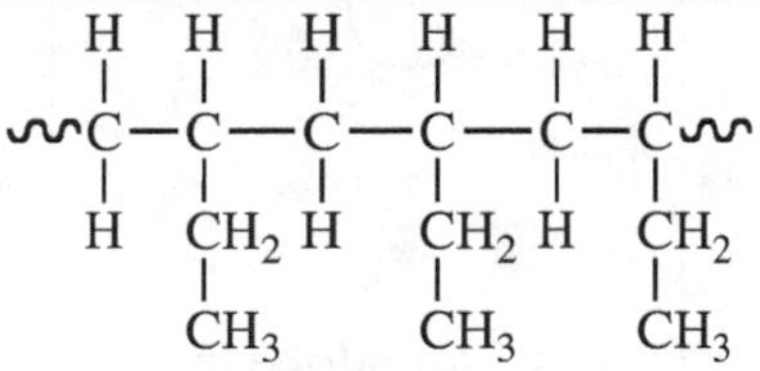

13.13
a. 1-pentyne
b. 4-bromo-1-butyne
c. 4,4-dimethyl-2-hexyne

13.14

$CH_3{-}CH_2{-}\underset{\displaystyle CH_3}{\underset{|}{CH}}{-}C{\equiv}CH$	$CH_3{-}\underset{\displaystyle Cl}{\underset{|}{CH}}{-}C{\equiv}C{-}CH_3$	$CH_3{-}\underset{\displaystyle CH_3}{\underset{|}{CH}}{-}\underset{\displaystyle CH_3}{\underset{|}{CH}}{-}C{\equiv}C{-}CH_3$
a. 3-methyl-1-pentyne	b. 4-chloro-2-pentyne	c. 4,5-dimethyl-2-hexyne

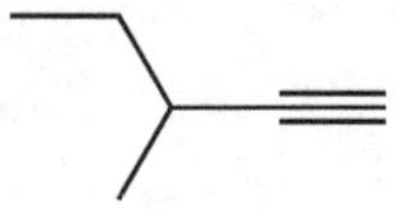

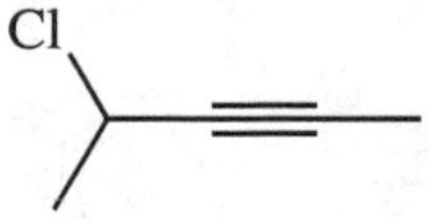

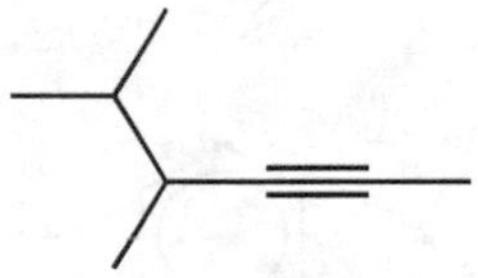

13.15

a. $$CH_3{-}CH_2{-}CH_2{-}C{\equiv}CH \; + \; 2Cl_2 \longrightarrow CH_3{-}CH_2{-}CH_2{-}CCl_2{-}CHCl_2$$

b. $$CH_3{-}CH_2{-}CH_2{-}C{\equiv}CH \; + \; 2HBr \longrightarrow CH_3{-}CH_2{-}CH_2{-}CBr_2{-}CH_3$$

c. $$CH_3{-}CH_2{-}CH_2{-}C{\equiv}CH \; + \; 1HCl \longrightarrow CH_3{-}CH_2{-}CH_2{-}\underset{\displaystyle Cl}{\underset{|}{C}}{=}CH_2$$

d. $$CH_3{-}CH_2{-}CH_2{-}C{\equiv}CH \; + \; 2H_2 \xrightarrow[\text{catalyst}]{\text{Ni}} CH_3{-}CH_2{-}CH_2{-}CH_2{-}CH_3$$

13.16
1. a. 1-bromo-2-ethylbenzene b. *o*-bromoethylbenzene
2. a. 1,4-dichlorobenzene b. *p*-dichlorobenzene
3. a. 1-bromo-3-iodobenzene b. *m*-bromoiodobenzene

13.17 a. 2-chlorotoluene or *o*-chlorotoluene b. 1,3-xylene or *m*-xylene
c. 3-bromotoluene or *m*-bromotoluene

13.18 a. 1-chloro-2-phenylpentane
b. 4-phenyl-2-pentene

13.19

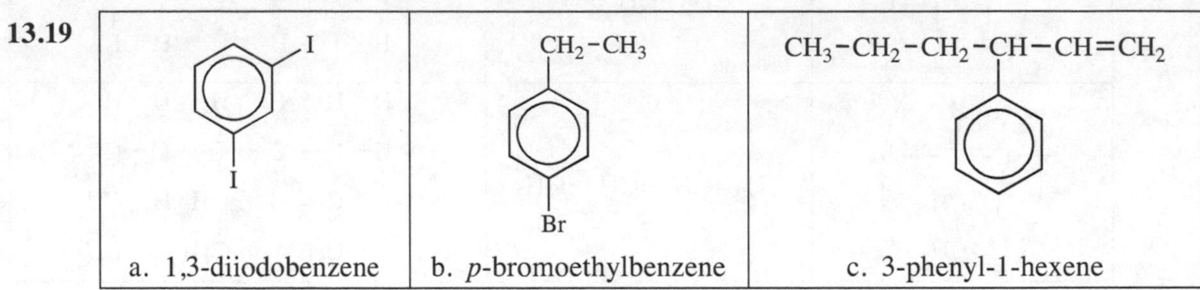

a. 1,3-diiodobenzene | b. *p*-bromoethylbenzene | c. 3-phenyl-1-hexene

13.20

a. benzene + CH_3Cl $\xrightarrow{AlCl_3}$ toluene (CH_3 on ring) + HCl

b. benzene + Br_2 $\xrightarrow{FeBr_3}$ bromobenzene (Br on ring) + HBr

c. benzene + Cl_2 $\xrightarrow{FeCl_3}$ chlorobenzene (Cl on ring) + HCl

d. benzene + CH_3CH_2Cl $\xrightarrow{AlCl_3}$ ethylbenzene (CH_2-CH_3 on ring) + HCl

13.21 a. E; b. B; c. F; d. D; e. A; f. C;
g. Structure A *trans*-2-butene
h. Structure B *cis*-2-butene
i. Structure C butane
j. Structure D 3-methylcyclohexene
k. Structure E 2-chlorotoluene
l. Structure F 1-butyne

Answers to Self-Test

The numbers in parentheses refer to sections in your textbook.
1. F; alkene (13.2) **2**. T (13.2) **3**. F; both carbon atoms (13.3) **4**. T (13.3) **5**. T (13.2)
6. F; acetylene (13.11) **7**. F; substitution (13.15) **8**. F; 2-butene (13.5) **9**. F; dehydrogenation (13.10)
10. T (13.6) **11**. F; 2,3-dibromobutane (13.9) **12**. T (13.9) **13**. F; no (13.10) **14**. d (13.9)
15. a (13.11) **16**. c (13.13) **17**. c (13.13) **18**. c (13.5) **19**. a (13.12) **20**. c (13.3, 13.11)
21. c (13.9) **22**. b (13.9, 13.15) **23**. d (13.16) **24**. b (13.7)

Alcohols, Phenols, and Ethers — Chapter 14

Chapter Overview

Most of the chemical reactions of organic molecules involve the functional groups on the hydrocarbon chains or rings. In this chapter you will consider the properties, chemical and physical, of some hydrocarbon derivatives with oxygen- and sulfur-containing functional groups. Alcohols, phenols, and ethers are compounds that are commonly encountered in naturally occurring substances.

In this chapter you will identify, name, and draw structures for alcohols, phenols, and ethers. You will compare the physical properties (melting point, boiling point, solubility in water) of these compounds to those of the hydrocarbons. You will write equations for the mild oxidation of alcohols and for other reactions that alcohols undergo.

Practice Exercises

14.1 An **alcohol** (Sec. 14.2) is a hydrocarbon derivative in which a **hydroxyl group** (–OH) (Sec. 14.2) is attached to a saturated carbon atom (a carbon atom bonded to four other atoms). The –OH functional group should not be confused with the hydroxide ion (OH^-).

The names of alcohols are derived from the hydrocarbons and given the ending *-ol*. Polyhydroxy alcohols have more than one hydroxyl group and are named as diols, triols, etc.

Give the IUPAC name for each of the following compounds (Sec. 14.3 in your textbook):

a. $CH_3-CH_2-CH_2-C(CH_3)_2-OH$ (C bearing two CH_3 groups, one above and one below)	b. $CH_3-CH_2-CH_2-CH_2-CH_2-CH_2-OH$ (OH on the last CH_2)
c. $CH_3-CH_2-CH(OH)-CH(OH)-CH_3$	d. CH_3–(cyclohexane ring)–OH (para positions)

14.2 Draw the condensed structural formula and line-angle structural formula for each of the following compounds:

a. 3-hexanol	b. 2,3-dimethyl-1-butanol
c. 1,2-butanediol	d. 2-methylcyclopentanol

14.3 Common names are sometimes used for alcohols in which –OH is attached to a simple alkyl group. In a common name, the alkyl group is named first and followed by the word *alcohol*. Draw the structural formulas for the four alcohols that have one, two, or three carbon atoms, and name them using common names.

14.4 Constitutional isomerism is possible for alcohols containing three or more carbon atoms. Draw condensed structural formulas for all of the alcohols that have the molecular formula $C_4H_{10}O$. Give the IUPAC name for each isomer. (Use the space below for the answers to Practice Exercises 14.4, 14.5, and 14.6.)

14.5 Alcohols are classified as **primary**, **secondary**, or **tertiary** alcohols (Sec. 14.8) on the basis of the number of carbon atoms attached to the hydroxyl-bearing carbon. Under each structure and IUPAC name in Practice Exercise 14.4, identify the alcohol as a primary, secondary, or tertiary alcohol.

14.6 Under each structure, IUPAC name, and structure type in Practice Exercise 14.4, give the common name for that alcohol.

14.7 The physical properties of alcohols are influenced by two factors: 1) hydrogen bonding (the strong dipole-dipole interaction between an electronegative oxygen atom and the partial positive charge on a hydrogen atom attached to an oxygen atom) and 2) the nonpolarity of the hydrocarbon part of the molecule.

Give the IUPAC name for each of these compounds below its structural formula:

$CH_3CH_2CH_2OH$ $CH_3CH_2CH_2CH_3$ $CH_3CH_2CH_2CH_2OH$

_______________ _______________ _______________

a. Write the names of the three compounds in order of increasing boiling point:

b. Write the names of the three compounds in order of increasing solubility in water:

14.8 As you have read in Section 13.10 of your textbook, one method for preparing alcohols is the hydration of alkenes in the presence of a sulfuric acid catalyst (Sec. 14.7). The predominant alcohol product can be determined from Markovnikov's rule.

Write an equation for the preparation of each of the following alcohols from the appropriate alkene.

a. 4-methyl-2-pentanol

b. cyclopentanol

14.9 Another method for synthesizing alcohols is the addition of hydrogen to a carbonyl group.

$$\gt C{=}O$$

The reaction is similar to addition of hydrogen to the double bond of an alkene. Complete the following equation for addition of hydrogen to acetaldehyde. Name the alcohol produced.

$$CH_3{-}\overset{\overset{\large O}{\|}}{C}{-}H \quad + \quad H_2 \xrightarrow{\text{Catalyst}}$$

14.10 Like the hydrocarbons, alcohols undergo combustion in air to produce CO_2 and H_2O. Write a balanced chemical equation for the complete combustion of propanol (C_3H_8O).

14.11 Alcohols undergo **dehydration** (Sec. 14.9), removal of a water molecule, when heated in the presence of a sulfuric acid catalyst. The product formed depends on the temperature of the reaction: At 180°C, intramolecular (within the molecule) dehydration produces an alkene. This is an **elimination reaction** (Sec. 14.9). According to **Zaitsev's rule** (Sec. 14.9), the alkene with the greatest number of alkyl groups attached to the double bond will be formed.

At a lower temperature (140°C), a primary alcohol can undergo a **condensation reaction** (Sec. 14.9) between two alcohol molecules, to produce an **ether** (Sec. 14.14) and water.

Complete the following equations by supplying the missing compounds:

a. $$CH_3{-}\underset{\underset{\large CH_3}{|}}{CH}{-}CH_2{-}OH \xrightarrow[H_2SO_4]{140^\circ C} \quad ? \quad + \quad ?$$

b. $$CH_3{-}CH_2{-}CH_2{-}\underset{\underset{\large OH}{|}}{CH}{-}CH_3 \xrightarrow[H_2SO_4]{180^\circ C} \quad ? \quad + \quad ?$$

c. $$? \xrightarrow[H_2SO_4]{140^\circ C} CH_3{-}CH_2{-}CH_2{-}O{-}CH_2{-}CH_2{-}CH_3 \quad + \quad H{-}OH$$

d. $$? \xrightarrow[H_2SO_4]{180^\circ C} CH_3{-}CH_2{-}CH_2{-}CH{=}\underset{\underset{\large CH_3}{|}}{C}{-}CH_3 \quad + \quad H{-}OH$$

14.12 Primary and secondary alcohols undergo **organic oxidation** (Sec. 14.9) in the presence of a mild oxidizing agent to produce compounds that contain a carbon-oxygen double bond. Oxidation is defined as an increase in the number of carbon-oxygen bonds or a decrease in the number of carbon-hydrogen bonds.

Primary alcohols may oxidize even further to form carboxylic acids. Tertiary alcohols cannot be oxidized in this way since the carbon attached to the hydroxyl group is attached to three other carbons and cannot form a carbon-oxygen double bond.

Complete the following equations by supplying the missing information:

a. **?** $\xrightarrow[\text{agent}]{\text{mild oxidizing}}$ $CH_3-CH_2-CH_2-CH_2-\underset{\underset{O}{\|}}{CH}$

b. 1-methylcyclohexanol (cyclohexane ring with OH and CH_3) $\xrightarrow[\text{agent}]{\text{mild oxidizing}}$ **?**

c. $CH_3-CH_2-CH_2-\underset{\underset{OH}{|}}{CH}-CH_3$ $\xrightarrow[\text{agent}]{\text{mild oxidizing}}$ **?**

14.13 Alcohols also undergo substitution reactions. An example is halogenation, the substitution of a halogen atom for a hydroxyl group to produce an alkyl halide. Complete the following equations for halogenation reactions.

a. $CH_3-\overset{\overset{OH}{|}}{\underset{\underset{H}{|}}{C}}-CH_3$ $\xrightarrow[\text{heat}]{PBr_3}$ **?**

b. **?** $\xrightarrow[\text{heat}]{PCl_3}$ $CH_3-\underset{\underset{CH_3}{|}}{CH}-CH_2-\underset{\underset{Cl}{|}}{CH}-CH_3$

14.14 A **phenol** (Sec. 14.10) is a compound in which a hydroxyl group is attached to a carbon atom in an aromatic ring system. The general formula for phenols is Ar–OH. Phenols are named in the same way as benzene derivatives, but with the parent name *phenol*.

Write the IUPAC name for each of the following compounds:

a. (benzene ring with OH and Br on adjacent carbons)	b. (benzene ring with OH; Cl adjacent to OH; CH_3 para to OH)

14.15 Draw a condensed structural formula for each of the following compounds:

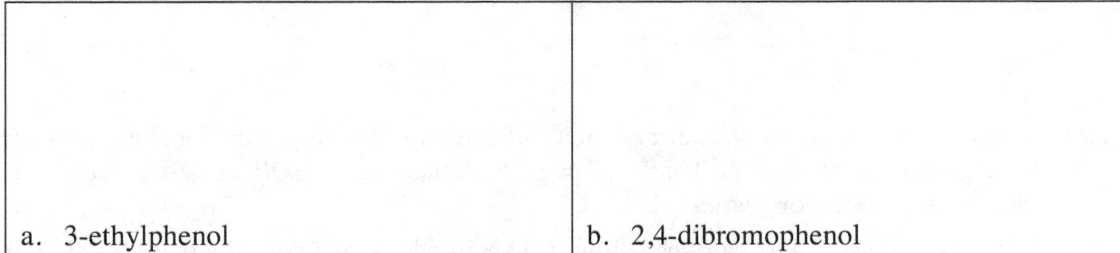

14.16 Phenols are not considered to be alcohols; the stable benzene ring alters the properties of the hydroxyl group. Phenols are weak acids; in aqueous solution the hydrogen attached to oxygen dissociates to form H_3O^+ and the phenoxide ion. Write an equation showing the reaction of phenol in aqueous solution with a strong base (NaOH).

14.17 An **ether** (Sec. 14.14) is a compound in which an oxygen atom is bonded to two carbon atoms by single bonds. According to the IUPAC system, ethers are named as substituted hydrocarbons. The smaller hydrocarbon and the oxygen atom are named as an **alkoxy group** (Sec. 14.15) and considered as a substituent on the larger hydrocarbon.

Write the IUPAC name for each of the following compounds:

$CH_3-CH_2-CH_2-CH_2-O-CH_3$ a.	(cyclohexane ring)$-O-CH_2-CH_2-CH_3$ b.

14.18 Ethers are often known by common names in which the hydrocarbon groups are named alphabetically followed by the word *ether*. Give the common name for each of the ethers in Practice Exercise 14.17.

a.

b.

14.19 Constitutional isomerism in ethers depends on the identity of the alkyl groups that are present. Only one ether is possible for the molecular formula C_3H_8O. Three ether constitutional isomers are possible for the molecular formula $C_4H_{10}O$ (Sec. 14.17).

Draw line-angle structural formulas for all possible ether constitutional isomers that have the formula $C_5H_{12}O$. (Hint: A butyl group has four isomeric forms; a propyl group has two.) Give the common name for each ether.

14.20 Ethers should be handled with caution because of two chemical properties:

1)

2)

14.21 **Thiols** (Sec. 14.20) are the sulfur analogs of alcohols. The thiol functional group is the **sulfhydryl group** (—SH). Thiols are named as alkanethiols (IUPAC system) or alkyl mercaptans (common names).

Thioethers (Sec. 14.21) are the sulfur analogs of ethers, and are named as alkylthioalkanes (IUPAC system) or as alkyl alkyl sulfides (common names).

Give the IUPAC name and the common name for each of the following sulfur-containing compounds.

$CH_3-CH_2-CH_2-CH_2-SH$	(cyclohexane ring)–SH	CH_3-CH_2-S-(cyclopentane ring)
a.	b.	c.

14.22 Write structural formulas for the following sulfur-containing compounds:

a. 2-methyl-1-butanethiol	b. *tert*-butyl thiol	c. diethyl sulfide

14.23 Many of the physical properties of an organic compound that has an oxygen-containing functional group are determined by the molecule's ability to form hydrogen bonds with a like molecule or with water molecules (Sec. 14.18).

For the hydrocarbons and their derivatives that you have studied so far (alkanes, alkenes, alkynes, alcohols, ethers, thiols, and thioethers), which of these types of molecules:

a. form hydrogen bonds between like molecules ______________________

b. form hydrogen bonds with water molecules ______________________

Indicate which compound in each of the following pairs would have the higher boiling point and the greater solubility in water. (Compounds in each pair have comparable molecular weights.)

Compounds	Higher boiling point	Greater water solubility
1-propanol and butane		
2-methyl-2-propanol and diethyl ether		
phenol and toluene		
1-butanol and 1-propanethiol		
1-propanethiol and pentane		

14.24 Use this identification exercise to review your knowledge of the structures introduced in this chapter. Using structures A through I, give the best match for the terms that follow:

$HO-CH_2-CH_2-OH$ A.	$CH_3-O-CH_2-CH_3$ B.	CH_3-S-CH_3 C.
$CH_3-CH_2-C(CH_3)_2-OH$ D.	(benzene ring)$-OH$ E.	$CH_3-CH_2-CH_2-OH$ F.
CH_3-CH_2-SH G.	$CH_3-O-O-CH_3$ H.	$CH_3-S-S-CH_3$ I.

Give the letter of the best match for the structures above:

a. 3° alcohol ____________
b. glycol ____________
c. peroxide ____________
d. ether ____________
e. phenol ____________
f. disulfide ____________
g. thioether ____________
h. thiol ____________
i. 1° alcohol ____________

Give the IUPAC names for structures A through G:

j. Structure A ____________
k. Structure B ____________
l. Structure C ____________
m. Structure D ____________
n. Structure E ____________
o. Structure F ____________
p. Structure G ____________

Self-Test

True-false: Indicate whether the following statements are true or false. If the statement is false, give the word or phrase that may be substituted for the underlined portion to make the statement true.

1. A compound in which a hydroxyl group is attached to a carbon atom in an aromatic ring system is called a thiol.
2. A secondary alcohol has two hydrogen atoms attached to the hydroxyl carbon.
3. Ether molecules cannot form hydrogen bonds with other ether molecules.
4. Diethyl ether and 1-butanol are structural isomers.
5. Oxygen has six valence electrons, which include two nonbonding electron pairs.
6. 2-Butanol is an example of a tertiary alcohol.
7. Alcohols are more soluble in water than alkanes of similar molecular mass.
8. A symmetrical ether can be prepared by the intramolecular dehydration of an alcohol.
9. Some phenols (BHA and BHT) are used as antioxidants because they are low-melting solids.
10. Furan is an example of a heterocyclic organic compound.
11. Storage of ethers can be hazardous because unstable peroxides may form.
12. Gasoline blends with up to 10% methanol are increasingly used in the United States.

13. Propylene glycol is a better automobile antifreeze than ethylene glycol because its oxidation product is nontoxic.
14. The strong odors of onion and garlic are due mainly to glycols.
15. Phenols can act as antioxidants because they react with oxidizing agents.

Multiple choice:

16. An alcohol may be prepared by:
    a. hydrogenation of an alkene
    b. hydration of an alkene
    c. dehydration of a carbonyl group
    d. dehydrogenation of an alkane
    e. none of these

17. Mercaptans is an older term for:
    a. thiols
    b. phenols
    c. glycols
    d. ethers
    e. none of these

18. The mild oxidation of a secondary alcohol can result in the production of a(n):
    a. acid
    b. aldehyde
    c. ketone
    d. alkene
    e. none of these

19. One possible functional group isomer of diethyl ether is:
    a. anisole
    b. 1-butanol
    c. 2-propanol
    d. dimethyl ether
    e. 1-methoxybutane

20. The common name for 2-propanol is:
    a. propyl alcohol
    b. 2-propyl alcohol
    c. isopropyl alcohol
    d. methyl ethyl alcohol
    e. none of these

21. Which of the following names is correct according to IUPAC rules?
    a. 2-ethyl-1-butene-2-ol
    b. 3-propylene-2-ol
    c. 3,4-butanediol
    d. 1,2-dimethylphenol
    e. none of these

22. According to Zaitsev's rule, the main product of the dehydration of 2-methyl-2-butanol would be:
    a. 2-methyl-2-butene
    b. 2-methyl-1-butene
    c. 1-methyl-2-butene
    d. 1-methyl-1-butene
    e. none of these

23. Rank butane, butanol, and diethyl ether by boiling points, lowest boiling point first and highest boiling point last:
    a. butane, butanol, diethyl ether
    b. butanol, butane, diethyl ether
    c. butane, diethyl ether, butanol
    d. butanol, diethyl ether, butane
    e. diethyl ether, butane, butanol

24. The intramolecular dehydration of 2-pentanol produces:
    a. 2-methoxypentane
    b. 2-pentene
    c. 1-pentene
    d. 2-methylpentene
    e. none of these

25. An example of a cyclic ether is:
    a. phenol
    b. methoxybenzene
    c. catechol
    d. pyran
    e. methyl *tert*-butyl ether

26. Mild reduction of a disulfide produces:
    a. two thiols b. a peroxide c. a thioether
    d. a cyclic ether e. none of these

Answers to Practice Exercises

14.1 a. 2-methyl-2-pentanol b. 1-hexanol
c. 2,3-pentanediol d. 4-methylcyclohexanol

14.2

$CH_3-CH_2-CH_2-CH(OH)-CH_2-CH_3$ a. 3-hexanol	$CH_3-CH(CH_3)-CH(CH_3)-CH_2-OH$ b. 2,3-dimethyl-1-butanol
$CH_3-CH_2-CH(OH)-CH_2(OH)$ c. 1,2-butanediol	d. 2-methylcyclopentanol

14.3

CH_3-OH
methyl alcohol

CH_3-CH_2-OH
ethyl alcohol

$CH_3-CH(OH)-CH_3$
isopropyl alcohol

$CH_3-CH_2-CH_2-OH$
propyl alcohol

14.4, 14.5, and 14.6

$CH_3-CH_2-CH_2-CH_2-OH$
1-butanol
primary
butyl alcohol

$CH_3-CH_2-CH(OH)-CH_3$
2-butanol
secondary
sec-butyl alcohol

$H_3C-CH(CH_3)-CH_2-OH$
2-methyl-1-propanol
primary
isobutyl alcohol

$CH_3-C(CH_3)_2-OH$
2-methyl-2-propanol
tertiary
tert-butyl alcohol

14.7 $CH_3CH_2CH_2OH$ (1-propanol) $CH_3CH_2CH_2CH_3$ (butane) $CH_3CH_2CH_2CH_2OH$ (1-butanol)

a. Increasing boiling point: butane, 1-propanol, 1-butanol
b. Increasing solubility in water: butane, 1-butanol, 1-propanol

14.8 a. $CH_3-CH(CH_3)-CH_2-CH=CH_2 + H-OH \xrightarrow{H_2SO_4} CH_3-CH(CH_3)-CH_2-CH(OH)-CH_3$

b. cyclopentene + $H-OH \xrightarrow{H_2SO_4}$ cyclopentyl$-OH$

14.9 $CH_3-C(=O)-H + H_2 \xrightarrow{\text{Catalyst}} CH_3-C(OH)(H)-H$ ethanol

14.10 $2C_3H_8O + 9O_2 \rightarrow 6CO_2 + 8H_2O$

14.11 a. $2\ CH_3-CH(CH_3)-CH_2-OH \xrightarrow[H_2SO_4]{140^oC} CH_3-CH(CH_3)-CH_2-O-CH_2-CH(CH_3)-CH_3 + H_2O$

b. $CH_3-CH_2-CH_2-CH(OH)-CH_3 \xrightarrow[H_2SO_4]{180^oC} CH_3-CH_2-CH=CH-CH_3 + H_2O$

c. $2\ CH_3-CH_2-CH_2-OH \xrightarrow[H_2SO_4]{140^oC} CH_3-CH_2-CH_2-O-CH_2-CH_2-CH_3 + H_2O$

d. $CH_3-CH_2-CH_2-CH_2-C(OH)(CH_3)-CH_3 \xrightarrow[H_2SO_4]{180^oC} CH_3-CH_2-CH_2-CH=C(CH_3)-CH_3 + H_2O$

14.12 a. $CH_3-CH_2-CH_2-CH_2-CH_2(OH) \xrightarrow[\text{agent}]{\text{mild oxidizing}} CH_3-CH_2-CH_2-CH_2-CH(=O)$

b. 1-methylcyclohexanol (ring with OH and CH_3) $\xrightarrow[\text{agent}]{\text{mild oxidizing}}$ No reaction, tertiary alcohol

c. $CH_3-CH_2-CH_2-CH(OH)-CH_3 \xrightarrow[\text{agent}]{\text{mild oxidizing}} CH_3-CH_2-CH_2-C(=O)-CH_3$

14.13

a. 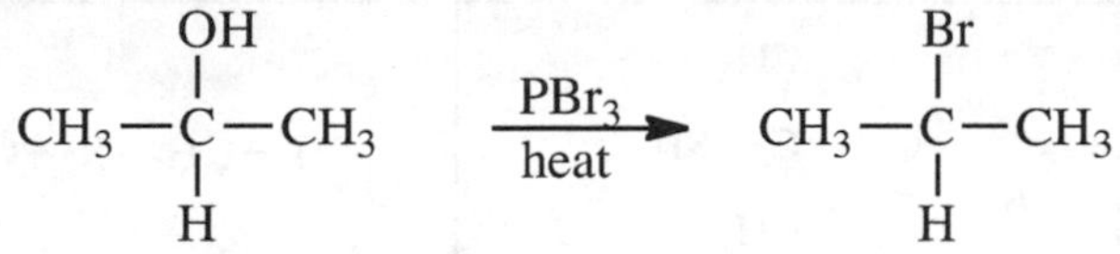 $CH_3-CH(OH)-CH_3 \xrightarrow[\text{heat}]{PBr_3} CH_3-CH(Br)-CH_3$

b. 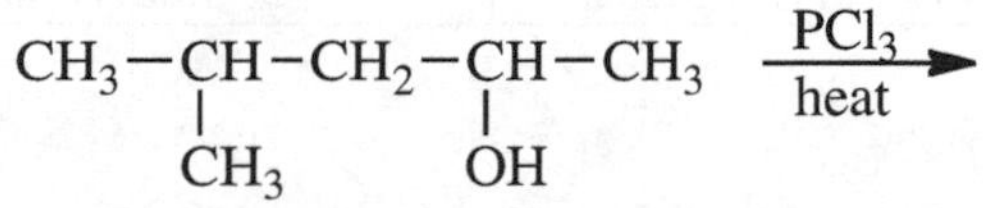 $CH_3-CH(CH_3)-CH_2-CH(OH)-CH_3 \xrightarrow[\text{heat}]{PCl_3} CH_3-CH(CH_3)-CH_2-CH(Cl)-CH_3$

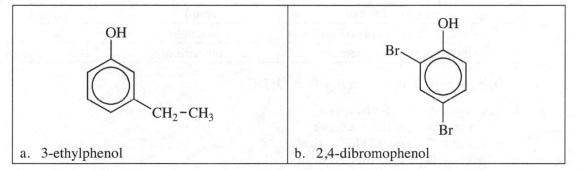

14.14 a. 2-bromophenol or *o*-bromophenol
b. 2-chloro-4-methylphenol

14.15

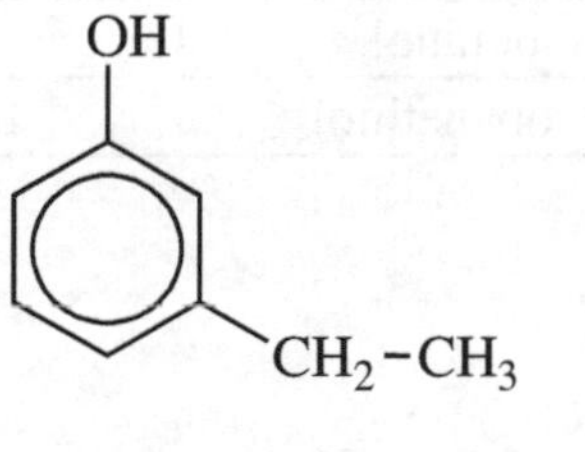

a. 3-ethylphenol	b. 2,4-dibromophenol

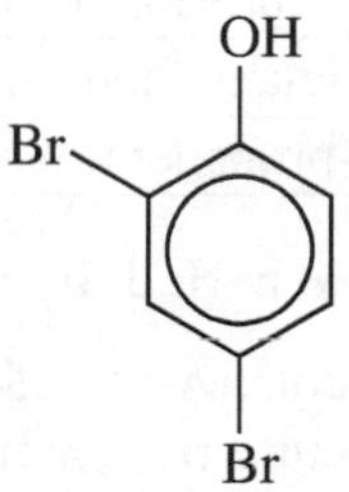

14.16 phenol (OH) + $NaOH_{(aq)}$ ⟶ phenoxide ($O^- Na^+$) + H_2O

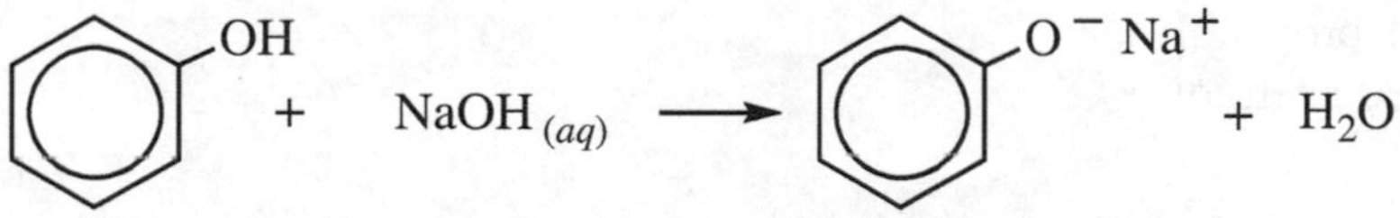

14.17 a. 1-methoxybutane b. propoxycyclohexane

14.18 a. butyl methyl ether b. cyclohexyl propyl ether

14.19

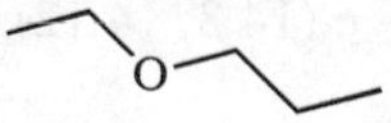

ethyl propyl ether

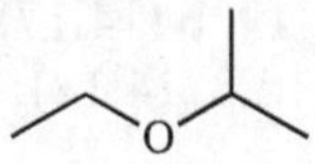

ethyl isopropyl ether

butyl methyl ether

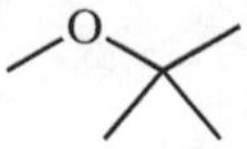

tert-butyl methyl ether

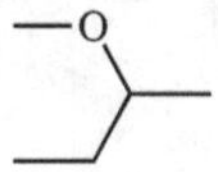

sec-butyl methyl ether

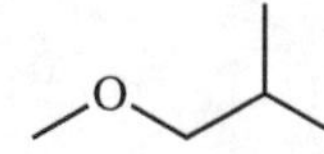

isobutyl methyl ether

14.20 1) Ethers are flammable. (Diethyl ether, bp = 35°C, is a flash-fire hazard.)
2) Ethers react slowly with oxygen from the air to form unstable (explosive) hydroperoxides and peroxides.

14.21 a. 1-butanethiol
butyl mercaptan

b. cyclohexanethiol
cyclohexyl mercaptan

c. ethylthiocyclopentane
ethyl cyclopentyl sulfide

14.22

$CH_3-CH_2-CH(CH_3)-CH_2-SH$	$CH_3-C(CH_3)_2-SH$	$CH_3-CH_2-S-CH_2-CH_3$
a. 2-methyl-1-butanethiol	b. *tert*-butyl thiol	c. diethyl sulfide

14.23 a. alcohols
b. alcohols and ethers

Compounds	Higher boiling point	Greater water solubility
1-propanol and butane	1-propanol	1-propanol
2-methyl-2-propanol and diethyl ether	2-methyl-2-propanol	2-methyl-2-propanol
phenol and toluene	phenol	phenol
1-butanol and 1-propanethiol	1-butanol	1-butanol
1-propanethiol and pentane	1-propanethiol	1-propanethiol

14.24 a. D; b. A; c. H; d. B; e. E; f. I; g. C; h. G; i. F

j. Structure A 1,2-ethanediol
k. Structure B methoxyethane
l. Structure C methylthiomethane
m. Structure D 2-methyl-2-butanol
n. Structure E phenol
o. Structure F 1-propanol
p. Structure G ethanethiol

Answers to Self-Test

The numbers in parentheses refer to sections in your textbook.
1. F; phenol (14.11) **2**. F; one (14.8) **3**. T (14.18) **4**. T (14.17) **5**. T (14.1) **6**. F; secondary (14.8) **7**. T (14.6) **8**. F; intermolecular (14.7) **9**. F; easily oxidized (14.13) **10**. T (14.19) **11**. T (14.18) **12**. F; 10% ethanol (14.5) **13**. T (14.5) **14**. F; sulfides and disulfides (14.5, 14.21) **15**. T (14.13) **16**. b (14.7) **17**. a (14.20) **18**. c (14.9) **19**. b (14.17) **20**. c (14.3) **21**. e (14.3, 14.12) **22**. a (14.9) **23**. c (14.6, 14.13) **24**. b (14.9) **25**. d (14.3, 14.12) **26**. a (14.20)

Aldehydes and Ketones — Chapter 15

Chapter Overview

The carbonyl functional group is very commonly found in nature. It contains an oxygen atom joined to a carbon atom by a double bond. Aldehydes and ketones contain the carbonyl functional group, as do several other oxygen-containing compounds that will be considered in the following chapters.

In this chapter you will learn to recognize, name, and write structural formulas for aldehydes and ketones and write equations for their preparation by oxidation of alcohols. You will compare the physical properties of aldehydes and ketones with those of other organic compounds. You will learn some tests used to distinguish aldehydes from ketones and will write equations for reactions involving addition to the carbonyl group of aldehydes and ketones.

Practice Exercises

15.1 The **carbonyl group** (Sec. 15.1) consists of a carbon atom and an oxygen atom joined by a double bond. The two single bonds of the carbon atom and the double bond are arranged in a trigonal planar structure with 120° bond angles.

Aldehydes (Sec. 15.1) are compounds in which the carbonyl carbon atom has at least one hydrogen atom directly attached to it. **Ketones** (Sec. 15.1) are compounds in which the carbonyl carbon atom has two other carbon atoms directly attached to it.

Classify each of the following structural formulas as an aldehyde, a ketone, or neither.

a. $CH_3-CH_2-CH_2-O-CH_3$	b. $CH_3-\overset{CH_3}{\overset{\vert}{CH}}-\overset{O}{\overset{\Vert}{C}}-H$
c. H_3C-⬡$=O$	d. $CH_3-\underset{\underset{O}{\Vert}}{C}-\overset{Cl}{\overset{\vert}{CH}}-\underset{\underset{CH_3}{\vert}}{CH}-CH_3$

15.2 In addition to aldehydes and ketones, there are other types of compounds (Sec. 15.2) that contain the carbonyl group in their functional groups. You will study some of these compounds in later chapters. Write the generic formulas for the carbonyl-containing compounds below. (Use R and R′ as generic alkyl groups.)

aldehyde	ketone	carboxylic acid	ester	amide

15.3 In naming aldehydes, change the *-e* ending of the hydrocarbon to *-al*, and name any substituents, starting the counting from the carbonyl carbon. No number is specified for the carbonyl group.

Give the IUPAC name for each of the following aldehydes:

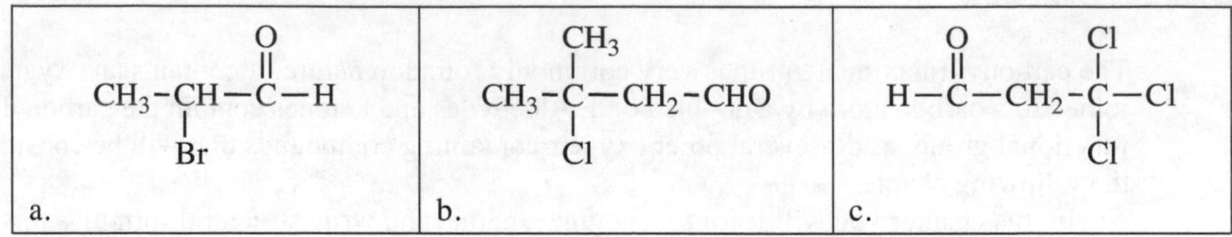

15.4 Draw both the condensed structural formula and the line-angle structural formula for each of the following aldehydes.

a. 4-chloro-3,3-dimethylbutanal	b. 4-bromo-3-methylpentanal

15.5 In naming ketones, change the *-e* ending of the hydrocarbon to *-one*. Give the carbonyl group the lowest possible number on the chain, and then name the substituents. The numbered position of the carbonyl group is included in the name.

Give the IUPAC names for the following ketones.

$CH_3-C(=O)-CH(Cl)-CH(CH_3)-CH_3$	$C_6H_5-CH(CH_3)-C(=O)-CH_3$	$CH_3-C_6H_{10}=O$ (4-methyl ring ketone)
a.	b.	c.

15.6 Draw both the condensed structural formula and the line-angle structural formula for each of the following ketones.

a. 6-chloro-4-methyl-3-hexanone	b. 3-ethyl-2-methylcyclohexanone

15.7 Common names (Sec. 15.5) are frequently used for some of the simpler aldehydes and ketones. Draw structural formulas for the following compounds:

a. acetone	b. acetaldehyde	c. formaldehyde	d. ethyl methyl ketone
e. benzaldehyde	f. butyraldehyde	g. propionaldehyde	h. methyl phenyl ketone

15.8 Aldehydes and ketones with the same number of carbon atoms and the same degree of saturation are functional group isomers. Given the molecular formula C_4H_8O, draw condensed structural formulas for the aldehyde and ketone isomers.

Are other functional group isomers possible for the molecular formula C_4H_8O? ________
____

15.9 Hydrogen bonding cannot occur between the molecules of an aldehyde or between the molecules of a ketone, because no hydrogen atoms are attached to the oxygen atoms. However, since the carbonyl group is polar, dipole-dipole attractions (weaker than hydrogen bonds) occur between molecules in aldehydes and ketones.

Which compound in each of the following pairs would have a higher boiling point? Explain your choice.

Compounds	Higher-boiling compound	Explanation
pentanal and hexane		
2-hexanone and 2-octene		
pentanal and 1-pentanol		

15.10 Aldehydes and ketones can be produced by the oxidation of primary and secondary alcohols, respectively (Sec. 15.9). Write an equation for the preparation of an aldehyde or a ketone by the mild oxidation of each of the following alcohols. Name the carbonyl compound.

a. 1-pentanol

b. cyclohexanol

c. 2-methyl-2-butanol

15.11 Aldehydes are easily oxidized to carboxylic acids; ketones are resistant to oxidation (Sec. 15.10). The Tollens test and the Benedict's test, which can distinguish between aldehydes and ketones, use metal ions (Ag^+ and Cu^{2+}) as oxidizing agents. The appearance of the reduced metal (Ag) or metal oxide (Cu_2O) indicates that an aldehyde is present.

Indicate whether each of the following compounds will give a positive test with Tollens or Benedict's reagent.

Compound	Benedict's reagent	Tollens reagent
a. 4-chlorocyclohexanone		
b. 2-methylhexanal		
c. 4-chloro-3-methyl-2-pentanone		
d. 3,3-dimethylpentane		
e. pentanal		

15.12 Aldehydes and ketones are easily reduced by hydrogen gas (H_2), in the presence of a catalyst (Ni, Pt, or Cu), to form alcohols. Aldehyde reduction produces a 1° alcohol; ketone reduction produces a 2° alcohol.

Write a chemical equation for the preparation of each of the following alcohols by the reduction of the appropriate aldehyde or ketone.

a. 3-chloro-1-butanol

b. 1-bromo-3-methyl-2-pentanol

15.13 Aldehydes and ketones easily undergo addition of one alcohol molecule to the double bond of the carbonyl group, forming a **hemiacetal** (Sec. 15.11).

Draw the condensed structural formula of the hemiacetal formed when one molecule of methanol, CH_3OH, reacts with one molecule of each of the following carbonyl compounds. Circle the hemiacetal carbon.

a. 2-bromopropanal + methanol	b. 4-methylcyclohexanone + methanol	c. 3-chloro-2-pentanone + methanol

15.14 When a hemiacetal molecule reacts with a second molecule of alcohol, an **acetal** (Sec. 15.11) is produced. An acetal is not an ether; an acetal has two alkoxy groups bonded to the same carbon atom while an ether has one alkoxy group bonded to a carbon atom.

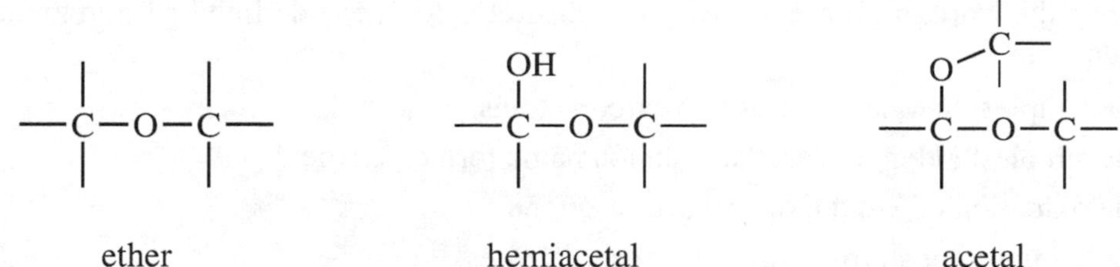

Draw the condensed structural formulas of the acetals formed when one additional molecule of methanol is added to each of the hemiacetal molecules in Practice Exercise 15.13. Circle the acetal carbon.

a.	b.	c.

15.15 The hydrolysis of an acetal or a hemiacetal produces the aldehyde or ketone and the alcohol that originally reacted to form the acetal or hemiacetal. Complete the following acid hydrolysis reactions.

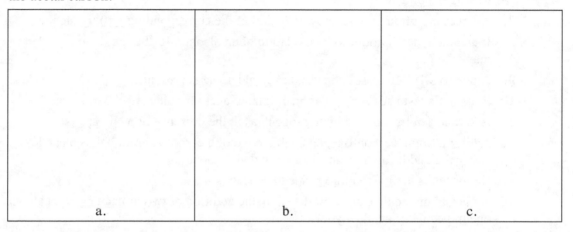

a. $CH_3-CH_2-CH(OH)-O-CH_2-CH_3 \xrightleftharpoons{\text{acid catalyst}} ? + ?$

b. $CH_3-CH(CH_3)-CH(O-CH_3)-O-CH_3 + H_2O \xrightleftharpoons{\text{acid catalyst}} ? + ?$

c. $C_5H_9-C(CH_3)(O-CH_2-CH_3)-O-CH_2-CH_3 + H_2O \xrightleftharpoons{\text{acid catalyst}} ? + ?$ (cyclopentyl ring attached to the central C)

Self-Test

True-false: Indicate whether the following statements are true or false. If the statement is false, give the word or phrase that may be substituted for the underlined portion to make the statement true.

1. The simplest ketone contains two carbon atoms.
2. The simplest aldehyde has the common name formaldehyde.
3. The oxidation of 2-butanol produces a ketone.
4. Many important steroid hormones are aldehydes.
5. Aldehydes and ketones have higher boiling points than the corresponding alcohols.
6. Low-molecular-mass aldehydes and ketones are water-soluble because water forms hydrogen bonds with them.
7. Ketones are often produced by oxidation of the corresponding tertiary alcohol.
8. If an aldehyde is produced by oxidation of an alcohol, further oxidation to a ketone may take place.
9. A positive Tollens test indicates that an aldehyde is present.
10. Benedict's test uses Ag^+ ion as an oxidizing agent for aldehydes.
11. A ketone can be reduced by hydrogen gas in the presence of a Ni catalyst.
12. Addition to the carbon-oxygen double bond of a carbonyl group takes place less easily than addition to a carbon-carbon double bond.
13. Hemiacetals are much more stable than acetals.
14. An acetal molecule can be produced by the addition of two molecules of an alcohol to one molecule of an aldehyde.
15. Replacement of the oxygen atom in a carbonyl compound with a sulfur atom produces a sulfoxide.
16. Acetaldehyde is a functional isomer of acetone.
17. The correct name for $CH_3-CH_2-\overset{\overset{\displaystyle O}{\|}}{C}-CH_3$ is 2-butanone.
18. A heterocyclic ring system in which an oxygen atom is part of the ring is known as a cyclic ketone.

Multiple choice:

19. Oxidation of ethanol cannot yield:

a. acetaldehyde b. acetone c. acetic acid
d. ethanal e. carbon dioxide

20. The IUPAC name for the compound $CH_3–CH_2–CH_2–CH_2–CHO$ is:

a. 1-pentanone b. 1-pentyl ketone c. 1-pentanal
d. pentanal e. none of these

21. A common name for propanone is:

a. acetone b. acetophenone c. ethyl methyl ketone
d. diethyl ketone e. none of these

22. A hemiacetal molecule is formed by the reaction between:
    a. two ketone molecules
    b. two aldehyde molecules
    c. a ketone molecule and an alcohol molecule
    d. a ketone molecule and an aldehyde molecule
    e. none of these

23. Aldehydes and ketones have boiling points lower than the corresponding alcohols because aldehydes and ketones:
    a. have stronger dipole-dipole attractions
    b. form more hydrogen bonds than alcohols do
    c. cannot form hydrogen bonds between molecules
    d. form bonds with water molecules
    e. none of these

24. How many ketones are isomeric with butanal?
    a. one   b. two   c. three   d. four   e. none of these

25. Acetaldehyde can be produced by the oxidation of:
    a. acetone   b. ethanol   c. dimethyl ether
    d. ethane   e. none of these

26. The reduction of ethyl methyl ketone produces which of the following:
    a. methanol   b. ethanol   c. 2-propanol
    d. propanal   e. 2-butanol

Answers to Practice Exercises

15.1 a. neither (an ether)   b. aldehyde   c. ketone   d. ketone

15.2

aldehyde	ketone	carboxylic acid	ester	amide
$R-\overset{\overset{O}{\|}}{C}-H$	$R-\overset{\overset{O}{\|}}{C}-R'$	$R-\overset{\overset{O}{\|}}{C}-OH$	$R-\overset{\overset{O}{\|}}{C}-O-R'$	$R-\overset{\overset{O}{\|}}{C}-NH_2$

15.3 a. 2-bromopropanal
b. 3-chloro-3-methylbutanal
c. 3,3,3-trichloropropanal

15.4

a. 4-chloro-3,3-dimethylbutanal	b. 4-bromo-3-methylpentanal
$CH_2Cl-C(CH_3)_2-CH_2-CHO$	$CH_3-CH(Br)-CH(CH_3)-CH_2-CHO$
Cl (skeletal structure), O	Br (skeletal structure), O

15.5 a. 3-chloro-4-methyl-2-pentanone;
b. 3-phenyl-2-butanone
c. 4-methylcyclohexanone

15.6

a. 6-chloro-4-methyl-3-hexanone	b. 3-ethyl-2-methylcyclohexanone
$CH_2Cl-CH_2-CH(CH_3)-C(=O)-CH_2-CH_3$	cyclohexanone ring (O) with CH_3 and CH_2-CH_3
Cl (skeletal structure), O	O (skeletal structure)

15.7

$CH_3-C(=O)-CH_3$	$CH_3-C(=O)-H$	$H-C(=O)-H$	$CH_3-CH_2-C(=O)-CH_3$
a. acetone	b. acetaldehyde	c. formaldehyde	d. ethyl methyl ketone
$C_6H_5-C(=O)-H$	$CH_3-CH_2-CH_2-C(=O)-H$	$CH_3-CH_2-C(=O)-H$	$C_6H_5-C(=O)-CH_3$
e. benzaldehyde	f. butyraldehyde	g. propionaldehyde	h. methyl phenyl ketone

15.8 $CH_3-CH_2-CH_2-C(=O)-H$ $\quad CH_3-CH_2-C(=O)-CH_3$ $\quad CH_3-CH(CH_3)-C(=O)-H$

The two aldehydes are skeletal isomers; each is a functional group isomer of the ketone.
Yes, there are other possible functional group isomers (ethers, unsaturated alcohols)

15.9

Compounds	Higher-boiling compound	Explanation
pentanal and hexane	pentanal	Dipole-dipole attractions occur between pentanal molecules.
2-hexanone and 2-octene	2-hexanone	Dipole-dipole attractions occur between 2-hexanone molecules.
pentanal and 1-pentanol	1-pentanol	The hydrogen-bonding between 1-pentanol molecules is stronger than the dipole-dipole attractions between pentanal molecules.

15.10

a. $CH_3-CH_2-CH_2-CH_2-CH_2(OH)$ $\xrightarrow[\text{agent}]{\text{mild oxidizing}}$ $CH_3-CH_2-CH_2-CH_2-C(=O)-H$

1-pentanol → pentanal

b. 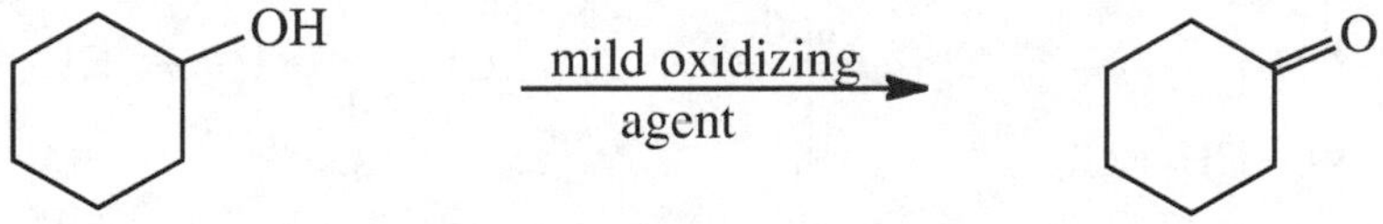 cyclohexanol $\xrightarrow[\text{agent}]{\text{mild oxidizing}}$ cyclohexanone

c. $CH_3-CH_2-C(CH_3)(OH)-CH_3$ $\xrightarrow[\text{agent}]{\text{mild oxidizing}}$ No reaction, tertiary alcohol.

2-methyl-2-butanol

15.11

Compound	Benedict's reagent	Tollens reagent
a. 4-chlorocyclohexanone	negative	negative
b. 2-methylhexanal	positive	positive
c. 4-chloro-3-methyl-2-pentanone	negative	negative
d. 3,3-dimethylpentane	negative	negative
e. pentanal	positive	positive

15.12

a. $CH_3-CH(Cl)-CH_2-C(=O)-H + H_2 \xrightarrow{Ni} CH_3-CH(Cl)-CH_2-CH_2-OH$

b. $CH_3-CH_2-CH(CH_3)-C(=O)-CH_2-Br + H_2 \xrightarrow{Ni} CH_3-CH_2-CH(CH_3)-CH(OH)-CH_2-Br$

15.13

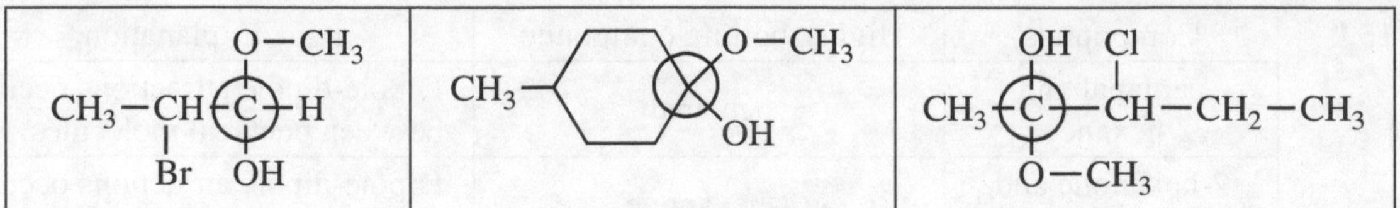

$CH_3-CH(Br)-C(O-CH_3)(OH)-H$	CH_3-cyclohexane-C$(O-CH_3)(OH)$	$CH_3-C(OH)(O-CH_3)-CH(Cl)-CH_2-CH_3$
a. 2-bromopropanal + methanol	b. 4-methylcyclohexanone + methanol	c. 3-chloro-2-pentanone + methanol

15.14

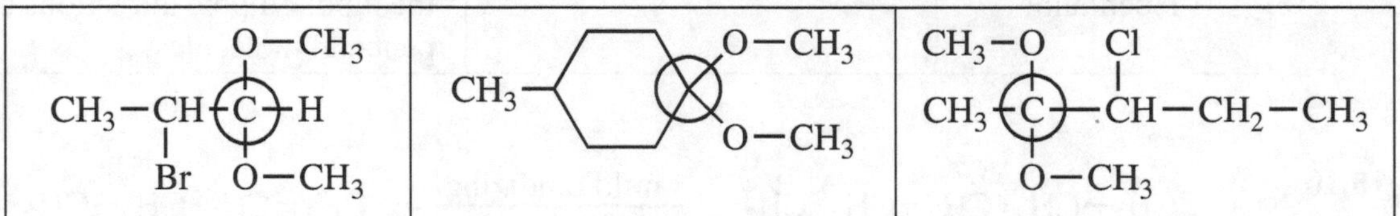

$CH_3-CH(Br)-C(O-CH_3)(O-CH_3)-H$	CH_3-cyclohexane-C$(O-CH_3)(O-CH_3)$	$CH_3-C(O-CH_3)(O-CH_3)-CH(Cl)-CH_2-CH_3$
a.	b.	c.

15.15 a. $CH_3-CH_2-CH(OH)(O-CH_2-CH_3) \xrightleftharpoons{\text{acid catalyst}} CH_3-CH_2-C(=O)-H + CH_2(OH)-CH_3$

b. $CH_3-CH(CH_3)-CH(O-CH_3)(O-CH_3) + H_2O \xrightleftharpoons{\text{acid catalyst}} CH_3-CH(CH_3)-C(=O)-H + 2\ CH_3-OH$

c. cyclopentyl$-C(O-CH_2-CH_3)(O-CH_2-CH_3)-CH_3 + H_2O \xrightleftharpoons{\text{acid catalyst}}$ cyclopentyl$-C(=O)-CH_3 + 2\ CH_2(OH)-CH_3$

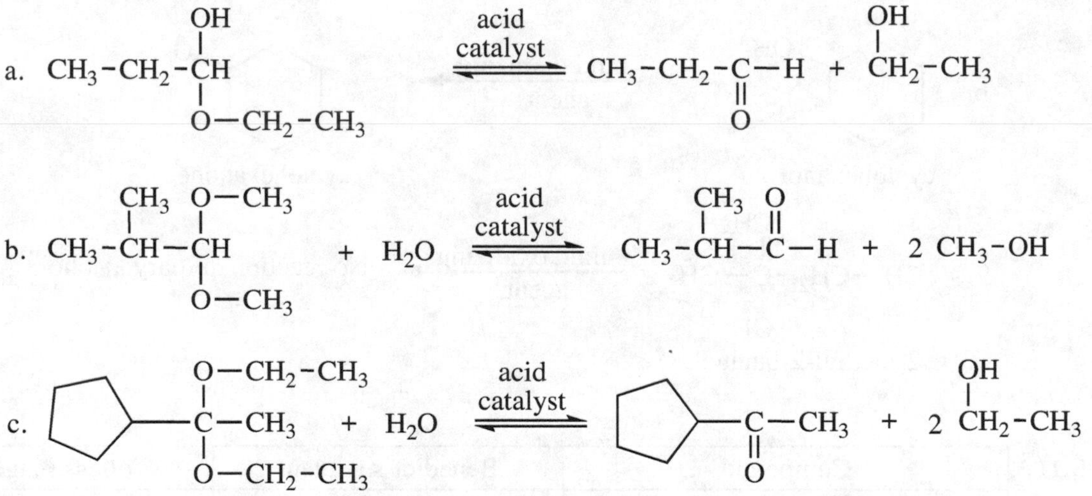

Answers to Self-Test

The numbers in parentheses refer to sections in your textbook.
1. F; three (15.3) **2**. T (15.4) **3**. T (15.9) **4**. F; ketones (15.7) **5**. F; lower (15.8)
6. T (15.8) **7**. F; secondary (15.9) **8**. F; a carboxylic acid (15.9) **9**. T (15.10)
10. F; Cu^{2+} (15.10) **11**. T (15.10) **12**. F; more easily (15.11) **13**. F; much less stable (15.11)
14. T (15.11) **15**. F; thiocarbonyl compound (15.13) **16**. F; propionaldehyde (Sec. 15.4)
17. F; butanone (15.5) **18**. F; cyclic ether (15.3) **19**. b (15.9, 15.10) **20**. d (15.4) **21**. a (15.5)
22. c (15.11) **23**. c (15.8) **24**. a (15.6) **25**. b (15.9) **26**. e (15.5, 15.10)

Carboxylic Acids, Esters, and Other Acid Derivatives — Chapter 16

Chapter Overview

Carboxylic acids participate in a variety of different reactions in organic chemistry and in biological systems. Some of their derivatives, soluble salts and esters, help to keep our world clean and sweet-smelling.

In this chapter you will learn to recognize and name carboxylic acids and their derivatives: acid salts, esters, acid chlorides and anhydrides. You will write equations for preparation of these compounds. You will compare their physical properties and write equations for some of their chemical reactions. You will identify thioesters and esters of phosphoric acid.

Practice Exercises

16.1 The generalized structures for carboxylic acids and their derivatives are characterized by an **acyl** (Sec. 16.1) group in which a carbonyl carbon atom is attached to an oxygen, a nitrogen, or a halogen atom. These compounds are called **acyl compounds** (Sec. 16.1). The C—Z bond in acyl compounds is polar; most reactions of acyl compounds involve the C—Z bond.

$$R-\overset{\overset{\displaystyle O}{\|}}{C}-$$

acyl group

$$R-\overset{\overset{\displaystyle O}{\|}}{C}-Z$$

acyl compound: C—Z = C—O, C—N, or C—Cl
carbonyl compound: C—Z = C—H, C—C

The compounds studied in the last chapter (aldehydes and ketones) have a nonpolar C–Z bond because the carbonyl carbon is attached to either a hydrogen atom or a carbon atom. This bond is nonpolar, so most reactions of aldehydes and ketones involve the carbonyl group. Aldehydes and ketones are called **carbonyl compounds** (Sec. 16.1).

For each structural formula below:

1. Circle the acyl group.
2. Indicate whether the C—Z bond is polar or nonpolar.
3. Classify the compound as an acyl compound or a carbonyl compound.
4. Identify the general family of compounds.

$R-\overset{\overset{\displaystyle O}{\|}}{C}-OH$	$R-\overset{\overset{\displaystyle O}{\|}}{C}-O-R'$	$R-\overset{\overset{\displaystyle O}{\|}}{C}-O-\overset{\overset{\displaystyle O}{\|}}{C}-R'$	$R-\overset{\overset{\displaystyle O}{\|}}{C}-NH_2$
2. 3. 4.	2. 3. 4.	2. 3. 4.	2. 3. 4.
$R-\overset{\overset{\displaystyle O}{\|}}{C}-Cl$	$R-\overset{\overset{\displaystyle O}{\|}}{C}-H$	$R-\overset{\overset{\displaystyle O}{\|}}{C}-R'$	
2. 3. 4.	2. 3. 4.	2. 3. 4.	

16.2 **Carboxylic acids** (Sec. 16.1) are acyl compounds in which the carbonyl carbon is attached to a hydroxyl group. The functional group is called the **carboxyl group** (Sec. 16.1). The naming of carboxylic acids is similar to that of aldehydes, with the *–al* ending replaced by *–oic acid*. **Dicarboxylic acids** (Sec. 16.2) are named by adding *–dioic acid* to the alkane name.

Give the IUPAC name for each of the following carboxylic acids.

a. $CH_3-CH(CH_3)-C(=O)-OH$	b. $I-C_6H_4-C(=O)-OH$	c. $HO-C(=O)-CH(CH_3)-CH_2-CH_2-C(=O)-OH$

16.3 Common names are often used for many of the smaller carboxylic acids and dicarboxylic acids. In common names, positions on the parent carbon chain are named relative to the carboxyl carbon: the α-carbon is carbon 2 (next to the carboxyl carbon), the β-carbon is carbon 3, etc. Compare the common name and the IUPAC name for some acids by completing the following table.

IUPAC name	Common name	Structural formula
butanoic acid		
	formic acid	
2-chloropentanoic acid		
	β-bromobutyric acid	
		HOOC—COOH
	adipic acid	

16.4 A carboxylic acid that contains one or more functional groups in addition to the one carboxyl group is a **polyfunctional carboxylic acid** (Sec. 16.4). Name the functional groups present in each of the following compounds:

a. pyruvic acid

b. glycolic acid

c. maleic acid

16.5 Carboxylic acids have very high boiling points because two molecules can form two hydrogen bonds with one another, from each of the double-bonded oxygens to the hydrogens of the –OH groups. Carboxylic acids also form hydrogen bonds with water molecules

Draw diagrams of hydrogen bonding between the following molecules. Use condensed structural formulas and show hydrogen bonds as dotted lines.

a. two molecules of propanoic acid	b. one molecule of propanoic acid and water

16.6 Carboxylic acids can be prepared by oxidation of a primary alcohol or an aldehyde. Aromatic acids can be prepared by oxidation of an alkyl side chain on a benzene derivative.

Write equations for the preparation of carboxylic acids from the indicated reactants, using a weak oxidizing agent (such as CrO_3 or $K_2Cr_2O_7$). Give the IUPAC name for each carboxylic acid formed.

a. 1-butanol

b. butanal

c. 4-bromo-1-ethylbenzene

16.7 Carboxylic acids are weak acids. In water, proton transfer from the carboxylic acid molecule to a water molecule occurs (usually less than 5%) to give a hydronium ion and a negative **carboxylate ion** (Sec. 16.7).

a. Write a chemical equation for the dissociation of propanoic acid in water.

Carboxylic acids react with strong bases to produce water and a **carboxylic acid salt** (Sec. 16.8). The negative ion of the salt is a carboxylate ion; it is named by dropping the *–ic acid* ending and adding *–ate*. Write equations (using structural formulas) for the reaction of each of these carboxylic acids with NaOH. Name the carboxylic acid salt that is formed.

b. propanoic acid

c. benzoic acid

16.8 A carboxylic acid salt can be converted to the carboxylic acid by reacting the salt with a strong acid such as HCl or H_2SO_4.

Write equations (using structural formulas) for the reaction of each of these carboxylic acid salts with hydrochloric acid.

a. potassium ethanoate

b. sodium butanoate

16.9 In living organisms, one reaction a carboxylic acid may undergo is a **decarboxylation reaction** (Sec. 16.9), in which the carboxylic acid is converted to a molecule of CO_2 and a smaller organic molecule.

Complete the following chemical equations for decarboxylation reactions.

a. $CH_3{-}CH_2{-}CH_2{-}C({=}O){-}OH \longrightarrow \qquad + \quad O{=}C{=}O$

b. $CH_3-CH_2-\underset{\underset{O}{\|}}{C}-C\overset{O}{\underset{OH}{\lessgtr}}$ $\longrightarrow$ $+\ O=C=O$

16.10 An **ester** (Sec. 16.10) is a carboxylic acid derivative in which the –OH portion of the carboxyl group has been replaced by –OR (an alkoxy group). The condensation reaction between a carboxylic acid and an alcohol, in the presence of a strong-acid catalyst, produces an ester. In this process, known as **esterification** (Sec. 16.11), the –OH from the carboxyl group combines with –H from the alcohol to form a molecule of water as a by-product.

Write in the structural formulas for the reactants or products that will complete the esterification reactions below:

a. $CH_3-CH_2-\overset{\overset{O}{\|}}{C}-OH + HO-CH_2-CH_3 \overset{H^+}{\rightleftharpoons}$

b. ? + ? $\overset{H^+}{\rightleftharpoons}$ $C_6H_5-\overset{\overset{O}{\|}}{C}-O-CH_3$

16.11 In naming an ester, it is useful to view the structure as having an "acid part" (the acyl group) and an "alcohol part" (the alkoxy group). Name the "alcohol part" first, followed by the "acid part" with an *-ate* ending. Complete the following table showing ester structures, IUPAC names, and the reactants that form the esters.

Acid and alcohol	Structure of ester formed	IUPAC name
ethanoic acid and methanol		
	$CH_3-CH_2-CH_2-\overset{\overset{O}{\|}}{C}-O-CH_3$	
		ethyl benzoate

16.12 Esters are often known by common names, which are based on the common names of the acid parts (Table 16.1) of the esters. Complete the following table comparing the IUPAC and common names of some esters.

IUPAC name	Common name
ethyl methanoate	
	methyl valerate
	propyl caproate

16.13 Constitutional isomerism (including positional, skeletal, and functional group isomerism) is possible for most carboxylic acids and esters. To practice differentiating between these types of isomerism, draw isomers as directed below.

a. Draw a functional group isomer of this ester.

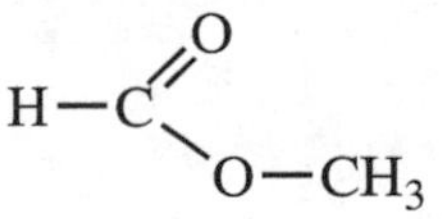

$$H-C(=O)-O-CH_3$$

b. Draw a positional isomer of this ester:

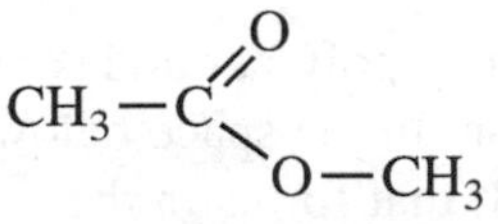

$$CH_3-C(=O)-O-CH_3$$

c. Draw a skeletal isomer of this ester

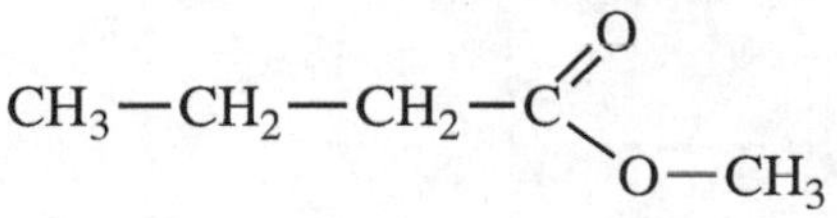

$$CH_3-CH_2-CH_2-C(=O)-O-CH_3$$

16.14 Ester molecules do not have a hydrogen atom bonded to an oxygen atom, so they cannot form hydrogen bonds with one another. They can, however, form hydrogen bonds with water molecules.

Which compound in each of the following pairs would have a higher boiling point? Explain your answer.

Compounds	Higher-boiling compound	Explanation
butanoic acid and methyl propanoate		
methyl propanoate and 1-butanol		
methyl hexanoate and decane		

16.15 Esters can be produced by the acid-catalyzed reaction of a carboxylic acid with an alcohol (esterification). Esters can undergo the reverse reaction, hydrolysis with an acid catalyst, to produce the acid and alcohol from which they were formed. Esters also undergo base-catalyzed hydrolysis, which is called **ester saponification** (Sec. 16.16). In this case the products are the alcohol and the salt of the carboxylic acid.

Give the IUPAC names of the products formed in the following reactions.

a. Ethyl butanoate undergoes basic hydrolysis with sodium hydroxide (saponification).

b. Ethyl butanoate undergoes acidic hydrolysis.

16.16 A thiol can react with a carboxylic acid to form a **thioester** (Sec. 16.17), a sulfur-containing analog of an ester. A molecule of water is formed as a by-product. Thioesters are named in the same way as esters, except that the prefix *thio* is added to the acid part of the name.

Complete the following reactions showing thioester formation, and give the IUPAC name for the thioester produced.

a. $HCOOH + CH_3-CH_2-SH \longrightarrow$?

b. ? + ? $\longrightarrow C_6H_5-C(=O)-S-CH_3 + H_2O$

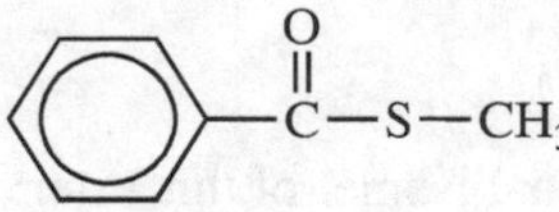

16.17 The condensation reaction by which difunctional monomers form a polymer and some small molecule as a by-product is called a condensation polymerization. In the space below draw a two monomer section of the **condensation polymer** (Sec. 16.18) that forms in the condensation reaction between terephthalic acid (a dicarboxylic acid) and ethylene glycol (a dialcohol).

a. $2\ HO-C(=O)-C_6H_4-C(=O)-OH + 2\ HO-CH_2-CH_2-OH \longrightarrow$

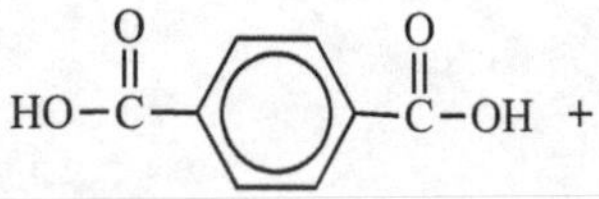

b. This polymer is one of a group of polymers called polyesters (16.18). Explain the name.

c. What is the by-product in this condensation polymerization?

16.18 Complete the following chemical equations for the preparation of an **acid chloride** (Sec. 16.19) or an **acid anhydride** (Sec. 16.19). Give the IUPAC name for the acid chloride (replace *–oic* with *–oyl chloride*) and for the acid anhydride (replace *acid* with *anhydride*).

a. $CH_3-C(=O)-OH \xrightarrow{SOCl_2}$

b. $CH_3-C(=O)-Cl + CH_3-C(=O)-O^- \longrightarrow$

Both acid chlorides and acid anhydrides react readily with water, undergoing hydrolysis reactions to produce the corresponding carboxylic acids. Complete the following hydrolysis reactions. Name the reactants and products.

c. $CH_3-CH_2-CH_2-CH_2-C(=O)-Cl + H_2O \longrightarrow$

d. $(CH_3-CH_2-C(=O))_2O + H_2O \xrightarrow{Heat}$

16.19 Tell whether each of the compounds below contains an acyl group (See Practice Exercise 16.1). If it does contain an acyl group, give the IUPAC name for that acyl group.

a. acetic anhydride ______________________________

b. methanoyl chloride ______________________________

c. diethyl ether ______________________________

d. propanal ______________________________

e. pentanoic acid ______________________________

f. methyl thioethanoate ______________________________

16.20 When compounds containing acyl groups react with an alcohol or a phenol, the acyl group is transferred to the oxygen atom of the alcohol or phenol. Write the equation for the **acyl transfer reaction** (Sec. 16.18), also called an acylation reaction, between ethanoyl chloride and phenol.

16.21 Inorganic acids react with alcohols to form esters; the reaction is similar to that for the formation of carboxylic acid esters. **Phosphate esters** (Sec. 16.20) are important in many biochemical reactions. Phosphoric acid has three hydroxyl groups and so can form phosphate esters with one, two, or three molecules of an alcohol. Draw the structural formula of the ester formed from each set of reactants named below.

a. One molecule of ethanol and one molecule of phosphoric acid

b. Two molecules of ethanol and one molecule of phosphoric acid

c. Three molecules of ethanol and one molecule of phosphoric acid

16.22 The ionic form (hydrogen ions removed) of a terminal phosphate group in an organic molecule is called a **phosphoryl group** (Sec. 16.20). A phosphoryl group has the formula $-PO_3^{2-}$.

A **phosphorylation reaction** (Sec. 16.20) is a chemical reaction in which a phosphoryl group is transferred from one molecule to another. This type of reaction will be encountered in later chapters dealing with metabolic reactions.

In the phosphorylation reaction below, draw a circle around the phosporyl group that is transferred.

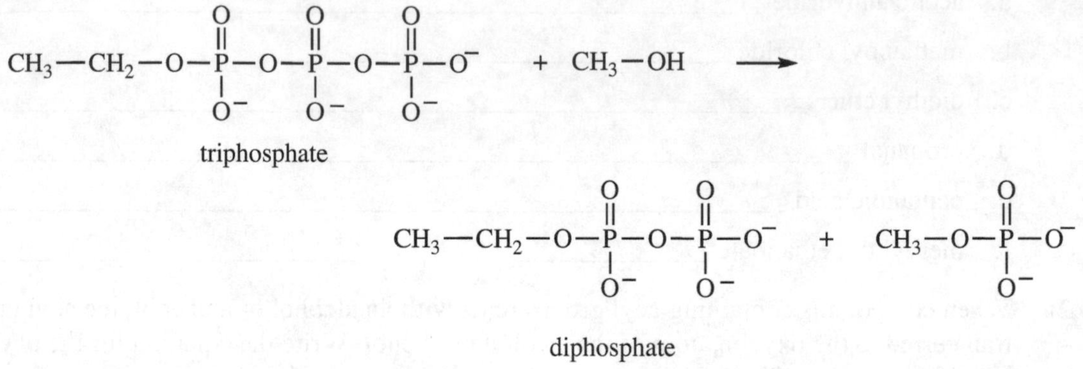

$$CH_3{-}CH_2{-}O{-}\overset{O}{\overset{\|}{\underset{O^-}{\underset{|}{P}}}}{-}O{-}\overset{O}{\overset{\|}{\underset{O^-}{\underset{|}{P}}}}{-}O{-}\overset{O}{\overset{\|}{\underset{O^-}{\underset{|}{P}}}}{-}O^- + CH_3{-}OH \longrightarrow$$

triphosphate

$$CH_3{-}CH_2{-}O{-}\overset{O}{\overset{\|}{\underset{O^-}{\underset{|}{P}}}}{-}O{-}\overset{O}{\overset{\|}{\underset{O^-}{\underset{|}{P}}}}{-}O^- + CH_3{-}O{-}\overset{O}{\overset{\|}{\underset{O^-}{\underset{|}{P}}}}{-}O^-$$

diphosphate

16.23 Use this exercise to review your knowledge of the functional groups in this chapter that contain the carbonyl group. Using letters A through I, give the best choice for each of the terms that follow.

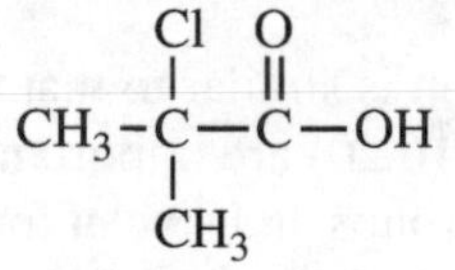

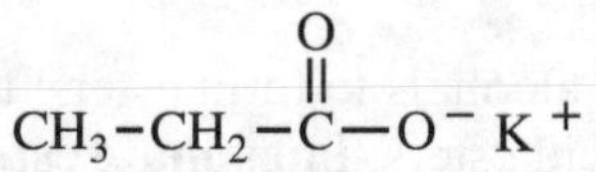

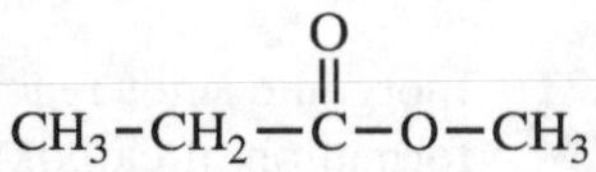

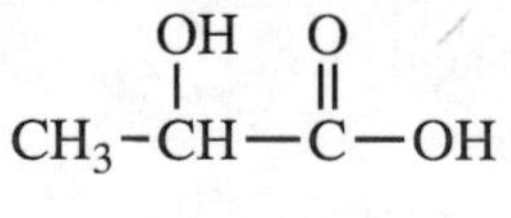

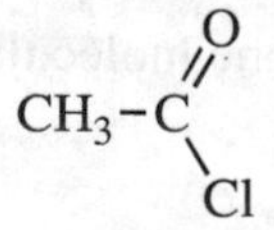

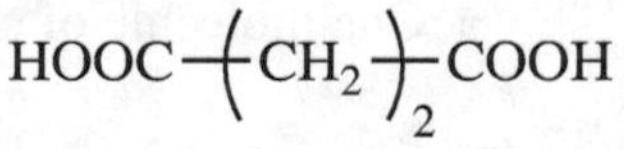

$CH_3{-}\overset{Cl}{\overset{|}{\underset{CH_3}{\underset{|}{C}}}}{-}\overset{O}{\overset{\|}{C}}{-}OH$ A.	$CH_3{-}CH_2{-}\overset{O}{\overset{\|}{C}}{-}O^-\ K^+$ B.	$CH_3{-}CH_2{-}\overset{O}{\overset{\|}{C}}{-}O{-}CH_3$ C.
$CH_3{-}\overset{OH}{\overset{|}{CH}}{-}\overset{O}{\overset{\|}{C}}{-}OH$ D.	$CH_3{-}C(=O)Cl$ E.	$HOOC{+}CH_2{+}_2COOH$ F.
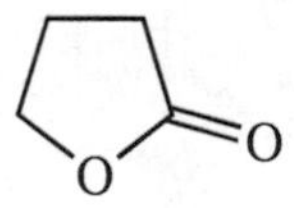 (five-membered ring lactone, ring O, C=O) G.	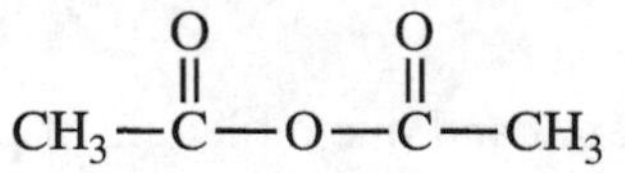 $CH_3{-}\overset{O}{\overset{\|}{C}}{-}O{-}\overset{O}{\overset{\|}{C}}{-}CH_3$ H.	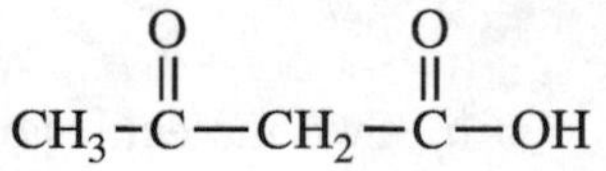 $CH_3{-}\overset{O}{\overset{\|}{C}}{-}CH_2{-}\overset{O}{\overset{\|}{C}}{-}OH$ I.

a. dicarboxylic acid____________
b. carboxylic acid ____________
c. β-keto carboxylic acid ______
d. lactone____________
e. potassium salt ____________
f. ester ____________
g. α-hydroxy carboxylic acid ____
h. acid chloride ____________
i. anhydride____________

Give IUPAC names for compounds A through H in the table above

A. ____________
B. ____________
C. ____________
D. ____________
E. ____________
F. ____________
G. ____________
H. ____________

Self-Test

True-false: Indicate whether the following statements are true or false. If the statement is false, give the word or phrase that may be substituted for the underlined portion to make the statement true.

1. Formic acid is the simplest carboxylic acid.
2. A carboxylic acid with six carbons in a straight chain is named 1-hexanoic acid.
3. Vinegar is made of glacial acetic acid.
4. Another name for 2-methylbutanoic acid is β-methylbutyric acid.
5. Glycolic acid is a dicarboxylic acid.
6. Oxidation of 2-butanol produces butanoic acid.
7. Because of their extensive hydrogen bonding, carboxylic acids have low boiling points.
8. Benzoic acid cannot be prepared by the oxidation of ethylbenzene.
9. A carboxylate ion is formed by the loss of an acidic hydrogen atom.
10. The reaction between sodium hydroxide and benzoic acid would produce sodium benzoate.
11. According to Le Châtelier's principle, adding an excess of alcohol to a carboxylic acid will decrease the amount of ester that is formed.
12. The pleasant fragrances of many flowers and fruits are produced by mixtures of carboxylic acids.
13. Aspirin has an acid functional group and an ester functional group.
14. The general molecular formula for an ester is $C_nH_{2n}O_2$.
15. The boiling points of esters are lower than those of alcohols and acids with comparable molecular mass.
16. Ester saponification is the base-catalyzed hydrolysis of an ester.
17. A polyester is an addition polymer with ester linkages.
18. Acetyl CoA is a biologically important thioester.

Multiple choice:

19. The reaction between methanol and ethanoic acid will produce:

a. methyl acetate	b. ethyl formate	c. methyl butyrate
d. ethyl ethanoate	e. none of these	

20. Rank these three types of compounds – alcohol, carboxylic acid, ester of carboxylic acid of comparable molecular mass – in order of boiling point, highest to lowest:

a. alcohol, acid, ester	b. ester, acid, alcohol	c. acid, alcohol, ester
d. ester, alcohol, acid	e. acid, ester, alcohol	

21. The products of the acid-catalyzed hydrolysis of an ester include:

a. a carboxylate ion	b. a ketone	c. an alcohol
d. an ether	e. none of these	

22. A thioester is formed by the reaction between:
    a. sulfuric acid and a carboxylate
    b. an alcohol and sulfuric acid
    c. a carboxylic acid and a thiol
    d. a carboxylic acid and a sulfate
    e. none of these

23. When a carboxylic acid and an alcohol react to form an ester:
    a. a molecule of water is incorporated in the ester
    b. a molecule of water is produced
    c. a molecule of oxygen is incorporated in the ester
    d. a molecule of hydrogen is produced
    e. none of these

24. A carboxylate ion is produced in which of these reactions?
    a. esterification
    b. saponification
    c. acid hydrolysis of an ester
    d. reaction of an alcohol with an inorganic acid
    e. none of these

25. An example of a dicarboxylic acid is:
    a. succinic acid
    b. butyric acid
    c. caproic acid
    d. lactic acid
    e. none of these

26. The hydrolysis of methyl butanoate in the presence of strong acid would yield:
    a. methanoic acid and butyric acid
    b. methanol and butanoic acid
    c. methanoic acid and 1-butanol
    d. sodium butanoate and methanol
    e. none of these

Answers to Practice Exercises

16.1

R—C(=O)—OH	R—C(=O)—O—R'	R—C(=O)—O—C(=O)—R'	R—C(=O)—NH_2
2. polar 3. acyl compound 4. carboxylic acid	2. polar 3. acyl compound 4. ester	2. polar 3. acyl compound 4. acid anhydride	2. polar 3. acyl compound 4. amide
R—C(=O)—Cl	R—C(=O)—H	R—C(=O)—R'	
2. polar 3. acyl compound 4. acid chloride	2. nonpolar 3. carbonyl compound 4. aldehyde	2. nonpolar 3. carbonyl compound 4. ketone	

16.2
a. 2-methylpropanoic acid
b. 4-iodobenzoic acid
c. 2-methylpentanedioic acid

16.3

IUPAC name	Common name	Structural formula
butanoic acid	butyric acid	$CH_3-(CH_2)_2-COOH$
methanoic acid	formic acid	$H-COOH$
2-chloropentanoic acid	α-chlorovaleric acid	$CH_3-CH_2-CH_2-CH(Cl)-COOH$
3-bromobutanoic acid	β-bromobutyric acid	$CH_3-CH(Br)-CH_2-COOH$
ethanedioic acid	oxalic acid	$HOOC-COOH$
hexanedioic acid	adipic acid	$HOOC-(CH_2)_4-COOH$

16.4 a. one carbonyl group, one carboxyl group
b. one hydroxyl group, one carboxyl group
c. one carbon-carbon double bond, two carboxyl groups

16.5 a. two molecules of propanoic acid

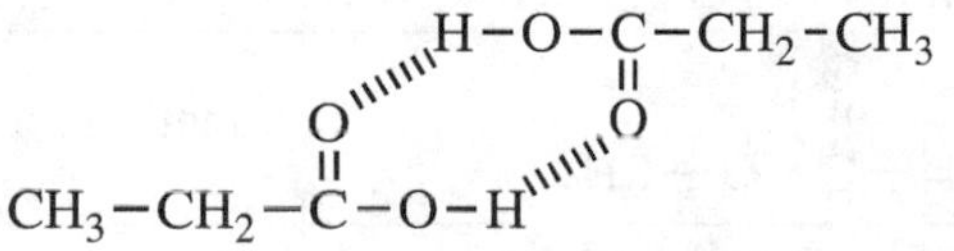

$H-O-C(=O)-CH_2-CH_3$ hydrogen-bonded (dimer) with $CH_3-CH_2-C(=O)-O-H$

b. one molecule of propanoic acid and water

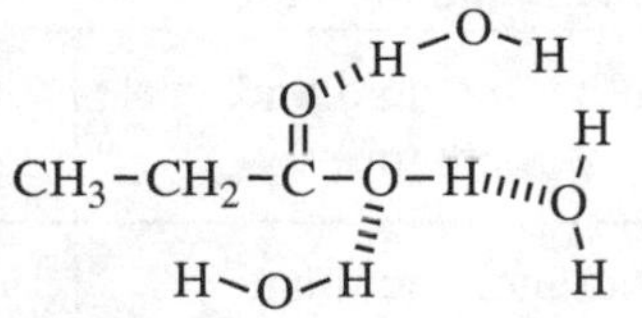

$CH_3-CH_2-C(=O)-O-H$ hydrogen-bonded with three $H-O-H$ molecules

16.6

a. $CH_3-CH_2-CH_2-CH_2-OH \xrightarrow[\text{(oxidizing agent)}]{CrO_3} CH_3-CH_2-CH_2-COOH$
butanoic acid

b. $CH_3-CH_2-CH_2-C(=O)-H \xrightarrow[\text{(oxidizing agent)}]{K_2Cr_2O_7} CH_3-CH_2-CH_2-COOH$
butanoic acid

c. 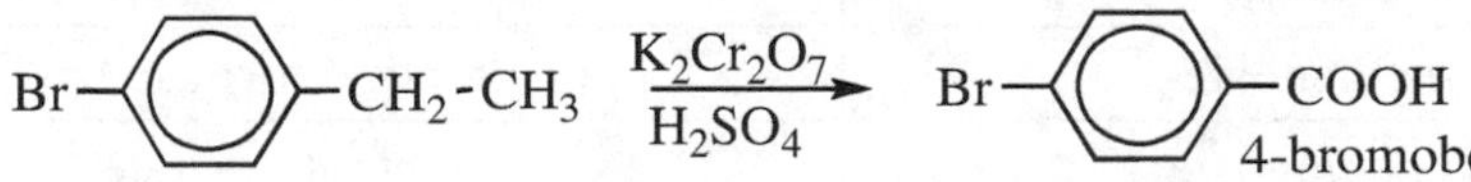 $Br-C_6H_4-CH_2-CH_3 \xrightarrow[H_2SO_4]{K_2Cr_2O_7} Br-C_6H_4-COOH$
4-bromobenzoic acid

16.7

a. 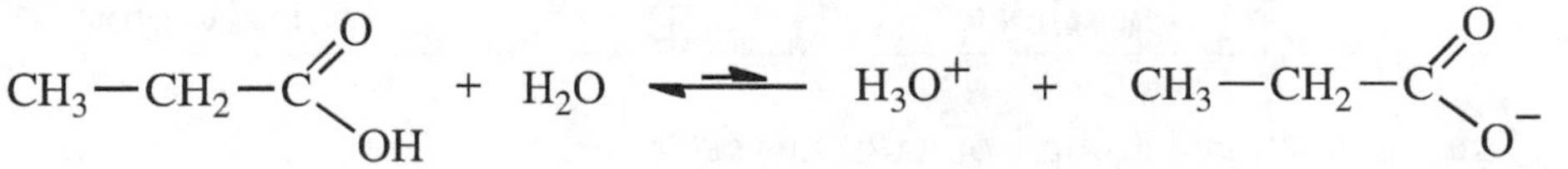 $CH_3-CH_2-C(=O)-OH + H_2O \rightleftharpoons H_3O^+ + CH_3-CH_2-C(=O)-O^-$

b. 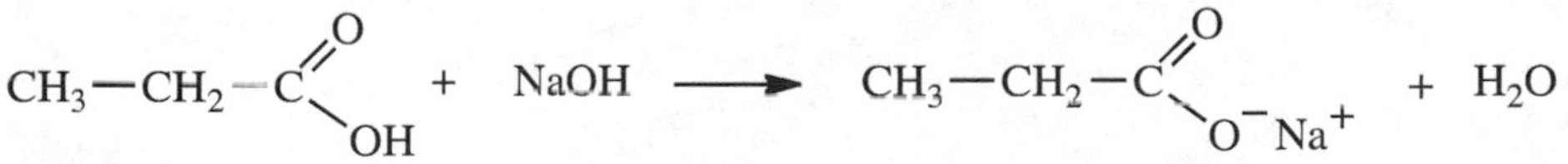 $CH_3-CH_2-C(=O)-OH + NaOH \longrightarrow CH_3-CH_2-C(=O)-O^-\ Na^+ + H_2O$
propanoic acid → sodium propanoate

c.  $C_6H_5-C(=O)-OH + NaOH \longrightarrow C_6H_5-C(=O)-O^-\ Na^+ + H_2O$
benzoic acid → sodium benzoate

16.8 a. $CH_3-C(=O)-O^-K^+ + HCl \longrightarrow CH_3-C(=O)-OH + KCl$

b. $CH_3-CH_2-CH_2-C(=O)-O^-Na^+ + HCl \longrightarrow CH_3-CH_2-CH_2-C(=O)-OH + NaCl$

16.9

a. $CH_3-CH_2-CH_2-C(=O)-OH \longrightarrow CH_3-CH_2-CH_3 + O=C=O$

b. $CH_3-CH_2-C(=O)-C(=O)-OH \longrightarrow CH_3-CH_2-C(=O)-H + O=C=O$

16.10 a. $CH_3-CH_2-C(=O)-OH + HO-CH_2-CH_3 \overset{H^+}{\rightleftharpoons} CH_3-CH_2-C(=O)-O-CH_2-CH_3 + H_2O$

b. 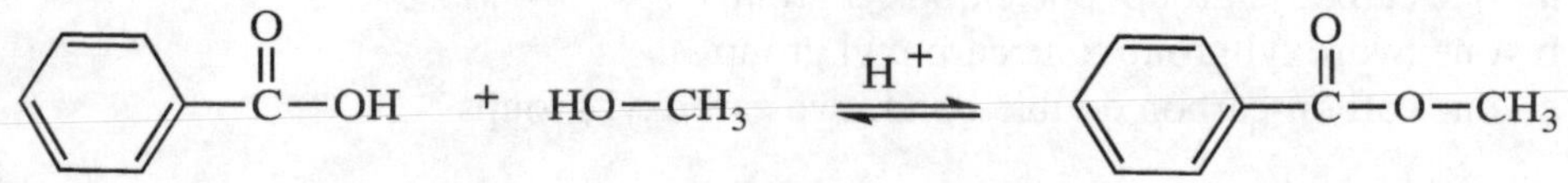 $C_6H_5-C(=O)-OH + HO-CH_3 \overset{H^+}{\rightleftharpoons} C_6H_5-C(=O)-O-CH_3$

16.11

Acid and alcohol	Structure of ester formed	IUPAC name
ethanoic acid and methanol	$CH_3-C(=O)-O-CH_3$	methyl ethanoate
butanoic acid and methanol	$CH_3-CH_2-CH_2-C(=O)-O-CH_3$	methyl butanoate
benzoic acid and ethanol	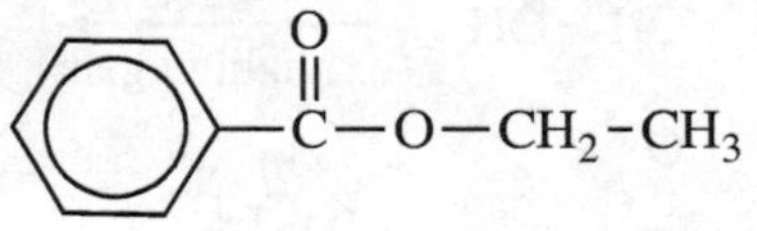 $C_6H_5-C(=O)-O-CH_2-CH_3$	ethyl benzoate

16.12

IUPAC name	Common name
ethyl methanoate	ethyl formate
methyl pentanoate	methyl valerate
propyl hexanoate	propyl caproate

16.13 a. Draw a functional group isomer of this ester.

$H-C(=O)-O-CH_3$ $\qquad$ $CH_3-C(=O)-OH$

b. Draw a positional isomer of this ester:

$CH_3-C(=O)-O-CH_3$ $\qquad$ $H-C(=O)-O-CH_2-CH_3$

c. Draw a skeletal isomer of this ester:

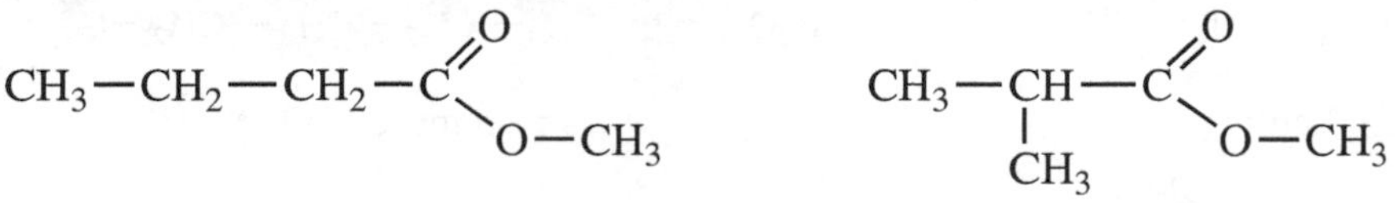

$CH_3-CH_2-CH_2-C(=O)-O-CH_3$ $CH_3-CH(CH_3)-C(=O)-O-CH_3$

16.14

Compounds	Higher-boiling compound	Explanation
butanoic acid, methyl propanoate	butanoic acid	hydrogen bonding between two molecules of the carboxylic acid (produces a dimer)
methyl propanoate, 1-butanol	1-butanol	hydrogen bonding between the alcohol molecules is stronger than dipole-dipole attraction between ester molecules
methyl hexanoate, decane	methyl hexanoate	dipole-dipole attraction between ester molecules

16.15 a. ethanol and sodium butanoate b. ethanol and butanoic acid

16.16 a. $HCOOH + CH_3-CH_2-SH \longrightarrow HC(=O)-S-CH_2-CH_3 + H_2O$
ethyl thiomethanoate

b. $C_6H_5-COOH + CH_3-SH \longrightarrow C_6H_5-C(=O)-S-CH_3 + H_2O$
methyl thiobenzoate

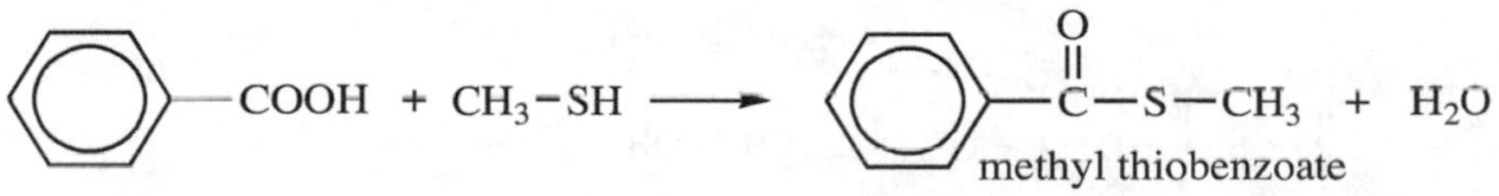

16.17 a. $2\ HO-C(=O)-C_6H_4-C(=O)-OH + 2\ HO-CH_2-CH_2-OH \longrightarrow$

$HO-C(=O)-C_6H_4-C(=O)-O-CH_2-CH_2-O-C(=O)-C_6H_4-C(=O)-O-CH_2-CH_2-OH + 2\ H_2O$
(ester linkage; ester linkage)

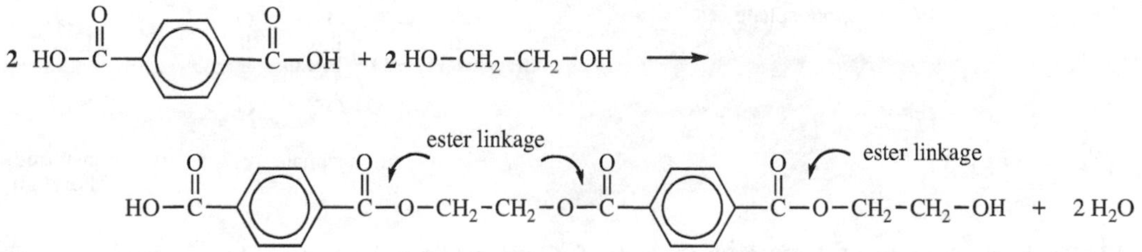

b. They are called polyesters because "poly" means many and "ester" refers to the ester linkage in the polymer.
c. The by-product is H_2O.

16.18 a. $CH_3-C(=O)-OH \xrightarrow{SOCl_2} CH_3-C(=O)-Cl$ + inorganic products
ethanoyl chloride

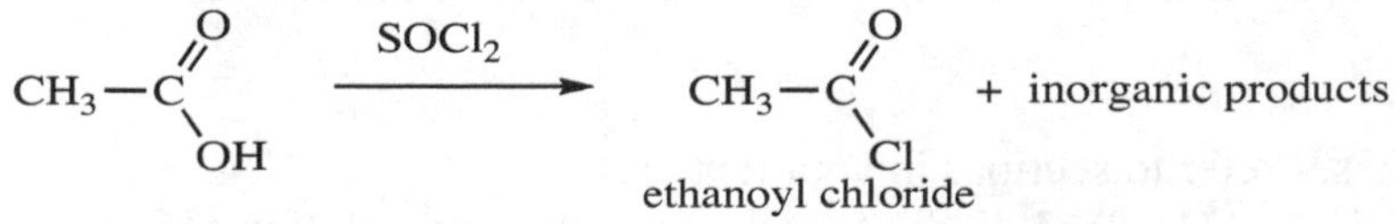

b. $CH_3-C(=O)-Cl + CH_3-C(=O)-O^- \longrightarrow CH_3-C(=O)-O-C(=O)-CH_3 + Cl^-$
ethanoic anhydride

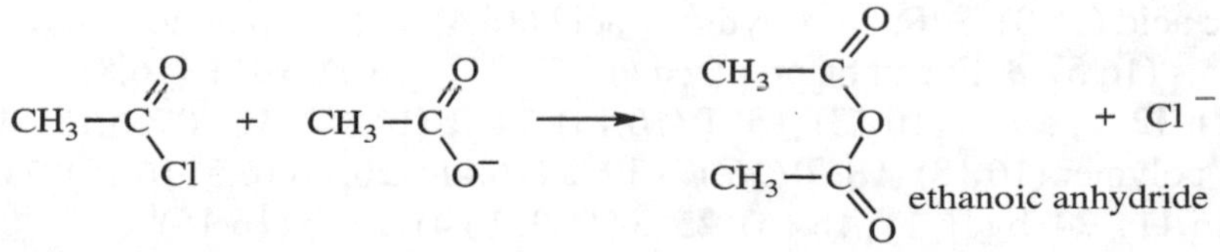

c. 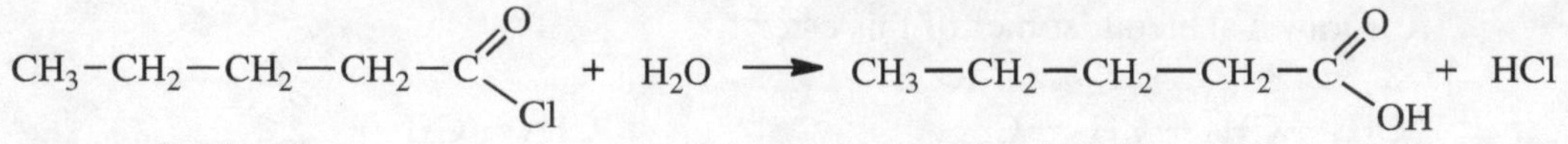 $CH_3-CH_2-CH_2-CH_2-C(=O)-Cl + H_2O \longrightarrow CH_3-CH_2-CH_2-CH_2-C(=O)-OH + HCl$

pentanoyl chloride → pentanoic acid

d. 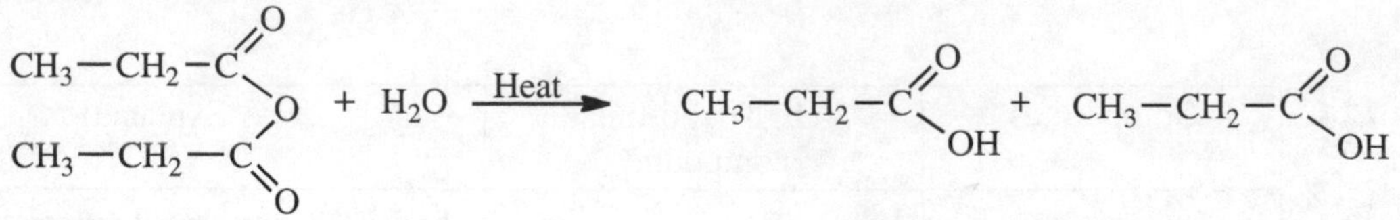 $(CH_3-CH_2-C(=O))_2O + H_2O \xrightarrow{\text{Heat}} CH_3-CH_2-C(=O)-OH + CH_3-CH_2-C(=O)-OH$

propanoic anhydride → 2 molecules of propanoic acid

16.19 a. acetic anhydride – yes; ethanoyl
b. methanoyl chloride – yes; methanoyl
c. diethyl ether – no
d. propanal – yes; propanoyl
e. pentanoic acid – yes; pentanoyl
f. methyl thioethanoate – yes; ethanoyl

16.20 $CH_3-C(=O)-Cl + HO-C_6H_5 \longrightarrow CH_3-C(=O)-O-C_6H_5 + HCl$

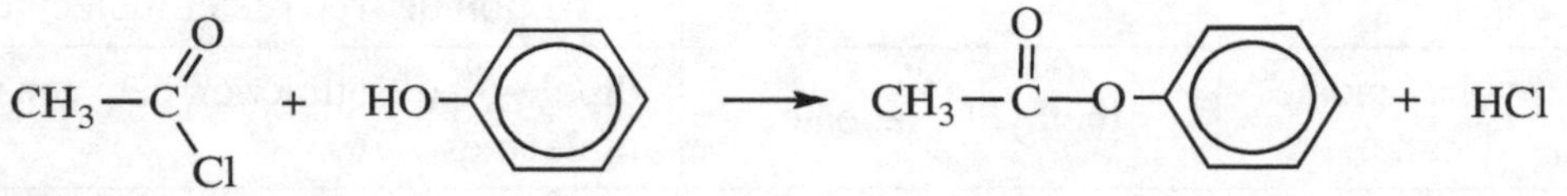

16.21 a. 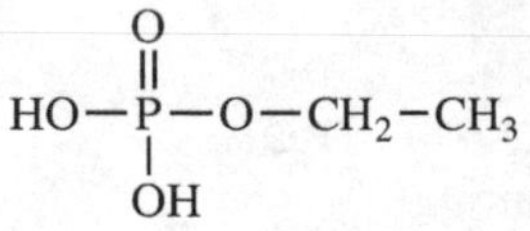 $HO-P(=O)(OH)-O-CH_2-CH_3$ b. 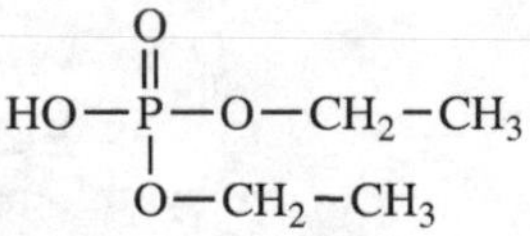 $HO-P(=O)(O-CH_2-CH_3)-O-CH_2-CH_3$ c. 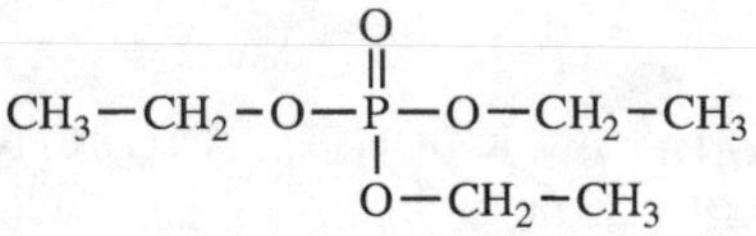 $CH_3-CH_2-O-P(=O)(O-CH_2-CH_3)-O-CH_2-CH_3$

16.22

$CH_3-CH_2-O-P(=O)(O^-)-O-P(=O)(O^-)-O-P(=O)(O^-)-O^- + CH_3-OH \longrightarrow$

triphosphate (circled terminal $P(=O)(O^-)-O^-$: phosphoryl group)

$CH_3-CH_2-O-P(=O)(O^-)-O-P(=O)(O^-)-O^- + CH_3-O-P(=O)(O^-)-O^-$

diphosphate; transferred phosphoryl group

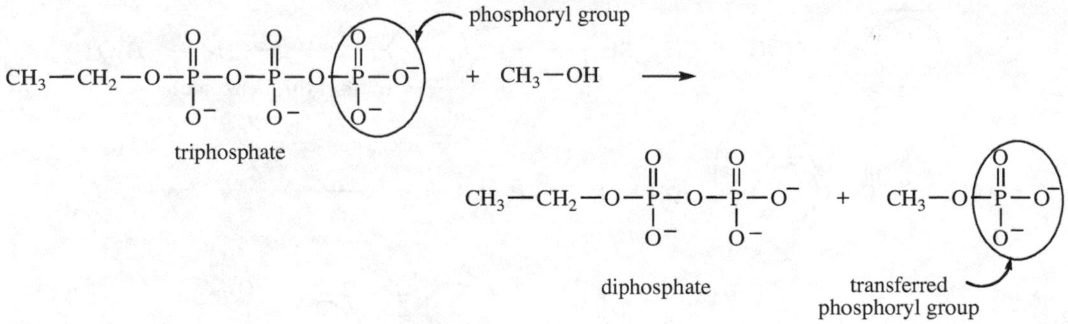

16.23 a. F; b. A (also D, I); c. I; d. G; e. B; f. C (also G); g. D; h. E; i. H

A. 2-chloro-2-methylpropanoic acid
B. potassium propanoate
C. methyl propanoate
D. 2-hydroxypropanoic acid
E. ethanoyl chloride
F. butanedioic acid
G. 4-butanolide
H. ethanoic anhydride

Answers to Self-Test

The numbers in parentheses refer to sections in your textbook.
1. T (16.1) **2**. F; hexanoic acid (16.2) **3**. F; dilute aqueous acetic acid solution (16.3) **4**. F; α-methylbutyric acid (16.3) **5**. F; an α-hydroxy acid (16.4) **6**. F; butanone (16.6) **7**. F; high boiling points (16.5) **8**. F; can be prepared (16.6) **9**. T (16.7) **10**. T (16.8) **11**. F; increase (16.11) **12**. F; esters (16.13) **13**. T (16.13) **14**. T (16.14) **15**. T (16.15) **16**. T (16.16) **17**. F; a condensation polymer (16.18) **18**. T (16.17) **19**. a (16.11) **20**. c (16.5, 16.15) **21**. c (16.16) **22**. c (16.17) **23**. b (16.11) **24**. b (16.11, 16.16) **25**. a (16.3, 16.4) **26**. b (16.16)

Amines and Amides — Chapter 17

Chapter Overview

Amines and amides are nitrogen-containing organic compounds that include many substances of importance in the human body as well as many natural and synthetic drugs.

In this chapter you will learn to recognize, name, and draw structural formulas for amines and amides. You will write equations for the preparation of amines and amides and for the hydrolysis of amides. You will learn the names and functions of some biologically important amines and amides.

Practice Exercises

17.1 Nitrogen has five valence electrons; it forms three covalent bonds to complete its octet of electrons. Draw a Lewis structure for ammonia (NH_3) showing the bonding and nonbonding electron pairs.

17.2 An **amine** (Sec. 17.2) is an organic derivative of ammonia in which alkyl groups replace one, two, or three of ammonia's hydrogen atoms. Amines are classified as primary (1°), secondary (2°), or tertiary (3°) on the basis of the number of alkyl groups attached to the nitrogen atom. Their general formulas are: RNH_2 (1° amines), R_2NH (2° amines), and R_3N (3° amines).

Draw structural diagrams for the following amines, and give their IUPAC and common names. See Sec. 17.3 in your textbook for rules for naming amines.

a. A 1° amine with one methyl group attached to the nitrogen atom.

b. A 2° amine with one ethyl group and one methyl group attached to the nitrogen atom.

c. A 3° amine with three ethyl groups attached to the nitrogen atom.

17.3 Amines (other than the simplest, methylamine) exist as constitutional isomers.

a. Draw line-angle structural formulas for the eight isomeric amines that have the molecular formula $C_4H_{11}N$. (Hint: There are four primary amines, three secondary amines, and one tertiary amine.)

b. Classify each amine as a 1°, 2°, or 3° amine.

c. Give the amine's IUPAC name and its common name (Sec. 17.3). One of the amines has been done as an example.

Line-angle structural formula	Classification	IUPAC name /Common name
(structure: CH$_3$–NH–CH$_2$CH$_2$CH$_3$, line-angle with H over N)	2°	*N*-methyl-1-propanamine; methylpropylamine

17.4 The simplest aromatic amine is called aniline. Draw structural diagrams for the two aniline derivatives named below.

(benzene ring)–NH_2 aniline	*N*-ethylaniline	2-ethylaniline

17.5 Amines are capable of forming hydrogen bonds between molecules; however, because nitrogen is less electronegative than oxygen, hydrogen bonds between molecules of amines are weaker than those between alcohol molecules. (Tertiary amines do not form hydrogen bonds between molecules; they have no hydrogen atoms attached to the nitrogen atom.)

Determine which compound would have a higher boiling point and explain your choice:

Compounds	Higher-boiling compound	Explanation
1-pentanamine or 1-pentanol		
1-pentanamine or hexane		
N,N-diethylethanamine or 1-hexanamine		

17.6 Amines are weak Brønsted-Lowry bases (nitrogen has a lone electron pair that can act as a proton acceptor). Reaction of an amine with an acid produces an **amine salt** (Sec. 17.7). An amine salt consists of a positive ion (the protonated amine) and the negative ion from the acid. The amine may be regenerated by deprotonation of the amine salt by a strong base.

Amine salts are named in the same way that other ionic compounds are: the positive ion, the **substituted ammonium ion** (Sec. 17.6), is named first, followed by the name of the negative ion. Complete the following equations and name the amine salts.

a. $CH_3-CH_2-CH_2-NH_2 \;+\; HCl \longrightarrow$

b. **?** + NaOH ⟶ (benzene ring)$-NH_2$ + NaBr + H_2O

17.7 An **alkylation reaction** (Sec. 17.8) is a reaction in which an alkyl group is transferred from one molecule to another. Amines can be prepared by the alkylation of ammonia with an alkyl halide in the presence of a strong base. The primary amine formed reacts further with more molecules of the alkyl halide to form secondary and tertiary amines, and the **quaternary ammonium salt** (Sec. 17.8).

Write the four reactions that take place when ammonia reacts with chloroethane in the presence of sodium hydroxide, a strong base. Name the primary, secondary, and tertiary amines and the quaternary ammonium salt formed.

a.

b.

c.

d.

17.8 **Heterocyclic amines** (Sec. 17.9) and their derivatives are common in naturally occurring compounds. Name the following heterocyclic amines.

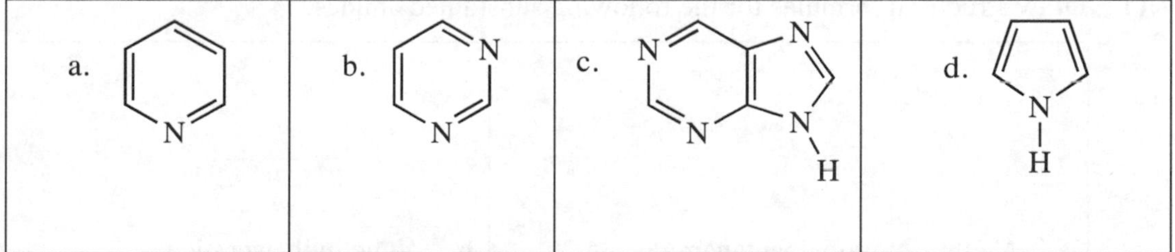

17.9 An **amide** (Sec. 17.12) is a carboxylic acid derivative; the carboxyl –OH group is replaced by an amino or substituted amino group. An amide can be prepared by an **amidification reaction** (Sec. 17.17), a condensation reaction at high temperature between a carboxylic acid and ammonia or a primary or secondary amine. (At room temperature, an acid-base reaction takes place. No amide is formed; the product is a carboxylate salt.)

Give the structure of the amide formed by the reaction of the following acids and amines.

a. 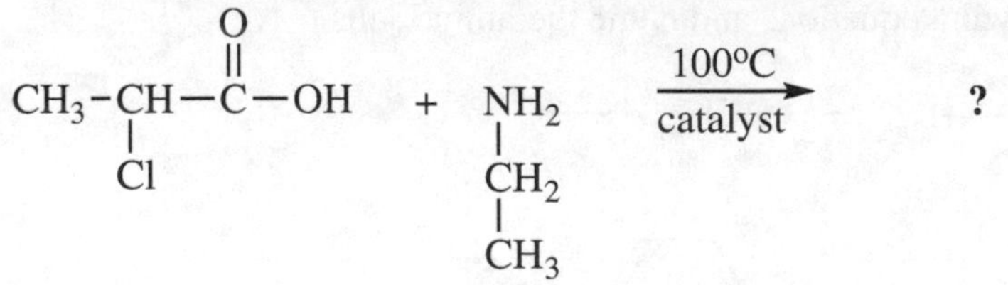 $CH_3-\underset{\underset{Cl}{|}}{CH}-\overset{\overset{O}{\|}}{C}-OH + \underset{\underset{\underset{CH_3}{|}}{\underset{CH_2}{|}}}{NH_2} \xrightarrow[\text{catalyst}]{100^\circ C}$?

b. 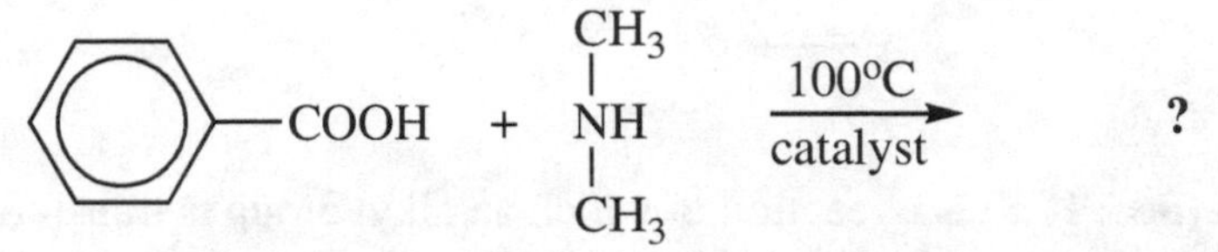 $C_6H_5-COOH + \underset{\underset{CH_3}{|}}{\overset{\overset{CH_3}{|}}{NH}} \xrightarrow[\text{catalyst}]{100^\circ C}$?

c. $CH_3-CH_2-CH_2-\overset{\overset{O}{\|}}{C}-OH + NH_3 \xrightarrow[\text{catalyst}]{100^\circ C}$ 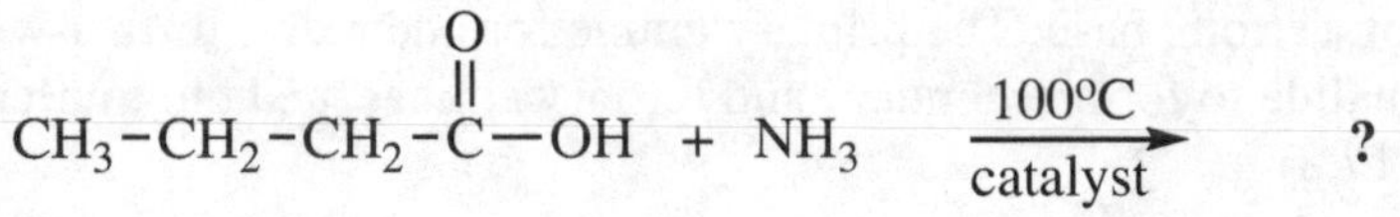 ?

17.10 Amides can be classified as primary, secondary, or tertiary on the basis of the number of carbon atoms attached to the nitrogen atom. Amide names in the IUPAC system use the name of the parent carboxylic acid with the ending *-amide*. Alkyl groups attached to the nitrogen atom are included as prefixes, using *N*- to locate them.

Classify each of the amides in Practice Exercise 17.9 as a primary, secondary, or tertiary amide, and give the IUPAC name for the amide produced.

a.

b.

c.

Common names for amides are similar to IUPAC names, but the common name of the parent carboxylic acid is used. Give the common names for the amides in Practice Exercise 17.9.

d.

e.

f.

17.11 Draw structural formulas for the following substituted amides.

a. *N*-ethyl-*N*-propylpentanamide	b. *N*-methylbenzamide

17.12 To review your understanding of nitrogen-containing functional groups, write generalized structural formulas for the following families of compounds. (Use R and R′ to represent alkyl groups.)

a. a primary amine	e. a primary amine chloride
b. a secondary amine	f. a primary amide
c. a tertiary amine	g. a secondary amide
d. a quaternary ammonium salt	h. a tertiary amide

17.13 Amides are the least reactive type of alkyl compounds. They will undergo hydrolysis under strenuous conditions. The products of hydrolysis depend on the catalyst that is used: an acid catalyst produces a carboxylic acid and an amine salt, and a basic catalyst produces a carboxylic acid salt and an amine. (Sec. 17.18)

Complete the following equations for the hydrolysis of each of these amides:

a. $CH_3-CH_2-CH_2-\overset{\overset{O}{\parallel}}{C}-\overset{\overset{CH_2-CH_3}{|}}{NH} + H_2O \xrightarrow[HCl]{heat} ? + ?$

b. $C_6H_5-\overset{\overset{O}{\parallel}}{C}-\underset{\underset{CH_2-CH_3}{|}}{\overset{\overset{CH_3}{|}}{N}} + H_2O \xrightarrow[HCl]{heat} ? + ?$

c. $CH_3-CH_2-CH_2-\overset{\overset{O}{\parallel}}{C}-\overset{\overset{CH_2-CH_3}{|}}{NH} + H_2O \xrightarrow[NaOH]{heat} ? + ?$

17.14 Give the IUPAC names for the organic products formed in Practice Exercise 17.13.

a.

b.

c.

17.15 Use this identification exercise to review your knowledge of common structures. Using structures A through I, give the best choice for each of the terms that follow.

 NH_2 (skeletal chain) A.	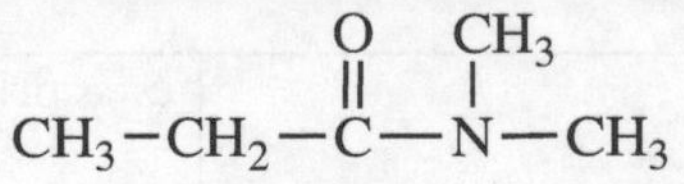 $CH_3-CH_2-\overset{O}{\overset{\Vert}{C}}-\overset{CH_3}{\overset{\vert}{N}}-CH_3$ B.	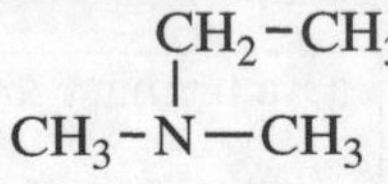 $CH_3-\overset{CH_2-CH_3}{\overset{\vert}{N}}-CH_3$ C.
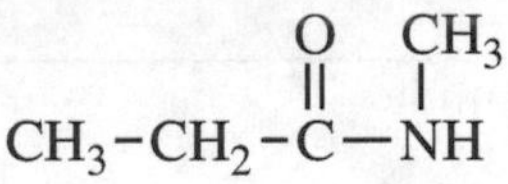 $CH_3-CH_2-\overset{O}{\overset{\Vert}{C}}-\overset{CH_3}{\overset{\vert}{NH}}$ D.	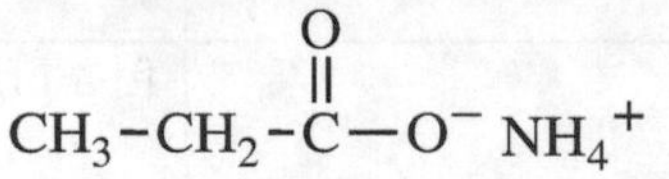 $CH_3-CH_2-\overset{O}{\overset{\Vert}{C}}-O^-\ NH_4^+$ E.	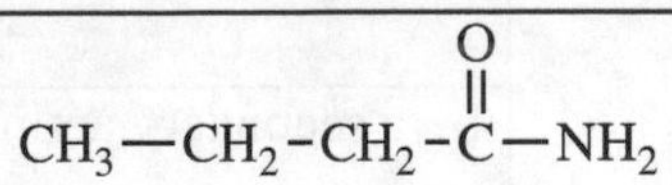 $CH_3-CH_2-CH_2-\overset{O}{\overset{\Vert}{C}}-NH_2$ F.
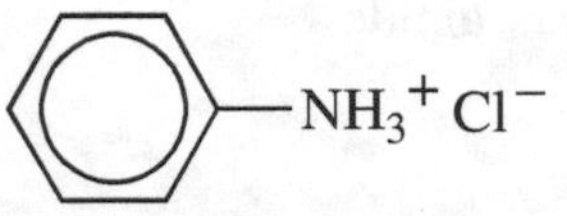 (benzene ring)$-NH_3^+\ Cl^-$ G.	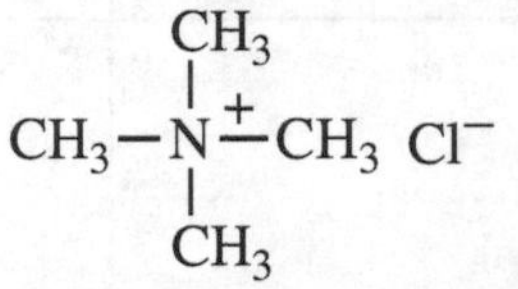 $CH_3-\underset{CH_3}{\underset{\vert}{\overset{CH_3}{\overset{\vert}{N^+}}}}-CH_3\ Cl^-$ H.	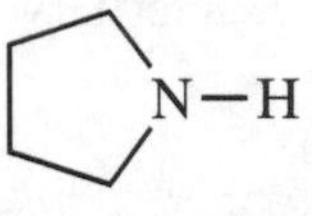 (five-membered ring)$N-H$ I.

Give the letter of the matching structural formula from the table above.

a. heterocyclic amine ________
b. salt of a carboxylic acid ________
c. quaternary ammonium salt ________
d. amine salt ________
e. primary amine ________
f. secondary amide ________
g. secondary amine ________
h. tertiary amine ________
i. primary amide ________
j. tertiary amide ________

Give the IUPAC names of compounds A - I.

A. ________
B. ________
C. ________
D. ________
E. ________
F. ________
G. ________
H. ________
I. ________

Self-Test

True-false: Indicate whether the following statements are true or false. If the statement is false, give the word or phrase that may be substituted for the underlined portion to make the statement true.

1. Nitrogen forms two covalent bonds to complete its octet of electrons.
2. *tert*-Butyl amine is a primary amine.
3. The IUPAC name for an amine having two methyl groups and one ethyl group attached to a nitrogen atom is ethyldimethylamine.
4. A benzene ring with an attached amino group is called aniline.
5. Amines are noted for their pleasant odors.
6. Hydrogen bonding between the molecules of an amine is stronger than hydrogen bonding between alcohol molecules.
7. Reaction of excess alkyl halide with ammonia produces an amine salt.
8. The nitrogen atom of a heterocyclic amine cannot be part of an aromatic system.
9. Urea is a naturally occurring diamine with only one carbon atom.
10. The various types of nylon are synthesized by the reactions of diamines with dicarboxylic acids.
11. In basic solution, an amine exists as an amine salt.

12. Amides <u>do not</u> exhibit basic properties in water as amines do.
13. Disubstituted amides have no hydrogens attached to nitrogen for hydrogen bonding and so have <u>low melting points</u>.
14. The symptoms of Parkinson's disease are caused by a deficiency of <u>serotonin</u>.
15. The nitrogen atom in an amide functional group <u>does not act</u> as a base.
16. The chemical reaction that is opposite to the process of amine salt formation is <u>protonation</u> of the amine salt.
17. Proteins are <u>polyamide</u> polymers.

Multiple choice:

18. An amine can be formed from its amine salt by treating the salt with:
    a. NaOH b. an alcohol c. H_2SO_4
    d. a carboxylic acid e. none of these

19. Which of the following is *not* a heterocyclic amine derivative?
    a. caffeine b. nicotine c. heme
    d. choline e. none of these

20. Alkaloids are a group of nitrogen-containing compounds obtained from plants. Which of the following compounds is *not* an alkaloid?
    a. quinine b. nicotine c. atropine
    d. dopamine e. morphine

21. Which of the following compounds is a secondary amine?
    a. 2-butanamine b. ethylmethylamine c. *N,N*-dimethylaniline
    d. 2-methylaniline e. none of these

22. *N*-methylpropanamide could be prepared as a product of the reaction between:
    a. *N*-propanamine and methanol b. propylamine and acetic acid
    c. methylamine and propanoic acid d. acetic acid and methylamine
    e. none of these

23. Basic hydrolysis of an amide produces:
    a. an amine salt and a carboxylic acid b. an amine salt and a carboxylic acid salt
    c. an amine and a carboxylic acid d. an amine and a carboxylic acid salt
    e. none of these

24. Mental depression may be caused by a deficiency of:
    a. histamine b. serotonin c. atropine
    d. porphyrin e. none of these

25. A correct name for the compound containing nitrogen bonded to two ethyl groups and one hydrogen is:
    a. diethylamine b. 2-ethylamine c. 2-ethyl amine
    d. diethyl amine e. none of these

26. Which of the following biologically active substances does not have phenylethylamine core structure?
    a. dopamine b. norepinephrine c. serotonin
    d. epinephrine e. none of these

Answers to Practice Exercises

17.1

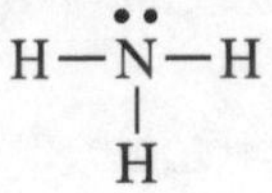

$$H-\ddot{N}-H \text{ with } H \text{ bonded to N (below)}$$

17.2

$H-N(H)-CH_3$ methanamine; methylamine

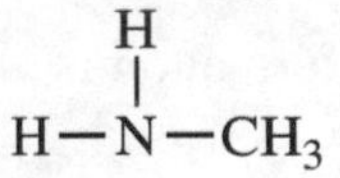

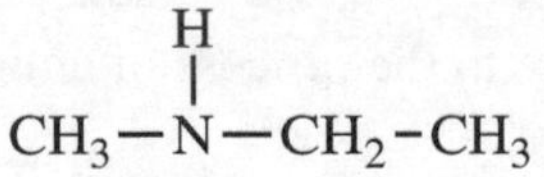

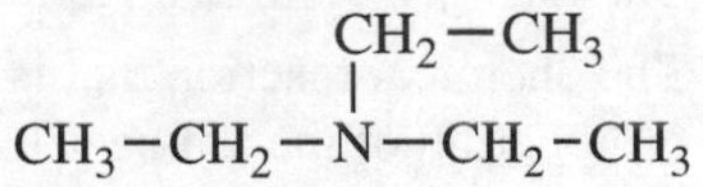

$CH_3-N(H)-CH_2-CH_3$ *N*-methylethanamine; ethylmethylamine

$CH_3-CH_2-N(CH_2-CH_3)-CH_2-CH_3$ *N*,*N*-diethylethanamine; triethylamine

17.3

Line-angle structural formula	Amine classification	IUPAC name / Common name
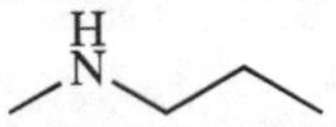 CH3-NH-CH2CH2CH3 (line-angle)	2°	*N*-methyl-1-propanamine; methylpropylamine
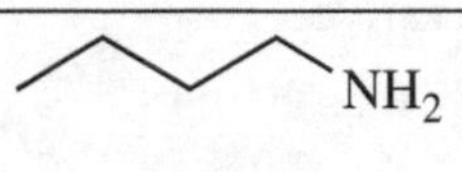 CH3CH2CH2CH2-NH2 (line-angle)	1°	1-butanamine; butylamine
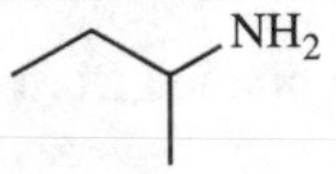 CH3CH2CH(NH2)CH3 (line-angle)	1°	2-butanamine; *sec*-butylamine
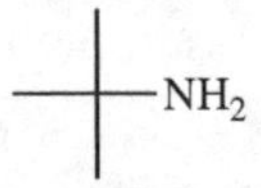 (CH3)3C-NH2 (line-angle)	1°	2-methyl-2-propanamine; *tert*-butylamine
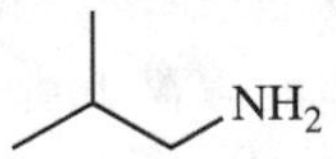 (CH3)2CHCH2-NH2 (line-angle)	1°	2-methyl-1-propanamine; isobutylpropylamine
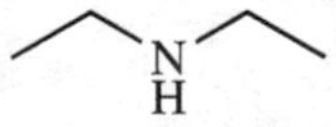 CH3CH2-NH-CH2CH3 (line-angle)	2°	*N*-ethylethanamine; diethylamine
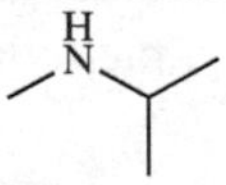 CH3-NH-CH(CH3)2 (line-angle)	2°	*N*-methyl-2-propanamine; methylisopropylamine
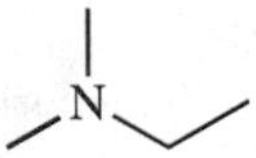 (CH3)2N-CH2CH3 (line-angle)	3°	*N*,*N*-dimethylethanamine; ethyldimethylamine

17.4

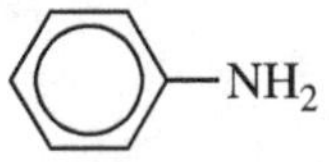 $C_6H_5-NH_2$	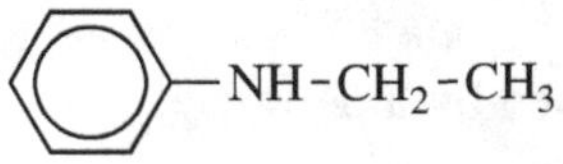 $C_6H_5-NH-CH_2-CH_3$	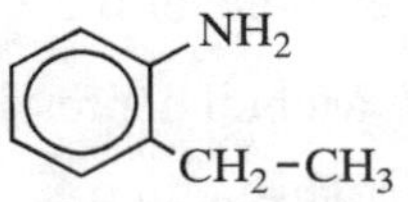 NH_2 and CH_2-CH_3 on adjacent ring carbons
aniline	*N*-ethylaniline	2-ethylaniline

17.5

Higher-boiling	Explanation
1-pentanol	-OH hydrogen bonds are stronger than -NH hydrogen bonds because oxygen is more electronegative than nitrogen
1-pentanamine	hydrogen bonding between primary amine molecules
1-hexanamine	hydrogen bonding between primary amine molecules but not between molecules of tertiary amines

17.6 a. $CH_3-CH_2-CH_2-NH_2 + HCl \longrightarrow CH_3-CH_2-CH_2-\overset{+}{N}H_3\ Cl^-$ propylammonium chloride

b. $C_6H_5-\overset{+}{N}H_3Br^- + NaOH \longrightarrow C_6H_5-NH_2 + NaBr + H_2O$ (anilinium bromide)

17.7 a. $NH_3 + NaOH + CH_3-CH_2-Cl \longrightarrow CH_3-CH_2-NH_2 + H_2O + NaCl$ primary; ethylamine

b. $CH_3-CH_2-NH_2 + CH_3-CH_2-Cl + NaOH \longrightarrow (CH_3-CH_2)(CH_3-CH_2)NH + H_2O + NaCl$ secondary; diethylamine

c. $(CH_3-CH_2)_2NH + CH_3-CH_2-Cl + NaOH \longrightarrow (CH_3-CH_2)_3N + H_2O + NaCl$ tertiary; triethylamine

d. $(CH_3-CH_2)_3N + CH_3-CH_2-Cl \xrightarrow{OH^-} (CH_3-CH_2)_4N^+\ Cl^-$ quaternary; tetraethylammonium chloride

17.8 a. pyridine b. pyrimidine
c. purine d. pyrrole

17.9 a. $CH_3-CH(Cl)-C(=O)-OH + NH_2-CH_2-CH_3 \xrightarrow[\text{catalyst}]{100^\circ C} CH_3-CH(Cl)-C(=O)-NH-CH_2-CH_3 + H_2O$

b. $C_6H_5-COOH + CH_3-NH-CH_3 \xrightarrow[\text{catalyst}]{100^\circ C} C_6H_5-C(=O)-N(CH_3)-CH_3 + H_2O$

c. $CH_3-CH_2-CH_2-C(=O)-OH + NH_3 \xrightarrow[\text{catalyst}]{100^\circ C} CH_3-CH_2-CH_2-C(=O)-NH_2 + H_2O$

17.10 a. secondary amide, 2-chloro-*N*-ethylpropanamide
b. tertiary amide, *N,N*-dimethylbenzamide
c. primary, butanamide
d. α-chloro-*N*-ethylpropionamide
e. *N,N*-dimethylbenzamide
f. butyramide

17.11

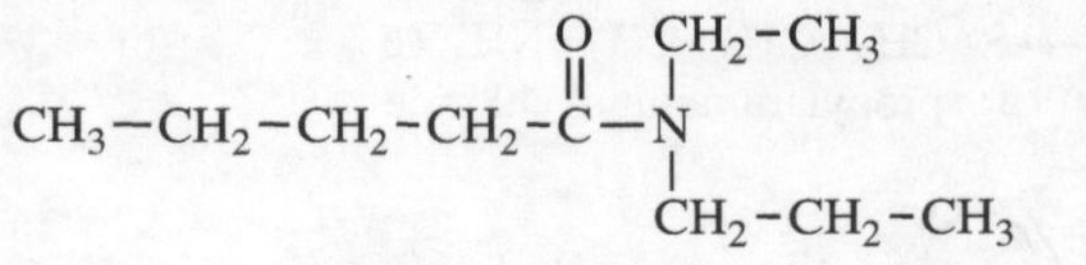 $CH_3-CH_2-CH_2-CH_2-\overset{O}{\overset{\Vert}{C}}-N(CH_2-CH_3)(CH_2-CH_2-CH_3)$ a. *N*-ethyl-*N*-propylpentanamide	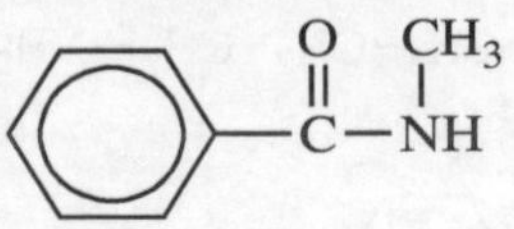 $C_6H_5-\overset{O}{\overset{\Vert}{C}}-NH-CH_3$ b. *N*-methylbenzamide

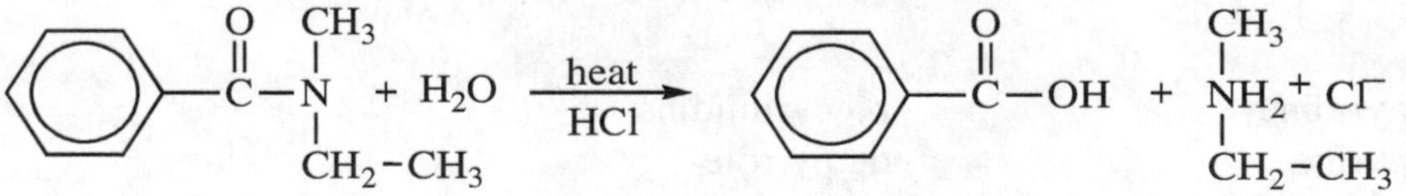

17.12

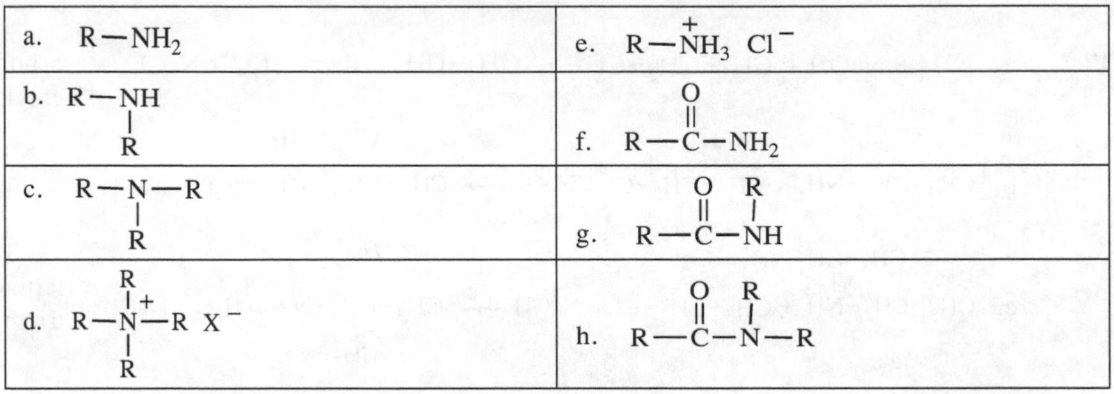

a. $R-NH_2$	e. $R-\overset{+}{N}H_3\ Cl^-$
b. $R-NH-R$	f. $R-\overset{O}{\overset{\Vert}{C}}-NH_2$
c. $R-N(R)-R$	g. $R-\overset{O}{\overset{\Vert}{C}}-NH-R$
d. $R-\overset{+}{N}(R)(R)-R\ X^-$	h. $R-\overset{O}{\overset{\Vert}{C}}-N(R)-R$

17.13 a. 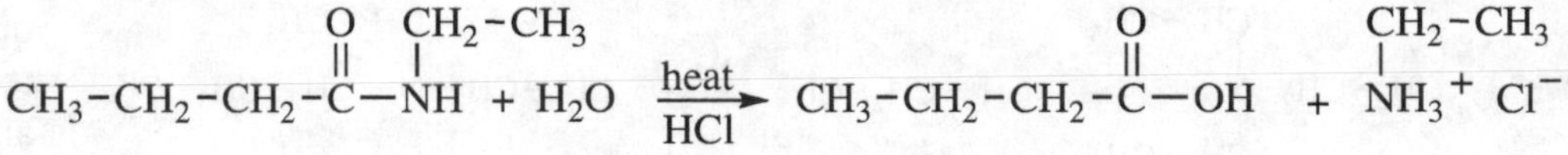 $CH_3-CH_2-CH_2-\overset{O}{\overset{\Vert}{C}}-NH(CH_2-CH_3) + H_2O \xrightarrow[HCl]{heat} CH_3-CH_2-CH_2-\overset{O}{\overset{\Vert}{C}}-OH + CH_3-CH_2-NH_3^+\ Cl^-$

b. $C_6H_5-\overset{O}{\overset{\Vert}{C}}-N(CH_3)(CH_2-CH_3) + H_2O \xrightarrow[HCl]{heat} C_6H_5-\overset{O}{\overset{\Vert}{C}}-OH + CH_3-NH_2^+(CH_2-CH_3)\ Cl^-$

c. $CH_3-CH_2-CH_2-\overset{O}{\overset{\Vert}{C}}-NH(CH_2-CH_3) + H_2O \xrightarrow[NaOH]{heat} CH_3-CH_2-CH_2-\overset{O}{\overset{\Vert}{C}}-O^-\ Na^+ + CH_3-CH_2-NH_2$

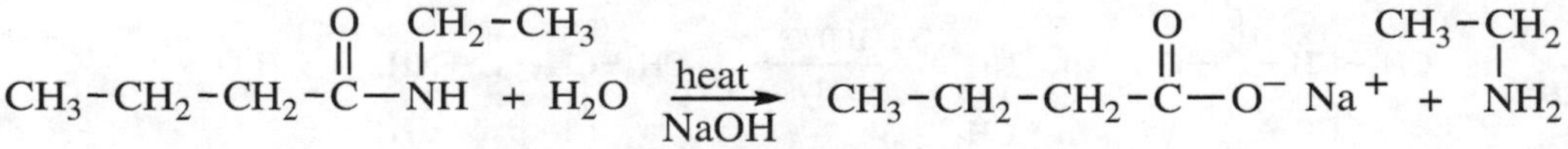

17.14 a. butanoic acid and ethylammonium chloride
b. benzoic acid and ethylmethylammonium chloride
c. sodium butanoate and ethanamine

17.15 a. I; b. E; c. H; d. G; e. A; f. D; g. I; h. C; i. F; j. B

A. 1-propanamine
B. *N,N*-dimethylpropanamide
C. *N,N*-dimethylethanamine
D. *N*-methylpropanamide
E. ammonium propanoate
F. butanamide
G. anilinium chloride
H. tetramethylammonium chloride
I. pyrrolidine

Answers to Self-Test

The numbers in parentheses refer to sections in your textbook.
1. F; three covalent bonds (17.1) **2**. T (17.2) **3**. F; *N,N*-dimethylethanamine (17.3) **4**. T (17.3) **5**. F; unpleasant or fishlike odors (17.5) **6**. F; weaker (17.5) **7**. T (17.7) **8**. F; can (17.9) **9**. F; diamide (17.14) **10**. T (17.19) **11**. F; the free amine (17.7) **12**. T (17.16) **13**. T (17.16) **14**. F; dopamine (17.10) **15**. T (17.15) **16**. F; deprotonation (17.7) **17**. T (17.19) **18**. a (17.7) **19**. d (17.9) **20**. d (17.11) **21**. b (17.3) **22**. c (17.17) **23**. d (17.18) **24**. b (17.10) **25**. a (17.3) **26**. c (17.10)

Carbohydrates — Chapter 18

Chapter Overview

The remaining chapters in the book are concerned with compounds that are important in the chemistry of biological systems. Bioorganic substances include carbohydrates, lipids, proteins, and nucleic acids. Bioinorganic substances include water and inorganic salts.

Carbohydrates are important energy-storage compounds. Their oxidation provides energy for the activities of living organisms. In this chapter you will identify various types of carbohydrates, write structural diagrams for some of the most common ones, and explain stereoisomerism in terms of chiral carbons. You will identify the structural features of some important disaccharides and polysaccharides and you will learn where in nature they are found and what functions they have.

Practice Exercises

18.1 A **monosaccharide** (Sec. 18.3) is a **carbohydrate** (Sec. 18.3) that contains a single polyhydroxy aldehyde or polyhydroxy ketone unit. Carbohydrates can be classified in terms of the number of monosaccharide units they contain. Give the names for carbohydrates having the following numbers of monosaccharide units.

a. 2 monosaccharide units ____________________

b. 3 to 10 monosaccharide units ____________________

c. 3 monosaccharide units ____________________

d. many (greater than 10) units ____________________

18.2 A left hand and a right hand are **mirror images** (Sec. 18.4) of one another; the two hands are not superimposable. Some molecules have the property of "handedness," which is a form of isomerism; these molecules have a "left-handed" and a "right-handed" form. A molecule that cannot be superimposed on its mirror image is a **chiral molecule**; a molecule that can be superimposed on its mirror image is **achiral** (Sec. 18.4).

An organic molecule is chiral if it contains a **chiral center** (Sec. 18.4), a single atom with four different atoms or groups of atoms attached to it by single bonds. Indicate whether the circled carbon in each structure below is a chiral center or is not.

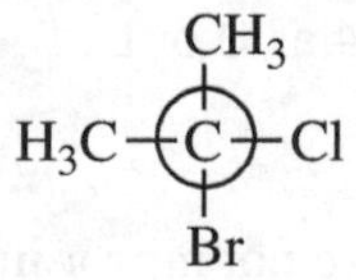 a.	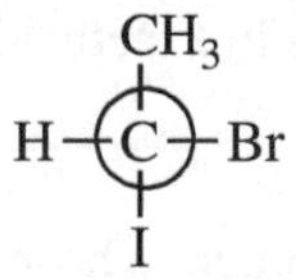 b.	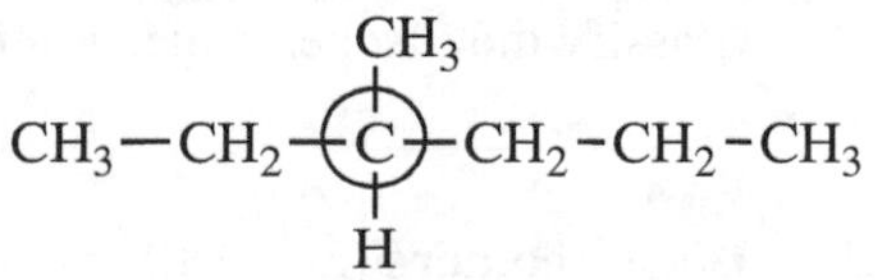 c.
CH_3 above, H_3C–(C)–Cl, Br below	CH_3 above, H–(C)–Br, I below	CH_3 above, CH_3-CH_2–(C)–$CH_2-CH_2-CH_3$, H below

18.3 Organic molecules may contain more than one chiral center. Circle the chiral centers in the following condensed structures.

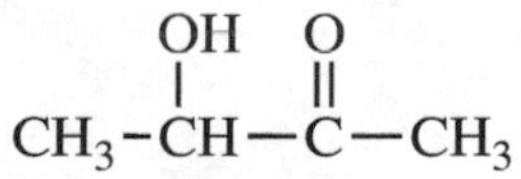

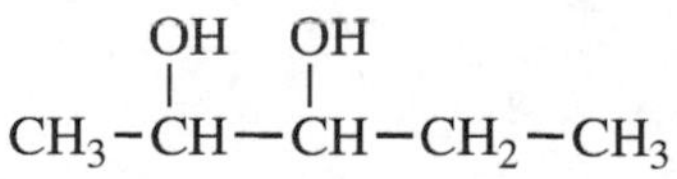

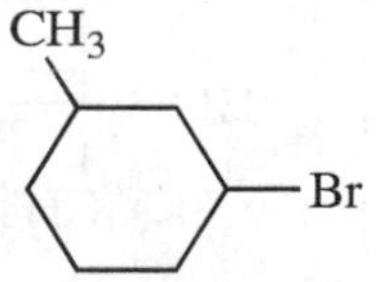

$CH_3-CH(OH)-C(=O)-CH_3$	$CH_3-CH(OH)-CH(OH)-CH_2-CH_3$	cyclohexane ring with CH_3 and Br substituents

18.4 The three-dimensional nature of chiral molecules can be represented by using **Fischer projection formulas** (Sec. 18.6). In a Fischer projection formula for a monosaccharide the carbon chain is positioned vertically with the carbonyl carbon at or near the top. The chiral carbon is shown as the intersection of horizontal and vertical lines, with vertical lines representing bonds directed into the page and horizontal lines representing bonds directed out of the page.

Convert the following Fischer projection formulas into condensed structural formulas.

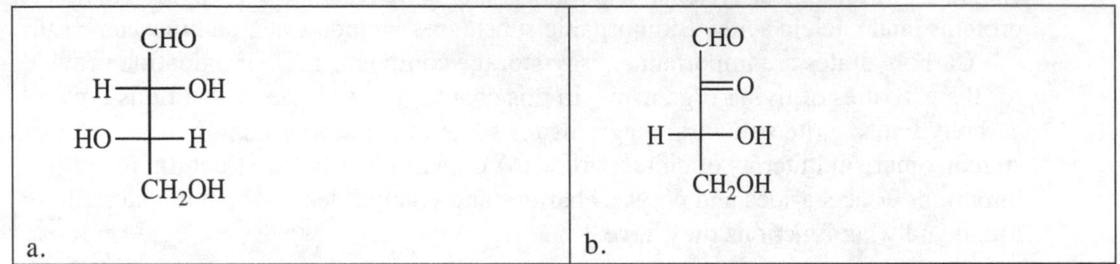

18.5 **Stereoisomers** (Sec. 18.5) are isomers that have the same molecular and structural formulas but a different orientation of their atoms in space. **Enantiomers** (Sec. 18.5) are stereoisomers whose molecules are mirror images of one another.

Using Fischer projection formulas, draw the enantiomers for each of the Fischer projection formulas in Practice Exercise 18.4.

a.	b.

18.6 Monosaccharides (Sec. 18.3) are the simplest kind of carbohydrates and the building blocks for more complex carbohydrates. Monosaccharides are almost always "right-handed." The "handedness" of a monosaccharide is determined by the highest numbered chiral carbon (the one farthest from the carbonyl group). If the –OH on this carbon is to the right, the isomer is, by definition, right-handed and has the designation D; if the –OH is to the left, the isomer is left-handed or L.

Classify the Fischer projection formulas in Practice Exercise 18.4 as D or L.

a. ____________ b. ____________

18.7 **Diastereomers** (Sec. 18.5) are stereoisomers whose molecules are not mirror images of each other. For the structure shown below: a. draw the enantiomer and b. draw one diastereomer.

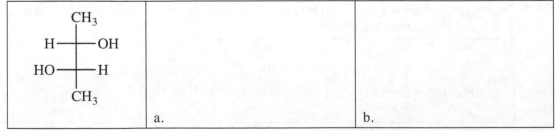

18.8 A monosaccharide is classified as an **aldose** or a **ketose** (Sec. 18.8) on the basis of the carbonyl group: an aldose contains an aldehyde group and a ketose contains a ketone group. Monosaccharides can be further classified by the number of carbons in each molecule. For example, an aldopentose is a five-carbon aldose, and a ketohexose is a six-carbon ketose.

a. Using the number of carbons and the type of carbonyl group present, classify the monosaccharides below.

b. A monosaccharide that has *n* chiral centers may exist in a maximum of 2^n stereoisomeric forms. Calculate the maximum number of stereoisomers for each monosaccharide below.

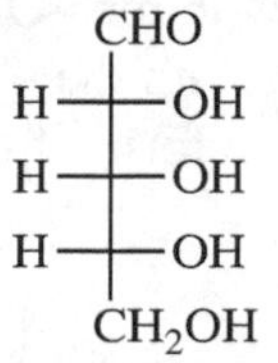 CHO H—OH H—OH H—OH CH_2OH a. b.	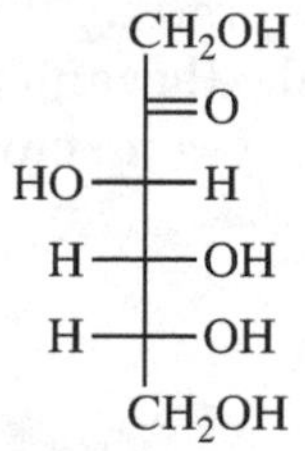 CH_2OH =O HO—H H—OH H—OH CH_2OH a. b.	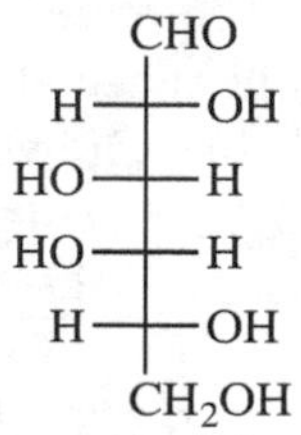 CHO H—OH HO—H HO—H H—OH CH_2OH a. b.

18.9 There are six monosaccharides (Sec. 18.9) that are important in the functioning of the human body; you should know their structures. Note that most biologically-important monosaccharides are D-isomers. Draw the Fischer projection formulas for each monosaccharide in the space provided.

D-glyceraldehyde	dihydroxyacetone	D-glucose	D-galactose	D-fructose	D-ribose

18.10 For monosaccharides containing five or more carbons, open-chain structures are in equilibrium with cyclic structures. In an intramolecular hemiacetal reaction, the carbonyl group reacts with the hydroxyl group on the highest numbered chiral center, and thus forms a cyclic hemiacetal. For the last four molecular structures in Practice Exercise 18.9:
(1) Number each carbon atom starting from the top of the Fischer projection formula.
(2) Tell whether the cyclic form of each sugar molecule will be a 5-membered or a 6-membered ring. (Hint: one atom of the ring will be an oxygen atom.)

a. D-glucose –
b. D-galactose –
c. D-fructose –
d. D-ribose –

18.11 The structural representations of the cyclic forms of monosaccharides are called **Haworth projections** (Sec. 18.11). Conventions for drawing Hayworth projection formulas are: 1) –OH groups drawn to the right in the Fischer projection are placed below the ring; –OH groups to the left are above the ring. 2) The –OH group formed from the carbonyl group may be above or below the plain of the ring, depending on how ring closure occurs.

a. For the sugar whose Haworth projection (cyclic form) is shown below, draw the Fischer projection (open-chain form) of the sugar. Use the conventions shown above.

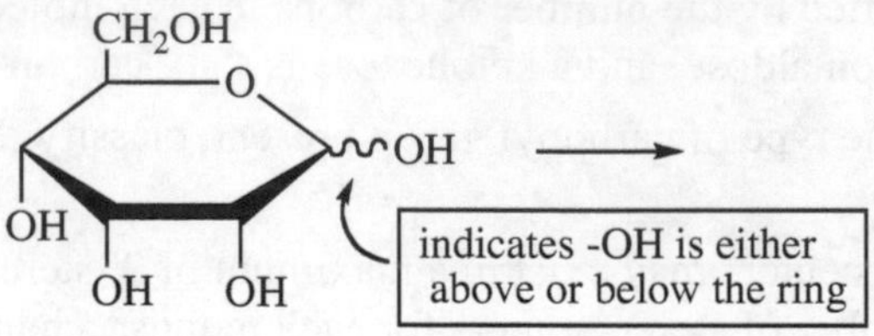

CH_2OH O OH → OH OH OH

indicates -OH is either above or below the ring

b. There are two possible cyclic forms for the sugar below: one with the –OH of the hemiacetal above the ring in the Haworth projection, the other with the –OH below the ring. Draw the two possible Hayworth projections for the cyclic form of the sugar below.

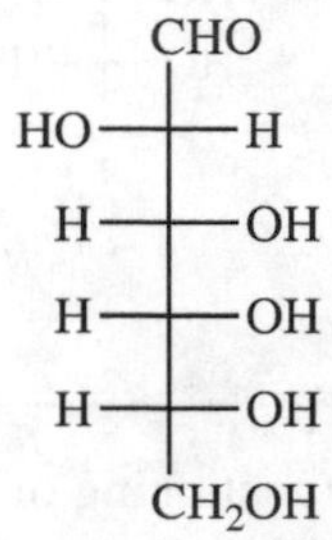

CHO
HO—H
H—OH
H—OH
H—OH
CH_2OH

18.12 **Anomers** (Sec. 18.10) are cyclic monosaccharides that differ only in the positions of the substituents on the **anomeric carbon atom** (the hemiacetal carbon atom). In naming the anomeric forms of a monosaccharide, the α- or β-configuration is determined by the position of the –OH formed from the carbonyl group, relative to the $–CH_2OH$ group (which is above the ring for D sugars): the –OH group is α when it is drawn in the "down" position (below the ring) and β when it is drawn in the "up" position (above the ring).

For the following structures: Circle the anomeric hemiacetal carbon and note whether the –OH group is α or β, and name each compound.

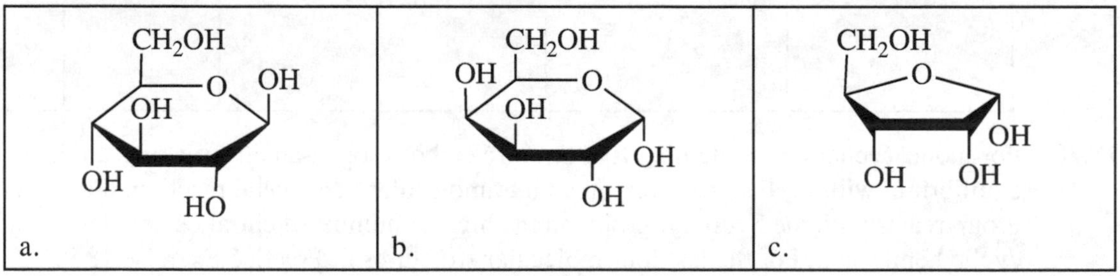

a. CH_2OH, O, OH, OH, OH, HO	b. CH_2OH, O, OH, OH, OH, OH	c. CH_2OH, O, OH, OH, OH

18.13 Monosaccharides and **disaccharides** (Sec. 18.3) are often called **sugars** (Sec. 18.8).

a. Weak oxidizing agents, such as Tollens and Benedict's solutions, oxidize the carbonyl group end of a monsaccharide to give an acid. Sugars that can be oxidized by these agents are called **reducing sugars** (Sec. 18.12). Complete the following reaction.

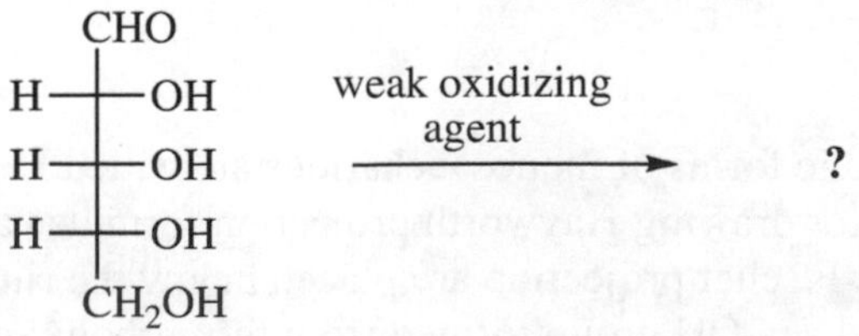

CHO
H—OH
H—OH
H—OH
CH_2OH

weak oxidizing agent → ?

b. The carbonyl group of sugars can be reduced to a hydroxyl group using H_2 as the reducing agent, producing a sugar alcohol or alditol. Reduction of D-ribose produces D-ribitol.

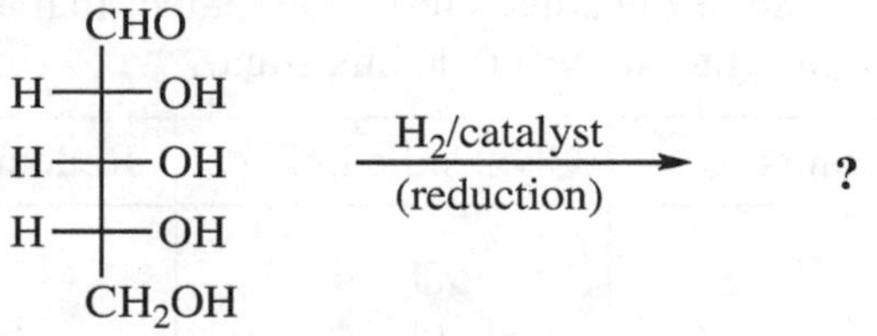

CHO
H—OH
H—OH
H—OH
CH_2OH

$\xrightarrow[\text{(reduction)}]{H_2\text{/catalyst}}$?

c. Another important reaction of monosaccharides is **glycoside** (Sec. 18.12) formation, reaction of the cyclic hemiacetal with an alcohol to form an acetal.

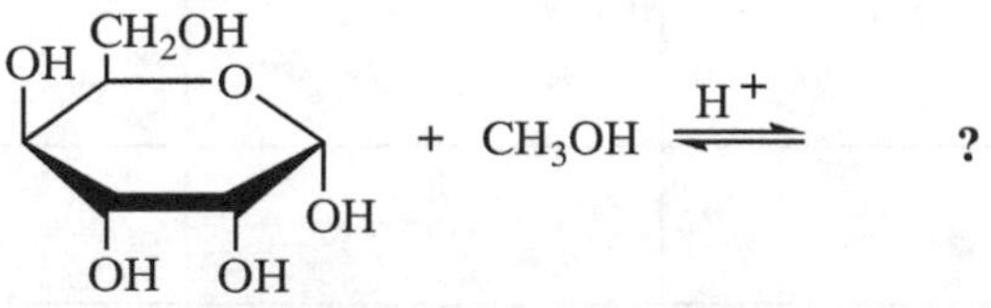

CH_2OH, OH, O, OH, OH, OH + CH_3OH $\overset{H^+}{\rightleftharpoons}$?

(Since this cyclic hemiacetal is glucose, the resulting acetal is called a glucoside.)

Two other reactions of monosaccharides are phosphate ester formation (at the –OH group on C-1 or C-6) and amino sugar formation (–NH_2 replaces –OH on C-2).

18.14 The two monosaccharides that form a disaccharide are joined by a **glycosidic linkage** (Sec. 18.13). In the glycoside (acetal) formation, one monosaccharide acts as the hemiacetal and the other as the alcohol. The configuration of the –OH group (α or β) of the hemiacetal carbon atom gives the configuration of the glycosidic linkage. The numbers of the carbon atoms linked together are specified. For example: (1→4) means the hemiacetal carbon is C-1 and the alcohol carbon is C-4.

Below are Haworth projections for four common disaccharides. In the diagrams circle the glycosidic bonds. Draw a box around any hemiacetal carbon atom (positive Benedict's test.)

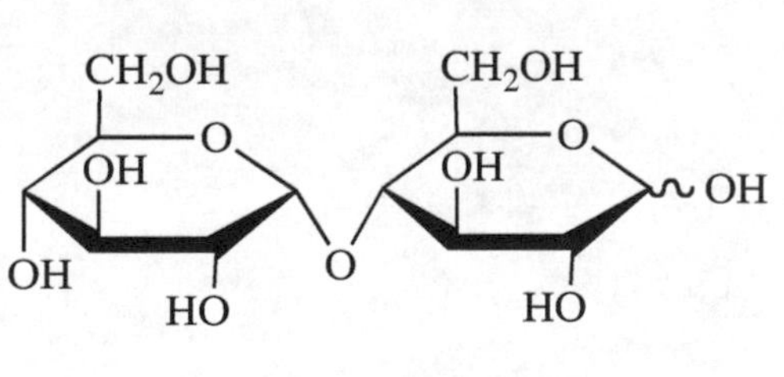

maltose	lactose
cellobiose	sucrose

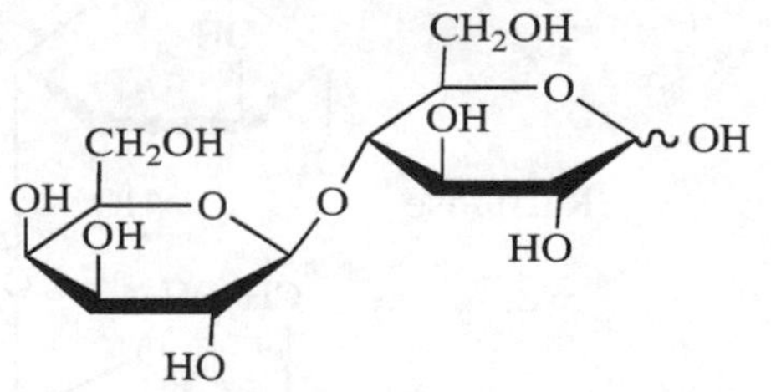

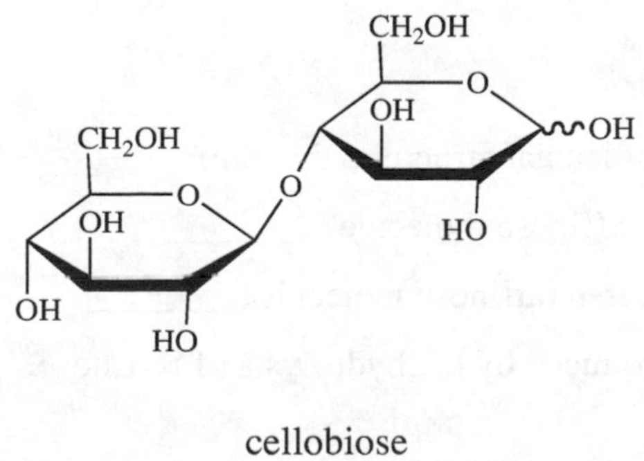

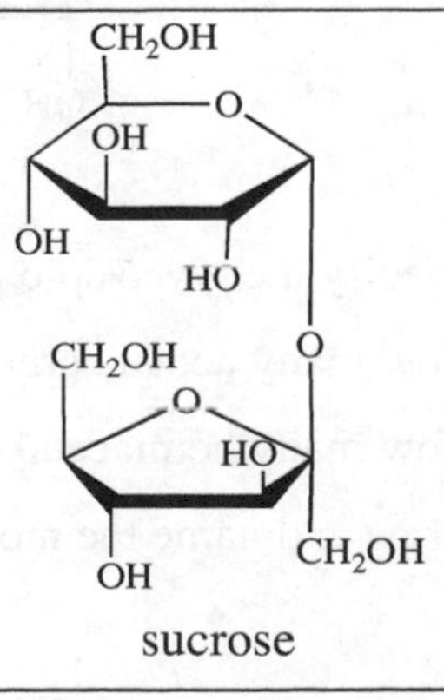

18.15 Use Practice Exercise 18.14 above to complete the following table on the four disaccharides. Name the monosaccharide units that are produced by the hydrolysis of each disaccharide. Specify the glycosidic linkage (one has been completed for you). Indicate whether the disaccharide is a reducing sugar (gives a positive Benedict's test). Remember that a disaccharide must have a hemiacetal carbon center to be a reducing sugar.

Disaccharide	Monosaccharide units	Glycosidic linkage	Reducing sugar?
maltose		$\alpha(1\rightarrow4)$	
cellobiose			
lactose			
sucrose			

18.16 **Oligosaccharides** (Sec. 18.14) contain three to ten monosaccharide units bonded to each other by glycosidic linkages. The structural diagram for raffinose, a naturally occurring oligosaccharide, is shown below:

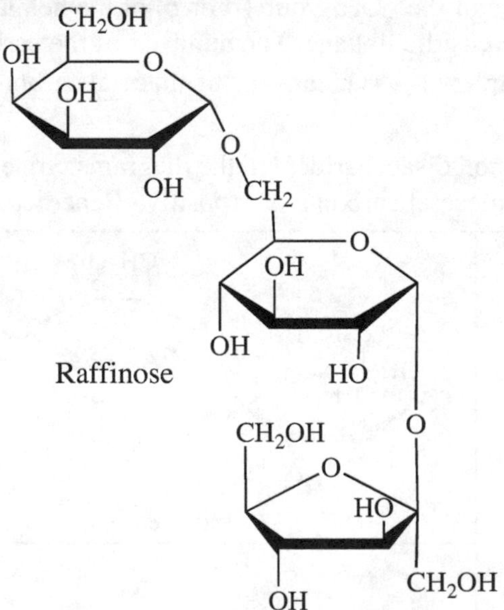

a. Specify the glycosidic linkages on the molecular structural diagram.

b. How many acetal carbons are there in a raffinose molecule? _______

c. How many hemiacetal carbons are there in a raffinose molecule? _______

d. Draw and name the monosaccharides produced by the hydrolysis of raffinose.

18.17 A **polysaccharide** (Sec. 18.15) contains many monosaccharide units bonded to one another by glycosidic linkages. **Homopolysaccharides** (Sec. 18.15) contain only one type of monosaccaride unit. Glycogen, amylose, and amylopectin, important **storage polysaccharides** (Sec. 18.16), and cellulose, a **structural polysaccharide** (Sec. 18.17), are all made up of D-glucose units.

These four homopolysaccharides in the table below differ in the type of glycosidic linkage (α or β) between monomers, and in size, degree of branching, and function. Use Sections 18.15 and 18.16 of your textbook to complete the table.

Polysaccharide	Linkage	Number of glucose units	Branching	Function
glycogen				
amylose				
amylopectin				
cellulose				

18.18 There are three common polysaccharides (considered in Sec. 18.17 and 18.18 of your textbook) whose monomers are not simple glucose units. Chitin is a homopolysaccharide; hyaluronic acid and heparin are **heteropolysaccharides** (Sec. 18.15). Complete the table below with the functions of the three polysaccharides and the monomeric units of which they are composed.

Polysaccharide	Function	Monosaccharide units
chitin		
hyaluronic acid		
heparin		

Self-Test

True-false: Indicate whether the following statements are true or false. If the statement is false, give the word or phrase that may be substituted for the underlined portion to make the statement true.

1. Bioorganic substances include carbohydrates, lipids, proteins, and <u>inorganic salts</u>.
2. Carbohydrates form part of the structural framework of <u>DNA and RNA</u>.
3. An oligosaccharide is a carbohydrate that contains <u>at least 20</u> monosaccharide units.
4. <u>A chiral molecule</u> is a molecule that is identical to its mirror image.
5. Naturally-occurring monosaccharides are almost always "<u>left-handed</u>" molecules.
6. The compound 2-methyl-2-butanol contains <u>one</u> chiral center.
7. In a Fischer projection, <u>horizontal lines</u> represent bonds to groups directed into the printed page.
8. Enantiomers are stereoisomers whose molecules <u>are not</u> mirror images of each other.

9. Diastereomers have identical boiling and freezing points.
10. A dextrorotatory compound rotates plane-polarized light in a clockwise direction.
11. The direction of rotation of plane-polarized light by sugar molecules is designated in the compound's name as D or L.
12. The responses of the human body to the two enantiomeric forms of a chiral molecule are identical.
13. A six-carbon monosaccharide with a ketone functional group is called an aldopentose.
14. Glycogen is the glucose storage polysaccharide known as animal starch.
15. The carbonyl group of a monosaccharide can be reduced to a hydroxyl group using Tollens solution as a reducing agent.
16. The bond between two monosaccharide units in a disaccharide is a glycosidic linkage.
17. Humans cannot digest cellulose because they lack an enzyme to catalyze hydrolysis of the $\alpha(1 \rightarrow 4)$ linkage.
18. Amylopectin molecules are more highly branched than amylose molecules are.
19. A homopolysaccharide is a polysaccharide in which more than one type of monosaccharide unit is present.
20. Disaccharides are classified as complex carbohydrates.

Multiple choice:

21. Which of the following is *not* required for the production of carbohydrates by photosynthesis?
    a. carbon dioxide b. oxygen c. water
    d. sunlight e. chlorophyll
22. Which of the following compounds contains only three chiral centers?
    a. glyceraldehyde b. glucose c. fructose
    d. galactose e. none of these
23. Ribose is an example of a(n):
    a. ketohexose b. aldohexose c. aldotetrose
    d. aldopentose e. none of these
24. Dextrose, or blood sugar, is:
    a. D-glucose b. D-galactose c. D-fructose
    d. D-ribose e. none of these
25. The sugar sometimes known as levulose is:
    a. D-ribose b. L-glucose c. L-galactose
    d. D-fructose e. none of these
26. Which of the following compounds is *not* made up solely of D-glucose units?
    a. maltose b. lactose c. cellulose
    d. starch e. glycogen
27. Which of the following sugars is *not* a reducing sugar?
    a. lactose b. maltose c. sucrose
    d. galactose e. all are reducing sugars

28. The main storage form of D-glucose in animal cells is:
    a. chitin b. amylopectin c. amylose
    d. sucrose e. none of these

29. Which of the following molecules has a chiral center?
    a. ethanol b. 1-chloro-1-bromoethane
    c. 1-chloro-2-bromoethane d. 1,2-dichloroethane
    e. none of these

30. Hydrolysis of sucrose yields:
    a. glucose and fructose b. glucose and galactose c. ribose and fructose
    d. ribose and glucose e. none of these

Answers to Practice Exercises

18.1 a. disaccharide b. oligosaccharide
c. trisaccharide d. polysaccharide

18.2 a. not a chiral center b. chiral center c. chiral center

18.3

a.	b.	c.
$CH_3-\underset{\text{OH}}{\underset{\vert}{CH}}-\overset{\text{O}}{\overset{\Vert}{C}}-CH_3$ (CH circled)	$CH_3-\underset{\text{OH}}{\underset{\vert}{CH}}-\underset{\text{OH}}{\underset{\vert}{CH}}-CH_2-CH_3$ (both CH circled)	ring with CH_3 and Br (two carbons circled)

18.4 a. $CH_2-CH-CH-CHO$ with OH on each of the first three carbons ($HOCH_2-CH(OH)-CH(OH)-CHO$)

b. $CH_2-CH-\overset{\text{O}}{\overset{\Vert}{C}}-CHO$ with OH on the first two carbons ($HOCH_2-CH(OH)-CO-CHO$)

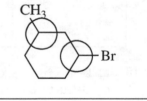

18.5

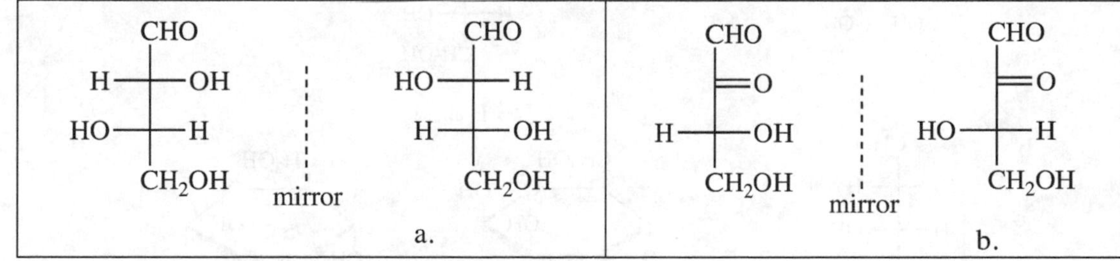

18.6 a. L b. D

18.7

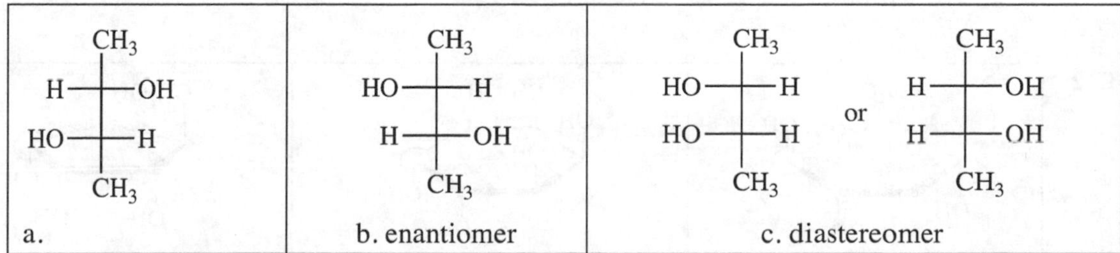

18.8

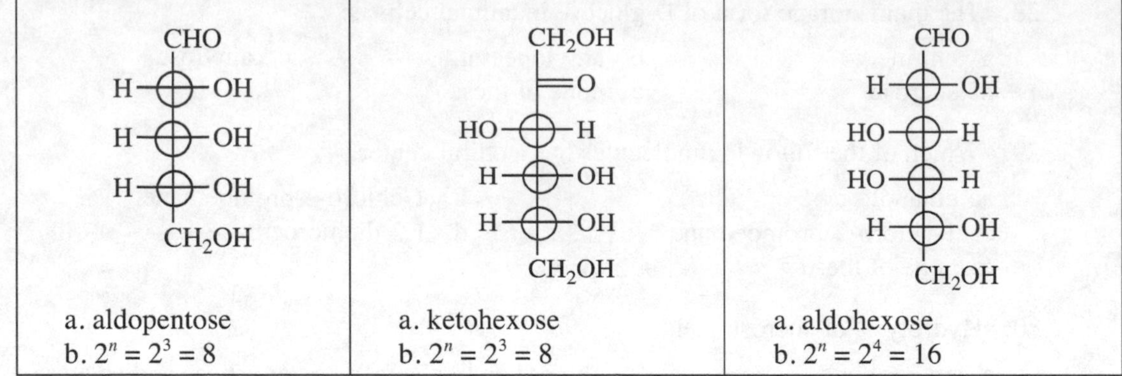

18.9

D-glyceraldehyde	dihydroxyacetone	D-glucose	D-galactose	D-fructose	D-ribose
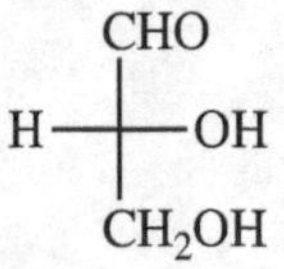 CHO H–C–OH CH_2OH	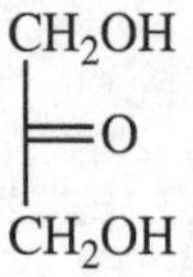 CH_2OH C=O CH_2OH	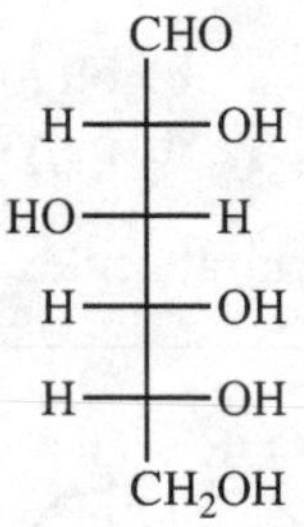 CHO H–C–OH HO–C–H H–C–OH H–C–OH CH_2OH	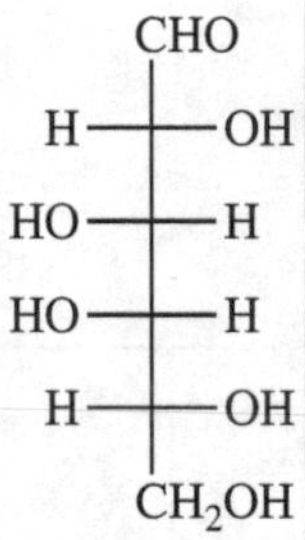 CHO H–C–OH HO–C–H HO–C–H H–C–OH CH_2OH	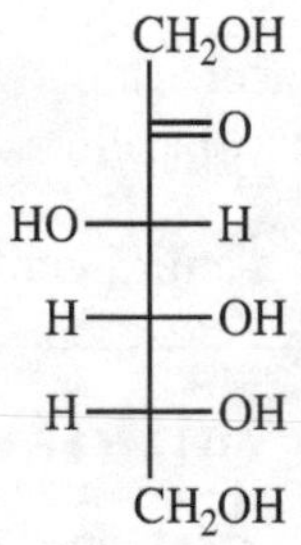 CH_2OH C=O HO–C–H H–C–OH H–C–OH CH_2OH	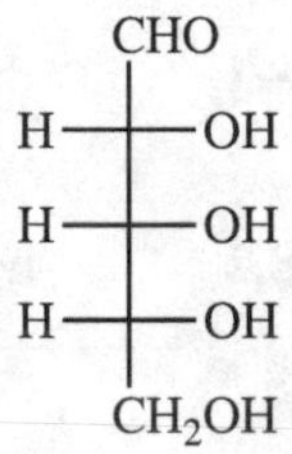 CHO H–C–OH H–C–OH H–C–OH CH_2OH

18.10 a. D-glucose – six-membered ring b. D-galactose – six-membered ring
c. D-fructose – five-membered ring d. D-ribose – five-membered ring

18.11 a.

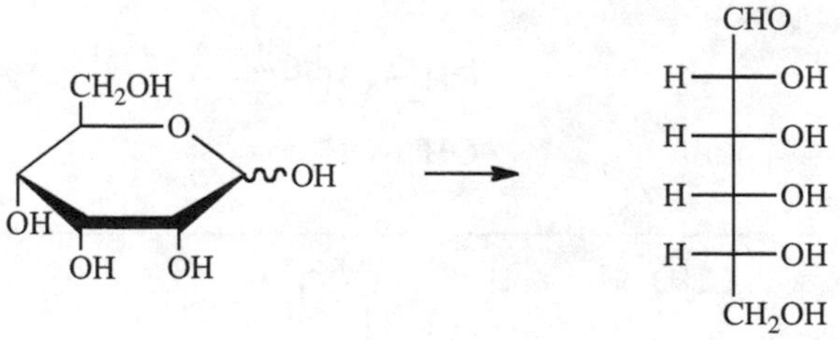

CHO
H–C–OH
H–C–OH
H–C–OH
H–C–OH
CH_2OH

D-allose

b.

CHO
HO–C–H
H–C–OH
H–C–OH
H–C–OH
CH_2OH

α–D-altrose

β–D-altrose

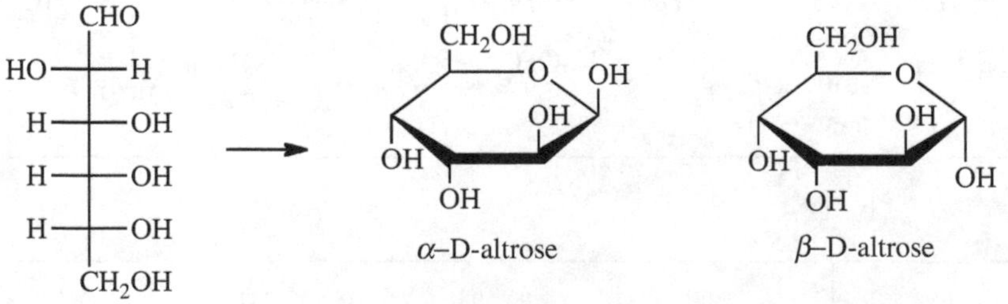

18.12

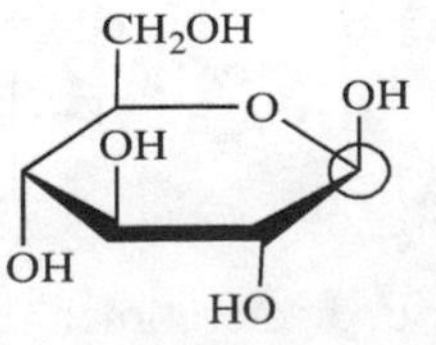 a. β-D-glucose	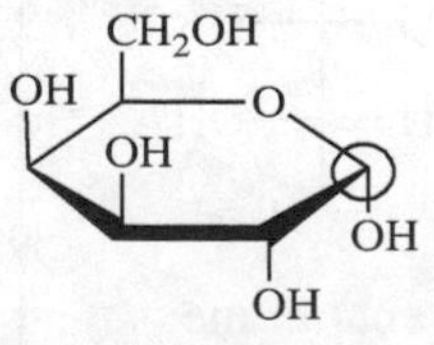 b. α-D-galactose	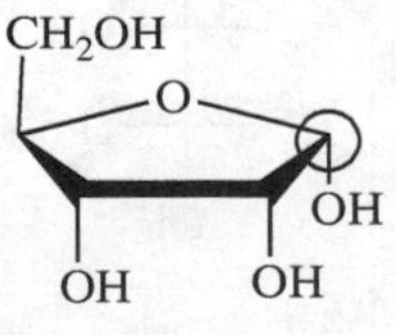 c. α-D-ribose

18.13

a. (Fischer projection: CHO / H–OH / H–OH / H–OH / CH_2OH) $\xrightarrow{\text{weak oxidizing agent}}$ (COOH / H–OH / H–OH / H–OH / CH_2OH)

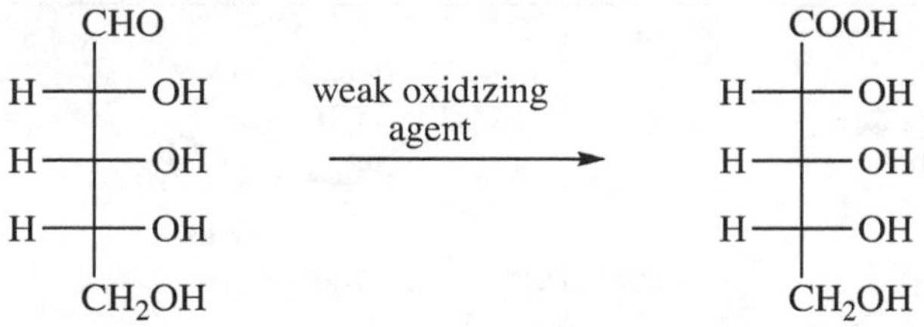

b. (CHO / H–OH / H–OH / H–OH / CH_2OH) $\xrightarrow[\text{(reduction)}]{H_2\text{/catalyst}}$ (CH_2OH / H–OH / H–OH / H–OH / CH_2OH)

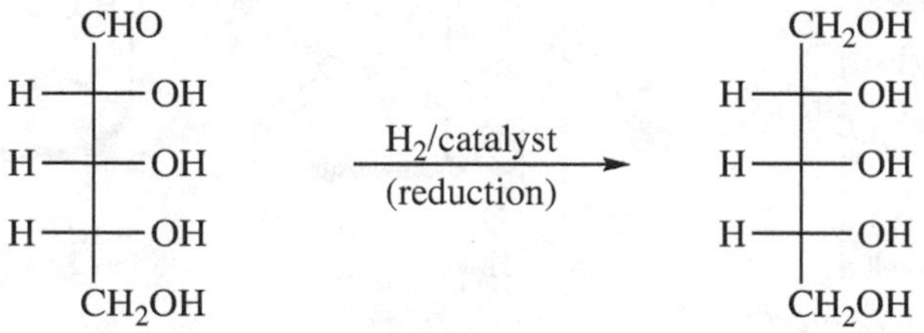

c. (cyclic sugar with CH_2OH, OH, OH, OH, OH) + CH_3OH $\overset{H^+}{\rightleftharpoons}$ (cyclic sugar with CH_2OH, OH, OH, OH, OCH_3) + H_2O

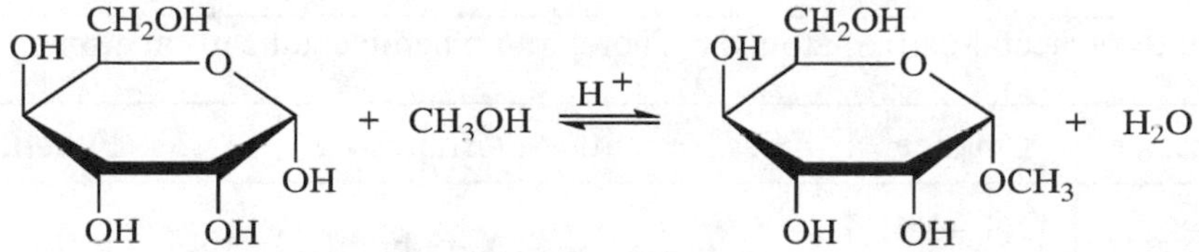

18.14

maltose

lactose

cellobiose

sucrose

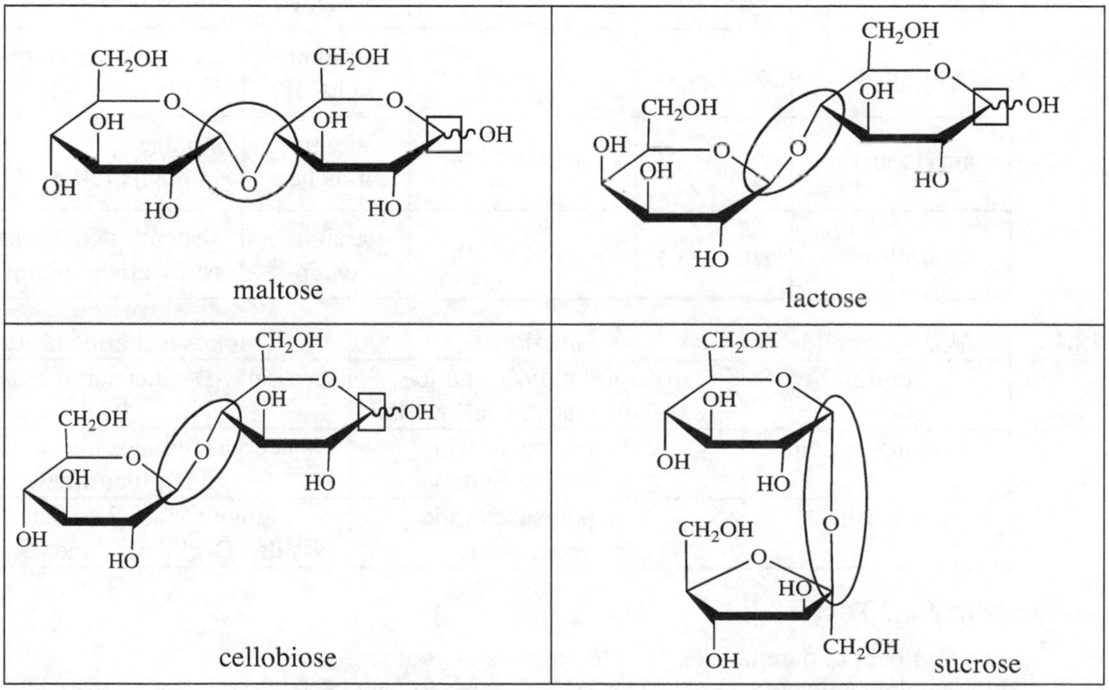

18.15

Disaccharide	Monosaccharide units	Glycosidic linkage	Reducing sugar?
maltose	D-glucose (one is α-D-glucose)	$\alpha(1 \to 4)$	yes
cellobiose	D-glucose (one is β-D-glucose)	$\beta(1 \to 4)$	yes
lactose	β-D-galactose and D-glucose	$\beta(1 \to 4)$	yes
sucrose	α-D-glucose and β-D-fructose	$\alpha,\beta(1 \to 2)$	no

18.16

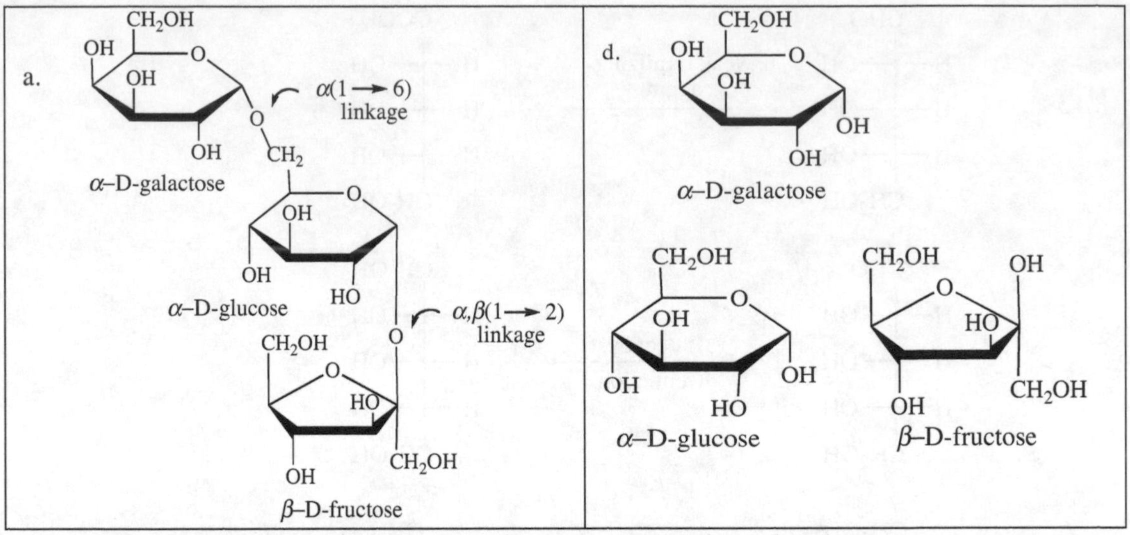

a. α-D-galactose, α(1→6) linkage, α-D-glucose, α,β(1→2) linkage, β-D-fructose

d. α-D-galactose, α-D-glucose, β-D-fructose

b. There are three acetal carbon atoms; c. There are no hemiacetal carbon atoms.

18.17

Polysaccharide	Linkage	Glucose units	Branching	Function
glycogen	$\alpha(1 \rightarrow 4)$ $\alpha(1 \rightarrow 6)$	up to 1,000,000	very highly branched	storage form of glucose in animals
amylose	$\alpha(1 \rightarrow 4)$	300 to 500	straight-chain	storage form of glucose in plants (15-20%)
amylopectin	$\alpha(1 \rightarrow 4)$ $\alpha(1 \rightarrow 6)$	up to 100,000	highly branched	storage form of glucose in plants (80-85%)
cellulose	$\beta(1 \rightarrow 4)$	about 5000	straight-chain	structural component of cell walls in plants

18.18

Polysaccharide	Function	Monosaccharide units
chitin	structural polysaccharide (exoskeletons and cell walls)	*N*-acetyl-D-glucosamine (NAG)
hyaluronic acid	acidic polysaccharide (lubricant in joints)	*N*-acetyl-D-glucosamine (NAG) and D-glucuronate
heparin	acidic polysaccharide (anticoagulant)	D-glucuronate-2-sulfate and N-sulfo-D-glucosamine -6-sulfate

Answers to Self-Test

The numbers in parentheses refer to sections in your textbook.
1. F; nucleic acids (18.1) **2**. T (18.2) **3**. F; three to ten (18.3)
4. F; An achiral molecule (18.4) **5**. F; right-handed (18.4) **6**. F; no (18.4)
7. F; vertical lines (18.6) **8**. F; are (18.5) **9**. F; enantiomers (18.5) **10**. T (18.7)
11. F; (+) or (–) (18.6, 18.7) **12**. F; often different (18.7) **13**. F; a ketohexose (18.8)
14. T (18.16) **15**. F; H_2 (18.12) **16**. T (18.13) **17**. F; $\beta(1 \rightarrow 4)$ linkage (18.17)
18. T (18.16) **19**. F; heteropolysaccharide (18.18) **20**. F; simple(18.19) **21**. b (18.2)
22. c (18.4) **23**. d (18.9) **24**. a (18.9) **25**. d (18.9) **26**. b (18.13, 18.16, 18.17)
27. c (18.13) **28**. e; glycogen (18.16) **29**. b (18.4) **30**. a (18.13)

Chapter Overview

Lipids are compounds grouped according to their common solubility in nonpolar solvents and insolubility in water, but there are also some structural features that help to identify them. They have a variety of functions in the body, acting as energy-storage compounds and chemical messengers and providing structure to cell membranes.

In this chapter you will define fats and oils and explain how they differ in molecular structure. You will identify various groups of lipids according to their structures and components, and you will learn some of the functions that they perform in biological systems.

Practice Exercises

19.1 There are two common methods of classifying **lipids** (Sec. 19.1): 1) classification by biochemical function (used in this chapter of your textbook) and 2) classification based on whether a lipid can be broken down into two or more smaller units by basic hydrolysis (saponification), discussed in Sec. 19.15.

Fatty acids (Sec. 19.2), which are components of many of the lipids studied in this chapter, are naturally occurring carboxylic acids that contain long, unbranched hydrocarbon chains, 12 to 26 carbon atoms in length. Three types of fatty acids are **saturated fatty acids**, SFAs, with no carbon-carbon double bonds; **monounsaturated fatty acids**, MUFAs, with one carbon-carbon double bond; and **polyunsaturated fatty acids**, PUFAs, with two or more carbon-carbon double bonds. (Sec. 19.2). A shorthand notation for the structure of fatty acids uses the ratio of the number of carbon atoms to the number of double bonds.

In the omega classification system, the first double bond is identified by its number on the carbon chain, counted from the methyl end of the chain. Omega-3 and omega-6 are the most common "omega" families.

Double-bond positioning may be specified by using a delta with the carbon numbers of the double bonds superscripted. For this notation, the double-bond position is counted from the carboxyl end of the chain.

Complete the table below using information from Table 19.1 in your textbook. The first fatty acid has been filled in as an example.

Name of fatty acid	Carbon atoms and double bonds	Type designation	"Omega" designation	"Delta" designation
linolenic acid	18:3	PUFA	omega-3	$\Delta^{9,12,15}$
palmitoleic acid				
arachidonic acid				
linoleic acid				
oleic acid				
stearic acid				

19.2 The solubility of carboxylic acids in water (Sec. 16.5) depends on the length of the hydrocarbon chain. Because of the nonpolar nature of the hydrocarbon chain, solubility decreases as chain length increases. Fatty acids, having hydrocarbon chains 12 to 26 carbon atoms in length, are very slightly soluble or insoluble in water.

Melting points of fatty acids depend on both the hydrocarbon chain length and the degree of unsaturation within the chains. As chain length increases, melting point increases. As degree of unsaturation increases, melting point decreases. For the following pairs of fatty acids, indicate which member of the pair has the greater melting point.

Fatty acid pair	Higher melting point?
lauric acid – 12:0 and myristic acid –14:0	
stearic acid – 18:0 and oleic acid – 18:2	
arachidic acid – 20:0 and arachidonic acid – 20:4	

19.3 The **energy-storage lipids** (Sec. 19.1) are the first of the lipid function groups to be studied. **Triacylglycerols** (Sec. 19.4) are produced by the esterification of three fatty acid molecules with a glycerol molecule.

Draw structural formulas for the triacylglycerols that have the following fatty acid residues: a. three molecules of oleic acid; b. two molecules of stearic acid and one of oleic acid; c. three molecules of stearic acid. Use line-angle representation for the hydrocarbon chains.

a.	b.	c.

19.4 **Fats** (Sec. 19.4) are triacylglycerols with a high percentage of saturated fatty acids; **oils** (Sec. 19.4) are triacylglycerols with a high percentage of unsaturated fatty acids. Tell whether each of the compounds in Practice Exercise 19.3 would be a fat or an oil, and explain.

Compound	Fat or oil?	Explanation
a.		
b.		
c.		

19.5 Triacylglycerols undergo several characteristic reactions. In this exercise you will work with four types of reactions and the three triacylglycerol molecules in Practice Exercise 19.3.

1. Acidic hydrolysis is the reverse of esterification. Complete acidic hydrolysis of a triacylglycerol molecule produces glycerol and three fatty acid molecules. Under each letter, name the products of the acidic hydrolysis of that triacylglycerol.

a.	b.	c.

2. Saponification (hydrolysis under alkaline conditions) of a triacylglycerol produces glycerol and the salts of the three fatty acids. Under each letter, name the products of the saponification of that triacylglycerol.

a.	b.	c.

3. Hydrogenation— the addition of H_2 molecules to the double bonds of the fatty acid residues. Under the letters below, tell how many molecules of hydrogen (H_2) would be used for the complete hydrogenation of each of the three triacylglycerol molecules.

a.	b.	c.

4. Oxidation, reaction with O_2, breaks the carbon-carbon double bonds and produces low-molecular-weight aldehydes and carboxylic acids that have unpleasant odors. Fats and oils oxidized in this way are said to be rancid. Under each letter write "yes" if the triacylglycerol is likely to become rancid, "no" if it is not likely.

a.	b.	c.

19.6 A second lipid function category is the **membrane lipid** (Sec. 19.1) category. Membrane lipids include the **phospholipids** (Sec. 19.7), with two subgroups (the **glycerophospholipids**, Sec. 19.7, and the **sphingophospholipids,** Sec. 19.7), and the **sphingoglycolipids** (Sec. 19.8). The phospholipids and the sphingoglycolipids can undergo hydrolysis in the same way that the triacylglycerols do.

All of these types of lipids have similar basic structures, but the component "building blocks" vary. In the block diagrams below, fill in the name of each structural subunit for each type of lipid named. Then follow the directions below the diagrams.

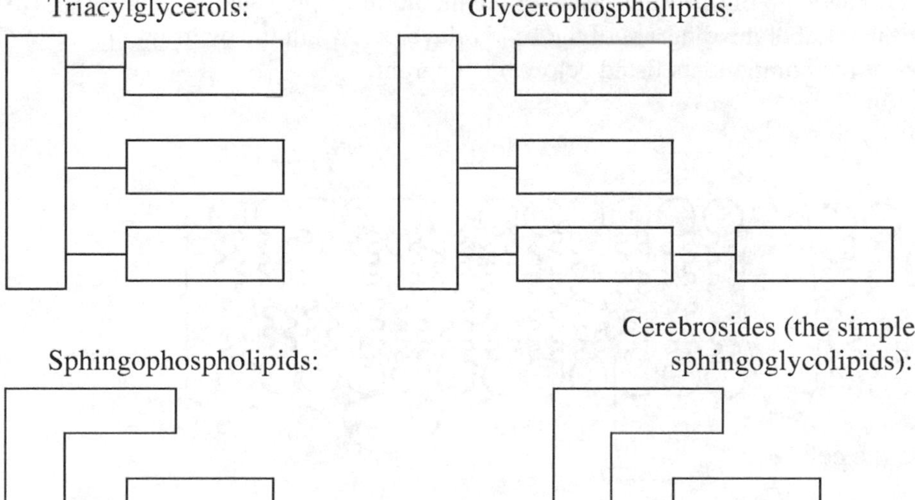

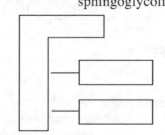

1. Label all ester linkages in the block diagrams with the letter A.
2. Label all amide linkages with the letter B.
3. Label all glycosidic linkages with the letter C.

19.7 Triacylglycerols and the three types of membrane lipids in Practice Exercise 19.6 have two general characteristics in common: they are all saponifiable (basic hydrolysis of the ester, amide, and glycosidic linkages) and they all have a polar head and one or two nonpolar tails. The polar head is **hydrophilic**; the nonpolar tails are **hydrophobic** (Sec. 19.7, 19.8).

Use the diagrams you completed in Practice Exercise 19.6 to help you identify the hydrophilic heads and the hydrophobic tails in these membrane lipids.

Lipids	Hydrophilic heads	Hydrophobic tails
triacylglycerols		
glycerophospholipids		
sphingophospholipids		
sphingoglycolipids		

19.8 Cholesterol is another membrane lipid. However, it cannot be hydrolyzed. It is a **steroid** (Sec. 19.9), a particular fused-ring system. Cholesterol and other steroid molecules do not have a large polar head (the –OH group is considered to be the head) and so are not very soluble in water. Cholesterol molecules are carried through the bloodstream as a part of a protein carrier system, a lipoprotein. Complete the information below for two common lipoproteins.

	Name	Function
LDL		
HDL		

19.9 The aqueous material within a living cell is separated from its surrounding aqueous environment by a **cell membrane** (Sec. 19.10). Most of the mass of the cell membrane is lipid material, in the form of a **lipid bilayer** (Sec. 19.10), in which the inner and outer surfaces are made up of the polar heads of the membrane lipids and the interior consists of the nonpolar tails. Label the diagram of the lipid bilayer shown in the diagram that follows, using the letters of the components listed below the diagram.

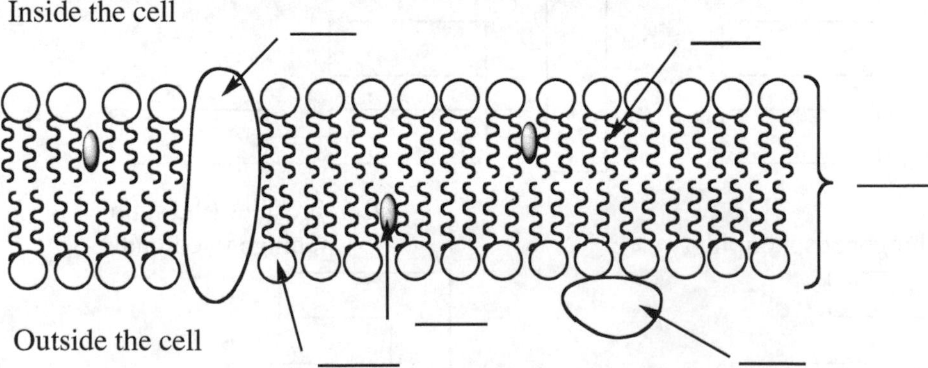

a. lipid bilayer
b. cholesterol molecule
c. polar heads of membrane molecules
d. nonpolar tails of membrane molecules
e. integral membrane protein
f. peripheral membrane protein

19.10 There are three methods of transport of materials across cell membranes: 1) passive transport, 2) facilitated transport, and 3) active transport.

Indicate (X) which type of cell transport would correspond to each of the descriptions below.

Description	Passive transport	Facilitated transport	Active transport
energy input needed			
no energy input			
closely related to osmosis			
requires aid of protein pump			
proteins serve as gates			
transport of Na^+, K^+, H_3O^+			
transport of O_2, N_2, urea			
transport of Cl^-, HCO_3^-, glucose			

19.11 Two more important groups of lipids are the **steroids** (structurally based on a fused-ring system, Sec. 19.9) and the **eicosanoids** (derivatives of a 20-carbon fatty acid, Sec. 19.13). Match the name of each of the lipids in the table below with the number of its specific function in biological systems.

Answer	Name of lipid	Physiological function of lipid
	thromboxanes	1. control reproduction, secondary sex characteristics
	mineralocorticoids	2. regulates body temperature, enhance inflammation
	sex hormones	3. starting material for synthesis of steroid hormones
	glucocorticoids	4. control glucose metabolism, reduce inflammation
	cholesterol	5. emulsifying agents, digestion of lipids
	complex bile acids	6. found in white blood cells, hypersensitivity response
	leukotrienes	7. promote formation of blood clots
	prostaglandins	8. control the balance of Na^+ and K^+ ions in cells

19.12 After each name in the table above, write A if the compound is a steroid or B if the compound is an eicosanoid.

19.13 As you have seen in this chapter, one of the principal ways to classify lipids is according to general biochemical function. Use the following letters to classify the lipids in the list below: A. energy-storage lipids, B. membrane lipids, C. emulsification lipids, D. messenger lipids, E. protective-coating lipids.

triacylglycerols ________
biological waxes ________
prostaglandins ________
cholic acid ________
adrenocorticoids ________
sphingomyelins ________

lecithins ________
androgens ________
thromboxanes ________
cerebrosides ________
cholesterol ________
steroid hormones ________

fats________
cephalins________
bile acids ________
estrogens ________
oils ________

19.14 In addition to classification of lipids according to their biological functions, lipids can also be classified as **saponifiable** or **nonsaponifiable** (Sec. 19.15). Classify each of the lipids or lipid groups in Practice Exercise 19.13 as saponifiable (S) or nonsaponifiable (N).

Self-Test

True-false: Indicate whether the following statements are true or false. If the statement is false, give the word or phrase that may be substituted for the underlined portion to make the statement true.

1. Fats contain a high proportion of unsaturated fatty acid chains.
2. Fats and oils become rancid when the double bonds on triacylglycerol side chains are oxidized.
3. The two essential fatty acids are linoleic acid and arachidonic acid.
4. Rancid fats contain low-molecular-mass acids.
5. Sphingophospholipids are complex lipids found in the brain and nerves.
6. In a cell membrane, the nonpolar tails of the phospholipids are on the outside surfaces of the lipid bilayer.
7. The presence of cholesterol molecules in the lipid bilayer of a cell membrane makes the bilayer more flexible.
8. Essential fatty acids cannot be synthesized within the human body from other substances.
9. Most naturally occurring triacylglycerols are formed with three identical fatty acid molecules.
10. Partial hydrogenation of vegetable oils produces a product of higher melting point than the original oil.
11. Glycerophospholipids are important components of cell membranes.
12. Bile acids play an important role in the human body in the process of digestion.
13. Cholesterol cannot be synthesized within the human body.
14. A fatty acid containing a *cis* double bond has a bent carbon chain.
15. Most fatty acids found in the human body have an odd number of carbon atoms.
16. The lipid molecules in the lipid bilayer of a cell membrane are held to one another by covalent bonds.
17. Waxes are esters of long-chain fatty acids and aromatic alcohols.
18. Some synthetic derivatives of adrenocorticoid hormones are used to control inflammatory diseases.
19. The eicosanoids are hormonelike substances that are transported in the bloodstream.
20. Eicosanoids are derivatives of polyunsaturated fatty acids that have 20 carbon atoms.
21. Facilitated transport involves movement of a substance across a membrane against a concentration gradient with the expenditure of cellular energy.

Multiple choice:

22. A triacylglycerol is prepared by combining glycerol and:

 a. long-chain alcohols b. fatty acids c. unsaturated hydrocarbons
 d. saturated hydrocarbons e. none of these

23. Which of the following fatty acids is saturated?

 a. palmitic acid b. oleic acid c. linoleic acid
 d. arachidonic acid e. none of these

24. An example of a glycerophospholipid is:

a. prostaglandin b. progesterone c. a lecithin
d. glycerol tristearate e. none of these

25. An example of a steroid hormone is:

a. progesterone b. cholesterol c. stearic acid
d. a cerebroside e. none of these

26. Cholesterol is the starting material within the human body for the synthesis of:

a. vitamin A b. vitamin B_1 c. vitamin C
d. vitamin D e. vitamin E

27. The most abundant steroid in the human body is:

a. estradiol b. testosterone c. progesterone
d. cholesterol e. estrogen

28. The two essential fatty acids that the human body must obtain from dietary sources are:

a. oleic and linoleic b. oleic and arachidonic c. arachidonic and linoleic
d. linoleic and oleic e. linolenic and linoleic

29. Consumption of a diet containing cold-water fish such as salmon and mackeral is beneficial because these fish contain high amounts of:

a. unsaturated fatty acids b. omega-3 fatty acids c. omega-6 fatty acids
d. saturated fatty acids e. polyunsaturated fatty acids

Answers to Practice Exercises

19.1

Name of fatty acid	Carbon atoms and double bonds	Type designation	"Omega" designation	"Delta" designation
linolenic acid	18:3	PUFA	omega-3	$\Delta^{9,12,15}$
palmitoleic acid	16:1	MUFA	omega-7	Δ^{9}
arachidonic acid	20:4	PUFA	omega-6	$\Delta^{5,8,11,14}$
linoleic acid	18:2	PUFA	omega-6	$\Delta^{9,12}$
oleic acid	18:1	MUFA	omega-9	Δ^{9}
stearic acid	18:0	SFA	–	–

19.2

Fatty acid pair	Higher melting point?
lauric acid – 12:0 and myristic acid –14:0	myristic acid
stearic acid – 18:0 and oleic acid – 18:2	stearic acid
arachidic acid – 20:0 and arachidonic acid – 20:4	arachidic acid

19.3 a.

O
‖
CH_2-O-C (H H, cis double bond)
O
‖
$CH-O-C$ (H H, cis double bond)
O
‖
CH_2-O-C (H H, cis double bond)

b.

O
‖
CH_2-O-C
O
‖
$CH-O-C$ (H H, cis double bond)
O
‖
CH_2-O-C

or

O
‖
CH_2-O-C (H H, cis double bond)
O
‖
$CH-O-C$
O
‖
CH_2-O-C

c.

O
‖
CH_2-O-C
O
‖
$CH-O-C$
O
‖
CH_2-O-C

19.4

	Fat or oil?	Explanation
a.	oil	All three fatty acid residues are unsaturated, so the molecules are bent and do not pack together closely, resulting in low melting point.
b.	fat	Only one fatty acid is unsaturated and two are saturated, so the molecules are less distorted than in the previous example and pack together more closely.
c.	fat	All three fatty acids are saturated and linear, resulting in close packing and high melting point.

19.5 1.

a.	b.	c.
glycerol, 3 molecules of oleic acid	glycerol, 1 molecule of oleic acid, 2 molecules of stearic acid	glycerol, 3 molecules of stearic acid

2.

a.	b.	c.
glycerol, 3 formula units of sodium oleate	glycerol, 1 formula unit of sodium oleate, 2 units of sodium stearate	glycerol, 3 formula units of sodium stearate

3.

a.	b.	c.
3	1	0

4.

a.	b.	c.
yes	yes	no

19.6 Triacylglycerols

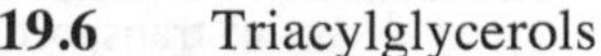

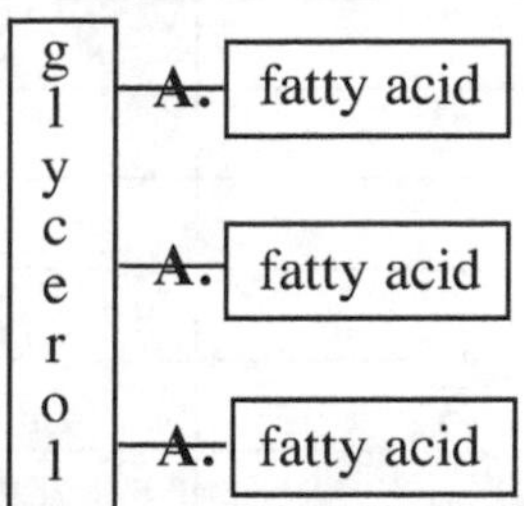

Glycerophospholipids

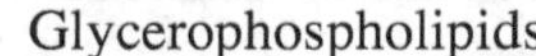

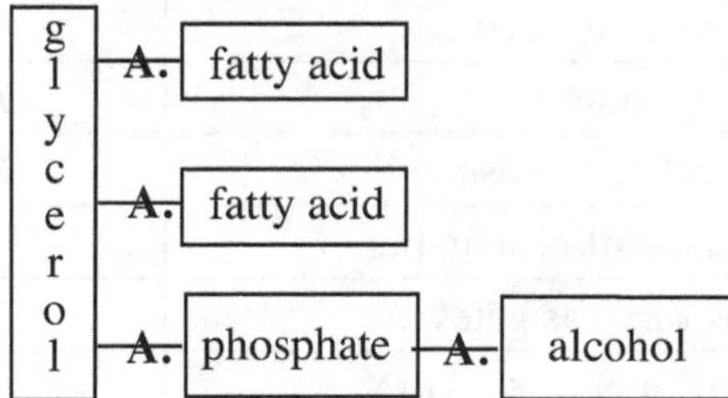

Sphingophospholipids

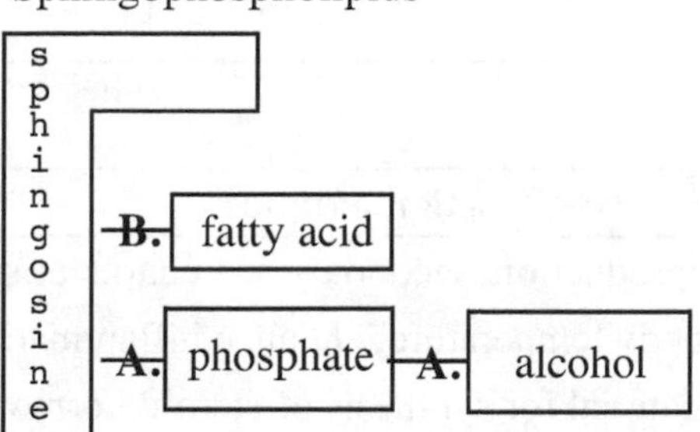

Cerebrosides (one type of glycolipid)

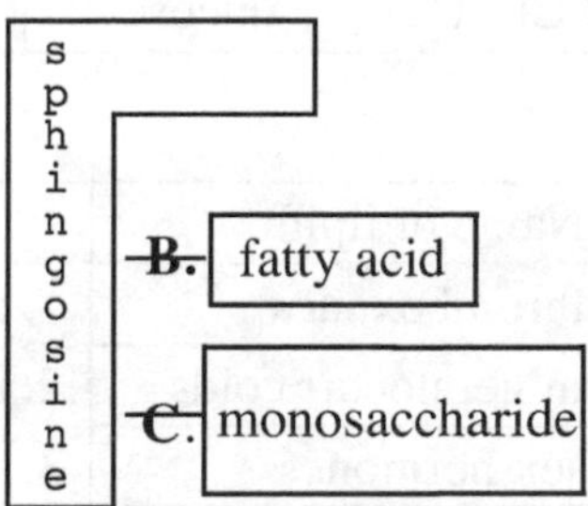

A. ≡ ester linkage **B.** ≡ amide linkage **C.** ≡ glycosidic linkage

19.7

Lipids	Hydrophilic heads	Hydrophobic tails
triacylglycerols	Glycerol ester	Fatty acid chain
glycerophospholipids	Alcohol phosphate	Fatty acid chains
sphingophospholipids	Alcohol phosphate	Sphingosine and fatty acid chain
sphingoglycolipids	monosaccharide	Sphingosine and fatty acid chain

19.8

	Name	Function
LDL	low-density lipoproteins	carry cholesterol from the liver to the various tissues in the body
HDL	high-density lipoproteins	carry excess cholesterol from tissues back to the liver

19.9

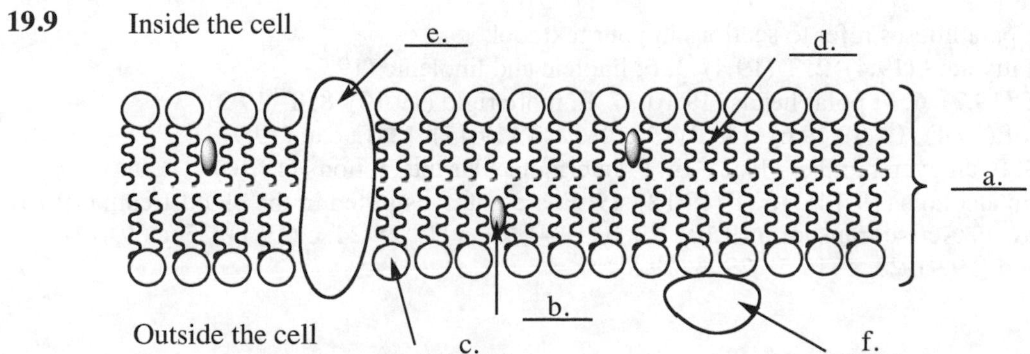

19.10

Description	Passive transport	Facilitated transport	Active transport
energy input needed			**X**
no energy input	**X**	**X**	
closely related to osmosis	**X**		
requires aid of protein pump			**X**
proteins serve as gates		**X**	
transport of Na^+, K^+, H_3O^+			**X**
transport of O_2, N_2, urea	**X**		
transport of Cl^-, HCO_3^-, glucose		**X**	

19.11

Answer	Name of lipid	Function of lipid
7.	thromboxanes	1. control reproduction, secondary sex characteristics
8.	mineralocorticoids	2. regulate body temperature, enhance inflammation
1.	sex hormones	3. starting material for synthesis of steroid hormones
4.	glucocorticoids	4. control glucose metabolism, reduce inflammation
3.	cholesterol	5. emulsifying agents, digestion of lipids
5.	complex bile acids	6. found in white blood cells, hypersensitivity response
6.	leukotrienes	7. promote formation of blood clots
2.	prostaglandins	8. control the balance of Na^+ and K^+ ions in cells

19.12 thromboxanes – B; mineralocorticoids – A; sex hormones – A; glucocorticoids – A; cholesterol – A; bile acids – A; leukotrienes – B; prostaglandins – B

19.13 and 19.14

triacylglycerols____A, S	lecithins_________B, S	fats_________A, S
biological waxes____E, S	androgens________D, N	cephalins_____B, S
prostaglandins______D, N	thromboxanes_____D, N	bile acids ____C, N
cholic acid_________C, N	cerebrosides______B, S	estrogens ____D, N
adrenocorticoids_____D, N	cholesterol _______B, N	oils__________A, S
sphingomyelins_____B, S	steroid hormones __D, N	

Answers to Self-Test

The numbers in parentheses refer to sections in your textbook.
1. F; saturated fatty acid (19.4) **2**. T (19.4) **3**. F; linoleic and linolenic (19.2)
4. T (19.6) **5**. T (19.7) **6**. F; polar heads (19.10) **7**. F; more rigid (19.10) **8**. T (19.2)
9. F; a mixture of (19.4) **10**. T (19.6) **11**. T (19.10) **12**. T (19.11) **13**. F; can (19.9)
14. T (19.2) **15**. F; an even number (19.2) **16**. F; intermolecular interactions (19.10)
17. F; long-chain alcohols (19.14) **18**. T (19.12) **19**. F; are not transported in the bloodstream (19.13)
20. T (19.13) **21**. F; Active transport (19.10) **22**. b (19.4) **23**. a (19.2) **24**. c (19.7) **25**. a (19.12)
26. d (19.9) **27**. d (19.9) **28**. e (19.5) **29**. b (19.5)

Proteins — Chapter 20

Chapter Overview

The functions of proteins in living systems are highly varied; they catalyze reactions, form structures, and transport substances. These functions depend on the properties of the small units that make up proteins (the amino acids) and on how these amino acids are joined together.

In this chapter you will identify the elements of which amino acids are made (C, H, O, and N), the basic structure of amino acids, and you will characterize amino acid side chains. You will draw Fischer projections for amino acids and explain why an amino acid exists as a zwitterion. You will define the peptide bond and write abbreviated names for peptides. You will compare the primary, secondary, tertiary, and quaternary structures of proteins.

Practice Exercises

20.1 A **protein** (Sec. 20.1) is a naturally-occurring, unbranched polymer in which the monomer units are amino acids. An ***α*-amino acid** (Sec. 20.2) contains an amino group and a carboxyl group, both attached to the same atom, the ***α*-carbon atom** (Sec. 20.2). Amino acids are classified as **nonpolar, polar neutral, polar basic,** or **polar acidic** (Sec. 20.2), depending on the nature of the side chain present.

Complete the following table for four of the 20 standard amino acids.

Name	Abbreviation	Structure	Classification
leucine			
aspartic acid			
lysine			
serine			

20.2 The α-carbon atom of an α-amino acid is a chiral center (except in glycine); naturally occurring amino acids are L-isomers. To draw the Fischer projection formula for an amino acid: put the –COOH group at the top and the R group at the bottom. The $-NH_2$ group will be to the left for the L-isomer, to the right for the D-isomer.

Using Fischer projection formulas, draw L-aspartic acid and the enantiomer of L-aspartic acid. (The enantiomer of the L-isomer is the D-isomer.)

$HO-\overset{\overset{O}{\|}}{C}-CH_2-\underset{\underset{NH_2}{\|}}{CH}-\overset{\overset{O}{\|}}{C}-OH$ aspartic acid	a. Fischer projection of L-aspartic acid	b. enantiomer of L-aspartic acid

20.3 Because an amino acid has both an acidic group (the carboxyl group) and a basic group (the amino group), it can exist as a **zwitterion** (Sec. 20.5), a molecule that has a positive charge on one atom and a negative charge on another atom. a. Draw the structural formula for alanine and b. Draw the alanine zwitterion.

a.	b.

20.4 In solution, three different amino acid forms can exist in equilibrium. The amount of each form present depends on the pH of the solution.
Draw the charged forms of alanine at each of the following pH values.

pH = 1	pH = 6	pH = 12

20.5 An amino acid's **isoelectric point** (Sec. 20.5) is the pH at which the amino acid exists primarily in its zwitterion form. Isoelectric points for neutral amino acids are in the pH range 4.8 – 6.3. Acidic or basic amino acids have isoelectric points that are lower or higher than those for amino acids with neutral side chains.

a. The isoelectric point for aspartic acid is 2.77. In the space below, draw the structural form for aspartic acid that predominates at its isoelectric point.

b. At what pH (relative to the isoelectric point) would you expect both carboxyl groups be protonated? ____________________ Draw the predominant form of aspartic acid at this pH.

a.	b.

c. The isoelectric point for lysine is 9.74. In the space below, draw the structural form for lysine that predominates at its isoelectric point.

d. At what pH (relative to the isoelectric point) would neither of the amino groups of lysine be protonated? ____________________ Draw the predominant form of lysine at this pH.

c.	d.

20.6 The standard amino acid cysteine (Sec. 20.6) has a side chain that contains a sulfhydry group (–SH). Cysteine dimerizes in the presence of a mild oxidizing agent to form cystine, which contains a disulfide bond. (In Sec. 20.12, you will see that cystine can link two peptide chains or two parts of the same chain and so is important in the structure and shape of proteins.)

Write an equation showing the dimerization of two cysteine molecules to form a cystine molecule.

20.7 A **peptide** (Sec. 20.7) is a chain of amino acids joined together by covalent bonds called **peptide bonds** (Sec. 20.7). The peptide bond is an amide linkage between the carboxyl group of one amino acid and the amino group of another. The end of the peptide with the free NH_3^+ is called the N-terminal end, and the end with the free COO^- is called the C-terminal end.

a. Draw the structural formulas of the two different dipeptides that could form from the amino acids valine (Val) and serine (Ser). Draw the N-terminal amino acid on the left and the C-terminal on the right. Label the N-terminal end and the C-terminal end.

b. Circle the peptide bond in each structure.

20.8 In naming a peptide, the names of the amino acids are changed from *–ine* or *–ic acid* to *–yl*, but the C-terminal amino acid keeps its name. Give the IUPAC name for each of the following tripeptides:

a. Phe–Asp–Asn

b. Asn–Phe–Asp

20.9 Peptides that contain the same amino acids but in different order are constitutional isomers; they are different molecules with different properties.

a. Using abbreviations for the amino acids, draw all possible tripeptides that could be formed from one molecule each of the following amino acids: Pro, Trp, and Lys. The N-terminal amino acid is written to the left, and the C-terminal amino acid is to the right.

b. Draw the abbreviated peptide formulas for all possible tripeptides that could be formed from two molecules of Pro and one molecule of Lys.

20.10 Many small peptides are biochemically active. For the following small peptides, give the number of amino acid residues each contains and the function of the peptide.

Peptide	Amino acid residues	Peptide function
oxytocin		
vasopressin		
enkephalins		
glutathione		

20.11 A **protein** (Sec. 20.9) is a polypeptide in which at least 40 amino acid residues are present. The shape of a protein molecule is the result of forces acting between various parts of the peptide chain.

a. The sequence of the amino acids in a protein is called its **primary structure** (Sec. 20.10). The first protein whose amino acid sequence was determined was ___________.

b. The **secondary structure** (Sec. 20.11) of a protein results from hydrogen bonding between a carbonyl oxygen atom in one peptide bond and a hydrogen atom attached to nitrogen in another peptide bond. Name two types of secondary structure common in proteins. ______________________________

c. The **tertiary structure** (Sec. 20.12) is the shape of the protein resulting from the interactions between amino acid side chains. The four types of stabilizing interactions are listed below. Next to each term, tell what kinds of amino acids are involved in that type of interaction.

Interaction	Kinds of amino acids
covalent disulfide bonds	
electrostatic interactions	
hydrogen bonds	
hydrophobic attractions	

d. A protein may be classified as a **monomeric protein** (Sec. 20.9), a protein in which only one peptide chain is present, or a **multimeric protein**, in which more than one peptide chain is present. The **quaternary structure** (Sec. 20.13) of a protein is the organization of the peptide subunits in a multimeric protein. What types of interactions are responsible for a multimeric protein's quaternary structure? ______________________________

____

20.12 Proteins and other peptides can undergo complete hydrolysis of their peptide bonds (Sec. 20.14) to yield the consituent amino acid residues. In partial protein hydrolysis only a few of the peptide bonds are hydrolyzed, yielding a mixture of smaller peptides. How many different di- and tripeptides could be present in a solution of partially hydrolyzed Arg-Arg-Cys-Lys? List them.

20.13 Proteins can be classified on the basis of general shape in three basic types: **fibrous**, **globular**, and **membrane** (Sec. 20.16). Describe the shape and solubility characteristics of each type of protein:

Protein type	Shape and solubility characteristics
fibrous	
globular	
membrane	

20.14 Proteins can be divided into 12 groups according to their functions (Sec. 20.17): 1) catalytic proteins, 2) defense proteins, 3) transport proteins, 4) messenger proteins, 5) contractile proteins, 6) structural proteins, 7) transmembrane proteins, 8) storage proteins, 9) regulatory proteins, 10) nutrient proteins, 11) buffer proteins, and 12) fluid balance proteins. Using these numbers, classify the following examples of proteins by functional type.

Protein	Type
insulin	
enzymes	
albumin	
α-keratin	
myoglobin	
hemoglobin	
casein	

Protein	Type
muscle proteins	
binding site for enzymes	
ferritin	
immunoglobulin	
glucagon	
collagen	
globulin	

20.15 Determine the primary structure of a heptapeptide containing six different amino acids, if the following smaller peptides are among the partial-hydrolysis products:

Tyr-Trp, Glu-Asp-Tyr, Trp-Asp-Val, Val-Pro, Tyr-Trp-Asp. The C-terminal end is proline.

Structure of heptapeptide ____

20.16 **Simple proteins** (Sec. 20.9) contain only amino acid residues. **Conjugated proteins** (Sec. 20.9) contain one or more non-amino acid entities, or **prosthetic groups** (Sec. 20.9). **Glycoproteins** (Sec. 20.18) are conjugated proteins containing carbohydrates as well as amino acids. Glycoproteins are important in several ways in biological systems. Match each term below with the statement that describes its role in a process involving glycoproteins.

Answer	Term	Statement
	collagen	1. glycoprotein produced as a protective response
	antigens	2. foreign substances that invade the body
	antibodies	3. aggregates of triple helix strands
	immunoglobulin	4. premilk substance containing immunoglobulins
	colostrum	5. converted to gelatin when boiled in water
	collagen fibrils	6. molecules that counteract specific antigens

20.17 There are four main classes of **plasma lipoproteins** (Sec. 20.19), proteins that are involved in the transport system for lipids in the bloodstream. List the four types of plasma lipoproteins in the table below in order of density, from least dense to most dense. Complete the table.

Plasma lipoproteins	Density	Percent protein

20.18 The four types of plasma lipoproteins in P.E. 20.17 have different functions in the body's transport system.

a. Which lipoproteins carry excess cholesterol to the liver to be broken down? ______________________

b. Which lipoproteins carry dietary triacylglycerols from the intestines to the liver and adipose tissue? ______________________

c. Which lipoproteins transport triacylglycerols synthesized in the liver to the adipose tissue? ______________________

d. Which lipoproteins transport cholesterol synthesized in the liver to cells throughout the body? ______________________

e. Which of the four types of plasma lipoproteins carries the highest content of cholesterol by mass? ______________________

Self-Test

True-false: Indicate whether the following statements are true or false. If the statement is false, give the word or phrase that may be substituted for the underlined portion to make the statement true.

1. Naturally occurring amino acids are generally in the L form.
2. A peptide bond is the bond formed between the amino group of an amino acid and the carboxyl group of the same amino acid.
3. Because amino acids have both an acidic group and a basic group, they are able to undergo internal acid-base reactions.
4. When the pH of an amino acid solution is lowered, the amino acid zwitterion forms more of the positively charged species.
5. The quaternary structure of a multimeric protein is maintained by covalent bonding between protein subunits.
6. Insulin cannot be given by mouth because the digestive process causes hydrolysis of the protein.
7. Vasopressin is a peptide in the human body that regulates uterine contractions and lactation.
8. The function of a protein is controlled by the protein's primary structure.
9. The alpha helix structure is a part of the primary structure of some proteins.
10. Denaturation of a protein involves changes in the protein's primary structure.
11. A protein in which more than one peptide chain is present is called a conjugated protein.
12. An antibody is a foreign substance that invades the body.
13. The peptides in a beta-pleated sheet are held in place by hydrogen bonds.

14. The nine amino acids that cannot be produced in the body and must be obtained from dietary protein are called <u>essential amino acids</u>.
15. A complete dietary protein is a protein that contains <u>all 20 of the standard amino acids</u>.

Multiple choice:

16. An example of a polar basic amino acid is:

a. lysine b. serine c. tryptophan
d. leucine e. none of these

17. A tripeptide is formed from two alanine molecules and one glycine molecule. The maximum number of different tripeptides that could be formed from this combination is:

a. two b. three c. four
d. five e. none of these

18. The peptide that regulates the excretion of water by the kidneys is:

a. glutathione b. insulin c. vasopressin
d. lysine e. none of these

19. The partial hydrolysis of the pentapeptide Val-Ala-Ala-Gly-Ser could yield which of the following peptides? (The amino acid on the left is the N-terminal amino acid.)

a. Val-Ala-Gly b. Ser-Gly-Ala c. Ala-Gly-Ser
d. Ala-Ala-Ser e. none of these

20. Which of the following amino acids would have a net charge of zero at a pH of 7?

a. arginine b. phenylalanine c. aspartic acid
d. lysine e. none of these

21. Which of the following attractive interactions does *not* affect the formation of a protein's tertiary structure?

a. disulfide bonds b. salt bridges c. hydrogen bonds
d. peptide bonds e. none of these

22. Proteins may be denatured by:

a. heat b. acid
c. ethanol d. a, b, and c
e. a and b only

23. Which of the following proteins is *not* a conjugated protein?

a. hemoglobin b. insulin c. collagen
d. immunoglobulin e. lipoprotein

24. Which of the following is a characteristic of globular proteins?

a. stringlike molecules b. water-insoluble
c. roughly spherical d. structural function in body
e. none of these

25. The most abundant protein in humans is:

a. myoglobin b. collagen c. hemoglobin
d. α-keratin e. none of these

Answers to Practice Exercises

20.1

Name	Abbreviation	Structure	Classification
leucine	Leu	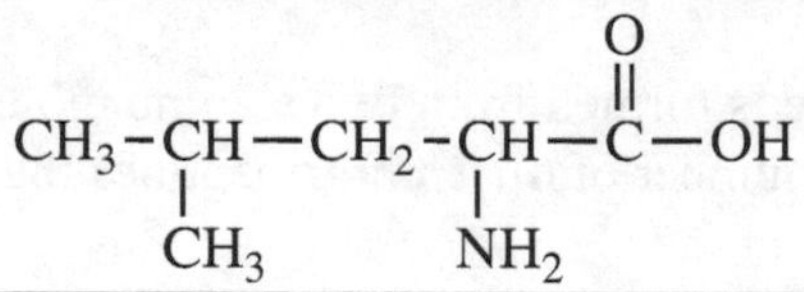 $CH_3-CH(CH_3)-CH_2-CH(NH_2)-C(=O)-OH$	nonpolar
aspartic acid	Asp	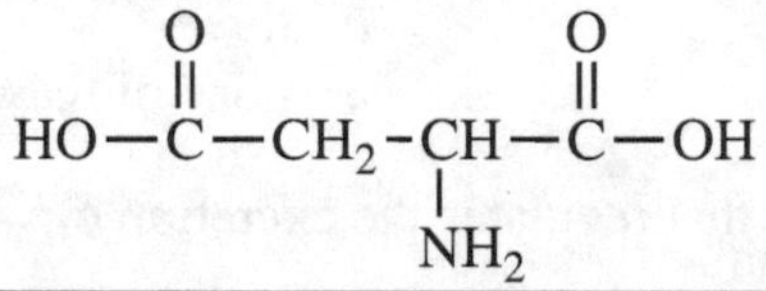 $HO-C(=O)-CH_2-CH(NH_2)-C(=O)-OH$	polar acidic
lysine	Lys	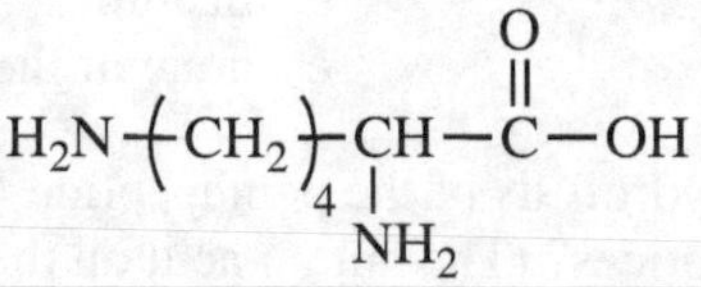 $H_2N-(CH_2)_4-CH(NH_2)-C(=O)-OH$	polar basic
serine	Ser	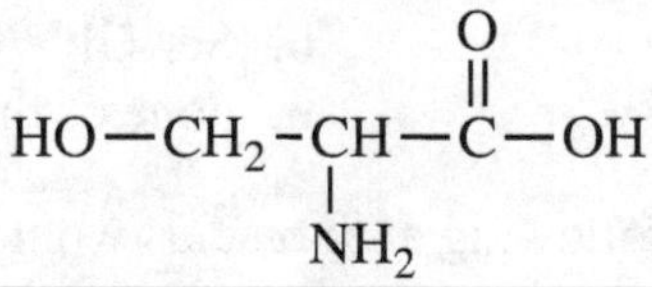 $HO-CH_2-CH(NH_2)-C(=O)-OH$	polar neutral

20.2

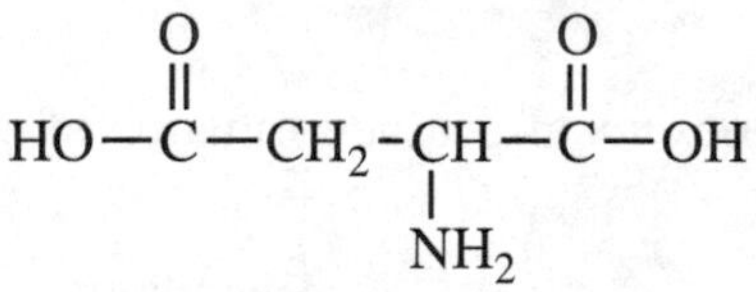 aspartic acid	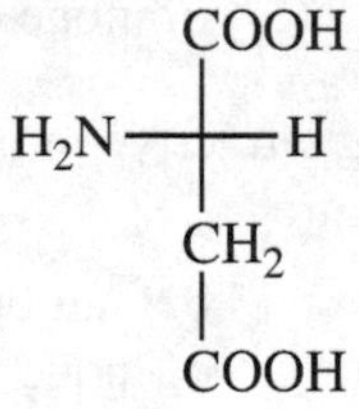 a. Fischer projection of L- aspartic acid	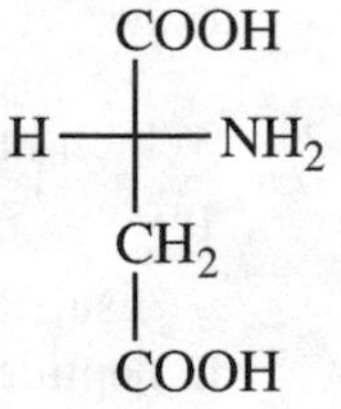 b. enantiomer of L-aspartic acid
$HO-C(=O)-CH_2-CH(NH_2)-C(=O)-OH$	COOH (top); H_2N (left), H (right); CH_2; COOH (bottom)	COOH (top); H (left), NH_2 (right); CH_2; COOH (bottom)

20.3

Structural formula of alanine	Alanine zwitterion
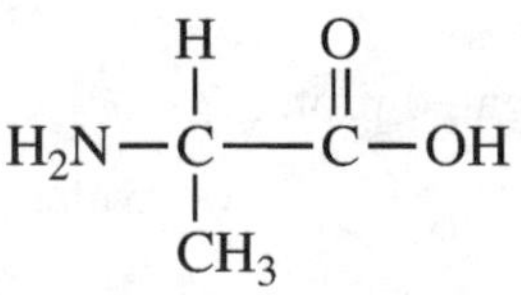 $H_2N-CH(CH_3)-C(=O)-OH$	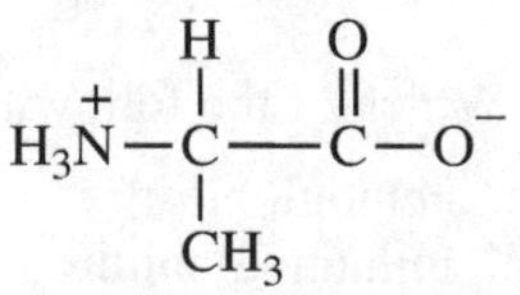 $H_3\overset{+}{N}-CH(CH_3)-C(=O)-O^-$

20.4

$H_3\overset{+}{N}-CH(CH_3)-C(=O)-OH$	$H_3\overset{+}{N}-CH(CH_3)-C(=O)-O^-$	$H_2N-CH(CH_3)-C(=O)-O^-$
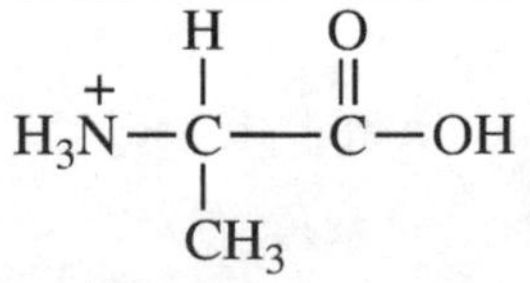 pH = 1	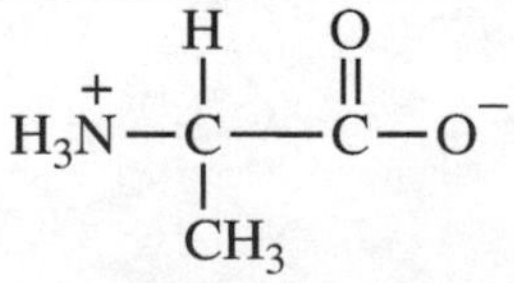 pH = 6	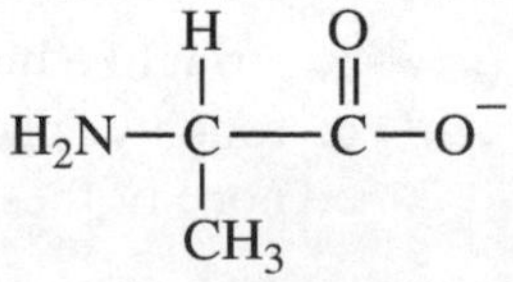 pH = 12

20.5 b. pH < 2.77

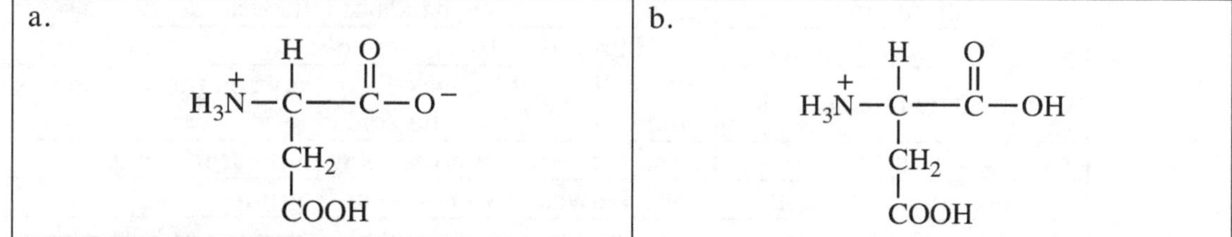

d. pH > 9.74

c.	d.
$H_3\overset{+}{N}-\underset{\underset{CH_2-CH_2-CH_2-NH_2}{CH_2}}{\overset{H}{C}}-\overset{O}{\overset{\|}{C}}-O^-$	$H_2N-\underset{\underset{CH_2-CH_2-CH_2-NH_2}{CH_2}}{\overset{H}{C}}-\overset{O}{\overset{\|}{C}}-O^-$

20.6

$$\underset{\text{Cysteine}}{H_3\overset{+}{N}-\underset{\underset{SH}{CH_2}}{CH}-COO^-} + \underset{\text{Cysteine}}{H_3\overset{+}{N}-\underset{\underset{SH}{CH_2}}{CH}-COO^-} \xrightarrow[\text{oxidizing agent}]{\text{mild}} \underset{\text{Cystine}}{\underset{\overset{+}{NH_3}}{\overset{COO^-}{CH}}-CH_2-S-S-CH_2-\underset{\overset{+}{NH_3}}{\overset{COO^-}{CH}}}$$

20.7

N-terminal end $H_3\overset{+}{N}-\underset{\underset{\underset{CH_3}{\vert}}{H_3C-CH}}{CH}-\overset{O}{\overset{\|}{C}}-NH-\underset{\underset{OH}{CH_2}}{\overset{H}{C}}-\overset{O}{\overset{\|}{C}}-O^-$ C-terminal end	N-terminal end $H_3\overset{+}{N}-\underset{\underset{OH}{CH_2}}{CH}-\overset{O}{\overset{\|}{C}}-NH-\underset{\underset{\underset{CH_3}{\vert}}{CH-CH_3}}{\overset{H}{C}}-\overset{O}{\overset{\|}{C}}-O^-$ C-terminal end

20.8 a. phenylalanylaspartylasparagine
b. asparagylphenylalanylaspartic acid

20.9 a. Pro-Trp-Lys, Pro-Lys-Trp, Trp-Pro-Lys, Trp-Lys-Pro, Lys-Trp-Pro, Lys-Pro-Trp
b. Pro-Pro-Lys, Pro-Lys-Pro, Lys-Pro-Pro

20.10

Peptide	Amino acid residues	Peptide function
oxytocin	9 (nonapeptide)	Hormone that regulates uterine contraction and lactation
vasopressin	9 (nonapeptide)	Antidiuretic hormone
enkephalins	5 (pentapeptides)	Neurotransmitters that reduce pain
glutathione	3 (tripeptide)	Antioxidant, protects cell contents from oxidizing agents

20.11 a. insulin b. alpha helix and beta pleated sheet

c.

Interaction	Kinds of amino acids
covalent disulfide bonds	–SH groups of two cysteine residues form a disulfide bond
electrostatic interactions	also called salt bridges, form between the side chains of acidic and basic amino acids
hydrogen bonds	form between amino acids with polar R-groups
hydrophobic attractions	form between two nonpolar R-groups

d. electrostatic interactions, hydrogen bonds, and hydrophobic attractions between peptide chains

20.12 Three dipeptides (Arg-Arg, Arg-Cys, and Cys-Lys) and two tripeptides (Arg-Arg-Cys and Arg-Cys-Lys)

20.13

Protein type	Shape and solubility characteristics
fibrous	elongated linear structure; water-insoluble
globular	spherical shape with hydrophobic R-groups oriented inward and hydrophilic R-groups outward; water-soluble
membrane	opposite of globular proteins, with hydrophobic R-groups oriented outwards; water-insoluble

20.14

Protein	Type	Protein	Type
insulin	4	muscle proteins	5
enzymes	1	binding sites for enzymes	9
albumin	12	ferritin	8
keratin	6	immunoglobulin	2
myoglobin	8	glucagons	4
hemoglobin	3, 11	collagen	6
casein	10	globulin	12

20.15 Structure of heptapeptide: Glu-Asp-Tyr-Trp-Asp-Val-Pro

20.16 collagen – 5; antigens – 2; antibodies – 6; immunoglobulin – 1; colostrums – 4; collagen fibrils – 3

20.17

Plasma lipoproteins	Density	Percent protein
chylomicrons	< 0.95 g/mL	1%
very-low-density lipoproteins (VLDL)	0.95 to 1.02 g/mL	8%
low-density lipoproteins (LDL)	1.02 to 1.06 g/mL	25%
high-density lipoproteins (HDL)	1.06 to 1.21 g/mL	45%

20.18 a. high-density lipoproteins b. chylomicrons c. very-low-density lipoproteins d. low-density lipoproteins e. low-density lipoproteins

Answers to Self-Test

The numbers in parentheses refer to sections in your textbook.
1. T (20.4) **2**. F; another (20.7) **3**. T (20.5) **4**. T (20.5) **5**. F; noncovalent interactions (20.13) **6**. T (20.14) **7**. F; Oxytocin (20.8) **8**. F; primary, secondary, tertiary, and quaternary (20.9) **9**. F; secondary (20.11) **10**. F; secondary, tertiary, and quaternary (20.14) **11**. F; multimeric protein (20.9) **12**. F; antigen (20.18) **13**. T (20.11) **14**. T (20.3) **15**. F; all of the essential amino acids (20.3) **16**. a (20.2) **17**. b (20.7) **18**. c (20.8) **19**. c (20.17) **20**. b (20.5) **21**. d (20.12) **22**. d (20.14) **23**. b (20.16) **24**. c (20.16) **25**. b (20.16)

Enzymes and Vitamins — Chapter 21

Chapter Overview

Enzymes provide the necessary "boost" for most of the chemical reactions of biological systems. Without the help of enzymes, most reactions would proceed so slowly that they would be ineffective. Vitamins are essential nutrients that often play a part in the activity of enzymes.

In this chapter you will learn the general characteristics of enzymes and how to predict the function of an enzyme from its name. You will describe the enzyme-substrate complex in terms of the lock-and-key model and the induced-fit model. You will learn how enzyme action can be inhibited and how it is controlled in biological systems, and you will identify the roles of vitamins in enzyme activity.

Practice Exercises

21.1 **Enzymes** (Sec. 21.1) are compounds that act as catalysts for biochemical reactions. They can be divided into two classes: **simple enzymes** and **conjugated enzymes** (Sec. 21.2). Other common terms used in describing enzymes are defined in Section 21.2 of your textbook. Match each of the following terms with the correct description.

Answer	Term	Description
	coenzyme	1. nonprotein portion of a conjugated enzyme
	cofactor	2. the combined apoenzyme and cofactor
	holoenzyme	3. protein portion of a conjugated enzyme
	apoenzyme	4. small organic molecule that serves as a cofactor

Using the terms from the table above, complete the following diagrams:

	=		+	
the entire conjugated enzyme		the protein portion		the nonprotein portion

cofactors:	metal ions	or	

21.2 Some enzymes are named by using the name of the **substrate** (Sec. 21.3), the compound undergoing change, and adding the ending *-ase*. Complete the table below, which gives examples of enzymes named for substrates:

Name of enzyme	Type of reaction
α-amylase	
	hydrolysis of peptide linkages
lipase	

More often, the enzyme name gives enzyme function. Enzymes are grouped into six major classes (and a number of subclasses) on the basis of the types of reactions they catalyze. For example: an isomerase catalyzes the conversion of the substrate to an isomer of itself.

Complete the table below, giving examples of enzyme classes and their functions.

Class of enzyme	Enzyme function
dehydrogenase	
	removal of CO_2 from a substrate
transaminase	
oxidase	
	conversion of D to L isomer, or L to D isomer

21.3 Before an enzyme-catalyzed reaction takes place, an **enzyme-substrate complex** (Sec. 21.4) is formed: the substrate binds to the **active site** (Sec. 21.4) of the enzyme.

a. What are two models that account for the specific way an enzyme selects a substrate?

b. What is the main difference between these two models?

21.4 **Enzyme specificity** (Sec. 21.5) is the extent to which an enzyme's activity is restricted to a specific substrate or type of chemical reaction. Types of specificity include: 1) absolute specificity, 2) group specificity, 3) linkage specificity, and 4) stereochemical specificity. In the table below, write the type of specificity exhibited by the given enzymes.

Enzyme name	Type of specificity
L-amino acid oxidase	
carboxypeptidase	
phosphatase	
catalase	

21.5 **Enzyme activity** (Sec. 21.6) is a measure of the rate at which an enzyme converts substrate to products. In the first column of the table below, name four factors that affect enzyme activity. In the second column, describe the effect on rate of reaction of an *increase* in each of the four factors named.

21.6 An **enzyme inhibitor** (Sec. 21.8) is a substance that slows or stops the normal catalytic function of an enzyme by binding to it. After reading the section on enzyme inhibition (Sec. 21.8) in your textbook, write a brief description of each of the three types of enzyme inhibitors, and give an example of each

Inhibitor type	Description	Example
irreversible enzyme inhibitor		
reversible noncompetitive enzyme inhibitor		
reversible competitive enzyme inhibitor		

21.7 Enzyme activity can also be changed by regulators produced within a cell (Sec. 21.4). Match the following terms with any descriptions that apply.

Answer	Term	Description
	allosteric enzyme	1. catalyzes the breaking of peptide bonds
	covalent modification	2. contains both active and regulator binding sites
	zymogen	3. inhibition of the first reaction in a sequence is controlled by the product of a sequence
	proteolytic enzyme	4. activation requires removal of some part of the proenzyme structure
	feedback control	5. inactive precursor of an enzyme
		6. a chemical group (usually phosphate) is removed from or attached to an enzyme
		7. regulator changes the shape of the active site

21.8 Several types of prescription drugs act by inhibiting enzyme activity. **Antibiotics** (Sec. 21.10) often act by inhibiting enzymes necessary to the growth of bacterial cells. Tell how each of the following types of drugs acts by inhibition of enzyme activity.

a. penicillin

b. sulfa drugs

c. ACE inhibitors

21.9 The human body needs small amounts of organic compounds called **vitamins** (Sec. 21.12). Vitamins cannot be synthesized by the body and so must be obtained in the diet.

a. In column I write the number of the matching function for each vitamin.
b. Vitamins are divided into two groups according to their solubility: **water-soluble vitamins** (Sec. 21.13, 21.14) and **fat-soluble vitamins** (Sec. 21.15). In column II, write W if the vitamin is water-soluble or F if it is fat-soluble.

I	II	Vitamin	Function in human body
		vitamin A	1. prevents oxidation of polyunsaturated fatty acids in lungs and blood cells
		vitamin C	2. helps form prothrombin and other proteins involved in blood clotting
		vitamin D	3. group of vitamins that are components of coenzymes
		vitamin B	4. cosubstrate in the formation of collagen; general antioxidant
		vitamin E	5. promotes deposition of calcium salts in bones; maintains normal blood levels of calcium
		vitamin K	6. helps to form the visual pigment rhodopsin; keeps skin and mucous membranes healthy

21.10 Vitamin C and the B vitamins are all water-soluble. The B vitamins have a biochemical function in common; they all serve as precursors for enzyme cofactors. Vitamin C does not. In the table below, give the preferred name for the numbered B vitamin and the coenzyme for which it is a precursor.

Vitamin	Preferred name	Coenzyme form
vitamin B_1		
vitamin B_2		
vitamin B_3		
vitamin B_5		
vitamin B_6		
vitamin B_7		
vitamin B_9		
vitamin B_{12}		

21.11 The four fat-soluble vitamins (vitamins A, D, E, and K) have structures that are more hydrocarbon-like than those of the water-soluble vitamins; therefore, they are less polar. Each contains at least one ring or ring system and at least one hydrocarbon chain. To help you remember their structural differences, complete the following matching exercise.

Answer	Vitamin	Structure
	vitamin A	1. Methylated naphthoquinone with long-chain R group
	vitamin D	2. Precursor is 7-dehydrocholesterol
	vitamin E	3. Retinoids, terpene-like structures
	vitamin K	4. Tocopherols, varying substituents on aromatic ring

Self-Test

True-false: Indicate whether the following statements are true or false. If the statement is false, give the word or phrase that may be substituted for the underlined portion to make the statement true.

1. In an enzyme-catalyzed reaction, the compound that undergoes a chemical change is called a substrate.
2. Enzyme names are usually based on the structure of the enzyme.
3. The protein portion of a conjugated enzyme is called the coenzyme.
4. The active site of an enzyme is the small part of an enzyme where catalysis takes place.
5. According to the lock-and-key model of enzyme action, the active site of the enzyme is flexible in shape.
6. A carboxypeptidase is an enzyme that is specific for one group of compounds.
7. A substance that binds to an enzyme's active site and is not released is called a noncompetitive enzyme inhibitor.
8. Enzymes undergo all the reactions of proteins except denaturation.
9. A cofactor is a protein part of an enzyme necessary for the enzyme's function.
10. Large doses of water-soluble vitamins may be toxic, because they can be retained in the body in excess of need.
11. Some of the body's vitamin C is produced in the skin with the help of sunlight.
12. Some vitamins act as cofactors in conjugated enzymes.
13. Tissue plasminogen activator is an enzyme used in the diagnosis of heart attacks.
14. The enzyme inhibition known as the "grapefruit effect" causes a decreased concentration of certain drugs in the bloodstream.

Multiple choice:

15. Enzymes assist chemical reactions by:

    a. increasing the rate of the reactions
    b. increasing the temperature of the reactions
    c. being consumed during the reactions
    d. all of these
    e. none of these

16. The enzyme that catalyzes the reaction of an alcohol to form an aldehyde would be:

    a. an oxidase
    b. a decarboxylase
    c. a dehydratase
    d. a reductase
    e. none of these

17. A competitive inhibitor of an enzymatic reaction:

    a. distorts the shape of the enzyme molecule
    b. attaches to the substrate
    c. blocks an enzyme's active site
    d. weakens the enzyme-substrate complex
    e. none of these

18. When the enzyme-substrate complex forms, the actual bond breaking and/or bond formation take place:

    a. at the active site
    b. at the regulator bonding site
    c. on the zymogen
    d. on the positive regulator
    e. none of these

19. α-Amylase does not catalyze the hydrolysis of β-glycosidic bonds because the enzyme α-amylase is:
    a. difficult to activate b. specific c. a proenzyme
    d. inhibited e. none of these

20. The antibiotic penicillin acts by:
    a. activating zymogens for bacterial enzymes
    b. cleaving peptide bonds in bacterial proteins
    c. catalyzing the hydrolysis of bacterial cell walls
    d. inhibiting bacterial enzymes that help form cell walls
    e. none of these

21. The vitamin that protects polyunsaturated fatty acids from oxidation is:
    a. vitamin B b. vitamin A c. vitamin D
    d. vitamin E e. none of these

22. An organism that contains enzymes active under conditions of high salinity is:
    a. an acidophile b. a xerophile c. a halophile
    d. a hyperthermophile e. none of these

Answers to Practice Exercises

21.1

Answer	Term	Description
4.	coenzyme	1. nonprotein portion of a conjugated enzyme
1.	cofactor	2. the combined apoenzyme and cofactor
2.	holoenzyme	3. protein portion of a conjugated enzyme
3.	apoenzyme	4. small organic molecule that serves as a cofactor

holoenzyme	=	apoenzyme	+	cofactor
the entire conjugated enzyme		the protein portion		the nonprotein portion

cofactors:		metal ions	or	coenzymes

21.2

Name of enzyme	Type of reaction
α-amylase	hydrolysis of α-linkages in starch molecules
peptidase or protease	hydrolysis of peptide linkages
lipase	hydrolysis of ester linkages in lipids

Class of enzyme	Enzyme function
dehydrogenase	removal of two atoms of hydrogen from a substrate
decarboxylase	removal of CO_2 from a substrate
transaminase	transfer of an amino group between substrates
oxidase	oxidation of a substrate
racemase	conversion of D to L isomer, or L to D isomer

21.3 a. The two models are the lock-and-key model and the induced-fit model.

b. The main difference between the two models is that according to the lock-and-key model, the active site of the enzyme is fixed and rigid, but in the induced-fit model, the active site can change its shape slightly to accommodate the shape of the substrate.

21.4

Enzyme name	Type of specificity
L-amino acid oxidase	stereochemical specificity
carboxypeptidase	group specificity
phosphatase	linkage specificity
catalase	absolute specificity

21.5

temperature	Increase in temperature causes rate of a reaction to increase. A temperature high enough to denature the protein enzyme causes rate to decrease.
pH	Each enzyme has an optimum pH for action; if the pH increases above this point, the reaction will slow down.
substrate concentration	An increase in substrate concentration causes the rate of reaction to increase until maximum enzyme capacity is reached; after this there is no further increase.
enzyme concentration	An increase in enzyme concentration causes the reaction rate to increase.

21.6

Inhibitor type	Description	Example
irreversible enzyme inhibitor	forms a strong covalent bond at the enzyme's active site	nerve gases, organo-phosphate insecticides
reversible noncompetitive enzyme inhibitor	binds to a site other than the enzyme's active site	the heavy metal ions Pb^{2+}, Ag^{+}, Hg^{2+}
reversible competitive enzyme inhibitor	.competes with substrate for the enzyme's active site	antihistamines

21.7

Answer	Term	Description
2. and 7.	allosteric enzyme	1. catalyzes the breaking of peptide bonds
6.	covalent modification	2. contains both active and regulator sites
4. and 5.	zymogen	3. inhibition of the first reaction in a sequence is controlled by the product of a sequence
1. and 4.	proteolytic enzyme	4. activation requires removal of some part of the proenzyme structure
3.	feedback control	5. inactive precursor of an enzyme
		6. a chemical group (usually phosphate) is removed from or attached to an enzyme
		7. regulator changes the shape of the active site

21.8 a. penicillin – inhibits transpeptidase, which is necessary for bacterial cell wall formation

b. sulfa drugs – inhibit bacterial enzymes that convert PABA to folic acid.

c. ACE inhibitors – inhibit ACE, the enzyme that converts angiotensinogen to angiotensin (a hormone that raises blood pressure).

21.9

I	II	Vitamin name	Function in human body
6.	F	vitamin A	1. prevents oxidation of polyunsaturated fatty acids in lungs and blood cells
4.	W	vitamin C	2. helps form prothrombin and other proteins involved in blood clotting
5.	F	vitamin D	3. a group of vitamins that are components of coenzymes
3.	W	vitamin B	4. cosubstrate in the formation of collagen; general antioxidant
1.	F	vitamin E	5. promotes deposition of calcium salts in bones; maintains normal blood levels of calcium
2.	F	vitamin K	6. helps to form the visual pigment rhodopsin; keeps skin and mucous membranes healthy

21.10

Vitamin	Preferred name	Coenzyme form
vitamin B_1	thiamin	thiamin pyrophosphate (TPP)
vitamin B_2	riboflavin	flavin adenine dinucleotide (FAD) and flavin mononucleotide (FMN)
vitamin B_3	niacin	nicotinamide adenine dinucleotide (NAD^+) and nicotinamide adenine dinuclotide phosphate ($NADP^+$)
vitamin B_5	pantothenic acid	coenzyme A and acyl carrier protein (ACP)
vitamin B_6	vitamin B_6	pyridoxine 5′phosphate (PNP), pyridoxal 5′phosphate (PLP), and pyridoxamine 5′phosphate (PMP)
vitamin B_7	biotin	amide formed between biotin's carboxyl group and the amine group of amino acid lysine.
vitamin B_9	folate	tetrahydrofolate (THF)
vitamin B_{12}	vitamin B_{12}	methylcobalamin

21.11

Answer	Vitamin	Structure
3	vitamin A	1. Methylated naphthoquinone with long-chain R group
2	vitamin D	2. Precursor is 7-dehydrocholesterol
4	vitamin E	3. Retinoids, terpene-like structures
1	vitamin K	4. Tocopherols, varying substituents on aromatic ring

Answers to Self-Test

The numbers in parentheses refer to sections in your textbook.
1. T (21.2) **2**. F; function (21.2) **3**. F; apoenzyme (21.3) **4**. T (21.4)
5. F; fixed and rigid (21.4) **6**. T (21.5) **7**. F; an irreversible (21.8) **8**. F; including (21.1)
9. F; nonprotein part (21.3) **10**. F; fat-soluble vitamins (21.15) **11**. F; vitamin D (21.15)
12. T (21.14) **13**. F; treatment (21.11) **14**. F; an increased concentration (21.10) **15**. a (21.1)
16. a (21.2) **17**. c (21.8) **18**. a (21.4) **19**. b (21.5) **20**. d (21.10) **21**. d (21.15) **22**. c (21.7)

Nucleic Acids   Chapter 22

Chapter Overview

Nucleic acids are the molecules of heredity. Every inherited trait of every living organism is coded in these huge molecules. The complexity of their structure is being unraveled, and this knowledge of the transmission of genetic information has led to the field of recombinant DNA technology.

In this chapter you will name and identify the structures of nucleotides and nucleic acids, the building blocks of DNA and RNA. You will write shorthand forms for nucleotide sequences in segments of DNA and RNA. You will identify the amino acid sequence coded by a given segment of DNA and will describe the processes of replication, transcription, and translation leading to protein synthesis. You will learn the basic ideas of recombinant DNA technology and gene therapy.

Practice Exercises

22.1 Two types of unbranched polymers called **nucleic acids** (Sec. 22.1) are found within the cells of living organisms. They are deoxyribonucleic acid (DNA) and ribonucleic acid (RNA). Complete the information about them below.

Nucleic acid	Location	Function
DNA		
RNA		

22.2 **Nucleotides** (Sec. 22.2) are the structural units from which nucleic acids are formed. A nucleotide is composed of a pentose sugar bonded to both a phosphate group and a nitrogen-containing heterocyclic base. The identities of the sugars and bases differ in RNA and DNA. Complete the table below identifying the pentoses and bases found in the nucleotides that make up RNA and DNA.

Nucleotide components	RNA	DNA
pentose		
purine bases		
pyrimidine bases		

22.3 The names of eight nucleotides, the monomer units making up RNA and DNA, are listed in Table 22.1 in your textbook. Both the names and the abbreviations of these names are commonly used. Complete this table showing the way that nucleotides are named.

Name of nucleotide	Abbreviation	Base	Sugar
deoxyadenosine 5′-monophosphate			
	dTMP		
		guanine	ribose
	CMP		

22.4 Nucleotide formation occurs in two steps: 1) The pentose sugar and the nitrogen-containing base form a **nucleoside** (Sec. 22.2) and 2) the nucleoside reacts with a phosphate group to form the nucleotide. Both reactions are condensation reactions; a molecule of water is given off in each step. Complete the table by naming the nucleoside formed from the given molecules, giving the abbreviation for the nucleotide formed, and indicating which nucleic acid the nucleotide would be a part of.

Base + Sugar	Nucleoside	Nucleotide abbreviation	DNA or RNA?
thymine + deoxyribose			
uracil + ribose			
adenine + ribose			
guanine + deoxyribose			

22.5 Nucleotide monomers can be linked to each other through sugar–phosphate bonds. Draw the structural formula for the dinucleotide that forms between dAMP and dTMP, so that dAMP is the 5′ end (the free phosphate group on the 5′ carbon atom of the sugar) on the left, and dTMP is the 3′ end (the free –OH group on the 3′ carbon atom of the sugar) on the right.

22.6 The **primary structure** (Sec. 22.4) of a nucleic acid consists of the sequence in which the nucleotides are linked together. Both RNA and DNA have an alternating sugar–phosphate backbone with the nitrogen-containing bases as side-chains.

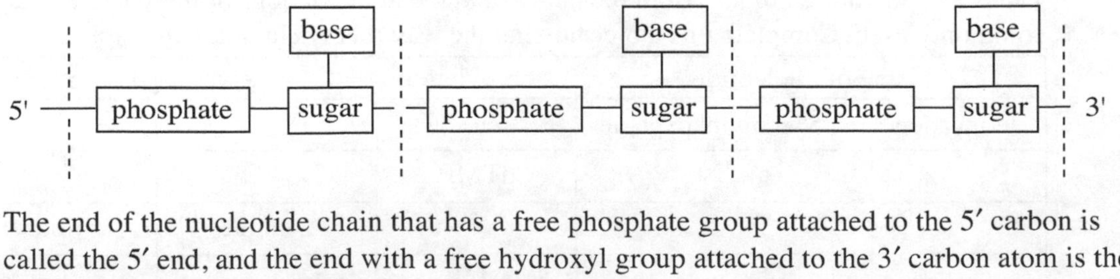

base base base
5' phosphate sugar phosphate sugar phosphate sugar 3'

The end of the nucleotide chain that has a free phosphate group attached to the 5′ carbon is called the 5′ end, and the end with a free hydroxyl group attached to the 3′ carbon atom is the 3′ end. The strand is read from the 5′ end to the 3′ end.

Using a structural block diagram similar to the one shown above, draw the structural formula for a trinucleotide that forms from dAMP, dTMP, and dCMP. In your diagram, dAMP will be on the 5′ end and to the left, and dTMP will be the 3′ end and to the right. Replace "base" and "sugar" with the specific names of the compounds.

22.7 The structure of DNA is that of a double helix, in which two strands of DNA are coiled around each other in a spiral. Two forces are important in stabilizing the double helix structure. Base-stacking interactions (Sec. 22.5) are hydrophobic interactions between the parallel planes of purine and pyrimidine bases in DNA.

The other important stabilizing force for the double helix is hydrogen bonding between two pairs of **complementary bases** (Sec. 22.5), A–T and G–C. The pairing of A with T and of G with C gives the maximum number of hydrogen bonds between DNA strands.

If, in a DNA molecule, the percentage of the base adenine is 20% of the total bases present, what would be the percentages of the bases thymine, cytosine, and guanine?

22.8 In the pairing of complementary bases, two hydrogen bonds form between adenine and thymine; three hydrogen bonds form between cytosine and guanine. In the DNA double helix, the two **complementary DNA strands** (Sec. 22.5) run in opposite directions, one in the 5′ to 3′ direction and the other 3′ to 5′. Complete the following segment of a DNA double helix. Write symbols for the missing bases. Indicate the correct number of hydrogen bonds between the bases in each pair.

```
5′ T – C –    – C –   – G –   – A  3′
   ||  |||
3′ A – G – T –    – T –   – A –    5′
```

22.9 During **DNA replication** (Sec. 22.6), the DNA molecule makes an exact duplicate of itself. The two strands unwind, and free nucleotides line up along each parent strand, with complementary base pairs attracted to one another by hydrogen bonding. Polymerization of each new strand occurs one nucleotide at a time. Each daughter strand is formed in the direction opposite to that of the parent strand.

a. Write the sequence of bases in a segment of a daughter strand formed by the replication of the DNA segment below:

5′ T–A–A–G–C–G–T–G–G 3′

b. A 3′ to 5′ base sequence can be converted to a 5′ to 3′ base sequence by reversing the order of the listed bases. Write the 5′ to 3′ base sequence for the segment of the daughter strand formed above.

22.10 The process of protein synthesis involves five types of RNA molecules (Sec. 22.8). Each type has a different function. Complete the table below on the different types of RNA.

Type of RNA	Abbreviation	Function
heterogeneous nuclear RNA		
small nuclear RNA		
messenger RNA		
ribosomal RNA		
transfer RNA		

22.11 During **transcription** (Sec. 22.9), one strand of a DNA molecule acts as the template for the formation of a molecule of hnRNA. The nucleotides that line up next to the DNA strand have ribose as a sugar; the same bases are present except that U is substituted for T.

DNA contains certain segments, called **introns** (Sec. 22.9), that do not convey genetic information, as well as **exons** (Sec. 22.9), which do carry the genetic code. Both exons and introns are transcribed to hnRNA. In post-transcription processing of hnRNA, deletion of introns and joining of exons (**splicing**, Sec. 22.9) takes place to form mRNA.

In the DNA template strand below, sections A, C, and E are exons, and B and D are introns. Write the base sequence for the hnRNA segment transcribed from this DNA strand. Below the hnRNA strand write the base sequence of the mRNA strand with the introns deleted.

A (G–C–C) B (T–G–T) C (A–C–T–T–C–G) D (A–T–T) E (G–G–A)

DNA 5′ G–C–C–T–G–T–A–C–T–T–C–G–A–T–T–G–G–A 3′

hnRNA 3′ 5′

mRNA 3′ 5′

22.12 During **translation** (Sec. 22.12) mRNA directs the synthesis of proteins by carrying the genetic code from the nucleus to a **ribosome** (Sec. 22.12), where the code is translated into the correct series of amino acids in the protein. A **codon** (Sec. 22.10) is a sequence of three nucleotides in an mRNA molecule that codes for a specific amino acid.

a. Complete the tables below with correct amino acid names and codons. The information can be obtained from Table 22.2 in your textbook.

Codon	Amino acid
UCA	
	asparagine
GAC	

Codon	Amino acid
	methionine
GAU	
	tryptophan

b. Are there any synonyms among the codons in the table in part a?

c. Why is ATC not listed in Table 22.2 as one of the codon sequences?

22.13 a. Write a base sequence for mRNA that codes for the tripeptide Gly-Pro-Leu.

b. Will there be only one answer? Explain.

22.14 An **anticodon** (Sec. 22.11) is a three-nucleotide sequence on tRNA that complements the mRNA sequence (codon) for the amino acid that bonds to that tRNA. Complete the table below for codons, their anticodons, and the amino acids that the codons specify.

Codon	5´ CAU 3´		
Anticodon		3´ GAG 5´	
Amino acid			Trp

22.15 The mRNA segment below is the one that was determined in Practice Exercise 22.11, and then reversed to the 5´ to 3´ direction. Use Table 22.2 in your textbook to determine the amino acid sequence of a tetrapeptide coded by the mRNA base sequence below.

5′ U–C–C–C–G–A–A–G–U–G–G–C 3′ mRNA

tetrapeptide

22.16 The two main processes of protein synthesis are transcription, in which DNA directs the synthesis of RNA molecules, and translation, in which RNA directs the synthesis of proteins. These two processes consist of various steps, which are reviewed in the table below. Tell what happens in each step and which molecules are involved.

Step	Process	Molecules involved
1. Formation of hnRNA		
2. Removal of introns		
3. mRNA to cytoplasm		
4. Activation of tRNA		
5. Initiation		
6. Elongation		
7. Termination		

22.17 A **mutation** (Sec. 22.13) is a change in the base sequence in a gene that is reproduced during DNA replications. In a **point mutation** (Sec. 22.13) one nucleotide is substituted for another; only one triplet is affected. Point mutation may change the primary protein structure, terminate protein synthesis, or there may be no change. A **frameshift mutation** (Sec. 22.13) inserts or deletes a base in a DNA sequence; the effect is to change all triplets that follow in the sequence.

Predict the effect on amino acid identity of each of the following point mutations:

Change in (3′ to 5′) DNA template strand	Change in DNA informational strand	mRNA (codon) change	Amino acid change
TTG → TTC			
AGT → AGC			
AAC → ATC			

Self-Test

True-false: Indicate whether the following statements are true or false. If the statement is false, give the word or phrase that may be substituted for the underlined portion to make the statement true.

1. In DNA the amount of adenine is equal to the amount of <u>guanine</u>.
2. In DNA strands, a phosphate ester bridge connects hydroxyl groups on the 3′ and 5′ positions of the <u>sugar</u> units.
3. Nearly all of the DNA of a cell occurs <u>within the cell nucleus</u>.
4. <u>Transfer RNA</u> molecules carry the genetic code from DNA to the ribosomes.
5. A <u>gene</u> is an individual DNA molecule bound to a group of proteins.
6. The two strands of a DNA molecule are connected to each other by <u>hydrogen bonds</u> between the base units.
7. The process by which a DNA molecule forms an exact duplicate of itself is called <u>transcription</u>.

8. Heterogeneous nuclear RNA is edited under the direction of snRNA and joined together to form messenger RNA.
9. Two different codons that specify the same amino acid are called synonyms.
10. Different species of organisms usually have the same genetic code for an amino acid.
11. A single mRNA molecule can serve as a codon sequence for the synthesis of one protein molecule at a time.
12. Mutagens are agents that cause a change in the structure of a gene.
13. Viruses are tiny disease-causing agents composed of a protein coat and a glycogen core.
14. Viruses can reproduce in water with dissolved organic nutrients.
15. A virus that contains RNA rather than DNA is called a retrovirus.
16. In DNA replication the two daughter strands form in opposite directions.
17. Antimetabolites are cancer drugs that inhibit RNA replication.
18. The concept of alternative splicing explains the large difference between the human genome and the human transcriptome.
19. The insertion or deletion of a base in a DNA sequence is called a point mutation.

Multiple choice:

20. The codon 5′ UGC 3′ would have as its anticodon:

 a. 5′ GCA 3′ b. 5′ UAG 3′ c. 5′ GAC 3′
 d. 5′ AUC 3′ e. none of these

21. Sections of DNA that carry noncoding base sequences are called:

 a. introns b. exons c. codons
 d. anticodons e. none of these

22. Fifteen nucleotide units in a DNA molecule can contain the code for no more than:

 a. 3 amino acids b. 5 amino acids c. 10 amino acids
 d. 15 amino acids e. none of these

23. Which of the following types of molecules does *not* carry information for protein synthesis?

 a. DNA b. ribosomal RNA c. messenger RNA
 d. transfer RNA e. none of these

24. A sequence of three nucleotides in an mRNA molecule is a(n):

 a. exon b. intron c. codon
 d. anticodon e. none of these

25. The intermediary molecules that deliver amino acids to the ribosomes are:

 a. snRNA b. tRNA c. rRNA
 d. mRNA e. none of these

26. The process of inserting recombinant DNA into a host cell is:

 a. translation b. transformation c. transcription
 d. translocation e. none of these

27. Cells that have descended from a single cell and have identical DNA are called:

a. mutagens b. mutations c. clones
d. plasmids e. none of these

Answers to Practice Exercises

22.1

Nucleic acid	Location	Function
DNA	cell nucleus	storage and transfer of genetic information
RNA	all parts of cells	synthesis of proteins

22.2

Nucleotide components	RNA	DNA
pentose	ribose	deoxyribose
purine bases	adenine, guanine	adenine, guanine
pyrimidine bases	uracil, cytosine	cytosine, thymine

22.3

Name of nucleotide	Abbreviation	Base	Sugar
deoxyadenosine 5′-monophosphate	dAMP	adenine	deoxyribose
deoxythymidine 5′-monophosphate	dTMP	thymine	deoxyribose
guanosine 5′-monophosphate	GMP	guanine	ribose
cytidine 5′-monophosphate	CMP	cytosine	ribose

22.4

Base + Sugar	Nucleoside	Nucleotide abbreviation	DNA or RNA
thymine + deoxyribose	deoxythymidine	dTMP	DNA
uracil + ribose	uridine	UMP	RNA
adenine + ribose	adenosine	AMP	RNA
guanine + deoxyribose	deoxyguanosine	dGMP	DNA

22.5

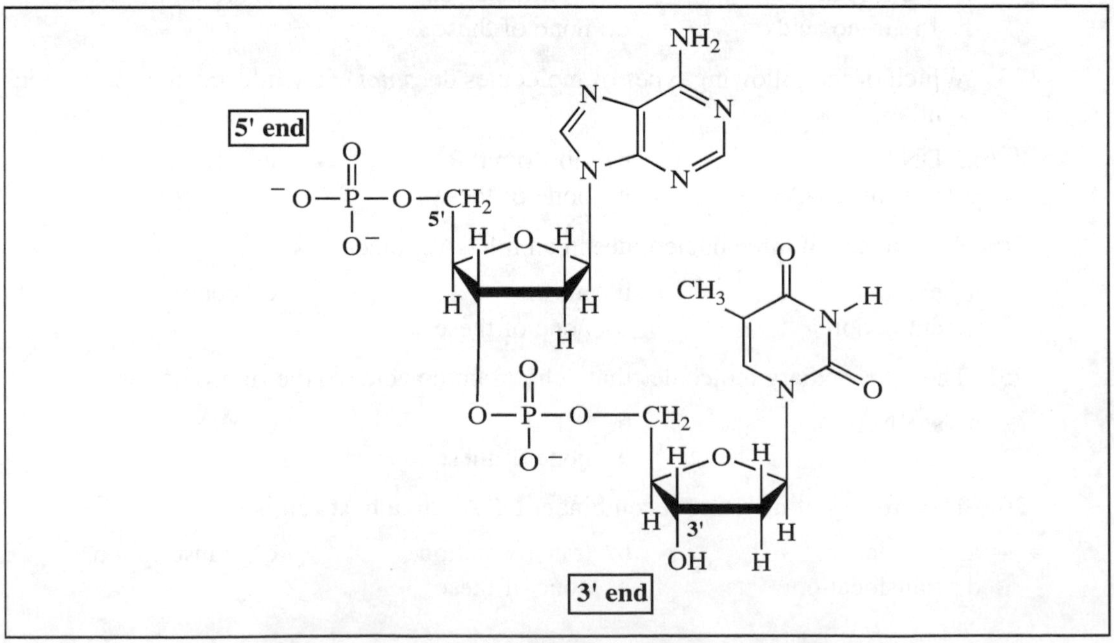

22.6 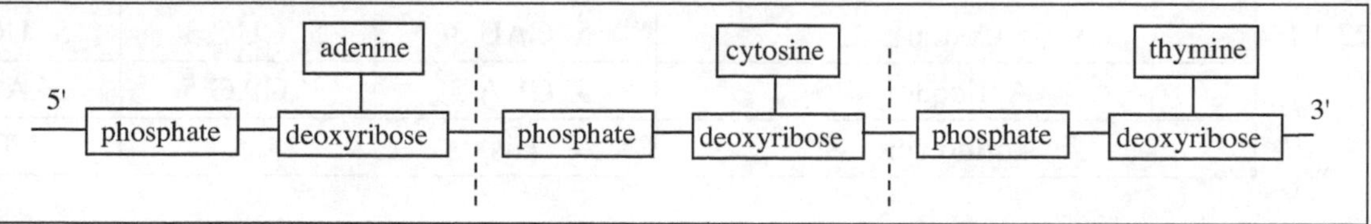

22.7 We know that %A = %T and %G = %C, because these bases are paired in DNA. If %A = 20%, then %T = 20%. %A + %T = 40%, so %G + %C = 60% and %G = %C = 30%.

22.8
```
5′ T – C – A – C – A – G – T – A  3′
   ‖   |||   ‖   |||   ‖   |||   ‖   ‖
3′ A – G – T – G – T – C – A – T  5′
```

22.9 a. 3′ A–T–T–C–G–C–A–C–C 5′ b. 5′ C–C–A–C–G–C–T–T–A 3′

22.10

Type of RNA	Abbreviation	Function
heterogeneous nuclear RNA	hnRNA	material from which mRNA is formed during transcription
small nuclear RNA	snRNA	governs the process of splicing hnRNA to form mRNA
messenger RNA	mRNA	carries genetic information from the nucleus to the ribosomes
ribosomal RNA	rRNA	combines with proteins to form ribosomes, the sites for protein synthesis
transfer RNA	tRNA	delivers individual amino acids to ribosomes for protein synthesis

22.11
```
     A     B        C         D     E
5′ G–C–C–T–G–T–A–C–T–T–C–G–A–T–T–G–G–A 3′        DNA
3′ C–G–G–A–C–A–U–G–A–A–G–C–U–A–A–C–C–U 5′        hnRNA
3′ C–G–G–U–G–A–A–G–C–C–C–U 5′                    mRNA
```

22.12

Codon	Amino acid
UCA	serine
AAU and AAC	asparagine
GAC	aspartic acid

Codon	Amino acid
AUG	methionine
GAU	aspartic acid
UGG	tryptophan

b. Yes; GAU and GAC both code for aspartic acid, and AAU and AAC both code for asparagine.

c. ATC is not listed as a codon because RNA does not contain thymine.

22.13 a. 5′ G–G–U–C–C–C–C–U–U 3′ is one possible answer.

b. No, because there is more than one codon for most amino acids.

22.14

Codon	5′ CAU 3′	5′ CUC 3′	5′ UGG 3′
Anticodon	3′ GUA 5′	3′ GAG 5′	3′ ACC 5′
Amino acid	His	Leu	Trp

22.15 5′ U–C–C–C–G–A–A–G–U–G–G–C 3′ mRNA
Ser–Arg–Ser–Gly tetrapeptide

22.16

Step	Process	Molecules involved*
1. Formation of hnRNA	DNA unwinds, acts as template for hnRNA formation	DNA, hnRNA, nucleotides
2. Removal of introns	hnRNA strand cut, introns removed, mRNA bonds form (splicing)	hnRNA, snRNA, mRNA
3. mRNA to cytoplasm	mRNA moves out of the nucleus into the cytoplasm	mRNA
4. Activation of tRNA	tRNA attaches to an amino acid and becomes energized	tRNA, amino acid, ATP
5. Initiation	mRNA attaches to ribosome, tRNA with amino acid moves to first codon	mRNA, tRNA with attached amino acid, rRNA
6. Elongation	more tRNA molecules move to next codons, polypeptide chain transfers to each new tRNA	mRNA, tRNA with attached amino acid, rRNA
7. Termination	stop codon appears on mRNA, peptide chain (protein) is cleaved from tRNA	mRNA, tRNA, protein

*Enzymes are also involved in each step of the processes.

22.17

Change in (3′ to 5′) DNA template strand	Change in DNA informational strand	mRNA (codon) change	Amino acid change
TTG → TTC	AAC → AAG	AAC → AAG	Asn → Lys
AGT → AGC	TCA → TCG	UCA → UCG	Ser → Ser (no change)
AAC → ATC	TTG → TAG	UUG → UAG	Leu → stop codon

Answers to Self-Test

The numbers in parentheses refer to sections in your textbook.
1. F; thymine (22.5) **2**. T (22.3) **3**. T (22.1) **4**. F; Messenger RNA (22.8)
5. F; chromosome (22.6, 22.9) **6**. T (22.5) **7**. F; replication (22.6) **8**. T (22.8) **9**. T (22.10)
10. T (22.10) **11**. F; many protein molecules at a time (22.12) **12**. T (22.13)
13. F; DNA or RNA core (22.14) **14**. F; only in cells of living organisms (22.14) **15**. T (22.14)
16. T (22.6) **17**. F; DNA replication (22.6) **18**. T (22.9) **19**. F; frameshift mutation (22.13)
20. a (22.12) **21**. a (22.9) **22**. b (22.10) **23**. b (22.8) **24**. c (22.10) **25**. b (22.8) **26**. b (22.15)
27. c (22.15)

Biochemical Energy Production — Chapter 23

Chapter Overview

The most important job of the body's cells is the production of energy to be utilized in carrying out all the complex processes known as life. The production and use of energy by living organisms involve an important intermediate called ATP.

In this chapter you will study the formation of acetyl CoA from the products of the digestion of food, as well as the further oxidation of acetyl CoA during the individual steps of the citric acid cycle. You will study the function and processes of the electron transport chain and the important role of ATP in energy transfer and release.

Practice Exercises

23.1 **Metabolism** (Sec. 23.1) is the sum total of all the chemical reactions of a living organism. It may be divided into **catabolism** (in which molecules are broken down) and **anabolism** (in which molecules are put together, Sec. 23.1). Complete the following table giving the type of reaction and classifying the process as anabolic or catabolic.

Chemical process	Reaction type	Anabolic or catabolic?
glucose → glycogen	esterification	
polypeptides → amino acids		
glucose → CO_2 + H_2O		
fatty acids → TAGs		
RNA → nucleotides		

23.2 Metabolic reactions take place in various locations within the cell. A eukaryotic cell, in which DNA is inside a membrane-enclosed nucleus, is shown below. Write the letters from the diagram next to the appropriate descriptions below the diagram.

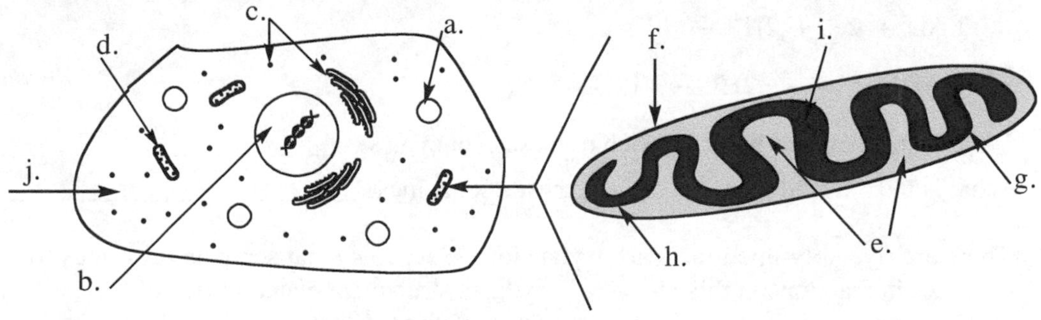

Cell:

ribosome ________	DNA replication ________
nucleus __________	protein synthesis ________
lysosome _________	energy production ________
mitochondrion_____	hydrolytic enzymes ______
cytosol __________	

Mitochondrion:

ATP synthase complex _____
matrix ____________________
inner membrane ___________
intermembrane space _______
outer membrane ___________

23.3 The catabolic reactions that convert food to energy involve several important nucleotide-containing compounds. To show how these compounds are structurally related to one another, complete the block diagrams below. Use information from Section 23.3 of your textbook.

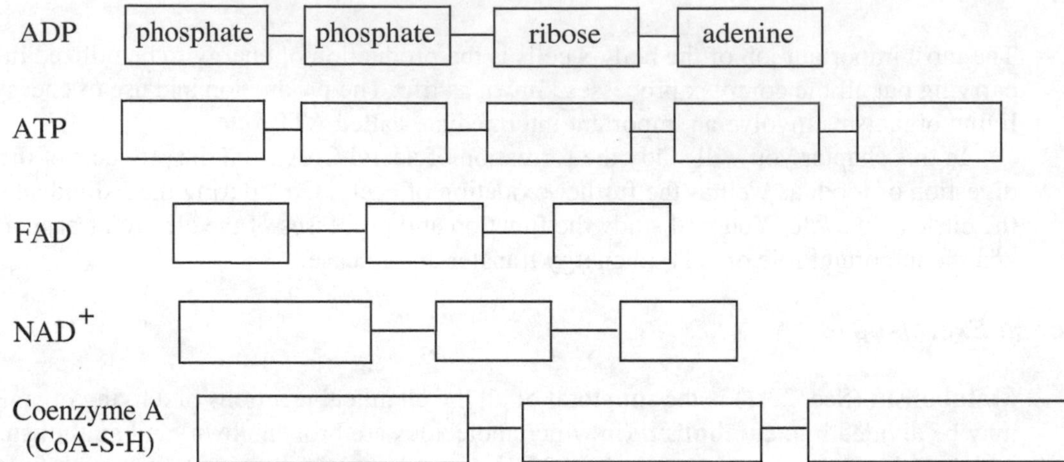

23.4 The bond between two phosphoryl (PO_3^{2-}) groups is a **phosphoanhydride bond** (Sec. 23.3). The two phosphoanhydride bonds in ATP and the one in ADP are very reactive bonds that release a large amount of free energy when they are broken by hydrolysis. Compounds that contain these very reactive bonds are referred to as **high energy compounds** (Sec. 23.3).

Complete the equations below showing the hydrolysis (energy-releasing reactions) of ATP.

ATP + H_2O → ? + P_i + energy

? + H_2O → AMP + P_i + ?

23.5 Two coenzymes, FAD and NAD^+, participate in redox reactions; they transport hydrogen atoms and electron pairs. Each of these coenzymes exists in an oxidized (low energy) form and a reduced (high energy) form. Complete the reactions below.

FAD + $2e^-$ + $2H^+$ →

NAD^+ + $2e^-$ + $2H^+$ →

What is the oxidized form for each of these coenzymes? ______________________________

What is the reduced form for each of these coenzymes? ______________________________

23.6 There are five polyfunctional carboxylate ions (Sec. 23.4) that serve as substrates for enzymes in reactions in this chapter. They have structures related to two dicarboxylic acids: three are related to succinic acid and two are related to glutaric acid. In the spaces below, draw the structures and name the carboxylate ions related to the two dicarboxylic acids.

$HOOC-CH_2-CH_2-COOH$ succinic acid			

$HOOC-CH_2-CH_2-CH_2-COOH$ glutaric acid		

23.7 Several phosphate-containing compounds found in metabolic pathways are known as **high-energy compounds** (Sec. 23.5). Structures of high-energy compounds are given in Table 23.1 in your textbook. How many high energy bonds do each of the following molecules have? ATP ______ ADP ________ AMP _________
Draw the structure of ATP (adenosine triphosphate) using R to represent adenosine and a squiggle (~) to represent each of the two high-energy (strained) bonds.

23.8 The catabolism of food begins with digestion. Products of digestion are absorbed and further broken down to release their stored energy. Complete the following table for the different stages (Sec. 23.6) of biochemical energy production by the catabolism of food.

Stage of catabolism	Where process occurs	Major products
1. digestion		
2. acetyl group formation		
3. citric acid cycle		
4. electron transport chain and oxidative phosphorylation		

The reactions in steps 3 and 4 above are the same for all types of foods; these reactions (the citric acid cycle, the electron transport chain and oxidative phosphorylation) are together referred to as the **common metabolic pathway** (23.6)

23.9 A metabolic pathway may be linear or cyclic. The **citric acid cycle** (Sec. 23.7) is the series of reactions in which the acetyl group of acetyl CoA is oxidized to CO_2, and $FADH_2$ and NADH are produced. Complete the following table summarizing the steps of the citric acid cycle. Use the discussion and equations in Section 23.7 of your textbook.

Step	Type of reaction	Final product(s)	# of C	Energy transfer intermediates
1.	condensation		C_6	none
2.		isocitrate		
3.				NADH
4.				
5.				
6.				
7.				
8.				

23.10 Four of the steps summarized in Practice Exercise 23.9 involve oxidation. For each of these steps, give the name and/or symbol for the substance oxidized and for the one reduced.

Step	Substance oxidized	Substance reduced
3.		
4.	α-ketoglutarate	
6.		FAD (to $FADH_2$)
8.		

23.11 The **electron transport chain** (Sec. 23.8) is a series of reactions in which electrons and hydrogen ions from NADH and $FADH_2$ are passed to intermediate carriers and ultimately react with molecular oxygen to produce water. Enzymes and electron carriers for the ETC are bound to the inner mitochondrial membrane in four distinct protein complexes. Coenzyme Q and cyt c move freely as electron carriers between the four complexes. Name the complexes in the spaces below.

Complex I: ______________________________

Complex II: ______________________________

Complex III: ______________________________

Complex IV: ______________________________

23.12 Each protein complex in the ETC has a different function. Use the summary of the electron transport chain given in Section 23.8 in your textbook.

a. Complete the following table.

Protein complex	Compound carrying electrons to complex	Compound transferring electrons out of complex
Complex I		
Complex II		
Complex III		
Complex IV		

b. What molecule is the final electron acceptor in the ETC? ____________________

c. What happens to O_2 from respiration that is not used in the ETC, and what problems can this cause? ______________________________

____

23.13 **Oxidative phophorylation** (Sec. 23.9 and 23.10) is the process by which ATP is synthesized from ADP using energy released in the ETC. Each NADH from the citric acid cycle produces 2.5 ATP molecules, each $FADH_2$ produces 1.5 ATP molecules, and each GTP produces 1 ATP molecule. Using Practice Exercise 23.9, add up the number of ATP moles per mole of acetyl CoA produced in one turn of the citric acid cycle (Steps 1 through 8).

Step	Energy-rich compound formed	Moles of ATP produced
3.		
4.		
5.		
6.		
8.		
	Total:	

23.14 There are four vitamins that function as coenzymes in the reactions of the common metabolic pathway. Name the four vitamins and give the names of the coenzymes formed.

Vitamin	Coenzyme

Self-Test

True-false: Indicate whether the following statements are true or false. If the statement is false, give the word or phrase that may be substituted for the underlined portion to make the statement true.

1. Catabolic reactions usually release energy.
2. Bacterial cells do not contain a nucleus and so are classified as eukaryotic.
3. Ribosomes are organelles that have a central role in the production of energy.
4. The active portion of the FAD molecule is the flavin subunit.
5. NAD^+ is the oxidized form of nicotine adenine dinucleotide.
6. The citric acid cycle occurs in the cytoplasm of cells.
7. In the first step of the citric acid cycle, oxaloacetate reacts with glucose.
8. The electron transport chain receives electrons and hydrogen ions from NAD^+ and FAD.
9. The electrons that pass through the electron transport chain gain energy in each transfer along the chain.
10. Cytochrome *c* molecules contain iron atoms that are reversibly oxidized and reduced.
11. ATP is produced from ADP using energy released in the electron transport chain.
12. Cytochrome *c* is the link between energy production and energy use in the cell.
13. Every acetyl CoA catabolized by the citric acid cycle ultimately produces 10 molecules of ATP.
14. The gas hydrogen cyanide (HCN) is deadly because it inhibits the enzyme phosphatase.
15. The bond between two phosphate groups in ATP and ADP is a phosphoanhydride bond.

16. A high-energy bond requires <u>less energy</u> to break than a normal bond.

Multiple choice:

17. The "fuel" for the citric acid cycle is:

a. acetyl CoA b. coenzyme Q c. cytochrome c
d. FAD e. none of these

18. Which of the following is *not* an organelle?

a. mitochondrion b. hemoglobin c. ribosome
d. lysosome e. none of these

19. Which of these molecules is *not* part of the electron transport chain?

a. coenzyme Q b. acetyl CoA c. cytochrome b
d. cytochrome c e. none of these

20. Energy used in cells is obtained directly from:

a. oxidation of NADH b. activation of acetyl CoA c. oxidation of $FADH_2$
d. hydrolysis of ATP e. none of these

21. The final acceptor of electrons in the electron transport chain is:

a. NAD^+ b. water c. oxygen
d. FAD e. none of these

22. The conversion of ATP to ADP is a part of:

a. the citric acid cycle b. oxidative phosphorylation
c. the electron transport chain d. energy use in the cells
e. none of these

23. Which B-vitamin is one of the subunits of coenzyme A?

a. niacin b. thiamine c. pantothenic acid
d. riboflavin e. biotin

Answers to Practice Exercises

23.1

Chemical process	Reaction type	Anabolic or catabolic?
glucose → glycogen	esterification	anabolic
peptides → amino acids	hydrolysis	catabolic
glucose → $CO_2 + H_2O$	oxidation	catabolic
fatty acids → TAGs	esterification	anabolic
RNA → nucleotides	hydrolysis	catabolic

23.2

Cell:

ribosome ______c.
nucleus _______b.
lysosome ______a.
mitochondrion__d.
cytosol _______j.

DNA replication ___b.
protein synthesis ___c.
energy production ___d.
hydrolytic enzymes __a.

Mitochondrion:

ATP synthase complex __g.
matrix __________i.
inner membrane ____h.
intermembrane space ____e.
outer membrane ______f.

23.3

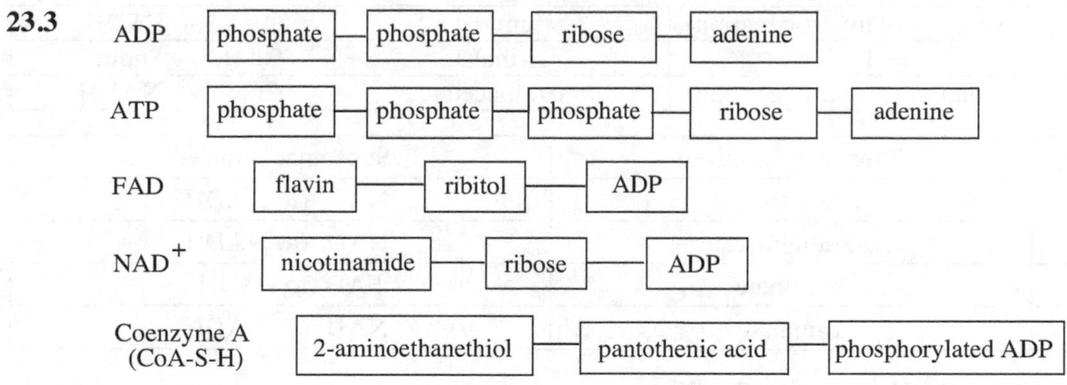

23.4 $ATP + H_2O \rightarrow ADP + P_i + \text{energy}$; $ADP + H_2O \rightarrow AMP + P_i + \text{energy}$

23.5 $FAD + 2e^- + 2H^+ \rightarrow FADH_2$ FAD – oxidized form; $FADH_2$ – reduced form.
$NAD^+ + 2e^- + 2H^+ \rightarrow NADH + H^+$ NAD^+ – oxidized form; NADH – reduced form

23.6

$HOOC-CH_2-CH_2-COOH$	$^-OOC-CH(OH)-CH_2-COO^-$	$^-OOC-C(=O)-CH_2-COO^-$	$H(^-OOC)C=C(H)COO^-$
succinic acid	malate	oxaloacetate	fumarate

$HOOC-CH_2-CH_2-CH_2-COOH$	$^-OOC-C(=O)-CH_2-CH_2-COO^-$	$^-OOC-CH_2-C(OH)(COO^-)-CH_2-COO^-$
glutaric acid	α-ketoglutarate	citrate

23.7 ATP – 2; ADP – 1; AMP – 0

$$R-O-\overset{O}{\overset{\|}{\underset{O^-}{\underset{|}{P}}}}-O\sim\overset{O}{\overset{\|}{\underset{O^-}{\underset{|}{P}}}}-O\sim\overset{O}{\overset{\|}{\underset{O^-}{\underset{|}{P}}}}-O^-$$

23.8

Stage of catabolism	Location	Major products
1. digestion	mouth, stomach, small intestine	glucose, fatty acids and glycerol, amino acids
2. acetyl group formation	cytoplasm of cell and inside mitochondria	acetyl CoA , NADH
3. citric acid cycle	inside mitochondria	CO_2, NADH, and $FADH_2$
4. ETC and oxidative phosphorylation	inside mitochondria	water and ATP

23.9

Step	Type of reaction	Final product(s)	# of C	Energy transfer intermediates
1.	condensation, hydrolysis	citrate, free coenzyme A	C_6	none
2.	isomerization	isocitrate	C_6	none
3.	oxidation, decarboxylation	α-ketoglutarate, CO_2	C_5	NADH
4.	oxidation, decarboxylation	succinyl CoA, CO_2	C_4	NADH
5.	phosphorylation	succinate, free coenzyme A	C_4	GTP

6.	oxidation (dehydrogenation)	fumarate	C_4	$FADH_2$
7.	hydration	L-malate	C_4	none
8.	oxidation (dehydrogenation)	oxaloacetate	C_4	NADH

23.10

Step	Substance oxidized	Substance reduced
3.	isocitrate	NAD^+ (to NADH)
4.	α-ketoglutarate	NAD^+ (to NADH)
6.	succinate	FAD (to $FADH_2$)
8.	L-malate	NAD^+ (to NADH)

23.11 Complex I: NADH-coenzyme Q reductase
Complex II: Succinate-coenzyme Q reductase
Complex III: Coenzyme Q-cytochrome c reductase
Complex IV: Cytochrome c oxidase

23.12 a.

Protein complex	Compound carrying electrons to complex	Compound transferring electrons out of complex
Complex I	NADH	$CoQH_2$
Complex II	$FADH_2$	$CoQH_2$
Complex III	$CoQH_2$	cyt c
Complex IV	cyt c	H_2O

b. O_2; c. O_2 can be converted to reactive oxygen species (ROS), such as hydrogen peroxide and oxygen radicals. The ROS react readily with biomolecules and can upset cellular activity.

23.13

Step	Energy-rich compound formed	Number of ATPs produced
3.	NADH	2.5
4.	NADH	2.5
5.	GTP	1
6.	$FADH_2$	1.5
8.	NADH	2.5
	Total:	10

23.14

Vitamin	Coenzyme
Niacin (nicotinamide)	NAD^+, NADH (nicotine adenine dinucleotide)
Riboflavin	FAD, $FADH_2$ (flavin adenine dinucleotide)
Thiamine	TPP (thiamine pyrophosphate)
Pantothenic Acid	CoA (Coenzyme A)

Answers to Self-Test

The numbers in parentheses refer to sections in your textbook.
1. T (23.1) **2**. F; prokaryotic (23.2) **3**. F; Mitochondria (23.2) **4**. T (23.3) **5**. T (23.3) **6**. F; mitochondria (23.6) **7**. F; acetyl CoA (23.7) **8**. F; NADH and $FADH_2$ (23.7) **9**. F; lose energy (23.8) **10**. F; copper (23.8) **11**. T (23.9) **12**. F; ATP (23.11) **13**. T (23.10) **14**. F; the ETC enzyme cytochrome *c* oxidase (23.8) **15**. T (23.3) **16**. T (23.5) **17**. a (23.3) **18**. b (23.2) **19**. b (23.8) **20**. d (23.11) **21**. c (23.8) **22**. d (23.11) **23**. c (23.13)

Carbohydrate Metabolism — Chapter 24

Chapter Overview

The complete oxidation of glucose supplies the energy needed by cells to carry out their many vital functions. This oxidation takes place in a number of steps that conserve the energy contained in the chemical bonds of glucose and transfer it efficiently.

In this chapter you will study the reactions of glycolysis to produce pyruvate, the pathways of pyruvate under aerobic and anaerobic conditions, and the pentose phosphate pathway. You will study the ways in which glycogen is synthesized in the body and broken down and the hormones that control these processes.

Practice Exercises

24.1 Carbohydrate **digestion** (Sec. 24.1) is the process in which carbohydrates are broken down through hydrolysis into units small enough to be absorbed into the bloodstream. Complete the table below showing the digestive processes for carbohydrates.

Digestion site	Enzymes/Process	Effect of enzyme action
mouth		
stomach		
small intestine		
instestinal mucosal cells		
villi		

24.2 **Glycolysis** (Sec. 24.2) is the metabolic pathway by which glucose is converted into two molecules of pyruvate and ATP and NADH are produced. Glycolysis is a linear pathway that functions in the cytosol of almost all cells. Using the steps of glycolysis discussed in Section 24.2 of your textbook, give the number(s) of the reaction steps for each of the following:

a. Where are ATPs used? __________

b. Where are ATPs produced? __________

c. Where is NAD^+ reduced? __________

d. Where is the carbon chain split? __________

e. Where is a ketone isomerized to an aldehyde? __________

f. Where are phosphate groups added to sugar molecules? __________

g. Where does **substrate level phosphorylation** (Sec. 24.2) take place? __________

h. Where is water lost? __________

i. What is the net gain in ATPs and in NADH for each glucose molecule converted to pyruvate? __________

24.3 The pyruvate produced by glycolysis is converted to three different products according to the availability of oxygen and the type of organism.

a. Complete the following table on the various fates of pyruvate.

Conditions	Name of process	Name of product	Number of NADH used or produced
1. aerobic			
2. anaerobic (humans)			
3. anaerobic (yeasts)			

b. Why do the two fermentation processes require NADH?

24.4 You have studied the many reactions involved in the complete oxidation of one glucose molecule. Below is a summary of the reactions in each of the four stages of this oxidation. Because of the complexity of the processes, the "equations" are unbalanced. To get an overview of the process, work with them as you would with ordinary balanced equations.

1. Add the equations together by crossing out molecules that occur on both the reactant side of one equation and the product side of another.
2. Write the molecules that remain as the final net equation for the entire process. Balance each type of atom in the final equation.
3. Sum up the ATPs produced.

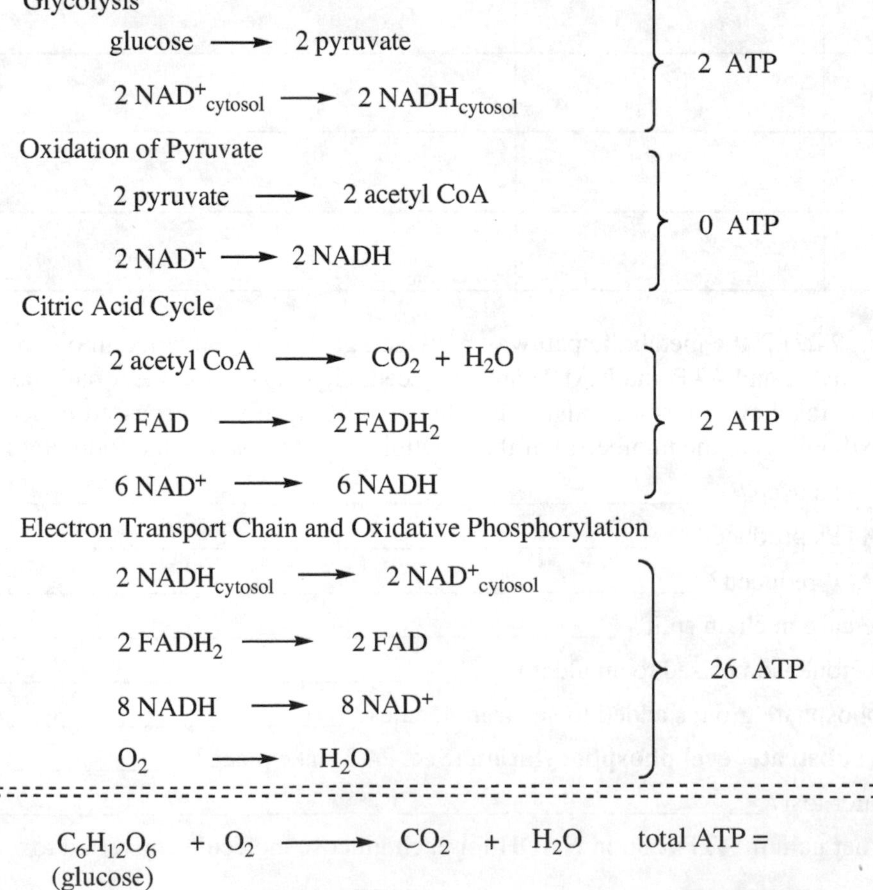

24.5 The very different metabolic processes of **glycogenesis** (Sec. 24.5), **glycogenolysis** (Sec. 24.5), and **gluconeogenesis** (Sec. 24.6) have like-sounding names. Fill in the following table showing the differences among these processes.

Name of process	What the process accomplishes	Where the process takes place	High-energy phosphate molecules used
glycogenesis			
glycogenolysis			
gluconeogenesis			

24.6 Another pathway by which glucose may be degraded is the **pentose phosphate pathway** (Sec. 24.8). What are the two main functions of the pentose phosphate pathway?

1.
2.

24.7 Three hormones affect carbohydrate metabolism. Complete the table below comparing these hormones.

Hormone	Source	Effect
insulin		
glucagon		
epinephrine		

24.8 Review the following terms introduced in this chapter by matching each with its correct description.

Answer	Term	Description
	insulin	1. lactate → pyruvate → glucose
	glycogenesis	2. stimulates: glycogen (liver) → glucose
	gluconeogenesis	3. glucose → glycogen
	glycolysis	4. stimulates: glycogen (muscle) → glucose-6-phosphate
	epinephrine	5. glycogen → glucose or glucose-6-phosphate
	glycogenolysis	6. stimulates: glucose (blood) → glucose (cells)
	glucagon	7. glucose → pyruvate

Self-Test

True-false: Indicate whether the following statements are true or false. If the statement is false, give the word or phrase that may be substituted for the underlined portion to make the statement true.

1. The main site for carbohydrate digestion is <u>the small intestine</u>.
2. Glycolysis takes place in the <u>cytosol</u> of cells.
3. The process by which glucose is degraded to ethanol is called <u>fermentation</u>.
4. The enzymes maltase, sucrase, and lactase convert <u>starch to disaccharides</u>.
5. Substrate-level phosphorylation is the direct transfer of <u>phosphate ions in solution</u> to ADP molecules to produce ATP.

6. The absorption of the monosaccharides glucose, galactose, and fructose through the intestinal walls into the bloodstream is by active transport.
7. Fructose and galactose are converted, in the liver, to intermediates that enter the glycolysis pathway.
8. In human metabolism, under anaerobic conditions, pyruvate is reduced to acetyl CoA.
9. The complete oxidation of glucose in skeletal muscle and nerve cells yields 30 ATP molecules per glucose molecule.
10. When glycogen stored in muscle and liver tissues is depleted by strenuous exercise, glucose molecules can be synthesized by the process of gluconeogenesis.
11. Glucagon speeds up the rate of glycogenolysis in the muscle cells.
12. The breakdown of glycogen in order to maintain normal glucose levels in the bloodstream is called glycolysis.
13. The pentose phosphate pathway metabolizes glucose to produce ribose and other sugars needed for biosynthesis.
14. Type 1 diabetes is the result of inadequate insulin production by the liver.
15. The process in which glucose is converted to lactate in muscle tissue, and the lactate is reconverted to glucose in the liver is called the Cori cycle.

Multiple choice:

16. The metabolic pathway converting glucose to two molecules of pyruvate is:
    a. glycogenolysis b. glycogenesis c. glycolysis
    d. gluconeogenesis e. none of these

17. The hormone that lowers blood glucose levels is:
    a. glucagon b. insulin c. epinephrine
    d. norepinephrine e. none of these

18. High blood glucose levels trigger the release of:
    a. epinephrine b. glucagon c. insulin
    d. adrenaline e. all of these

19. The lactate produced during strenuous exercise is converted to pyruvate in the:
    a. kidneys b. liver c. muscles
    d. small intestine e. none of these

20. One of the control mechanisms of glycolysis is feedback inhibition of hexokinase by:
    a. glucose b. glucose 6-phosphate c. glucose 1-phosphate
    d. fructose 6-phosphate e. none of these

21. During the complete oxidation of glucose, the dihydroxyacetone phosphate-glycerol 3-phosphate shuttle system provides transport from the cytosol into the mitochondrial membrane for:
    a. electrons from NADH b. electrons from NAD^+ c. electrons from FAD
    d. electrons from $FADH_2$ e. none of these

22. The two B-vitamins that are *not* involved in carbohydrate metabolism are:
    a. biotin and folate b. vitamin B_{12} and folate c. vitamin B_6 and biotin
    d. vitamin B_6 and thiamine e. thiamine and vitamin B_{12}

Answers to Practice Exercises

24.1

Digestion site	Enzymes/Process	Effect of enzyme action
mouth	salivary α-amylase	some hydrolysis of α-glycosidic linkages in starch and glycogen
stomach	no enzymes for carbohydrate digestion	no change
small intestine	pancreatic α-amylase	polysaccharides hydrolyzed to disaccharides (lactose, sucrose, maltose)
instestinal mucosal cells	maltase, sucrase, lactase	disaccharides hydrolyzed to glucose, galactose, fructose (monosaccharides)
villi	active transport across intestinal lining (ATP is used)	monosaccharides enter the bloodstream

24.2 a. Steps 1 and 3; b. Steps 7 and 10; c. Step 6; d. Step 4; e. Step 5; f. Steps 1, 3, and 6; g. Steps 7 and 10; h. Step 9; i. two ATPs and two NADHs

24.3 a.

Conditions	Name of process	Product	Number of NADH
1. aerobic	oxidation	acetyl CoA	1 produced
2. anaerobic (humans)	lactate fermentation	lactate	1 used
3. anaerobic (yeasts)	ethanol fermentation	ethanol	1 used

b. Fermentation is a reduction, and NADH is the reducing agent.

24.4

Net ATPs

Glycolysis

glucose ⟶ 2 pyruvate

$2\ NAD^{+}_{cytosol} \longrightarrow 2\ NADH_{cytosol}$

} 2 ATP

Oxidation of Pyruvate

2 pyruvate ⟶ 2 acetyl CoA

$2\ NAD^+ \longrightarrow 2\ NADH$

} 0 ATP

Citric Acid Cycle

2 acetyl CoA $\longrightarrow CO_2 + H_2O$

$2\ FAD \longrightarrow 2\ FADH_2$

$6\ NAD^+ \longrightarrow 6\ NADH$

} 2 ATP

Electron Transport Chain and Oxidative Phosphorylation

$2\ NADH_{cytosol} \longrightarrow 2\ NAD^{+}_{cytosol}$

$2\ FADH_2 \longrightarrow 2\ FAD$

$8\ NADH \longrightarrow 8\ NAD^+$

$O_2 \longrightarrow H_2O$

} 26 ATP

$C_6H_{12}O_6$ (glucose) $+ 6\ O_2 \longrightarrow 6\ CO_2 + 6\ H_2O$ total ATP = 30 ATP

24.5

Name of Process	What the process accomplishes	Where the process takes place	High-energy phosphate molecules used
glycogenesis	synthesis of glycogen from glucose 6-phosphate	muscle and liver tissue	2 ATP
glycogenolysis	breakdown of glycogen to glucose 6-phosphate	muscle and brain (glucose-6-phosphate) and liver (free glucose)	none
gluconeogenesis	glucose synthesis from noncarbohydrates	liver	4 ATP, 2 GTP

24.6
1. production of NADPH (a coenzyme needed in lipid synthesis)
2. production of ribose 5-phosphate for synthesis of nucleic acids and coenzymes

24.7

Hormone	Source	Effect
insulin	beta cells of pancreas	increases uptake and utilization of glucose by cells; lowers blood glucose
glucagon	alpha cells of pancreas	increases blood glucose by speeding up glycogenolysis in liver
epinephrine	adrenal glands	stimulates glycogenolysis in muscle cells; gives quick energy to muscle cells

24.8

Answer	Term	Description
6.	insulin	1. lactate → pyruvate → glucose
3.	glycogenesis	2. stimulates: glycogen (liver) → glucose
1.	gluconeogenesis	3. glucose → glycogen
7.	glycolysis	4. stimulates: glycogen (muscle) → glucose-6-phosphate
4.	epinephrine	5. glycogen → glucose or glucose-6-phosphate
5.	glycogenolysis	6. stimulates: glucose (blood) → glucose (cells)
2.	glucagon	7. glucose → pyruvate

Answers to Self-Test

The numbers in parentheses refer to sections in your textbook.
1. T (24.1) **2**. T (24.2) **3**. T (24.3) **4**. F; disaccharides to monosaccharides (24.1)
5. F; high-energy phosphate groups from substrate molecules (24.2) **6**. T (24.1) **7**. T (24.2)
8. F; lactate (24.3) **9**. T (24.4) **10**. T (24.6) **11**. F; Epinephrine (24.9)
12. F; glycogenolysis (24.5) **13**. T (24.8) **14**. F; beta cells of the pancreas (24.9)
15. T (24.6) **16**. c (24.2) **17**. b (24.9) **18**. c (24.9) **19**. b (24.3) **20**. b (24.2)
21. a (24.4) **22**. b (24.10)

Lipid Metabolism — Chapter 25

Chapter Overview

Lipids are the most efficient energy-storage compounds of the body. They are also important materials in membranes. This chapter discusses the biosynthesis and storage of lipids and their degradation to produce energy.

In this chapter you will study the processes of digestion of triacylglycerols and their absorption into the bloodstream, their storage in the body, and their degradation by means of the β-oxidation pathway. You will define ketone bodies and learn the conditions under which they are produced. You will compare fatty acid synthesis to fatty acid oxidation.

Practice Exercises

25.1 Most dietary lipids are triacylglycerols (TAGs). Complete the following diagram showing the stages of digestion and absorption of TAGs (Sec. 25.1). In each box, write the processes that take place at that location.

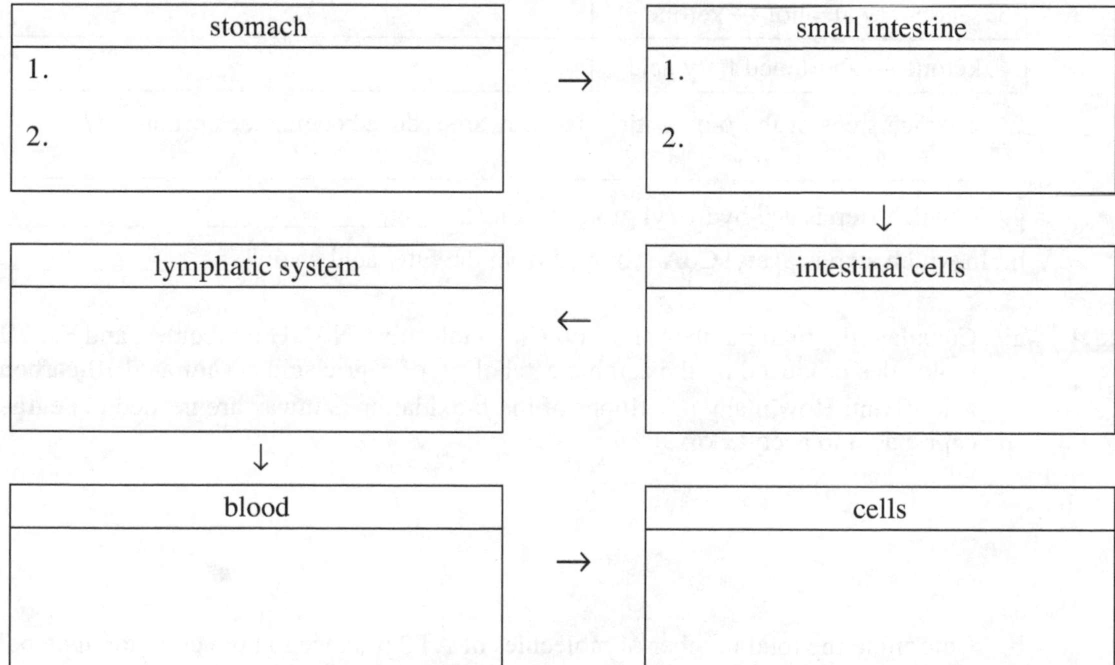

25.2 TAG molecules can be stored in **adipose tissue** (Sec. 25.2) until needed for energy production.

a. What is the process of **triacylglycerol mobilization** (Sec. 25.2)?

b. What happens to the products of triacylglycerol mobilization?

c. In glycerol metabolism, dihydroxyactone is produced. Which two biochemical processes can make use of the dihyroxyacetone?

25.3 The breakdown of fatty acids to produce energy takes place in three stages. Fatty acids are:
1) activated by bonding to coenzyme A.
2) transported through the inner mitochondrial membrane.
3) oxidized in two-carbon segments for each repetition of the ***β*-oxidation pathway** (Sec. 25.4).

Answer the following questions about this degradation process:

a. Where does fatty acid activation take place? ________________

b. How many high-energy phosphate bonds are used in activating the fatty acid? ________

c. What is the activated molecule called? ________________

d. Why is the acyl group transferred from coenzyme A to carnitine? ________________

e. Where does the *β*-oxidation pathway take place? ________________
Complete the table below for the reactions in the *β*-oxidation pathway:

Step	Type of reaction	Coenzyme oxidizing agent
1. alkane → alkene		
2. alkene → secondary alcohol		
3. secondary alcohol → ketone		
4. ketone → shortened fatty acid		

f. In which steps of the *β*-oxidation pathway are reduced coenzymes produced?

____

g. In which step is a *β*-hydroxyl group oxidized? ________________

h. In which step is acetyl CoA removed from the fatty acid chain? ________________

25.4 a. Calculate the total number of acetyl CoA molecules, NADH molecules, and $FADH_2$ molecules produced by the aerobic catabolism of capric acid, a saturated 10-carbon fatty acid. (Hint: How many repetitions of the *β*-oxidation pathway are needed to degrade capric acid to acetyl CoA?)

b. Determine the total number of molecules of ATP produced in the common metabolic pathway by the total oxidation of capric acid to carbon dioxide and water. (Include ATPs produced in the ETC by the molecules of NADH and $FADH_2$. Remember also that ATPs are used up in the activation of the free fatty acid to acyl CoA.)

25.5 Acetyl CoA from the β-oxidation pathway is usually oxidized further through the citric acid cycle; however, under some conditions there is not enough oxaloacetate produced to react with acetyl CoA in the first step of the citric acid cycle. The excess acetyl CoA undergoes **ketogenesis** (Sec. 25.6) producing **ketone bodies** (Sec. 25.6).

a. Name the three ketone bodies produced in the human body.

b. Under what conditions do ketone bodies form?

c. What is **ketosis** (Sec. 25.6)?

d. Which ketone bodies could produce a lowered blood pH (acidosis)?

25.6 **Lipogenesis** (Sec. 25.7) is the synthesis of fatty acids from acetyl CoA.

a. Since lipogenesis takes place in the cytosol, acetyl CoA is transported from the mitochondria to the cytosol in a three-step process involving citrate. Briefly describe each step of this citrate-malate shuttle.

1.

2.

3.

b. What are the two ACP (acyl carrier protein) complexes needed to start lipogenesis? What is the starting material for each?

c. Lipogenesis is not simply the reverse of the β-oxidation pathway for degradation of fatty acids. Complete the table below to compare the two processes.

Comparisons	Degradation of fatty acids	Lipogenesis
reaction site in cell		
carbons lost/gained per turn		
intermediate carriers		
types of enzymes		
coenzymes for energy transfer		

25.7 The chain elongation process of lipogenesis consists of four reactions that occur in a cyclic pattern. Complete the following table summarizing these four reactions.

Step	Type of reaction	Coenzyme reducing agent
1.		
2.		
3.		
4.		

25.8 Answer the following questions about the biosynthesis of capric acid, a 10-carbon saturated fatty acid, from acetyl CoA molecules (Sec. 25.7).

a. How many rounds of the fatty acid biosynthesis pathway are needed? ______________

b. How many molecules of malonyl ACP must be used? ______________

c. How many high-energy ATP bonds are consumed? ______________

d. How many NADPH molecules are needed? ______________

25.9 There are some notable similarities between the C_4 intermediates formed in the first cycle of lipogenesis and the C_4 intermediates in the last four steps of the citric acid cycle (See Fig. 25.13 in your textbook). A major difference is that the lipogenesis intermediates are monoacids rather than diacids as in the citric acid cycle.

a. In the table below, fill in the names of the comparable intermediates in the two cycles.

b. Number the intermediates in each column from 1 to 4 according to the order in which they are produced in their respective cycles (Sec. 25.8).

Type of molecule	Lipogenesis	Citric acid cycle
C_4 keto acid		
C_4 saturated acid		
C_4 unsaturated acid		
C_4 hydroxy acid		

25.10 Acetyl CoA is a key intermediate for many metabolic processes (Sec. 25.9). It is used up in some processes and produced in others. Complete the table below showing the role of acetyl CoA in some of the processes of lipid metabolism.

Process	Substances entering process	Substances produced
lipogenesis		
ketogenesis		
β-oxidation pathway		
citric acid cycle		
glycolysis		

Acetyl CoA cannot be used for the net synthesis of glucose. Why? ______________

____

Self-Test

True-false: Indicate whether the following statements are true or false. If the statement is false, give the word or phrase that may be substituted for the underlined portion to make the statement true.

1. Bile is released into the small intestine where it acts as <u>a hydrolytic enzyme</u> for lipids.
2. A meal high in triacylglycerols will cause the concentration of <u>fatty acid micelles</u> in the blood and lymph to peak in 4 to 6 hours.
3. Adipocytes are triacylglycerol storage cells found mainly in <u>liver tissue</u>.
4. Chylomicrons consist of lipoproteins combined with <u>fatty acids</u> in the intestinal cells.
5. The first stage of glycerol metabolism takes place in the <u>liver and kidneys</u>.
6. Glycerol is metabolized to <u>dihydroxyacetone phosphate</u> before entering glycolysis.
7. A stearic acid molecule produces <u>4 times</u> as much energy as a glucose molecule.
8. Fatty acid activation takes place within the <u>mitochondrial matrix</u>.
9. <u>*β*-Hydroxybutyrate</u> is classified as a ketone body but is not a ketone.
10. The B vitamin that makes up part of the structure of acetyl CoA is <u>biotin</u>.
11. Acetyl CoA molecules can be further oxidized in the <u>*β*-oxidation pathway</u>.
12. Cholesterol biosynthesis takes place in the <u>liver</u>.
13. When glucose is not available to the body, acetyl CoA molecules are used to manufacture <u>pyruvate</u>.
14. The liver uses <u>fatty acids</u> as the preferred fuel.
15. The last turn of the fatty acid spiral produces <u>two</u> acetyl CoA molecules.
16. In <u>ketoacidosis</u>, elevated levels of acetoacetate and *β*–hydroxybutyrate can cause a significant decrease in blood pH.

Multiple choice:

17. Triacylglycerol mobilization is the process in which triacylglycerols are:
    a. oxidized b. hydrolyzed c. synthesized
    d. hydrogenated e. none of these

18. Acetyl CoA may *not* be used to:
    a. synthesize ketone bodies b. synthesize fatty acids c. synthesize glucose
    d. synthesize cholesterol e. none of these

19. Diabetic ketosis results in:
    a. ketonemia b. ketonuria c. metabolic acidosis
    d. a, b, and c e. none of these

20. One repetition of the *β*–oxidation pathway produces:
    a. one NADH and one FAD b. two NADH and one FAD
    c. one NADH and two $FADH_2$ d. one NADH and one $FADH_2$
    e. none of these

21. Which of the following does *not* aid in the digestion of lipids?
    a. pancreatic lipases b. cholecystokinin
    c. salivary enzymes d. churning action of stomach
    e. all of the above are used

22. A fatty acid micelle produced by hydrolysis of lipids contains:
    a. triacylglycerols b. diacylglycerols c. monoacylglycerols
    d. lipoproteins e. none of these

23. A 20-carbon saturated fatty acid will produce 10 acetyl CoA molecules from the β–oxidation pathway after:
    a. 1 turn b. 9 turns c. 10 turns d. 20 turns e. none of these

24. The net number of ATPs produced in the body by complete oxidation of an 18-carbon saturated fatty acid is:
    a. 120 b. 118 c. 152 d. 98 e. none of these

Answers to Practice Exercises

25.1

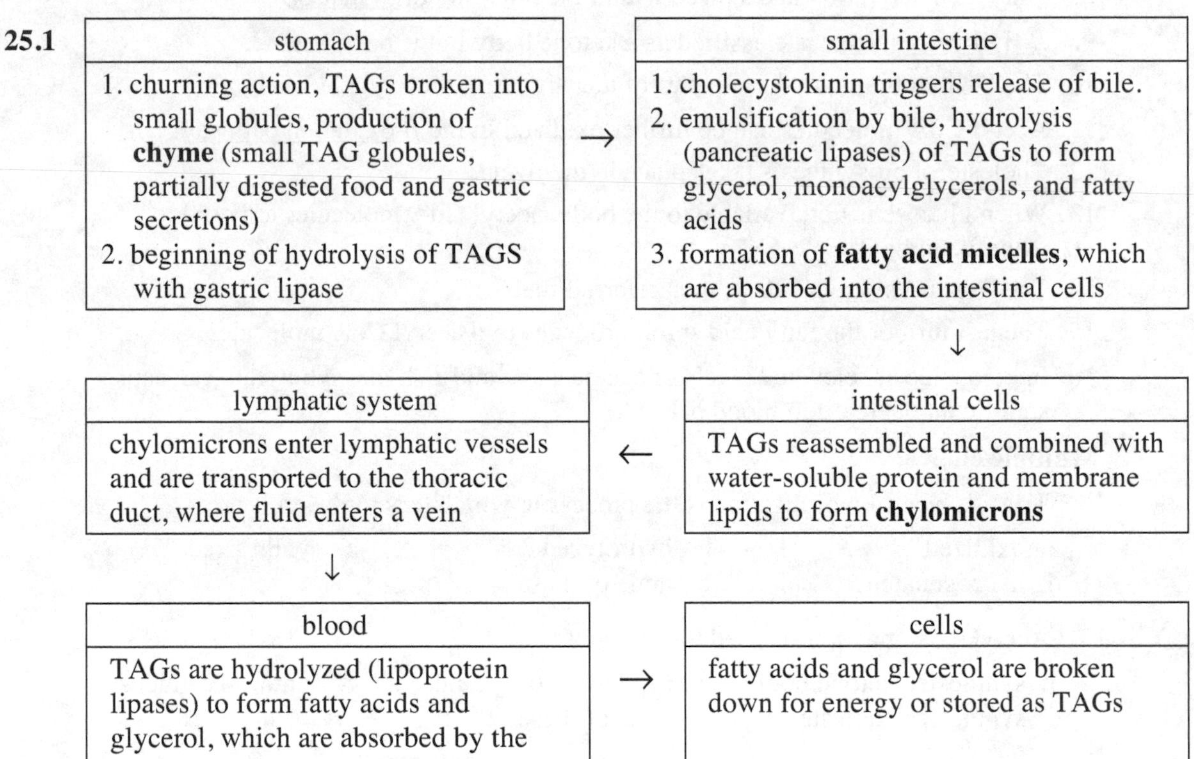

stomach
1. churning action, TAGs broken into small globules, production of **chyme** (small TAG globules, partially digested food and gastric secretions)
2. beginning of hydrolysis of TAGS with gastric lipase
→
small intestine
1. cholecystokinin triggers release of bile.
2. emulsification by bile, hydrolysis (pancreatic lipases) of TAGs to form glycerol, monoacylglycerols, and fatty acids
3. formation of **fatty acid micelles**, which are absorbed into the intestinal cells
↓
intestinal cells
TAGs reassembled and combined with water-soluble protein and membrane lipids to form **chylomicrons**
←
lymphatic system
chylomicrons enter lymphatic vessels and are transported to the thoracic duct, where fluid enters a vein
↓
blood
TAGs are hydrolyzed (lipoprotein lipases) to form fatty acids and glycerol, which are absorbed by the cells
→
cells
fatty acids and glycerol are broken down for energy or stored as TAGs

25.2
a. Triacylglycerol mobilization is the hydrolysis of stored TAGs. The fatty acids and glycerol produced enter the bloodstream.

b. The fatty acids and glycerol are oxidized to produce energy; the fatty acids are oxidized through the β-oxidation pathway, and the glycerol is oxidized in a two-step process that produces dihydroxyacetone phosphate.

c. The two processes are glycolysis and gluconeogenesis.

25.3 a. Fatty acid activation takes place in the outer mitochondrial membrane.
b. Two high-energy phosphate bonds are used. (ATP → AMP)
c. The activated molecule is called acyl CoA.
d. The acyl group is transferred to carnitine because the acyl CoA molecule is too large to pass through the inner mitochondrial membrane.
e. The β-oxidation pathway takes place in the mitochondrial matrix.

Step	Type of reaction	Oxidizing agent
1. alkane → alkene	first dehydrogenation (oxidation)	FAD
2. alkene → secondary alcohol	hydration	none
3. secondary alcohol → ketone	second dehydrogenation (oxidation)	NAD^+
4. ketone → shortened fatty acid	thiolysis (chain cleavage)	none

f. Reduced coenzymes are produced in Step 1 ($FADH_2$) and Step 3 (NADH).
g. A β-hydroxyl group is oxidized in Step 3 (hence, the name "β-oxidation pathway").
h. Acetyl CoA is removed from the fatty acid chain in Step 4.

25.4 a. In four times through the β-oxidation pathway, a 10-carbon fatty acid produces: 5 acetyl CoA molecules, 4 NADH, and 4 $FADH_2$.

b. Citric acid cycle: 5 acetyl CoA x 10 ATP/ 1 acetyl CoA = 50 ATP
Electron transport chain: 4 NADH x 2.5 ATP/ 1 NADH = 10 ATP
4 $FADH_2$ x 1.5 ATP/ 1 $FADH_2$ = 6 ATP
Activation of the fatty acid: –2 ATP (ATP → AMP)
Total: 64 ATP

25.5 a. Acetoacetate, β-hydroxybutyrate, acetone
b. Ketone bodies form from acetyl CoA when there is insufficient oxaloacetate being formed from pyruvate. This would occur when dietary intakes are high in fat and low in carbohydrates, when the body cannot adequately process glucose (as in diabetes), and during prolonged fasting or starvation.
c. Ketosis is the accumulation of ketone bodies in the blood and urine.
d. Acetoacetate and β-hydroxybutyrate are acids and could produce a lowered blood pH.

25.6 a. 1. Mitochondrial acetyl CoA reacts with oxaloacetate to produce citrate.
2. The citrate produced is transported through the inner mitochondrial membrane.
3. In the cytosol, the citrate regenerates the acetyl CoA and oxaloacetate.
b. The two ACP complexes are acetyl-ACP and malonyl-ACP; the starting material for both is cytosolic acetyl CoA.
c.

Comparisons	Fatty acid degradation	Lipogenesis
reaction site in cell	mitochondrial matrix	cytosol
carbons lost/gained per turn	two carbons lost	two carbons gained
intermediate carriers	CoA–SH	ACP–SH (acyl carrier protein)
types of enzymes	independent enzymes	fatty acid synthase complex
coenzymes – energy transfer	FAD and NAD^+	NADPH

25.7

Step	Type of reaction	Coenzyme reducing agent
1.	condensation	none
2.	hydrogenation	NADPH
3.	dehydration	none
4.	hydrogenation	NADPH

25.8 a. 4 rounds b. 4 malonyl ACP molecules
c. 4 ATP bonds d. 8 NADPH molecules

25.9

Type of molecule	Lipogenesis		Citric acid cycle	
C_4 keto acid	acetoacetate	1	oxaloacetate	4
C_4 saturated acid	butyrate	4	succinate	1
C_4 unsaturated acid	crotonate	3	fumarate	2
C_4 hydroxy acid	β-hydroxybutyrate	2	malate	3

25.10

Process	Substances entering process	Substance produced
lipogenesis	acetyl-CoA	fatty acids, cholesterol, etc.
ketogenesis	acetyl-CoA	ketone bodies
β-oxidation pathway	fatty acids	acetyl-CoA, NADH, $FADH_2$
citric acid cycle	acetyl-CoA	CO_2, NADH, $FADH_2$
glycolysis	glucose	pyruvate → acetyl-CoA, NADH

Acetyl CoA cannot be used for the net synthesis of glucose because humans and animals do not possess the enzymes needed to convert acetyl CoA to pyruvate, the starting material for gluconeogenesis.

Answers to Self-Test

The numbers in parentheses refer to sections in your textbook.
1. F; an emulsifier (25.1) **2**. F; chylomicrons (25.1) **3**. F; adipose tissue (25.2)
4. F; triacylglycerols (25.1) **5**. T (25.3) **6**. T (25.3) **7**. T (25.5)
8. F; outer mitochondrial membrane **9**. T (25.6) **10**. F; pantothenic acid (25.11)
11. F; citric acid cycle (25.4) **12**. T (25.9) **13**. F; ketone bodies (25.6) **14**. T (25.5)
15. T (25.4) **16**. T (25.6) **17**. b (25.2) **18**. c (25.10) **19**. d (25.6) **20**. d (25.3) **21**. c (25.1)
22. c (25.1) **23**. b (25.4) **24**. a (25.5)

Protein Metabolism — Chapter 26

Chapter Overview

Like carbohydrates and lipids, proteins and amino acids are metabolized by the body. Although the primary use of amino acids in biological systems is in biosynthesis of proteins and other nitrogen-containing compounds, they can also be used as sources of energy for the body.

In this chapter you will study the digestion of proteins, the absorption of amino acids and their breakdown in the body, and the entry of their degradation products into the metabolic pathways of carbohydrates and fatty acids and into the urea cycle to rid the body of toxic ammonium ions. You will study the biosynthesis of the nonessential amino acids.

Practice Exercises

26.1 The digestion of proteins produces amino acids, which enter the bloodstream for distribution throughout the body. In the diagram below, write down the digestive processes that take place in each area and the enzymes that are involved.

dietary protein → stomach [ ]

↓

small intestine [ ]

amino acid pool ←

26.2 The total supply of available amino acids in the body is called the **amino acid pool** (Sec. 26.2). What are the four ways in which amino acids from the amino acid pool are utilized in the human body?

1.
2.
3.
4.

26.3 The degradation of amino acids (Sec. 26.3) takes place in stages that involve **transamination** reactions and **oxidative deamination** reactions.

a. In transamination, the α-amino group of an amino acid is interchanged with the keto group of an α-keto acid (usually α-ketoglutarate, less often oxaloacetate). Complete the equation below, drawing a structural formula for the missing product. Name the product.

$$\underset{\text{an amino acid}}{R-\overset{\overset{+}{N}H_3}{\overset{|}{C}H}-COO^-} + \underset{\alpha\text{-ketoglutarate}}{^-OOC-CH_2-CH_2-\overset{O}{\overset{\|}{C}}-COO^-} \xrightarrow[\text{transferase}]{\text{amino-}} \underset{\text{an } \alpha\text{-keto acid}}{R-\overset{O}{\overset{\|}{C}}-COO^-} + \ ?$$

b. There are two pathways for further processing the glutamate molecules produced by transamination. One pathway is a second transamination reaction that produces the amino acid aspartate. Complete the chemical equation below and name the products.

$$\underset{\text{glutamate}}{^{-}OOC-CH_2-CH_2-\overset{\overset{+}{NH_3}}{\overset{|}{C}H}-COO^-} + \underset{\text{oxaloacetate}}{^{-}OOC-CH_2-\overset{\overset{O}{||}}{C}-COO^-} \xrightarrow{\text{aminotransferase}} \quad + \ ^{-}OOC-CH_2-\overset{\overset{+}{NH_3}}{\overset{|}{C}H}-COO^-$$

c. In the second pathway, oxidative deamination, ammonia is removed from the glutamate formed in transamination. Complete the chemical equation below and name the product.

$$\underset{\text{glutamate}}{^{-}OOC-CH_2-CH_2-\overset{\overset{+}{NH_3}}{\overset{|}{C}H}-COO^-} + \quad ? \quad \longrightarrow NH_4^+ + \quad + NADH + H^+$$

26.4 The ammonium ion produced by oxidative deamination is toxic; it is either used by the human body in biosynthesis reactions or removed from the body as urea in the **urea cycle** (Sec. 26.4). Answer the following questions about the urea cycle.

a. One of the "fuels" for the urea cycle is carbamoyl phosphate. What are the reactants that form this "fuel"? ____________________

b. What molecule is used in Step 1 and regenerated in Step 4? ____________________

c. Where does the other "fuel," (aspartate from the transamination of glutamate) enter the cycle? ____________________

d. Where is urea removed from the cycle? ____________________

e. Where is ATP used? ____________________

f. How much ATP is consumed in the production of one urea molecule? ____________

g. Complete the net reaction for the urea cycle:

$$NH_4^+ + CO_2 + 3ATP + 2H_2O + \text{aspartate} \longrightarrow$$

26.5 Transamination or oxidative deamination of the 20 amino acids produces α-keto acids; these compounds undergo a sequence of degradations, producing a total of seven different degradation products. (The product depends on the carbon skeleton of the original amino acid.) The seven degradation products enter several metabolic pathways. The original amino acids are classified in terms of the pathways taken by their degradation products.

Classification of original amino acids	Products of amino acid degradation	Further metabolic reactions of degradation products
glucogenic (Sec. 26.5)		
ketogenic (Sec. 26.5)		
glucogenic and ketogenic		

26.6 The starting materials for the biosynthesis of the nonessential amino acids can be synthesized by the human body. The starting materials are intermediates in glycolysis and the citric acid cycle. Three of the nonessential amino acids can be biosynthesized by transamination of an α-keto acid. In the example below, draw the structure of the amino acid produced and name the amino acid.

$$\underset{\text{pyruvate}}{CH_3-\overset{\overset{O}{||}}{C}-COO^-} \xrightarrow{\text{aminotransferase}}$$

26.7 Hemoglobin (Sec. 26.7) is the conjugated protein responsible for the oxygen-carrying ability of red blood cells. Complete the sentences below about the catabolism of hemoglobin.

a. Hemoglobin breaks down into ___________ and ___________________.

b. Globin is hydrolyzed to produce _________________

c. Heme's ring opens to produce an iron atom and ___________________, which is converted by reduction of a methylene bridge to form _______________, which is excreted from the liver in ________________ into the small intestine. There, other ___________________ are formed and excreted from the body.

26.8 Two of the standard amino acids, cysteine and methionine, contain an atom of sulfur in their side chains. Complete the following discussion of cysteine.

a. The biosynthesis of cysteine (Sec. 26.8) involves the sulfhydrylation of _____________ with H_2S. What is the source of the H_2S? ______________________________

b. Biodegradation of cysteine involves the release of sulfur as _________________.

c. Name three actions of H_2S in the human body.

1.

2.

3.

26.9 Complete the table below summarizing some of the metabolism of the three groups of nutrients you have been studying.

Class of nutrient	Basic structural units	Storage compounds	Degradation pathway
carbohydrates			
lipids			
proteins			

Self-Test

True-false: Indicate whether the following statements are true or false. If the statement is false, give the word or phrase that may be substituted for the underlined portion to make the statement true.

1. Protein digestion begins in the stomach.
2. Proteins are denatured in the stomach by hydrochloric acid.
3. The pH of the pancreatic juice that aids in protein digestion is between 1.5 and 2.0.
4. Transamination of an amino acid releases the amino group as an ammonium ion.
5. Transamination of an amino acid requires the presence of the coenzyme pyridoxal phosphate.
6. Oxidative deamination is the conversion of an amino acid into a keto acid with the release of urea.
7. The effect of transamination is to convert a variety of amino acids to two amino acids, glutamate and serine.
8. Fumarate produced in the urea cycle is also an intermediate in the citric acid cycle.
9. Glucogenic amino acids are converted to acetyl CoA.

10. All of the nonessential amino acids can be synthesized in the body.
11. Two of the essential amino acids can be synthesized in the body.
12. A positive nitrogen balance means that more nitrogen is excreted than is taken in.
13. Phenylketonuria (PKU) is a lack of the enzyme needed to synthesize phenylalanine.
14. During periods of negative nitrogen balance, the body uses amino acids obtained from the amino acid pool.
15. Only two amino acids are purely ketogenic: leucine and lysine.
16. When heme is degraded, its iron atom becomes part of biliverdin, an iron-storage protein.
17. All eight B vitamins are needed as cofactors in the degradation of the 20 amino acid "carbon skeletons."
18. The digestive enzyme pepsin effects the hydrolysis of approximately 90% of peptide bonds present in proteins.

Multiple choice:

19. The urea cycle is important because it:
    a. removes ammonium ion from the body
    b. regenerates α-ketoglutarate by oxidative deamination
    c. results in transamination of amino acids
    d. all of these
    e. none of these
20. Before entering the urea cycle, ammonia is converted to carbamoyl phosphate. This requires how much ATP?
    a. 1 molecule b. 2 molecules c. 3 molecules
    d. 4 molecules e. none of these
21. Which of the following amino acids can be synthesized by the human body?
    a. valine b. lysine c. alanine
    d. tryptophan e. phenylalanine
22. The degradation products of the globin portion of hemoglobin contribute to the formation of:
    a. bile pigments b. ferritin molecules c. tetrapyrrole
    d. the amino acid pool e. none of these
23. The net effect of the transamination of amino acids is to produce the single amino acid:
    a. glutamate b. aspartate c. pyridoxine
    d. oxaloacetate e. none of these
24. The amino acids whose carbon skeletons are degraded to citric acid cycle intermediates are classified as:
    a. essential amino acids b. nonessential amino acids
    c. ketogenic amino acids d. glucogenic amino acids
    e. both glucogenic and ketogenic
25. The oxygen-carrying complex present in hemoglobin is formed by the interaction of oxygen with:
    a. biliverdin b. globin c. pyrrole
    d. iron e. none of these

Answers to Practice Exercises

26.1

stomach

dietary protein → | The enzyme gastrin causes release of pepsinogen and HCl, denaturation of proteins by HCl, pepsinogen changes to pepsin, hydrolysis of some peptide bonds by the enzyme pepsin. |

↓

small intestine

amino acid pool ← | Secretin stimulates pancreatic production of HCO_3^- (neutralizes HCl), hydrolysis of peptides by trypsin, chymotrypsin and carboxypeptidase in the pancreatic juice and aminopeptidase from the intestinal mucosal cells; absorption of free amino acids by active transport across the intestinal wall. |

26.2 Four ways in which the amino acids from the amino acid pool are utilized:
1. biosynthesis of functional proteins
2. biosynthesis of nonprotein nitrogen-containing compounds
3. biosynthesis of nonessential amino acids
4. energy production

26.3 a.

$$\underset{\text{an amino acid}}{R-\overset{\overset{+}{NH_3}}{\underset{|}{C}}H-COO^-} + \underset{\alpha\text{-ketoglutarate}}{{}^-OOC-CH_2-CH_2-\overset{\overset{O}{\|}}{C}-CHOOH_2} \xrightarrow{\text{aminotransferase}} \underset{\alpha\text{-keto acid}}{R-\overset{\overset{O}{\|}}{C}-COO^-} + \underset{\text{glutamate}}{{}^-OOC-CH_2-CH_2-\overset{\overset{+}{NH_3}}{\underset{|}{C}}H-COO^-}$$

b.

$$\underset{\text{glutamate}}{{}^-OOC-CH_2-CH_2-\overset{\overset{+}{NH_3}}{\underset{|}{C}}H-COO^-} + \underset{\text{oxaloacetate}}{{}^-OOC-CH_2-\overset{\overset{O}{\|}}{C}-COO^-} \xrightarrow{\text{aminotransferase}} \underset{\alpha\text{-ketoglutarate}}{{}^-OOC-CH_2-CH_2-\overset{\overset{O}{\|}}{C}-COO^-} + \underset{\text{aspartate}}{{}^-OOC-CH_2-\overset{\overset{+}{NH_3}}{\underset{|}{C}}H-COO^-}$$

c.

$$\underset{\text{glutamate}}{{}^-OOC-CH_2-CH_2-\overset{\overset{+}{NH_3}}{\underset{|}{C}}H-COO^-} + NAD^+ + H_2O \xrightarrow{\text{glutamate dehydrogenase}} \overset{+}{N}H_4 + \underset{\alpha\text{-ketoglutarate}}{{}^-OOC-CH_2-CH_2-\overset{\overset{O}{\|}}{C}-COO^-} + NADH + H^+$$

26.4
a. ammonium ion, carbon dioxide, water, and two ATP molecules
b. ornithine
c. Step 2
d. Step 4
e. in the formation of carbamoyl phosphate (2 ATP) and in Step 2 of the urea cycle (2 high-energy bonds)

f. the equivalent of 4 ATP molecules (2 ATP → 2 ADP and 1 ATP → 1 AMP)

g. $NH_4^+ + CO_2 + 3ATP + 2H_2O + \text{aspartate} \longrightarrow$

$\text{urea} + 2ADP + AMP + PP_i + 2P_i + \text{fumarate}$

26.5

Classification of original amino acids	Products of amino acid degradation	Further metabolic reactions of degradation products
glucogenic	citric acid cycle intermediates, glucose precursors	glucose formation by gluconeogenesis, ATP production
ketogenic	acetyl CoA, acetoacetyl CoA	production of ketone bodies or fatty acids, ATP production
glucogenic and ketogenic	Pyruvate (and any of the above)	any of the above

26.6

$$CH_3-\overset{O}{\overset{\|}{C}}-COO^- \xrightarrow{\text{aminotransferase}} CH_3-\overset{\overset{+}{NH_3}}{\overset{|}{CH}}-COO^-$$

pyruvate → alanine

26.7 a. heme and globin; b. amino acids; c. biliverdin, bilirubin, bile, bile pigments

26.8 a. serine; sulfate assimilation and conversion to the sulfide
b. H_2S
c. smooth muscle relaxant and vasodilator (cardioprotective); active in the brain (in Alzheimer's disease H_2S concentration is low); toxic to pancreas (excess H_2S leads to death of pancreatic β-cells, reducing insulin production (type 1 diabetes)

26.9

Nutrient class	Basic structural units	Storage compounds	Degradation pathway
carbohydrates	monosaccharides (glucose)	glycogen	glycolysis, citric acid cycle
lipids (TAGs)	fatty acids, glycerol	triacylglycerols	β-oxidation pathway, citric acid cycle
proteins	amino acids	functional proteins (amino acids are not "stored")	transamination, oxidative deamination, citric acid cycle

Answers to Self-Test

The numbers in parentheses refer to sections in your textbook.
1. T (26.1) **2**. T (26.1) **3**. F; 7 and 8 (26.1) **4**. F; oxidative deamination (26.3)
5. T (26.3) **6**. F; ammonium ion (26.3) **7**. F; glutamate and aspartate (26.3) **8**. T (26.5)
9. F; ketogenic (26.5) **10**. T (26.6) **11**. F; none (26.6) **12**. F; negative (26.2) **13**. F; oxidize (26.6)
14. F; degradation of functional proteins (26.2) **15**. T (26.5) **16**. F; ferritin (26.7) **17**. T (26.10)
18. F; 10% (26.1) **19**. a (26.4) **20**. b (26.4) **21**. c (26.6) **22**. d (26.7) **23**. a (26.3) **24**. d (26.3)
25. d (26.7)

Basic Concepts About Matter — Chapter 1

Solutions to Selected Problems

1.1 All samples of matter have (1) mass and (2) occupy space.

1.3 Air, pizza, and gold are matter; each has mass and occupies space. Sound is a form of energy.
a. matter b. matter c. energy d. matter

1.5 a. Liquids have an indefinite shape; solids have a definite shape.
b. Gases have an indefinite volume; liquids have a definite volume.

1.7 a. Copper wire does not take the shape of its container; yes, it does have a definite volume.
b. Oxygen gas does not have a definite shape nor a definite volume. It takes the shape and volume of its container.
c. Each granule of sugar has its own definite shape; however, the granules are small, so the general shape of the mass of crystals does take the shape of the container. Yes, granulated sugar has a definite volume.
d. Liquid water takes the shape of its container; yes, it has a definite volume.

1.9 a. The state of a substance is a <u>physical</u> property.
b. Ignition on heating with air produces a new substance, so this is a <u>chemical</u> property.
c. A new substance is produced (hydrogen gas), so this is a <u>chemical</u> property.
d. Density is a <u>physical</u> property.

1.11 a. Chemical. The key word is "reacting," which indicates that a new substance is formed.
b. Physical. Red color can be observed without the formation of a new substance.
c. Chemical. The toxicity of beryllium indicates that it produces a change of substances in the human body.
d. Physical. Pulverizing a substance changes its shape.

1.13 a. Chemical. The word "reaction" indicates the inability to form a new substance.
b. Chemical. The word "reacts" indicates that a new substance is formed.
c. Physical. Change of state is a physical property.
d. Physical. Change of state is a physical property.

1.15 a. Physical change. The leaf changes shape, but the crushed leaf is not a new substance.
b. Physical change. The metal has changed shape, but it is not a new chemical substance.
c. Chemical change. Burning is always a chemical change; new substances are formed.
d. Physical change. The ham changes shape, but it is still ham; no new substance is formed.

1.17 a. Physical. Mechanical changes are physical.
b. Physical. A change of state, from liquid to solid in this case, is always a physical change.
c. Chemical. The smell of sour milk indicates that a new substance has been formed.
d. Physical. Breaking or cutting is always a physical change.

1.19 a. Physical. Ice and liquid water are the same substance in two different forms.
b. Physical. Crushed ice and ice chips are both forms of water; no new substance is formed.
c. Chemical. Burning a newspaper produces a change in its chemical identity; the gases, charred paper, etc., formed are new substances.
d. Physical. Pulverizing a sugar cube produces a new shape, but the substance is still sugar.

1.21 a. Chemical change. Burning a newspaper produces new substances.
b. Chemical property. Metallic copper reacts with chlorine gas to produce a new substance.
c. Physical change. Ice melting is a change of state. No new substance is produced.
d. Physical property. The state (solid) of a substance (gold) is a physical property.

1.23 a. False. A heterogeneous mixture contains two or more substances.
b. True. A pure substance contains only one substance, and so has a definite composition.
c. False. Substances maintain their identity in all mixtures; they are physically mixed, not chemically combined.
d. True. Most substances in the "everyday world" are mixtures.

1.25 a. Heterogeneous mixture. "Two substances" makes this a mixture; "two phases" shows that it is not uniformly mixed.
b. Homogeneous mixture. "One phase" indicates that the mixture of two substances has uniform properties throughout.
c. Pure substance. The two phases present represent two forms of a single substance (for example, ice and liquid water).
d. Heterogeneous mixture. The existence of three substances in three different phases indicates that the substances are not mixed uniformly.

1.27 a. Homogeneous mixture, one phase. The word "dissolved" indicates that the salt is uniformly distributed throughout the salt-water mixture.
b. Heterogeneous mixture, two phases. Undissolved sand is not uniformly mixed with water.
c. Heterogeneous mixture, three phases. The three phases present are ice (solid), liquid water, and liquid oil. The two liquids are not soluble in one another; they are separate phases.
d. Heterogeneous mixture, two phases. The water solution (water and dissolved sugar) is one phase; the undissolved sugar is the other phase.

1.29 a. Compound. A single substance (A) made up of two elements is a compound.
b. Compound. B must contain more than one element to decompose chemically, so it is a compound.
c. Classification is not possible because not enough information is given.
d. Classification is not possible. Melting is a physical change that both elements and compounds can undergo.

1.31 a. For A and B, classification is not possible. C is a compound because it contains the elements in A and B.
b. D is a compound because it breaks down into simpler substances. It is not possible to classify E, F, and G.

1.33 a. True. An element contains one kind of atoms and so is a single pure substance. A compound has a definite, constant composition, so it is also a pure substance.

b. False. A compound results from the chemical combination of two or more elements.

c. False. In order for matter to be heterogeneous, at least two substances (either elements or compounds) must be present.

d. False. Both compounds and elements have a definite composition.

1.35 The first box is an analogy for a mixture; the nuts and bolts are not evenly distributed. The second box is an analogy for a compound; each bolt is attached to a nut.

1.37 a. Compound. A compound is a substance that contains more than one element.

b. Mixture. Two substances are "mixed" together; this is a mixture.

c. Element. A substance that cannot be decomposed by chemical means is an element.

d. Mixture. Since the composition is variable, this is not a compound; it is a mixture.

1.39 a. Homogeneous mixture. The sample is a single phase (homogeneous); when it is boiled away (evaporated), it reveals that there are two substances (a mixture).

b. Heterogeneous mixture. A "cloudy" mixture is not homogeneous; it has two phases.

c. Heterogeneous mixture. The sample is nonuniform; it is heterogeneous.

d. Compound. The sample is not a mixture; it cannot be separated by physical means. It reacts with magnesium (an element) to form two compounds containing different elements. Therefore, the sample contains more than one element; it is a compound.

1.41 a. True. Variable composition is one indication of a mixture.

b. True. The two elements are physically combined and retain their physical properties.

c. False. The elements in a compound are chemically combined and cannot be separated by physical methods.

d. True. Elements in a compound are chemically combined.

1.43 a. True. There are a number of heavier elements that do not occur naturally.

b. False. There are 118 known elements.

c. False. Any elements discovered in the future will be highly unstable and thus not naturally occurring.

d. False. There are 88 naturally occurring elements.

1.45 a. True. Oxygen is the most abundant element and silicon is the second most abundant element in Earth's crust.

b. True. Oxygen is the most abundant element in Earth's crust.

c. False. Hydrogen and helium are the two most abundant elements in the universe as a whole.

d. True. 60.1% of all elemental particles (atoms) within Earth's crust are oxygen atoms.

1.47 In Earth's crust:
a. Silicon (20.1%) is more abundant than aluminum (6.1%).
b. Calcium (2.6%) is less abundant than hydrogen (2.9%).
c. Iron (2.2%) is less abundant than oxygen (60.1%).
d. Sodium (2.1%) is more abundant than potassium (<1.5%).

1.49 a. N is the symbol for nitrogen. b. Ni is the symbol for nickel.
c. Pb is the symbol for lead. d. Sn is the symbol for tin.

1.51 a. Al is the symbol for aluminum. b. Ne is the symbol for neon.
c. H is the symbol for hydrogen. d. U is the symbol for uranium.

1.53 The elements and symbols (Table 1.1 of your textbook) are as follows:
a. Na – sodium, S – sulfur b. Mg – magnesium, Mn – manganese
c. Ca – calcium, Cd – cadmium d. As – arsenic, Ar – argon

1.55 a. No. The symbols are Mg, N, and P. b. Yes. The symbols are Br, Fe, and Ca.
c. Yes. The symbols are Al, Cu, and Cl. d. No. The symbols are B, Ba, and Be.

1.57 a. Heteroatomic. Two kinds of atoms. b. Heteroatomic. Two kinds of atoms.
c. Homoatomic. One kind of atom. d. Heteroatomic. Two kinds of atoms.

1.59 a. Triatomic. The molecule contains three atoms.
b. Triatomic. The molecule contains three atoms.
c. Diatomic. Two oxygen atoms are present.
d. Diatomic. One carbon atom and one oxygen atom are present.

1.61 a. Compound. Two kinds of atoms are present.
b. Compound. Two kinds of atoms are present.
c. Element. All atoms are of the same kind.
d. Compound. Two kinds of atoms are present.

1.63 a. A triatomic molecule <u>could be an element or a compound</u>.
b. A molecule containing one kind of atom is an <u>element</u>.
c. A heteroatomic molecule contains more than one kind of molecule and so is a <u>compound</u>.
d. A homoatomic molecule contains one kind of atom and <u>must be an element</u>.

1.65 a. True. Atoms are the fundamental building blocks of both elements and compounds.
b. False. Triatomic molecules may contain one, two, or three kinds of atoms.
c. True. A compound is made up of two or more elements; its molecules must contain more than one kind of atom.
d. False. Both heteratomic and homoatomic molecules may contain three or more atoms.

1.67 a. Diagram I. The molecules present must all be triatomic, identical (a pure substance), and heteroatomic (a compound).

b. Diagram III. Two kinds of molecules must be present (a mixture), and the molecules must all be heteroatomic (a compound).

c. Diagram II. Two kinds of diatomic molecules are present.

d. Diagrams I and IV. All molecules must be identical, but can be homoatomic (an element) or heteroatomic (a compound).

1.69 a. Element. Because one kind of homoatomic molecule is present, this is an element.

b. Mixture. Because two phases are present, this is a mixture.

c. Mixture. Because two kinds of molecules are present, this is a mixture.

d. Mixture. Because two kinds of molecules are present, this is a mixture.

1.71 a. Changes 3 and 4 are physical changes; no new molecules are formed.

b. In change 1 two elements combine to form a compound; a new substance is formed.

1.73 Chemical formulas for the substances represented by the models are:

a. H_2O The molecule contains two hydrogen atoms and one oxygen atom.

b. CO_2 The molecule contains one carbon atom and two oxygen atoms.

c. O_2 The molecule contains two oxygen atoms.

d. CO The molecule contains one carbon atom and one oxygen atom.

1.75 a. Chemical formula of a compound. Six atoms are present in one formula unit.

b. Chemical formula of a compound. Five atoms are present in one formula unit.

c. Chemical symbol of an element. One atom is present in one formula unit.

d. Chemical formula of a compound. Two atoms are present in one formula unit.

1.77 The chemical formulas are:

a. $C_{12}H_{22}O_{11}$ b. $C_8H_{10}O_4N_2$

1.79 The chemical formulas are: XZ_4, X_2Z_3, X_3Z_2, X_4Z

1.81 a. HCN. Since 3 atoms are present and 3 elements are present there can be only one atom of each element present.

b. H_2SO_4. Subtracting 2 atoms of hydrogen and 1 atom of sulfur from the 7 atom total means that 4 atoms of O are present.

1.83 a. The first and second formulas contain the same number of total atoms.
HN_3 (1H + 3N = 4 atoms) and NH_3 (1N + 3H = 4 atoms) Each contains four atoms.

b. The first formula contains more total atoms. $CaSO_4$ (1Ca + 1S + 4O = 6 atoms); $Mg(OH)_2$ [1Mg + 2(1O + 1H) = 5 atoms]. The parentheses around the OH group indicate that each atom in the group is multiplied by the group's subscript.

c. The first formula contains the same number of total atoms as the second.
$NaClO_3$ (1Na + 1Cl + 3O = 5 atoms); $Be(CN)_2$ [1Be + 2(1C + 1N) = 5 atoms]

d. The first formula contains fewer total atoms. $Be_3(PO_4)_2$ [3Be + 2(1P + 4O) = 13 atoms]; $Mg(C_2H_3O_2)_2$ [1Mg + 2(2C + 3H + 2O) = 15 atoms]

1.85 a. $1 + 2 + x = 6; x = 3$ b. $2 + 3 + 3x = 17; x = 4$

c. $1 + x + x = 5; x = 2$ d. $x + 2x + x = 8; x = 2$

1.87 a. In a mixture of N_2, N_2H_4, NH_3, CH_4, and CH_3Cl, two kinds of molecules contain four or fewer atoms (N_2 – 2 atoms; NH_3 – 4 atoms).

b. There are four kinds of atoms (N, H, C, Cl) in the mixture.

c. There are 110 atoms in a mixture containing five molecules of each substance. $5(2 + 6 + 4 + 5 + 5) = 110$

d. There are 56 hydrogen atoms in a sample containing four molecules of each substance. $4(0 + 4 + 3 + 4 + 3) = 56$

Measurements in Chemistry — Chapter 2

Solutions to Selected Problems

2.1 It is easier to use because it is a decimal unit system.

2.3 a. The metric prefix *giga,* abbreviated as G, has a value of 10^9.
b. The metric prefix *nano,* abbreviated as n, has a value of 10^{-9}.
c. The metric prefix *mega,* abbreviated as M, has a value of 10^6.
d. The metric prefix *micro,* abbreviated as μ, has a value of 10^{-6}.

2.5 a. A kilogram, abbreviated as kg, measures mass.
b. A megameter, abbreviated as Mm, measures length.
c. A nanogram, abbreviated as ng, measures mass.
d. A milliliter, abbreviated as mL, measures volume.

2.7 The meaning of a metric system prefix is independent of the base unit it modifies. The lists, arranged from smallest to largest, are:
a. nanogram, milligram, centigram
b. kilometer, megameter, gigameter
c. picoliter, microliter, deciliter
d. microgram, milligram, kilogram

2.9 a. Measure the thickness of a chemistry textbook in centimeters.
b. Measure the mass of a cantaloupe seed in milligrams.
c. Measure the capacity of a car's gasoline tank in liters.
d. Measure the length of a man's tie in decimeters.

2.11 60 minutes is a counted (exact) number and 60 feet is a measured (inexact) number.

2.13 An exact number has no uncertainty associated with it; an inexact number has a degree of uncertainty. Whenever defining a quantity or counting, the resulting number is exact. Whenever a measurement is made, the resulting number is inexact.
a. 32 is an exact number of chairs
b. 60 is an exact number of seconds.
c. 3.2 pounds is an inexact weight.
d. 323 is an exact number of words.

2.15 Measurement results in an inexact number; counting and definition result in exact numbers.
a. The length of a swimming pool is an inexact number because it is measured.
b. The number of gummi bears in a bag is an exact number; gummi bears are counted.
c. The number of quarts in a gallon is exact because it is a defined number.
d. The surface area of a living room rug is an inexact number because it is calculated from two inexact measurements of length.

2.17 The last digit of a measurement is estimated.
a. The estimated digit is the 4.
b. The estimated digit is the zero.
c. The estimated digit is the 4.
d. The estimated digit is the 4.

2.19 The magnitude of the uncertainty is indicated by a 1 in the last measured digit.
a. The magnitude of the uncertainty is ±1
b. The magnitude of the uncertainty is ±0.1
c. The magnitude of the uncertainty is ±0.001
d. The magnitude of the uncertainty is ±0.00001

2.21 Only one estimated digit is recorded as part of a measurement.
a. Temperature recorded using a thermometer marked in degrees is recorded to 0.1 degree.
b. The volume from a graduated cylinder with markings of tenths of milliliters is recorded to 0.01 mL.
c. Volume using a volumetric device with markings every 10 mL is recorded to 1 mL.
d. Length using a ruler with a smallest scale marking of 1 mm is recorded to 0.1 mm.

2.23 a. 0.1 cm; since the ruler is marked in ones units, the estimated digit is tenths
b. 0.1 cm; since the ruler is marked in ones units, the estimated digit is tenths

2.25 a. 2.70 cm; the value is very close to 2.7, with the estimated value being 2.70
b. 27 cm; the value is definitely between 20 and 30, with the estimated value being 27

2.27 a. ruler 4; since the ruler is marked in ones units it can be read to tenths
b. ruler 1 or 4; since both rulers are marked in ones units they can be read to tenths
c. ruler 2; since the ruler is marked in tenths units it can be read to hundredths
d. ruler 3; since the ruler is marked in tens units it can be read to ones

2.29 Significant figures are the digits in a measurement that are known with certainty plus one digit that is uncertain. In a measurement, all nonzero numbers, and some zeros, are significant.
a. 0.444 has three significant figures.
b. 0.00444 has three significant figures.
c. 0.04040 has four significant figures.
d. 0.00004 has one significant figure.

2.31 a. 275.00 has 5 significant figures. Trailing zeros are significant when a decimal point is present.
b. 27,500 has 3 significant figures. Trailing zeros are not significant if the number lacks an explicit decimal point.
c. 6,720,000 has 3 significant figures. Trailing zeros are not significant if the number lacks an explicit decimal point.
d. 6,720,100 has 5 significant figures. Trailing zeros are not significant if the number lacks an explicit decimal point. Confined zeros are significant

2.33 a. 11.01 and 11.00 have the same number (four) of significant figures. All of the zeros are significant because they are either confined or trailing with an explicit decimal point.
b. 2002 has four significant figures, and 2020 has three. The last zero in 2020 is not significant because there is no explicit decimal point.
c. 0.000066 and 660,000 have the same number (two) of significant figures. None of the zeros in either number are significant because they are either leading zeros or trailing zeros with no explicit decimal point.
d. 0.05700 and 0.05070 have the same number (four) of significant figures. The trailing zeros are significant because there is an explicit decimal point.

2.35 a. Yes, 11.01 and 11.00 have the same uncertainty.
b. No, 2002 and 2020 do not have the same uncertainty.
c. No, 0.000066 and 660,000 do not have the same uncertainty.
d. Yes, 0.05700 and 0.05070 have the same uncertainty.

2.37 a. For 5371: number of significant figures is 4; estimated digit is 1; uncertainty is ± 1.
b. For 0.41: number of significant figures is 2; estimated digit is 1; uncertainty is ± 0.01.
c. For 3200: number of significant figures is 2; estimated digit is 2; uncertainty is ± 100.
d. For 5050: number of significant figures is 3; estimated digit is last 5; uncertainty is ± 10.

2.39 The estimated number of people is 50,000.
a. If the uncertainty is 10,000, the low and high estimates are 40,000 to 60,000.
b. If the uncertainty is 1000, the low and high estimates are 49,000 to 51,000.
c. If the uncertainty is 100, the low and high estimates are 49,900 to 50,100.
d. If the uncertainty is 10, the low and high estimates are 49,990 to 50,010.

2.41 When rounding numbers, if the first digit to be deleted is 4 or less, drop it and the following digits; if it is 5 or greater, drop that digit and all of the following digits and increase the last retained digit by one.

a. 0.350763 to four s.f. is 0.3508
b. 13.43 to three s.f. is 13.4
c. 22.4555 to two s.f. is 22
d. 0.030303 to three s.f. is 0.0303

2.43 To obtain a number with three significant figures:
a. 3567 becomes 3570
b. 323,200 becomes 323,000
c. 18 becomes 18.0
d. 2,345,346 becomes 2,350,000

2.45 Rounding:
a. 1.8828: to four significant figures is 1.883; to two significant figures is 1.9
b. 24,233: to four significant figures is 24,230; to two significant figures is 24,000
c. 0.51181: to four significant figures is 0.5118; to two significant figures is 0.51
d. 7.4500: to four significant figures is 7.450; to two significant figures is 7.5

2.47 In multiplication and division, the number of significant figures in the answer is the same as the number of significant figures in the measurement that contains the fewest significant figures. (s.f. stands for significant figures)
a. 10,300 (three s.f.) × 0.30 (two s.f.) × 0.300 (three s.f.) Since the least number of significant figures is two, the answer will have two significant figures.
b. 3300 (two s.f.) × 3330 (three s.f.) × 333.0 (four s.f.) The lowest number of significant figures is two, so the answer will have two significant figures.
c. 6.0 (two s.f.) ÷ 33.0 (three s.f.) The answer will have two significant figures.
d. 6.000 (four s.f.) ÷ 33 (two s.f.) The answer will have two significant figures.

2.49 In multiplication and division of measured numbers, the answer has the same number of significant figures as the measurement with the fewest significant figures. (s.f. stands for significant figures.)
a. 2.0000 (five s.f.) × 2.00 (three s.f.) × 0.0020 (two s.f.) = 0.0080 (two s.f.)
b. 4.1567 (five s.f.) × 0.00345 (three s.f.) = 0.0143 (three s.f.)
c. 0.0037 (two s.f.) × 3700 (two s.f.) × 1.001 (four s.f.) = 14 (two s.f.)
d. 6.00 (three s.f.) ÷ 33.0 (three s.f.) = 0.182 (three s.f.)
e. 530,000 (two s.f.) ÷ 465,300 (four s.f.) = 1.1 (two s.f.)
f. 4670 (four s.f.) × 3.00 (three s.f.) ÷ 2.450 (four s.f.) = 5720 (three s.f.)

2.51 In addition and subtraction of measured numbers, the answer has no more digits to the right of the decimal point than are found in the measurement with the fewest digits to the right of the decimal point.
a. 12 + 23 + 127 = 162 (no digits to the right of the inferred decimal point)
b. 3.111 + 3.11 + 3.1 = 9.3 (one digit to the right of the decimal point)
c. 1237.6 + 23 + 0.12 = 1261 (no digits to the right of the inferred decimal point)
d. 43.65 – 23.7 = 20.0 (one digit to the right of the decimal point)

2.53 a. The uncertainty of 12.37050 rounded to 6 significant figures is 0.0001
b. The uncertainty of 12.37050 rounded to 4 significant figures is 0.01
c. The uncertainty of 12.37050 rounded to 3 significant figures is 0.1
d. The uncertainty of 12.37050 rounded to 2 significant figures is 1

2.55 Scientific notation is a numerical system in which a decimal number is expressed as the product of a number between 1 and 10 (the coefficient) and 10 raised to a power (the exponential term). To convert a number from decimal notation to scientific notation, move the decimal point to a position behind the first nonzero digit. The exponent in the exponential term is equal to the number of places the decimal point was moved.
a. 0.0123 expressed in scientific notation is 1.23×10^{-2}. Negative exponent.
b. 375,000 expressed in scientific notation is 3.75×10^{5}. Positive exponent.
c. 0.100 expressed in scientific notation is 1.00×10^{-1}. Negative exponent.
d. 68.75 expressed in scientific notation is 6.875×10^{1}. Positive exponent.

2.57 To convert a number from decimal notation to scientific notation, move the decimal point to a position behind the first nonzero digit. The exponent in the exponential term is equal to the number of places the decimal point was moved.
a. For 0.0123 move the decimal 2 places to the right.
b. For 375,000 move the decimal 5 places to the left.
c. For 0.100 move the decimal 1 place to the right.
d. For 68.75 move the decimal 1 place to the left.

2.59 In scientific notation, only significant figures become part of the coefficient.
a. For 0.0123 there will be 3 significant figures in the scientific notation.
b. For 375,000 there will be 3 significant figures in the scientific notation.
c. For 0.100 there will be 3 significant figures in the scientific notation.
d. For 68.75 there will be 4 significant figures in the scientific notation.

2.61 To convert a number from decimal notation to scientific notation, move the decimal point to a position behind the first nonzero digit. The exponent in the exponential term is equal to the number of places the decimal point was moved.
a. 120.7 expressed in scientific notation is 1.207×10^{2}
b. 0.0034 expressed in scientific notation is 3.4×10^{-3}
c. 231.00 expressed in scientific notation is 2.3100×10^{2}
d. 23,100 expressed in scientific notation is 2.31×10^{4}

2.63 To convert a number from scientific notation to decimal notation, move the decimal point in the coefficient to the right for a positive exponent or to the left for a negative exponent. The number of places the decimal point is moved is specified by the exponent.

a. 2.34×10^{2} expressed in decimal notation is 234
b. 2.3400×10^{2} expressed in decimal notation is 234.00
c. 2.34×10^{-3} expressed in decimal notation is 0.00234
d. 2.3400×10^{-3} expressed in decimal notation is 0.0023400

2.65 To multiply numbers expressed in scientific notation, multiply the coefficients and add the exponents in the exponential terms. To divide numbers expressed in scientific notation, divide the coefficients and subtract the exponents.

a. $(3.20 \times 10^7) \times (1.720 \times 10^5) = 5.504 \times 10^{12} = 5.50 \times 10^{12}$ The coefficient in the answer is expressed to three significant figures because one of the numbers being multiplied has only three significant figures.
b. $(1.00 \times 10^3) \times (5.00 \times 10^3) \times (3.0 \times 10^{-3}) = 15 \times 10^3 = 1.5 \times 10^4$ To express the answer in correct scientific notation, the decimal point in the coefficient was moved one place to the left, and the exponent was increased by 1.
c. $(3.0 \times 10^{-5}) \div (1.5 \times 10^2) = 2.0 \times 10^{-7}$
d. $(2.2 \times 10^6) \times (2.3 \times 10^{-6}) \div (1.2 \times 10^{-3}) \div (3.5 \times 10^{-3}) = 1.2 \times 10^6$

2.67 a. 10^2; the uncertainty in the coefficient is 10^{-2} and multiplying this by the power of ten gives $10^{-2} \times 10^4 = 10^2$
b. 10^4; $10^{-2} \times 10^6 = 10^4$
c. 10^4; $10^{-1} \times 10^5 = 10^4$
d. 10^{-4}; $10^{-1} \times 10^{-3} = 10^{-4}$

2.69 To convert a number from decimal notation to scientific notation, move the decimal point to a position behind the first nonzero digit. The exponent in the exponential term is equal to the number of places the decimal point was moved.
a. 0.00300300 to three significant figures becomes 3.00×10^{-3}
b. 936,000 to two significant figures becomes 9.4×10^5
c. 23.5003 to three significant figures becomes 2.35×10^1
d. 450,000,001 to six significant figures becomes 4.50000×10^8

2.71 Conversion factors are derived from equations (equalities) that relate units. They always come in pairs, one member of the pair being the reciprocal of the other.
a. 1 day = 24 hours The conversion factors derived from this equality are:

$$\frac{1 \text{ day}}{24 \text{ hours}} \quad \text{or} \quad \frac{24 \text{ hours}}{1 \text{ day}}$$

b. 1 century = 10 decades The conversion factors derived from this equality are:

$$\frac{1 \text{ century}}{10 \text{ decades}} \quad \text{or} \quad \frac{10 \text{ decades}}{1 \text{ century}}$$

c. 1 yard = 3 feet The conversion factors derived from this equality are:

$$\frac{1 \text{ yard}}{3 \text{ feet}} \quad \text{or} \quad \frac{3 \text{ feet}}{1 \text{ yard}}$$

d. 1 gallon = 4 quarts The conversion factors derived from this equality are:

$$\frac{1 \text{ gallon}}{4 \text{ quarts}} \quad \text{or} \quad \frac{4 \text{ quarts}}{1 \text{ gallon}}$$

2.73 The conversion factors are derived from the definitions of the metric system prefixes.

a. $1\ \text{kL} = 10^3\ \text{L}$ The conversion factors are: $\dfrac{1\ \text{kL}}{10^3\ \text{L}}$ or $\dfrac{10^3\ \text{L}}{1\ \text{kL}}$

b. $1\ \text{mg} = 10^{-3}\ \text{g}$ The conversion factors are: $\dfrac{1\ \text{mg}}{10^{-3}\ \text{g}}$ or $\dfrac{10^{-3}\ \text{g}}{1\ \text{mg}}$

c. $1\ \text{cm} = 10^{-2}\ \text{m}$ The conversion factors are: $\dfrac{1\ \text{cm}}{10^{-2}\ \text{m}}$ or $\dfrac{10^{-2}\ \text{m}}{1\ \text{cm}}$

d. $1\mu\text{sec} = 10^{-6}\ \text{sec}$ The conversion factors are: $\dfrac{1\ \mu\text{sec}}{10^{-6}\ \text{sec}}$ or $\dfrac{10^{-6}\ \text{sec}}{1\ \mu\text{sec}}$

2.75 Exact numbers occur in definitions, counting and simple fractions. Inexact numbers result when a measurement is made.

a. 1 dozen = 12 objects This is a definition; the conversion factors are exact numbers.

b. 1 kilogram = 2.20 pounds This equality is measured; the conversion factors are inexact numbers.

c. 1 minute = 60 seconds The equality is derived from a definition; the conversion factors are exact numbers.

d. 1 millimeter = 10^{-3} meters The equality is derived from a definition; the conversion factors are exact numbers.

2.77 Using dimensional analysis: (1) identify the given quantity and its unit, and the unknown quantity and its unit; and (2) multiply the given quantity by a conversion factor that allows cancellation of any units not desired in the answer.

a. 1.6×10^3 dm is the given quantity. The unknown quantity will be in meters. The equality is $1\ \text{dm} = 10^{-1}\ \text{m}$, and the conversion factors are: $\dfrac{1\ \text{dm}}{10^{-1}\ \text{m}}$ or $\dfrac{10^{-1}\ \text{m}}{1\ \text{dm}}$

The second of these will allow the cancellation of decimeters and leave meters.

$$1.6 \text{ x } 10^3\ \text{dm} \times \left(\frac{10^{-1}\ \text{m}}{1\ \text{dm}}\right) = 1.6 \times 10^2\ \text{m}$$

b. Convert 24 nm to meters. The equality is $1\ \text{nm} = 10^{-9}\ \text{m}$.

$$24\ \text{nm} \times \left(\frac{10^{-9}\ \text{m}}{1\ \text{nm}}\right) = 2.4 \times 10^{-8}\ \text{m}$$

c. Convert 0.003 km to meters. The equality is $1\ \text{km} = 10^3\ \text{m}$.

$$0.003\ \text{km} \times \left(\frac{10^3\ \text{m}}{1\ \text{km}}\right) = 3\ \text{m}$$

d. Convert 3.0×10^8 mm to meters. The equality is $1\ \text{mm} = 10^{-3}\ \text{m}$.

$$3.0 \times 10^8\ \text{mm} \times \left(\frac{10^{-3}\ \text{m}}{1\ \text{mm}}\right) = 3.0 \times 10^5\ \text{m}$$

2.79 Convert 2500 mL to liters. The equality is $1\ \text{mL} = 10^{-3}\ \text{L}$.

$$2500\ \text{mL} \times \left(\frac{10^{-3}\ \text{L}}{1\ \text{mL}}\right) = 2.5\ \text{L}$$

2.81 Convert 1550 g to pounds. Some conversion factors relating the English and Metric Systems of measurement can be found in Table 2.2 of your textbook.

$$1550 \text{ g} \times \left(\frac{1.00 \text{ lb}}{454 \text{ g}}\right) = 3.41 \text{ lb}$$

2.83 Convert 25 mL to gallons. For this conversion, use two conversion factors, one derived from the defined relationship of mL and L, and the other, relating gallons and liters, from Table 2.2.

$$25 \text{ mL} \times \left(\frac{10^{-3} \text{ L}}{1 \text{ mL}}\right) \times \left(\frac{0.265 \text{ gal}}{1.00 \text{ L}}\right) = 0.0066 \text{ gal}$$

2.85 Convert 83.2 kg to pounds. See Table 2.2 in your textbook for the conversion factor relating kilograms and pounds.

$$83.2 \text{ kg} \times \left(\frac{2.20 \text{ lb}}{1.00 \text{ kg}}\right) = 183 \text{ lb}$$

Convert 1.92 m to feet. Use two conversion factors: the relationship between inches and meters from Table 2.2, and the defined relationship between feet and inches.

$$1.92 \text{ m} \times \left(\frac{39.4 \text{ in.}}{1.00 \text{ m}}\right) \times \left(\frac{1 \text{ ft}}{12 \text{ in.}}\right) = 6.30 \text{ ft}$$

2.87 Exact numbers occur in definitions (1 foot = 12 inches). Therefore the answer will have the same number of significant figures as the measurement.

a. 4.3 feet – two significant figures
b. 3.09 feet – three significant figures
c. 0.33030 feet – five significant figures
d. 5.12310 feet – six significant figures

2.89 The conversion factor is obtained from a measurement (1 inch = 2.540 cm). Therefore, the answer will have the same number of significant figures as the number in the measurement or the conversion factor with the least number of significant figures.

a. 4.3 cm – two significant figures
b. 3.09 cm – three significant figures
c. 0.33030 cm – four significant figures
d. 5.12310 cm – four significant figures

2.91 Density is the ratio of the mass of an object to the volume occupied by that object. To calculate the density of mercury, substitute the given mass and volume values into the defining formula for density.

$$\text{Density} = \text{mass/volume} = \frac{524.5 \text{ g}}{38.72 \text{ cm}^3} = 13.55 \frac{\text{g}}{\text{cm}^3}$$

2.93 Use the reciprocal of the density of acetone, 0.791 g/mL, as a conversion factor to convert 20.0 g of acetone to milliliters.

$$20.0 \text{ g} \times \left(\frac{1 \text{ mL}}{0.791 \text{ g}}\right) = 25.3 \text{ mL}$$

2.95 Use the density of homogenized milk, 1.03 g/mL, as a conversion factor to convert 236 mL of homogenized milk to grams.

$$236 \text{ mL} \times \left(\frac{1.03 \text{ g}}{1 \text{ mL}}\right) = 243 \text{ g}$$

2.97 An object or a water-insoluble substance will float in water if its density is less than that of water, 1.0 g/cm^3.

a. Paraffin wax will float in water because its density, 0.90 g/cm^3, is less than that of water.

b. Limestone will sink in water because its density, 2.8 g/cm^3, is greater than that of water.

2.99 Density = mass/volume The answer will have the same number of significant figures as the measurement with the least number of significant figures.

a. Density = 1.0 g ÷ 2.0 cm^3 = 5.0 × 10^{-1} g/cm^3

b. Density = 1.000 g ÷ 2.00 cm^3 = 5.00 × 10^{-1} g/cm^3

c. Density = 1.0000 g ÷ 2.0000 cm^3 = 5.0000 × 10^{-1} g/cm^3

d. Density = 1.000 g ÷ 2.0000 cm^3 = 5.000 × 10^{-1} g/cm^3

2.101 Calculate the volume of the given mass of substance by using density as a conversion factor.

a. Gasoline: $75.0 \text{ g} \times \frac{1.0 \text{ mL}}{0.56 \text{ g}} = 1.3 \times 10^2 \text{ mL}$

b. Sodium metal: $75.0 \text{ g} \times \frac{1.0 \text{ cm}^3}{0.93 \text{ g}} \times \frac{1.0 \text{ mL}}{1.0 \text{ cm}^3} = 81 \text{ mL}$

c. Ammonia gas: $75.0 \text{ g} \times \frac{1.00 \text{ L}}{0.759 \text{ g}} \times \frac{1000 \text{ mL}}{1.00 \text{ L}} = 9.88 \times 10^4 \text{ mL}$

d. Mercury: $75.0 \text{ g} \times \frac{1.00 \text{ mL}}{13.6 \text{ g}} = 5.51 \text{ mL}$

2.103 The relationship between the Fahrenheit and Celsius temperature scales can be stated in the form of an equation:

$$°\text{F} = \frac{9}{5}(°\text{C}) + 32 \quad \text{or} \quad °\text{C} = \frac{5}{9}(°\text{F} - 32)$$

To find the temperature for baking pizza in degrees Celsius, substitute the degrees Fahrenheit in the appropriate form of the equation and solve for °C.

$$\frac{5}{9}\left(525° - 32°\right) = 274°\text{C}$$

2.105 Convert the freezing point of mercury, –38.9°C, to degrees Fahrenheit using the appropriate equation.

$$\frac{9}{5}\left(-38.9°\right) + 32.0° = -38.0°\text{F}$$

2.107 Convert one of the temperatures to the other temperature scale.

$$\frac{9}{5}\left(-10°\right) + 32° = 14°\text{F}; \; -10°\text{C is higher}$$

Atomic Structure and the Periodic Table — Chapter 3

Solutions to Selected Problems

3.1
a. An electron possesses a negative electrical charge.
b. A neutron has no electrical charge.
c. A proton has a mass slightly less than a neutron.
d. A proton has a positive electrical charge.

3.3
a. False. The nucleus of an atom is positively charged because it contains proton(s).
b. False. The nucleus of an atom contains protons and neutrons.
c. False. A nucleon is any subatomic particle found in the nucleus. Protons and neutrons are both found in the nucleus and so are both nucleons.
d. True. Neutrons and protons are in the nucleus, and both particles have much more mass than electrons.

3.5 An atom's atomic number (Z) is equal to the number of protons and also the number of electrons in the atom. Its mass number (A) is the sum of the number of protons and the number of neutrons in the nucleus of the atom.

a. $Z = 7$, so there are 7 protons and 7 electrons. $A = 15$, so there are 8 neutrons.
b. $Z = 20$, so there are 20 protons and 20 electrons. $A = 40$, so there are 20 neutrons.
c. $Z = 11$, so there are 11 protons and 11 electrons. $A = 23$, so there are 12 neutrons.
d. $Z = 35$, so there are 35 protons and 35 electrons. $A = 79$, there are 44 neutrons.

3.7 Atomic number (Z) is equal to the number of protons in an atom and also the number of electrons. Mass number (A) is the sum of the number of protons and the number of neutrons in the nucleus of an atom.

a. Z (atomic number) = protons = electrons = 50 protons = 50 electrons
Number of neutrons = $A - Z = 118 - 50 = 68$ neutrons.
b. Z (atomic number) = protons = electrons, so there are 78 protons and 78 electrons.
Number of neutrons = $A - Z = 195 - 78 = 117$ neutrons
c. Z = number of protons = number of electrons = 31 electrons = 31 protons.
$A - Z$ = number of neutrons = $72 - 31 = 41$ neutrons
d. $Z = 12$ protons = 12 electrons; $A - Z = 26 - 12 = 14$ neutrons

3.9 In a neutral atom: atomic number = number of protons = number of electrons;
mass number = protons + neutrons

a. atomic number = protons = 5
b. atomic number = protons = 7
c. atomic number = protons = 13
d. atomic number = protons = 20

3.11 Mass number = protons + neutrons

a. mass number = 5 + 6 = 11
b. mass number = 7 + 8 = 15
c. mass number = 13 + 14 = 27
d. mass number = 20 + 28 = 48

3.13 The number of nucleons in an atom is the sum of the protons and neutrons.

a. nucleons = 5 + 6 = 11
b. nucleons = 7 + 8 = 15
c. nucleons = 13 + 14 = 27
d. nucleons = 20 + 28 = 48

3.15 Protons and electrons are the charged particles in an atom; therefore the total number of charged particles = protons + electrons

a. charged particles = 5 + 5 = 10
b. charged particles = 7 + 7 = 14
c. charged particles = 13 + 13 = 26
d. charged particles = 20 + 20 = 40

3.17 Total nuclear charge = number of protons

a. +5
b. +7
c. +13
d. +20

3.19 In a neutral atom: atomic number = number of protons;
mass number (superscript) = protons + neutrons

	Symbol	Atomic number	Mass number	Number of protons	Number of neutrons
	$^{37}_{17}Cl$	17	37	17	20
a.	$^{232}_{94}Pu$	94	232	94	138
b.	$^{32}_{16}S$	16	32	16	16
c.	$^{56}_{26}Fe$	26	56	26	30
d.	$^{40}_{20}Ca$	20	40	20	20

3.21 In a neutral atom: number of protons = number of electrons; number of nucleons = number of protons + number of neutrons; total number of particles = number of protons + number of neutrons + number of electrons

a. Protons and electrons carry a charge. 11 + 11 = 22
b. Neutrons carry no charge. 14
c. Protons and neutrons are found in the nucleus. 11 + 14 = 25
d. Electrons are not found in the nucleus. 11

3.23 a. 34; the total number of subatomic particles is given by the sum of the atomic number and the mass number
b. 23; the total number of subatomic particles in the nucleus is given by the mass number
c. 23; the total number of nucleons is the same as the total number of subatomic particles present in the nucleus
d. +11; the total positive charge present on the nucleus is determined by the number of protons present; the number of protons present is given by the atomic number

3.25 In a neutral atom: number of protons (subscript) = number of electrons
mass number (superscript) = number of nucleons = number of protons + number of neutrons;
total number of particles = number of protons + number of neutrons + number of electrons

a. (3) The two atoms, $^{13}_{7}C$ and $^{13}_{6}N$, have the same number of nucleons.
b. (4) The two atoms, $^{37}_{17}Cl$ and $^{36}_{18}Ar$, have the same total number of subatomic particles.
c. (2) The two atoms, $^{35}_{17}Cl$ and $^{37}_{17}Cl$, have the same number of protons.
d. (1) The two atoms, $^{18}_{8}O$ and $^{19}_{9}F$, have the same number of neutrons

3.27 carbon-12, $^{12}_{6}C$; carbon-13, $^{13}_{6}C$; carbon-14, $^{14}_{6}C$

3.29 a. The mass number would not be the same for two different isotopes, because mass number is the total number of protons and neutrons; isotopes have the same number of protons but a different number of neutrons.

b. The number of electrons would be the same for two different isotopes; isotopes differ only in the number of neutrons.

c. The isotopic mass would not be the same for two different isotopes because isotopes differ in the number of neutrons in the nucleus.

d. This chemical property would be the same for two isotopes; chemical properties depend on the number of electrons in an atom.

3.31 An element's atomic mass is calculated by multiplying the relative mass of each isotope by its fractional abundance and then totaling the products.

a. 0.0742×6.01 amu = 0.446 amu
0.9258×7.02 amu = 6.50 amu
6.946 amu = 6.95 amu (answer is limited to the hundredths place)

b. 0.7899×23.99 amu = 18.95 amu
0.1000×24.99 amu = 2.499 amu
0.1101×25.98 amu = 2.860 amu
24.309 amu = 24.31 amu (answer is limited to the hundredths place)

3.33 An element's atomic mass is calculated by multiplying the relative mass of each isotope by its fractional abundance and then totaling the products.

a. False. Different isotopes have different masses.
b. False. Atomic mass is an average.
c. True.
d. True.

3.35 a. They are not isotopes; the atomic numbers differ.
b. They are not isotopes; the atomic numbers differ.
c. They are isotopes; they have the same atomic number.

3.37 a. False. Both sodium isotopes have the same number of electrons, 11.
b. False. Both sodium isotopes have the same number of protons, 11, but a different number of neutrons (12 and 13).
c. True. The total number of subatomic particles for ^{23}Na is 34; for ^{24}Na, it is 35.
d. True. These two isotopes of sodium have the same atomic number, 11.

3.39 a. Two isotopes of an element have the same atomic number and the same number of protons.
b. Two isotopes of an element have a different number of neutrons and nucleons.
c. The atomic number for an element is always the same; they have the same atomic number.
d. $A + Z$ = mass number + atomic number = (protons + neutrons) + protons
For two isotopes the number of neutrons is different; $A + Z$ is different.

3.41 We know that Cr has an atomic number of 24, so it has 24 protons.

a. $^{54}_{24}Cr$ is the isotope with 30 more neutrons than protons.

b. $^{50}_{24}Cr$ is the isotope with two fewer subatomic particles than $^{52}_{24}Cr$

c. $^{54}_{24}Cr$ is the isotope with the same number of neutrons ($A - Z$) as $^{55}_{25}Mn$

d. $^{65}_{24}Cr$ is the isotope with the same number of subatomic particles ($A + Z$) as $^{60}_{29}Cu$

3.43 In the periodic table, a period is a horizontal row of elements and a group is a vertical column of elements. Use the periodic table on the inside cover of your textbook.

a. $_{4}Be$ is in Period 2, Group IIA
b. $_{15}P$ is in Period 3, Group VA
c. $_{19}K$ is in Period 4, GroupIA
d. $_{53}I$ is in Period 5, Group VIIA

3.45 Use the periodic table on the inside cover of your textbook to determine these numbers.

a. The atomic number of carbon (C) is 6.
b. The atomic mass of silicon (Si) is 28.09 amu.
c. The element whose atomic mass is 88.91 has an atomic number of 39.
d. The element located in Period 2 and Group IIA has an atomic mass of 9.01 amu.

3.47 Elements in the same group in the periodic table have similar chemical properties. The following pairs of elements would be expected to have similar chemical properties.

a. K and Rb are both found in Group IA.
b. P and As are found in Group VA.
c. F and I are found in Group VIIA.
d. Na and Cs are found in Group IA.

3.49 a. The halogens are in Group VII; the Period 2 halogen is F.
b. The alkaline earths are in Group IIA; the Period 2 alkaline earth is Be.
c. The alkali metals are in Group IA; the Period 2 alkali metal is Li.
d. The noble gases are in Group VIIIA; the Period 2 noble gas is Ne.

3.51 Check your periodic table to find the number of elements in these groups with an atomic number less than 40.

a. Three elements. Halogens are found in Group VIIA.
b. Four elements. Noble gases are found in Group VIIIA.
c. Four elements. Alkali metals are found in Group IA.
d. Four elements. Alkaline earth metals are found in Group IIA.

3.53 a. blue element b. yellow element c. yellow element d. green element

3.55 a. $^{12}_{4}Be$ is in period 2 and group IIA and has four more neutrons than protons.

b. $^{132}_{54}Xe$ is in period 5 and group VIIIA and has a mass number of 132.

c. $^{16}_{8}O$ is in period 2 and group VIA and has an equal number of the two kinds of nucleons (protons = neutrons).

d. $^{40}_{20}Ca$ is in period 4 and group IIA and has an equal number of all three kinds of subatomic particles (protons = neutrons = electrons).

3.57 Figure 3.6 shows the location of metals and nonmetals in the periodic table.

a. No. Cl and Br are in Group VIIA and are nonmetals.

b. No. Al (Group IIIA) is a metal and Si (Group IVA) is a nonmetal.

c. Yes. Cu and Mo are both found in the metals section of the periodic table.

d. Yes. Zn and Bi are both metals.

3.59 Figure 3.6 shows the location of metals and nonmetals in the periodic table.

a. S is a nonmetal; Na and K are Group IA metals.

b. P is a nonmetal. Cu (Group IB) and Li (Group (IA) are both metals.

c. I is a nonmetal (Group VIIA). Be and Ca are both metals.

d. Cl is a nonmetal (Group VIIA). Fe and Ga are both metals.

3.61 a. Ductility is one of the properties of a metal.

b. Nonmetals have low electrical conductivity.

c. Metals have high thermal conductivity.

d. Nonmetals are good heat insulators.

3.63 a. metal b. nonmetal

c. poor conductor of electricity d. good conductor of heat

3.65 Compare the six elements (nitrogen, beryllium, argon, aluminum, silver, and gold) using information from the periodic table.

a. For Be and Al the period number and Roman numeral group number are numerically equal.

b. Be, Al, Ag, and Au are all metals and readily conduct electricity.

c. Each of these elements (N, Be, Ar, Al, Ag, and Au) has an atomic mass greater than its atomic number.

d. Ag and Au have a nuclear charge that is greater than +20.

3.67 Electron configurations are written as number-letter combinations in which the number is the number of the shell and the letter is the name of the subshell.

a. The electron shell for a 2*s* electron is 2. b. The electron shell for a 4*s* electron is 4.

c. The electron shell for a 3*d* electron is 3. d. The electron shell for a 5*p* electron is 5.

3.69 An *s*-subshell can accommodate a maximum of 2 electrons; a *p*-subshell can accommodate a maximum of 6 electrons; a *d*-subshell can accommodate a maximum of 10 electrons.

a. 2 electrons b. 2 electrons c. 10 electrons d. 6 electrons

3.71 An electron orbital can accommodate a maximum of 2 electrons.

a. 2 electrons b. 2 electrons c. 2 electrons d. 2 electrons

3.73 For the subshell types listed in Problem 3.67,

a. the 2*s* subshell contains one electron orbital.

b. the 4*s* subshell contains one electron orbital.

c. the 3*d* subshell contains 5 electron orbitals.

d. the 5*p* subshell contains 3 electron orbitals.

3.75 a. True. The shape and size of an electron orbital are related to the energy of the electrons it accommodates.

b. True. Orbitals in a subshell have the same energy although they differ in orientation.

c. False. An *s*-subshell can have a maximum of two electrons, while an *f*-subshell can have up to 14 electrons.

d. True. A *p*-subshell has three orbitals.

3.77 An electron configuration is a statement of how many electrons an atom has in each of its electron subshells. Subshells containing electrons are listed in order of increasing energy using number-letter combinations. A superscript indicates the number of electrons in that subshell. Fig. 3.12 shows the order for filling electron subshells.

a. Carbon has 6 electrons: $1s^2 2s^2 2p^2$
b. Sodium has 11 electrons: $1s^2 2s^2 2p^6 3s^1$
c. Sulfur has 16 electrons: $1s^2 2s^2 2p^6 3s^2 3p^4$
d. Argon has 18 electrons: $1s^2 2s^2 2p^6 3s^2 3p^6$

3.79 a. Oxygen. Adding the numbers of electrons in the electron configuration (add the superscripts) gives a total of 8 electrons, which corresponds to the atomic number of oxygen.

b. Neon. The total number of electrons is 10, which corresponds to the atomic number of neon.

c. Aluminum. The total number of electrons is 13, and this is the atomic number of aluminum.

d. Calcium. The total number of electrons is 20, and this is the atomic number of calcium.

3.81 Electron configurations are written as number-letter combinations in which the number is the number of the shell and the letter is the name of the subshell. A superscript to the right of the letter gives the number of electrons in the subshell. The filling order of electron subshells is given in Fig. 3.11.

a. $1s^2 2s^2 2p^6 3s^2 3p^5$
b. $1s^2 2s^2 2p^6 3s^2 3p^6 4s^2 3d^{10} 4p^6 5s^2 4d^7$
c. $1s^2 2s^2 2p^6 3s^2 3p^6 4s^2$
d. $1s^2 2s^2 2p^6 3s^2 3p^6 4s^2 3d^1$

3.83 An orbital diagram is a statement of how many electrons an atom has in each of its electron orbitals. Electrons will occupy equal-energy orbitals singly to the maximum extent possible before any orbital acquires a second electron (as in parts a., c., and d. below).

	1*s*	2*s*	2*p*	3*s*	3*p*	4*s*	3*d*
a.	[↓↑]	[↓↑]	[↑] [↑] [ ]				
b.	[↓↑]	[↓↑]	[↓↑] [↓↑] [↓↑]	[↓↑]			
c.	[↓↑]	[↓↑]	[↓↑] [↓↑] [↓↑]	[↓↑]	[↑] [↑] [↑]		
d.	[↓↑]	[↓↑]	[↓↑] [↓↑] [↓↑]	[↓↑]	[↓↑] [↓↑] [↓↑]	[↓↑]	[↓↑] [↓↑] [↑] [↑] [↑]

3.85 As shown in the electron configurations in Problem 3.83, the number of unpaired electrons is:
a. 2 b. 0 c. 3 d. 3

3.87 Since two isotopes of the same element have the same number of electrons, their electron configurations are the same: $1s^22s^22p^4$

3.89 a. The group III element in the same period as $_4$Be is $_5$B; its electron configuration is $1s^22s^22p^1$.
b. The period 3 element in the same group as $_5$B is $_{13}$Al; its electron configuration is $1s^22s^22p^63s^23p^1$.
c. The lowest-atomic-numbered *metal* in group IIA is $_4$Be; its electron configuration is $1s^22s^2$.
d. The two period 3 elements that have no unpaired electrons are $_{12}$Mg and $_{18}$Ar. The electron configuration for $_{12}$Mg is $1s^22s^22p^63s^2$; for $_{18}$Ar it is $1s^22s^22p^63s^23p^6$.

3.91 a. No. From the electron configuration, we can see that the first element has one valence electron and is in Group IA. The second contains two valence electrons and is in Group IIA.
b. Yes. Both elements have six valence electrons (Group VIA).
c. No. The first element has three valence electrons, and the second has five valence electrons.
d. Yes. Both elements have six valence electrons and are in Group VIA.

3.93 The distinguishing electron is the last electron added to the electron configuration for an element. Figure 3.12 relates area of the periodic table in which an element is found to the distinguishing electron of its atom.
a. The *s* area. The distinguishing electron for magnesium ($1s^22s^22p^63s^2$) is in the 3*s* orbital.
b. The *d* area. The distinguishing electron for copper ($1s^22s^22p^63s^23p^64s^23d^9$) is in the 3*d* orbital.
c. The *p* area. The distinguishing electron for bromine ($1s^22s^22p^63s^23p^64s^23d^{10}4p^5$) is in the 4*p* orbital.
d. The *d* area. The distinguishing electron for iron ($1s^22s^22p^63s^23p^64s^23d^6$) is in the 3*d* orbital.

3.95 Figure 3.12 and the periodic table will give you the area and group to which the element belongs.
a. p^1. Aluminum is found in Group IIIA, which is the first column of the *p* area, which makes the distinguishing electron p^1.
b. d^3. Vanadium is found in Group VB, which is the third column of the *d* area, which makes the distinguishing electron d^3.
c. s^2. Calcium is found in Group IIA, which is the second column of the *s* area, which makes the distinguishing electron s^2.
d. p^6. Krypton is found in Group VIIIA, which is the sixth column of the *p* area, which makes the distinguishing electron p^6.

3.97 a. *s* area b. *d* area c. p^4 element d. s^2 element

3.99 Figure 3.13 gives a classification scheme for the elements according to their position in the periodic table.
a. Phosphorus is a representative element.
b. Argon is a noble gas.
c. Gold is a transition element.
d. Uranium is an inner transition element.

3.101 a. 4; (2 in the *s* area and 2 in the first five columns of the *p* area)
b. 1; (the last column of the *p* area)
c. 2; (H in the *s* area and one element in the *p* area)
d. 6; (everything other than H and two elements in the *p* area)

3.103 Totaling the electrons in the electron configuration gives the atomic number of each element. Use Figure 3.13 to classify the element in the correct block of the periodic table.
a. Noble gas. The atom has 10 electrons; the element is neon.
b. Representative element. The atom has 16 electrons; the element is sulfur.
c. Transition element. The atom has 21 electrons; the element is scandium.
d. Representative element. The atom has 20 electrons; the element is calcium.

Chemical Bonding: The Ionic Bond Model — Chapter 4

Solutions to Selected Problems

4.1 The mechanism for ionic bond formation is electron transfer and that for covalent bond formation is electron sharing.

4.3 A valence electron is an electron in the outermost electron shell of a representative element or a noble-gas element.

a. Two valence electrons in the 2*s* subshell (shown by the superscript 2)
b. Two valence electrons, in the 3*s* subshell
c. Three valence electrons, two in the 2*s* subshell and one in the 3*p* subshell
d. Four valence electrons, two in the 4*s* subshell and two in the 4*p* subshell

4.5 The group number of an element is found above the vertical column of that element in the Periodic Table. A valence electron is an electron in the outermost electron shell of an element.

a. ${}_{7}N$ is in Group VA and has 5 valence electrons.
b. ${}_{11}Na$ is in Group IA and has 1 valence electron.
c. ${}_{16}S$ is in Group VIA and has 6 valence electrons.
d. ${}_{19}K$ is in Group IA and has 1 valence electron.

4.7 a. A Period 2 element with 4 valence electrons is found in Group IVA; this element is carbon: $1s^22s^22p^2$

b. A Period 2 element with 7 valence electrons is found in Group VIIA; this element is fluorine: $1s^22s^22p^5$

c. A Period 3 element with 2 valence electrons is found in Group IIA; this element is magnesium: $1s^22s^22p^63s^2$

d. A Period 3 element with 5 valence electrons is found in Group VA; this element is phosphorus: $1s^22s^22p^63s^23p^3$

4.9 A valence electron is an electron in the outermost shell of a representative element.

a. Mg has two valence electrons, which is <u>more</u> than the one valence electron that Na has.
b. P and As both have the <u>same</u> number of valence electrons, five.
c. O has six valence electrons, which is <u>more</u> than the five valence electrons that N has.
d. Si has four valence electrons, which is <u>fewer</u> than the seven valence electrons that Cl has.

4.11 The number of valence electrons for representative elements is the same as the Roman numeral periodic-table group number.

a. There is one highlighted element with five valence electrons (Group VA).
b. There is one highlighted element with one valence electron (Group IA).
c. There is one highlighted element with seven valence electrons (Group VIIA).
d. There are two highlighted elements with two valence electrons (Group IIA).

4.13 A Lewis symbol is the chemical symbol of an element surrounded by dots equal in number to the number of valence electrons (electrons in the outermost shell) in atoms of the element. The number of valence electrons can be determined from the element's group number in the periodic table.

a. 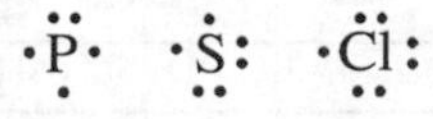

b. Na· K· Rb·

4.15 The number of valence electrons an atom has corresponds to its group number. Count the valence electrons given in the Lewis symbol. Find the Period 2 element that is in that group.
a. Li b. F c. Be d. N

4.17 A Lewis symbol is the chemical symbol of an element surrounded by dots equal in number to the number of valence electrons present in atoms of the element.

a. · Be · is the Lewis structure; Be has the electron configuration $1s^22s^2$.

b. ·Mg· is the Lewis structure; Mg has the electron configuration $1s^22s^22p^63s^2$.

c. :Ṡ̤: is the Lewis structure; S has the electron configuration $1s^22s^22p^63s^23p^4$.

d. · Ġ̣e · is the Lewis structure; Ge has the electron configuration $1s^22s^22p^63s^23p^64s^23d^{10}4p^2$.

4.19 Noble gases are the most unreactive of all elements.

4.21 They lose, gain, or share electrons in such a way that they achieve a noble-gas electron configuration.

4.23 An ion is an atom (or group of atoms) that is electrically charged because it has lost or gained electrons.
a. The symbol for the ion is O^{2-}. The atom has gained two electrons and so has a –2 charge.
b. The symbol for the ion is Mg^{2+}. The atom has lost two electrons and so has a +2 charge.
c. The symbol for the ion is F^-. The atom has gained one electron and so has a –1 charge.
d. The symbol for the ion is Al^{3+}. The atom has lost three electrons and so has a +3 charge.

4.25 The number of protons in each ion gives the atomic number, and thus the atomic symbol, of the element. Since electrons are negative and protons are positive, the difference between the numbers of protons and electrons gives the charge on the ion and its magnitude.
a. The chemical symbol is Ca^{2+}. The charge is +2; there are two more protons than electrons.
b. The chemical symbol is O^{2-}. The charge is –2; there are two more electrons than protons.
c. The chemical symbol is N^{3-}. The charge is –3; there are three more electrons than protons.
d. The chemical symbol is Be^{2+}. The charge is +2; there are two more protons than electrons.

4.27 The charge on an ion and its magnitude is equal to the number of protons minus the number of electrons.
a. The ion N^{3-} has 7 protons.
b. The ion Mg^{2+} has 12 protons, 10 electrons, and has lost two electrons.
c. The ion F^- has 10 electrons.
d. The ion Li^+ has lost one electron.

4.29 Metal atoms containing one, two, or three valence electrons tend to form ions by losing valence electrons; nonmetal ions containing five, six, or seven valence electrons tend to form ions by gaining valence electrons.

	Chemical symbol	Ion formed	Number of electrons in ion	Number of protons in ion
	Ca	Ca^{2+}	18	20
a.	Be	Be^{2+}	2	4
b.	I	I^-	54	53
c.	Al	Al^{3+}	10	13
d.	S	S^{2-}	18	16

4.31 Atoms containing one, two, or three valence electrons (Group IA, IIA, or IIIA) tend to form ions by losing valence electrons; atoms containing five, six, or seven valence electrons (Group VA, VIA, or VIIA) tend to form ions by gaining electrons.

a. An ion (X^{2+}) with 10 electrons is in period 3, group IIA; it is Mg.
b. An ion (X^{2+}) with 18 electrons is in period 4, group IIA; it is Ca.
c. An ion (X^{3-}) with 18 electrons is in period 3, group VA; it is P.
d. An ion (X^{-}) with 10 electrons is in period 2, group VIIA; it is F.

4.33 a. An ion with two more protons than electrons has a +2 charge; it would be in group IIA. In period 2 this is the ion Be^{2+}.
b. An ion with two fewer protons than electrons has a –2 charge; it would be in group VIA. In period 2 this is the ion O^{2-}.
c. An ion with three more protons than electrons has a +3 charge; it would be in group IIIA. In period 2 this is the ion B^{3+}.
d. An ion with four fewer protons than electrons has a –4 charge; it would be in group IVA. In period 2 this is the ion C^{4-}.

4.35 A Lewis symbol for an ion is the chemical symbol of an element surrounded by dots equal in number to the number of valence electrons present in that ion. The Lewis structure for each of these negatively charged ions shows a complete octet.

a. $[:\ddot{\underset{..}{P}}:]^{3-}$ b. $[:\ddot{\underset{..}{N}}:]^{3-}$ c. $[:\ddot{\underset{..}{C}}:]^{4-}$ d. $[:\ddot{\underset{..}{F}}:]^{-}$

4.37 Atoms tend to gain or lose electrons until they have obtained an electron configuration that is the same as that of a noble gas.

a. Mg loses two electrons to gain the electron configuration of neon; the ion has a +2 charge.
b. N gains three electrons to gain the electron configuration of neon; the ion has a –3 charge.
c. K loses one electron to gain the electron configuration of argon; the ion has a +1 charge.
d. F gains one electron to the electron configuration of neon; the ion has a –1 charge.

4.39 When atoms form ions, they tend to lose or gain the number of electron that will give them the electron configuration of a noble gas. Find each element's nearest noble gas in the periodic table.

a. Two electrons are lost when Be forms an ion.
b. One electron is gained when Br forms an ion.
c. Two electrons are lost when Sr forms an ion.
d. Two electrons are gained when Se forms an ion.

4.41 a. Ne; there are 10 electrons present in an O^{2-} ion, the same number as in Ne.
b. Ar; there are 18 electrons present in a P^{3-} ion, the same number as in Ar.
c. Ar; there are 18 electrons present in a Ca^{2+} ion, the same number as in Ar.
d. Ar; there are 18 electrons present in a K^{+} ion, the same number as in Ar.

4.43 Isoelectronic means the same electron configuration; thus, the answers are the same as in Problem 4.33.

a. Ne b. Ar c. Ar d. Ar

4.45 Isoelectronic means the atoms/ions have the same electron configuration.

a. Yes, Ca^{2+} and Ar are isoelectronic. b. Yes, Mg^{2+} and F^{-} are isoelectronic.
c. No, Ar and Kr are not isoelectronic, d. Yes, O^{2-} and N^{3-} are isoelectronic.

4.47 a. Four of the highlighted elements will form a positively-charged ion.
b. Four of the highlighted elements will form an ion through loss of electrons.
c. One of the highlighted elements forms an ion having a charge magnitude of two.
d. One of the highlighted elements will form an ion by gaining two or more electrons.

4.49 Aluminum (13 electrons) forms a +3 ion, which means that the atom has lost three electrons to form the ion (10 electrons).

a. Aluminum atom: $1s^22s^22p^63s^23p^1$ b. Aluminum ion: $1s^22s^22p^6$

4.51 A valence electron is an electron in the outermost shell of a representative element. Each of these ions has the electron configuration of the nearest gas, an octet of electrons. In the case of Li^+, an "octet" is two electrons.

a. N^{3-} has 8 valence electrons. b. P^{3-} has 8 valence electrons.
c. Al^{3+} has 8 valence electrons. d. Li^+ has 2 valence electrons.

4.53 A Lewis structure is a combination of Lewis symbols that represents either the transfer or the sharing of valence electrons in chemical bonds.

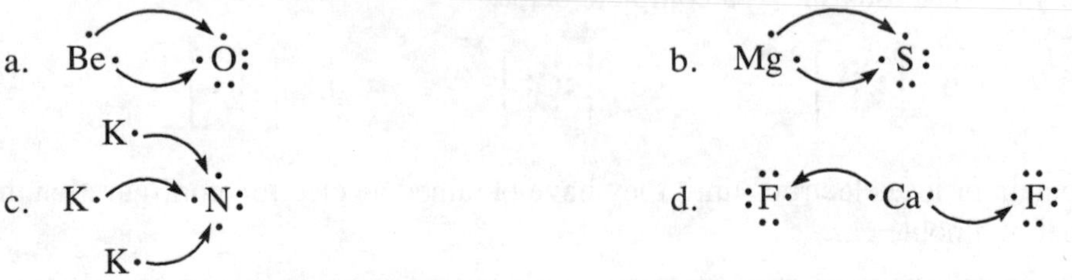

a. Be· → ·O: b. Mg· → ·S:

c. K· K· K· → ·N: d. :F· ← ·Ca· → ·F:

4.55 a. 2 extra electrons, –2 charge; S has 6 valence electrons, and the ion has 8 valence electrons.
b. 1 extra electron, –1 charge; F has 7 valence electrons, and the ion has 8 valence electrons.
c. 3 extra electrons, –3 charge; N has 5 valence electrons, and the ion has 8 valence electrons.
d. 2 extra electrons, –2 charge; Se has 6 valence electrons, and the ion has 8 valence electrons.

4.57 The chemical formula for an ionic compound combines ions in a ratio that achieves a balance of positive and negative charges.

a. Na_2S One S^{2-} ion combines with two Na^+ ions to give an uncharged chemical formula.
b. CaI_2 Two I^- ions combine with one Ca^{2+} ion to give an uncharged chemical formula.
c. Li_3N One N^{3-} ion combines with three Li^+ ions to give an uncharged chemical formula.
d. $AlBr_3$ Three Br^- ions combine with one Al^{3+} ion to give an uncharged chemical formula.

4.59 To write chemical formulas for these combining ions, balance the positive and negative charges of the ions.

		F^-	O^{2-}	N^{3-}	C^{4-}
	Na^+	NaF	Na_2O	Na_3N	Na_4C
a.	Ca^{2+}	CaF_2	CaO	Ca_3N_2	Ca_2C
b.	Al^{3+}	AlF_3	Al_2O_3	AlN	Al_4C_3
c.	Ag^+	AgF	Ag_2O	Ag_3N	Ag_4C
d.	Zn^{2+}	ZnF_2	ZnO	Zn_3N_2	Zn_2C

4.61 a. $BeCl_2$; Be forms a +2 ion and Cl forms a –1 ion.
b. BaI_2; Ba forms a +2 ion and I forms a –1 ion.
c. Na_2O; Na forms a +1 ion and O forms a –2 ion.
d. AlN; Al forms a +3 ion and N forms a –3 ion.

4.63 When forming ions or compounds, metals tend to lose electrons to reach the electron configuration of the nearest noble gas.
a. The chemical symbol for the potassium ion is K^+.
b. The chemical symbol for the calcium ion is Ca^{2+}.
c. The chemical symbol for the magnesium ion is Mg^{2+}.
d. The chemical symbol for the aluminum ion is Al^{3+}.

4.65 The chemical formula for an ionic compound combines ions in a ratio that achieves a balance of positive and negative charges.
a. XZ_2 X will lose 2 electrons, each Z will gain 1 electron. The combining ratio is 1:2
b. X_2Z X will lose 1 electron, Z will gain 2 electrons. The combining ratio is 2:1
c. XZ X will lose 3 electrons, Z will gain 3 electrons. The combining ratio is 1:1
d. ZX X will gain 2 electrons, Z will lose 2 electrons. The combining ratio is 1:1. The metal ion (the ion losing the electrons) is written first.

4.67 The chemical formula for an ionic compound combines ions in a ratio that achieves a balance of positive and negative charges.
a. LiCl b. MgF_2 c. CaS d. K_3N

4.69 a. Representative elements in Group VIA form ions with a minus two charge.
b. The Lewis symbol is: $\cdot\ddot{\underset{\cdot\cdot}{X}}\colon$
c. Since X is in Group VIA, it has six valence electrons.
d. CaX is the chemical formula that achieves a balance of positive and negative charges.

4.71 In the ionic compound K_2S:
a. Each K atom loses one electron.
b. The S atom gains two electrons.
c. The K^+ ions have a +1 charge.
d. The S^{2-} ions have a –2 charge.

4.73 A solid-state ionic compound is an extended array of alternating positive and negative ions.

4.75 A formula unit of an ionic compound is the smallest whole-number ratio of ions present.

4.77 Ions of opposite charge (a metal and a nonmetal) must be present in an ionic compound.

4.79 A binary ionic compound forms between a metal (positive ion) and a nonmetal (negative ion).
a. Al_2O_3 is an ionic compound; Al is a metal and O is a nonmetal.
b. H_2O_2 is not an ionic compound; both H and O are nonmetals.
c. K_2S is an ionic compound; K is a metal and S is a nonmetal.
d. N_2H_4 is not an ionic compound; both N and H are nonmetals.

4.81 In naming binary ionic compounds, name the metallic element first, followed by a separate word containing the stem of the nonmetallic element name and the suffix *–ide*.
a. KI – potassium iodide
b. BeO – beryllium oxide
c. AlF_3 – aluminum fluoride
d. Na_3P – sodium phosphide

4.83 In a binary ionic compound, oxide ion always has a charge of –2. The charge on the metal ion must balance the negative charge of the oxygen ions present.

a. The charge on Au is +1. Since one oxide ion carries a –2 charge, this must be balanced by a +2 charge on two Au ions, or a +1 for each Au.
b. The charge on Cu is +2. One oxide ion carries a –2 charge, which is balanced by one Cu^{2+}.
c. The charge on Sn is +4. Two oxide ions carrying a –4 charge are balanced by one Sn^{+4}.
d. The charge on Sn is +2. One oxide ion carrying a –2 charge is balanced by one Sn^{+2}.

4.85 a. Au_2O is gold (I) oxide. b. NiO is nickel (II) oxide.
c. SnO_2 is tin (IV) oxide. d. SnO is tin (II) oxide.

4.87 When naming a binary ionic compound containing a variably-charged metal ion, the charge on the metal ion must be included in the name by using a Roman numeral after the metal name, copper (II) oxide.

4.89 Figure 4.8 shows which metals have fixed ionic charges. The charge on a variably-charged metal ion is specified by a Roman numeral after the metal name in the compound name.
a. AuCl is named gold(I) chloride. b. KCl is named potassium chloride.
c. AgCl is named silver chloride. d. $CuCl_2$ is named copper(II) chloride.

4.91 When naming a binary ionic compound containing a variably-charged metal ion, the charge on the metal ion is incorporated in the name by using a Roman numeral after the metal name.
a. PbO is named lead (II) oxide; PbO_2 is named lead (IV) oxide.
b. $MgCl_2$ is named magnesium chloride; $FeCl_2$ is named iron (II) chloride.
c. Na_2O is named sodium oxide; MgO is named magnesium oxide.
d. Cu_2S is named copper (I) sulfide; CuS is named copper (II) sulfide.

4.93 The chemical formulas of binary ionic compounds must be balanced in terms of ionic charge. Table 4.2 gives the names of some common nonmetallic ions and their charges, and Figure 4.8 gives charges for metallic elements with fixed ionic charges. The chemical formula
a. for potassium bromide is KBr. b. for silver oxide is Ag_2O.
c. for beryllium fluoride is BeF_2. d. for barium phosphide is Ba_3P_2.

4.95 Since the charges on the nonmetallic ions are fixed, and we know the charges on the metal ions from their names, we can balance the chemical formulas in terms of ionic charge.
a. CoS is the chemical formula for cobalt(II) sulfide.
b. Co_2S_3 is the chemical formula for cobalt(III) sulfide.
c. SnI_4 is the chemical formula for tin(IV) iodide.
d. Pb_3N_2 is the chemical formula for lead(II) nitride.

4.97 In naming binary ionic compounds, name the metallic element first, followed by a separate word containing the stem of the nonmetallic element name and the suffix *–ide*. The ionic charges in the chemical formulas are balanced choosing the numbers of positive and negative ions that will give the chemical formula a charge of zero.
a. The name of the compound is magnesium nitride.
b. The chemical formula for the compound is Mg_3N_2.
c. There are five ions present in a formula unit (3 Mg^{+2} + 2 N^{3-}).
d. Six electrons are transferred per formula unit.

4.99 In naming binary ionic compounds, name the metallic element first, followed by a separate word containing the stem of the nonmetallic element name and the suffix *–ide*.

	Period number	Electron-dot symbols	Formula of compound	Name of compound
	2	·X· and :Y·	BeO	Beryllium oxide
a.	3	X· and :Y:	NaCl	Sodium chloride
b.	4	X· and :Y·	K_2Se	Potassium selenide
c.	2	·X· and :Y:	BeF_2	Beryllium fluoride
d.	3	·X· and :Y·	AlN	Aluminum nitride

4.101 a. Sulfate ion (SO_4^{2-}) has a –2 charge.
b. Nitrate ion (NO_3^-) has a –1 charge.
c. Phosphate ion (PO_4^{3-}) has a –3 charge.
d. Carbonate ion (CO_3^{2-}) has a –2 charge.

4.103 a. The hydronium ion (H_3O^+) contains 1 oxygen atom.
b. The cyanide ion (CN^-) contains no oxygen atoms.
c. The hydroxide ion (OH^-) contains 1 oxygen atom.
d. The ammonium ion (NH_4^+) contains no oxygen atoms.

4.105 a. HCO_3^- is named hydrogen carbonate ion.
b. $H_2PO_4^-$ is named dihydrogen phosphate ion.
c. H_3O^+ is named hydronium ion.
d. NH_4^+ is named ammonium ion.

4.107 The ionic charges in the chemical formulas are balanced choosing the numbers of positive and negative ions that will give the chemical formula a charge of zero.
a. NaOH The +1 charge on one sodium ion balances the –1 charge on one hydroxide ion.
b. $Fe(HSO_4)_3$ The +3 charge on one iron(III) ion balances the –1 charge on each of the three hydrogen sulfate ions.
c. $Ba(NO_3)_2$ The +2 charge on one barium ion balances the –1 charge on each of the two nitrate ions.
d. $Al_2(CO_3)_3$ The +3 charge on each of the two aluminum ions balances the –2 charge on each of the three carbonate ions.

4.109 The ionic charges in the chemical formulas are balanced choosing the numbers of positive and negative ions that will give the chemical formula a charge of zero.

		CN^-	**NO^{3-}**	**HCO^{3-}**	**SO_4^{2-}**
	Na^+	NaCN	$NaNO_3$	$NaHCO_3$	Na_2SO_4
a.	**Al^{3+}**	$Al(CN)_3$	$Al(NO_3)_3$	$Al(HCO_3)_3$	$Al_2(SO_4)_3$
b.	**Ag^+**	AgCN	$AgNO_3$	$AgHCO_3$	Ag_2SO_4
c.	**Ca^{2+}**	$Ca(CN)_2$	$Ca(NO_3)_2$	$Ca(HCO_3)_2$	$CaSO_4$
d.	**NH_4^+**	NH_4CN	NH_4NO_3	NH_4HCO_3	$(NH_4)_2SO_4$

4.111 The number of total ions present is the sum of the number of positive ions and the number of negative ions.
a. There are two ions present per formula unit in $LiHCO_3$, one Li^+ and one HCO_3^-.
b. There are two ions present per formula unit in $BeCO_3$, one Be^{2+} and one CO_3^{2-}.
c. There are five ions present per formula unit in $Al_2(SO_4)_3$, two Al^{3+} and three SO_4^{2-}.
d. There are two ions present per formula unit in NH_4NO_3, one NH_4^+ and one NO_3^-.

4.113 In naming ionic compounds containing negative polyatomic ions, give the metal ion name first and then the name of the polyatomic ion.
a. $MgCO_3$ is named magnesium carbonate. b. $ZnSO_4$ is named zinc sulfate.
c. $Be(NO_3)_2$ is named beryllium nitrate. d. Ag_3PO_4 is named silver phosphate.

4.115 When naming ionic compounds containing variable-charge metal ions and polyatomic ions, it is necessary to find the magnitude of the charge on each metal ion. To do this, balance the total negative charges against the total positive charges.
a. $Fe(OH)_2$ is named iron(II) hydroxide because the –2 charge on the hydroxide ions is balanced by a +2 charge on the iron ion (the variable-charge metal ion).
b. $CuCO_3$ is named copper(II) carbonate because the –2 charge on the carbonate ion is balanced by a +2 charge on the copper ion.
c. AuCN is named gold(I) cyanide because the –1 charge on the cyanide ion is balanced by a +1 charge on the gold ion.
d. $Mn_3(PO_4)_2$ is named manganese(II) phosphate because the –3 charge on each of the two phosphate ions is balanced by a +2 charge on each of the three manganese ions.

4.117 Balance the chemical formulas in terms of ionic charge.
a. Potassium bicarbonate – $KHCO_3$ One potassium ion is balanced by one bicarbonate ion.
b. Gold(III) sulfate – $Au_2(SO_4)_3$ Two gold(III) ions are balanced by three sulfate ions.
c. Silver nitrate – $AgNO_3$ One silver ion is balanced by one nitrate ion.
d. Copper(II) phosphate – $Cu_3(PO_4)_2$ Three copper(II) ions are balanced by two phosphate ions.

4.119 Balance the chemical formulas in terms of ionic charge.
a. The chemical formula for sodium sulfide is Na_2S.
b. The chemical formula for sodium sulfate is Na_2SO_4.
c. The chemical formula for sodium carbonate is Na_2CO_3.
d. The chemical formula for sodium hydrogen carbonate is $NaHCO_3$.

4.121 a. In naming ionic compounds containing negative polyatomic ions, give the metal ion name first and then the name of the polyatomic ion. The name of the compound formed from Al^{3+} and CO_3^{2-} is aluminum carbonate.
b. Balance the chemical formulas in terms of ionic charge. The chemical formula of the compound is $Al_2(CO_3)_3$.
c. There are three polyatomic ions present in one formula unit of $Al_2(CO_3)_3$, three CO_3^{2-}.
d. There are two monoatomic ions present in one formula unit of $Al_2(CO_3)_3$, two Al^{3+}.

4.123 Balance the chemical formulas in terms of ionic charge. In naming ionic compounds, give the metal ion name first and then the name of the nonmetal ion with the appropriate suffix (*-ide*, *-ite*, or *–ate*).

	Positive ion	Negative ion	Chemical formula	Name
	Mg^{2+}	OH^-	$Mg(OH)_2$	magnesium hydroxide
a.	Ba^{2+}	Br^-	$BaBr_2$	barium bromide
b.	Zn^{2+}	NO_3^-	$Zn(NO_3)_2$	zinc nitrate
c.	Fe^{3+}	CO_3^{2-}	$Fe_2(CO_3)_3$	iron (III) carbonate
d.	Pb^{4+}	O^{2-}	PbO_2	lead (IV) oxide

Chemical Bonding: The Covalent Bond Model

Chapter 5

Solutions to Selected Problems

5.1 An ionic bond involves a metal and a nonmetal: a covalent bond involves two nonmetals.

5.3 An ionic compound structural unit is an extended array of alternating positive and negative ions; a molecular compound structural unit is a molecule.

5.5 Covalent bond formation occurs between similar or identical atoms. Most often two nonmetals are involved.

a. Yes, a covalent bond forms between two N atoms.
b. Yes, a covalent bond forms between an H atom and an F atom.
c. Yes, a covalent bond forms between two Cl atoms.
d. No, a covalent bond does not form between a Cu atom and an O atom.

5.7 Lewis structures for molecular compounds are drawn so that each atom has an octet of electrons. Bromine, chlorine, and fluorine are in Group VIIA of the periodic table with seven valence electrons each; each atom needs one covalent bond (one shared pair of electrons) to have an octet of electrons. In part b., hydrogen shares its one electron to form a covalent bond with one of chlorine's electrons; for hydrogen, an "octet" is only two electrons.

a. $:\ddot{\underset{..}{Cl}}:\ddot{\underset{..}{Cl}}:$ b. $H:\ddot{\underset{..}{Cl}}:$ c. $:\ddot{\underset{..}{Br}}:\ddot{\underset{..}{Cl}}:$ d. $:\ddot{\underset{..}{Cl}}:\ddot{\underset{..}{F}}:$

5.9 a. 8 b. 4 c. zero d. 6

5.11 Use the octet rule (each atom in the molecule shares in an octet of electrons) to determine the number of atoms of each type in the simplest compound.

a. The two red atoms are H and S. S has 6 valence electrons. The addition of one electron from each of two hydrogens will form an octet. The chemical formula is H_2S.
b. The blue atoms are C (4 valence electrons) and Br (7 valence electrons). Each of the four C electrons is shared with a Br atom, completing its octet. The chemical formula is CBr_4.
c. The yellow atoms are O (6 valence electrons) and F (7 valence electrons). Each of two F atoms shares one electron with one O atom, completing octets for all three atoms. The chemical formula is OF_2.
d. The green atoms are N (5 valence electrons) and Cl (7 valence electrons). Each Cl atom needs one more electron and N needs 3 more electrons; one N atom shares 1 electron with each of three Cl atoms, completing octets for all four atoms. The chemical formula is NCl_3.

5.13 In a single covalent bond, two atoms share one pair of electrons. In a double covalent bond, two atoms share two pairs of electrons. In a triple covalent bond, two atoms share three pairs of electrons.

a. N_2 has one triple bond and two pairs of nonbonding electrons.
b. H_2O_2 has three single bonds and four pairs of nonbonding electrons.
c. H_2CO has one double bond, two single bonds, and two pairs of nonbonding electrons.
d. C_2H_4 has one double bond, four single bonds, and no nonbonding electrons.

5.15 Replace each pair of bonding electrons with a line; nonbonding electrons are written as dots.

a. :N≡N:

b. H—Ö—Ö—H (two nonbonding pairs on each O)

c.
```
H—C—H
  ||
 :O:
```

d.
```
H—C=C—H
  |  |
  H  H
```

5.17 a. normal; O should form two bonds.
b. normal; N should form three bonds.
c. not normal; C should form four bonds instead of two.
d. normal; O should form two bonds.

5.19 a. Nitrogen. An element that forms three single bonds has five valence electrons (three octet vacancies). In Period 2 this is the Group VA element, nitrogen.
b. Carbon. An element that forms four single bonds has four valence electrons (four octet vacancies). In Period 2 this is the Group IVA element, carbon.
c. Nitrogen. An element that forms one single bond and one double bond has three octet vacancies (five valence electrons). In Period 2 this is the Group VA element, nitrogen.
d. Carbon. An element that forms two single bonds and one double bond has four octet vacancies (four valence electrons). In Period 2 this is the Group IVA element, carbon.

5.21 a. An element that forms 3 single bonds has 5 valence electrons (3 octet vacancies).
b. An element that forms 2 double bonds has 4 valence electrons (4 octet vacancies).
c. An element that forms 1 single bond and 1 double bond has 5 valence electrons (3 octet vacancies).
d. An element that forms 2 single bonds and 1 double bond has 4 valence electrons (4 octet vacancies).

5.23 A coordinate covalent is a covalent bond in which both electrons of a shared pair come from one of the two atoms involved in the bond. Atoms participating in coordinate covalent bonds generally deviate from the common bonding pattern for that type of atom. The "hint" from the Lewis structure that coordinate covalency is involved is that oxygen forms three bonds instead of the normal two.

5.25 a. A nitrogen–oxygen bond; nitrogen has four bonds (not normal) and oxygen has one bond (not normal).
b. A coordinate covalent bond is not present.
c. An oxygen–chlorine bond; oxygen has one bond (not normal) and chlorine has two bonds (nor normal).
d. Two oxygen–bromine bonds; oxygen has one bond (not normal) and bromine has three bonds (not normal).

5.27 a. CBr_4 has 32 valence electrons (4 from C and 7 from each Br).
b. SF_2 has 20 valence electrons (6 from S and 7 from each F).
c. PH_3 has 8 valence electrons (5 from P and 1 from each H).
d. OF_2 has 20 valence electrons (6 from O and 7 from each F).

5.29 a. The Lewis structure for CBr_4 has 32 electron dots (4 from C and 7 from each Br).
b. The Lewis structure for SF_2 has 20 electron dots (6 from S and 7 from each F).
c. The Lewis structure for PH_3 has 8 electron dots (5 from P and 1 from each H).
d. The Lewis structure for OF_2 has 20 electron dots (6 from O and 7 from each F).

5.31 The central atom in each molecule will be the atom that has the most octet vacancies and will form the most covalent bonds. Remember that one bond line equals two electron dots.

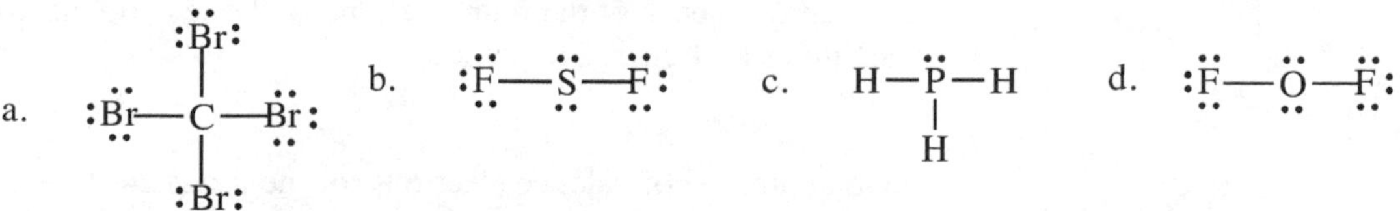

5.33 In some molecules there are not enough electrons to give the central atom an octet. Then we use one or more pairs of nonbonding electrons on the atoms bonded to the central atom to form double or triple bonds. Remember to count the electrons around each atom to make sure that the octet rule is followed.

a. H:C::C::C:H (with H below each terminal C) — A central C atom shares an electron pair with each of the other carbons to complete octets around each atom.

b. :F:N::N:F: — The central N atoms share a pair of electrons from each N atom to give a double bond.

c. H:C:C:::N: (with H above and below the first C) — The triple bond (three pairs of electrons) between the C and N atoms is formed from three electrons from the N atom and three from the C atom.

d. H:C:C:::C:H (with H above and below the first C) — The triple bond between the two C atoms is formed from three electrons from each of them.

5.35 To determine the number of electrons in the Lewis structure for a polyatomic ion, sum the numbers of the valence electrons of the atoms in the ion and add one electron for each negative charge or subtract one electron for each positive charge.

a. The Lewis structure for ClO^- has 14 "electron dots." (7 + 6 + 1 = 14)
b. The Lewis structure for ClO_2^- has 20 "electron dots." (7 + 6 + 6 + 1 = 20)
c. The Lewis structure for S_2^{2-} has 14 "electron dots." (6 + 6 + 2 = 14)
d. The Lewis structure for NH_4^+ has 8 "electron dots." (5 + 4 – 1 = 8)

5.37 Lewis structures for polyatomic ions are written in the same way as those for molecules except that the total number of electrons must be adjusted (increased or decreased) to take into account ion charge. The octet rule around each of the atoms must be followed.

a. [:O:H]⁻ — The total number of valence electrons for the ion is eight: six from O, one from H, and one extra electron, giving a –1 charge to the ion.

b. [H:Be:H (with H above and below Be)]²⁻ — The total number of valence electrons for the ion is eight: two from Be, one from each of the four H atoms, and two extra electrons, which give a –2 charge to the ion.

c. $\left[\begin{array}{c} :\ddot{Cl}: \\ :\ddot{Cl}:\ddot{Al}:\ddot{Cl}: \\ :\ddot{Cl}: \end{array} \right]^-$

The total number of valence electrons for the ion is 32: three from Al, seven from each of the four Cl atoms, and one extra electron, which gives a –1 charge to the ion.

d. $\left[\begin{array}{c} :\ddot{O}:N:\ddot{O}: \\ :\ddot{O}: \end{array} \right]^-$

The total number of valence electrons for the ion is 24: five from N, six from each of the three O atoms, and one extra electron, which gives a –1 charge to the ion.

5.39 The Lewis structure for a polyatomic ion is drawn in the same way as the Lewis structure for a molecule, except that the charge on the ion must be taken into account in calculating the number of electrons. The positive and negative ions for the ionic compounds are treated separately to show that they are not linked by covalent bonds.

a. $Na^+ [:C:::N:]^-$

The polyatomic ion CN^- has 10 valence electrons: four from C, five from N, and one extra electron, giving the ion a –1 charge. The –1 charge is balanced by a +1 charge on the sodium ion.

b. $3[K^+] \left[\begin{array}{c} :\ddot{O}: \\ :\ddot{O}:P:\ddot{O}: \\ :\ddot{O}: \end{array} \right]^{3-}$

The polyatomic ion PO_4^{3-} has a total of 32 electrons: five from P, 6 from each of the three O atoms, and three extra electrons, giving the ion a –3 charge. The –3 charge is balanced by three potassium ions, each having a +1 charge.

5.41 Calculate the number of valence electrons for each structure (sum of the valence electrons of the atoms plus one electron for each negative charge). Draw the molecular structure with covalent bonds between bonded atoms. Add nonbonding electron pairs to give each atom an octet of electrons.

$\begin{array}{c} :\ddot{O}:S::\ddot{O} \\ :\ddot{O}: \end{array}$

a. Lewis structure for SO_3

$\left[\begin{array}{c} :\ddot{O}:\ddot{S}:\ddot{O}: \\ :\ddot{O}: \end{array} \right]^{2-}$

b. Lewis structure for SO_3^{2-}

5.43 Single covalent bonds are not adequate to explain covalent bonding in all molecules. Sometimes two atoms must share two or three pairs of electrons (double bonds or triple bonds) to provide a complete octet of electrons.

a. HCl and HI have the same bond multiplicity; each has a single bond.
b. S_2 and Cl_2 have different bond multiplicities; S_2 has a double bond and Cl_2, a single bond.
c. CO and NO^+ have the same bond multiplicity; each has a triple bond.
d. OH^- and HS^- have the same bond multiplicity; each has a single bond.

5.45 According to VSEPR theory, electrons in the valence shell of a central atom are arranged in a way that minimizes the repulsions between negatively-charged electron groups. In order to predict the molecular geometry of a simple molecule, count the number of single, double, or triple bonds around the central atom, and assign a molecular geometry according to the chart in Chemistry at a Glance in Sec. 5.8. The molecular geometry for:

a. 4 bonding groups and 0 nonbonding electron groups is tetrahedral.
b. 2 bonding groups and 2 nonbonding electron groups is angular.
c. 3 bonding groups and 0 nonbonding electron groups is trigonal planar.

5.47 a. linear (atoms are in a straight line)
b. angular (atoms are not in a straight line)
c. tetrahedral (a central atom with four other atoms attached to it that lie at the corners of a tetrahedron)
d. linear (atoms are in a straight-line)

5.49 According to VSEPR theory, electrons in the valence shell of a central atom are arranged in a way that minimizes the repulsions between negatively-charged electron groups. (An electron group can be: a single bond, a double bond, a triple bond, or a nonbonding electron pair.)

In order to predict the molecular geometry of a simple molecule, count the number of VSEPR electron groups around the central atom, and assign a molecular geometry (see Chemistry at a Glance in Sec. 5.8).

a. Angular. The central S atom in this molecule has four electron groups around it, two single bonds and two nonbonding electron pairs. These four groups are in a tetrahedral arrangement, giving H_2S an angular molecular geometry.
b. Angular. The central O atom has four electron groups around it, two single bonds and two nonbonding electron pairs. The four electron groups have a tetrahedral arrangement; the three atoms have an angular geometry.
c. Angular. The central O atom has three electron groups: one double bond, one single bond, and one nonbonding pair of electrons. The three electron groups have a trigonal planar arrangement; the three atoms have an angular geometry.
d. Linear. The central N atom has two electron groups, two double bonds. The two electron groups have a linear arrangement, and the three atoms have a linear geometry.

5.51 In each of the following molecules: count the number of VSEPR electron groups around the central atom, and assign a molecular geometry (see Chemistry at a Glance in Sec. 5.8).

a. Trigonal pyramidal. The central N atom has four electron groups around it: three single bonds and one nonbonding electron pair. The four electron groups have a tetrahedral arrangement, and the four atoms have a trigonal pyramidal geometry.
b. Trigonal planar. The central C atom has three electron groups: two single bonds and a double bond. The three electron groups have a trigonal planar arrangement, and the four atoms have a trigonal planar geometry.
c. Tetrahedral. The central P atom has four electron groups around it: four single bonds. The four electron groups have a tetrahedral arrangement, and the five atoms of the molecule have a tetrahedral molecular geometry.
d. Tetrahedral. The central C atom has four electron groups around it: four single bonds. The four electron groups have a tetrahedral arrangement, and the five atoms of the molecule have a tetrahedral molecular geometry.

5.53 Each of the molecules in this problem has two central atoms. For a given molecule, consider each central atom separately, and then combine the results.

a. Trigonal planar about each carbon atom. Each central carbon atom has three electron groups around it (two single bonds and one double bond), a trigonal planar arrangement of electron groups, and a trigonal planar geometry around each carbon atom.
b. Tetrahedral about the carbon atom and angular about the oxygen atom. The carbon atom has four electron groups (four single bonds), a tetrahedral arrangement of electron groups, and a tetrahedral geometry. The oxygen atom has four electron groups (two single bonds and two nonbonding electron pairs), a tetrahedral arrangement of electron groups, and the three atoms (carbon, oxygen, and hydrogen) have an angular geometry.

5.55 In order to predict the molecular geometry of a simple molecule: 1) Draw a Lewis structure for the molecule, 2) count the number of VSEPR electron groups around the central atom, and 3) assign a molecular geometry (see Chemistry at a Glance in Sec. 5.8).

a. Trigonal pyramidal. The central N atom has four electron groups around it: three single bonds and a nonbonding pair of electrons. The four electron groups have a tetrahedral arrangement, and the four atoms of the molecule have trigonal pyramidal geometry.

b. Tetrahedral. The central Si atom has four electron groups around it: four single bonds. The four electron groups have a tetrahedral arrangement, and the five atoms of the molecule have a tetrahedral molecular geometry.

c. Angular. The central Se atom has four electron groups around it: two single bonds and two nonbonding electron groups. The four electron groups have a tetrahedral arrangement, and the three atoms of the molecule have an angular molecular geometry.

d. Angular. The central S atom has four electron groups around it: two single bonds and two nonbonding electron groups. The four electron groups have a tetrahedral arrangement, and the three atoms of the molecule have an angular molecular geometry.

5.57 According to VSEPR theory, electrons in the valence shell of a central atom are arranged in a way that minimizes the repulsions between negatively-charged electron groups. (An electron group can be: a single bond, a double bond, a triple bond, or a nonbonding electron pair.)

To predict the electron group geometry of a simple molecule, count the number of VSEPR electron groups around the central atom. To predict the molecular geometry, count the bonds but not the nonbonding electron pairs. Assign a VSEPR electron group geometry and a molecular geometry using information from Chemistry at a Glance in Sec. 5.8.

a. SH_4 has four VSEPR electron group and four single bonds. VSEPR electron group geometry is tetrahedral; molecular geometry is also tetrahedral.

b. NH_2^- has four VSEPR electron group, two single bonds and two nonbonding electron pairs. VSEPR electron group geometry is tetrahedral; molecular geometry is angular.

c. ClNO has three VSEPR electron group: one single bond, one double bond, and one nonbonding electron pair. VSEPR electron group geometry is trigonal planar; molecular geometry is angular.

d. CO_3^{2-} has three VSEPR electron group: two single bonds, one double bond, and no nonbonding electron pairs. VSEPR electron group geometry is trigonal planar; molecular geometry is trigonal planar.

5.59 In the periodic table, electronegativity values increase from left to right across periods and from bottom to top within groups.

a. True.

b. False. Elements 4 and 5 are in the same period, but element 4 is to the left of 5.

c. False. Elements 3 and 8 are in the same period, but 3 is to the left of 8.

d. False. Element 6 is to the right and above element 7, so 6 is more electronegative.

5.61 Electronegativity is a measure of the relative attraction that an atom has for the shared electrons in a bond. In the periodic table, electronegativity values increase from left to right across periods and from bottom to top within groups.

a. Na, Mg, Al, P. All four are in Period 3 (electronegativity increases from left to right).
b. I, Br, Cl, F. All four are halogens (Group VIIA); electronegativity increases bottom to top.
c. Al, P, S, O. The first three elements in the series are in Period 3 (left to right). O is above S and is more electronegative than S.
d. Ca, Mg, C, O. The most electronegative atom is O, to the right and/or above the other three. C is to the left of O (less electronegative than O). Mg and Ca are to the left and below C (less electronegative than C), and Ca is below Mg (same group, less electronegative).

5.63 Figure 5.11 shows electronegativity values for selected elements.

a. Br, Cl, N, O, F have values greater than that of C.
b. Na, K, Rb have values less than 1.0.
c. Cl, N, O, F are the four most electronegative elements listed in Figure 5.11.
d. Period 2 elements differ sequentially by 0.5 units.

5.65 In the periodic table, electronegativity values increase from left to right across periods and from bottom to top within groups. The bonded atom with the greater electronegativity will have the partial negative charge.

a. $\overset{\delta+}{\text{B}}$—$\overset{\delta-}{\text{N}}$ b. $\overset{\delta+}{\text{Cl}}$—$\overset{\delta-}{\text{F}}$ c. $\overset{\delta-}{\text{N}}$—$\overset{\delta+}{\text{C}}$ d. $\overset{\delta-}{\text{F}}$—$\overset{\delta+}{\text{O}}$

5.67 The polarity of a bond increases as the numerical value of the electronegativity difference between the two bonded atoms increases. The electronegativity values for the bonded atoms are found in Figure 5.11. For example, the elcctronegativity differences in part a. are calculated as follows: Cl–H (3.0 – 2.1 = 0.9), Br–H: (2.8 – 2.1 = 0.7), O–H (3.5 – 2.1 = 1.4). For this sample calculation, the more electronegative atom in the bond has been placed first.

a. H–Br (0.7), H–Cl (0.9), H–O (1.4)
b. O–F (0.5), P–O (1.4), Al–O (2.0)
c. Br–Br (0.0), H–Cl (0.9), B–N (1.0)
d. P–N (0.9), S–O (1.0), Br–F (1.2)

5.69 The electronegativity differences between the bonded atoms in three types of bonds are: nonpolar covalent bonds, 0.4 or less; polar covalent bonds, 0.4 to 1.5; ionic bonds, greater than 2.0.

a. polar covalent (electronegativity difference 1.0)
b. ionic (electronegativity difference 2.1)
c. nonpolar covalent (electronegativity difference 0.0)
d. ionic (electronegativity difference 1.5), between a metal and a nonmetal

5.71 The electronegativity difference between the bonded atoms in three types of bonds are: nonpolar covalent bonds, 0.4 or less; polar covalent bonds, 0.4 to 1.5; ionic bonds, greater than 2.0. The electronegativity values for the elements are found in Section 5.9. Find the difference betwccn the electronegativities of the two atoms forming the bond.

a. 3.0 – 2.5 = 0.5 The bond is polar covalent.
b. 3.5 – 2.5 = 1.0 The bond is polar covalent.
c. 2.5 – 2.1 = 0.4 The bond is nonpolar covalent.
d. 2.5 – 2.5 = 0.0 The bond is nonpolar covalent.

5.73 In the periodic table, electronegativity values increase from left to right across periods and from bottom to top within groups. The bonded atom with the greater electronegativity will have the partial negative charge. The electronegativity difference between the bonded atoms in three types of bonds are: nonpolar covalent bonds, 0.4 or less; polar covalent bonds, 0.4 to 1.5; ionic bonds, greater than 2.0.

	Bond identity	More electronegative atom	Atom with negative partial charge (δ^-)	Bond type
	N–P	N	N	polar covalent
a.	Se–P	Se	Se	nonpolar covalent
b.	O–P	O	O	polar covalent
c.	Br–P	Br	Br	polar covalent
d.	F–P	F	F	polar covalent

5.75 The electronegativies of A, B, C, and D are: A = 3.8, B = 3.3, C = 2.8, D = 1.3.
The electronegativity differences are: A – B = 0.5, A – D = 2.5, B – D = 2.0, A – C = 1.0

a. A greater difference in electronegativity means formation of a more ionic bond. The order of increasing ionic character is: BA, CA, DB, DA.
b. A greater difference in electronegativity means formation of a less covalent bond. The order of decreasing covalent bond character is : BA, CA, DB, DA.

5.77 A polar molecule is a molecule in which there is an unsymmetrical distribution of electronic charge. Molecular polarity depends on two factors: bond polarity and molecular geometry.

a. Nonpolar. Since the molecule is symmetrical, the effects of the two identical polar bonds are cancelled (the electron distribution is symmetrical).
b. Polar. The molecule is linear but not symmetrical, so the polar bond makes it polar.
c. Polar. In an angular molecule the bond polarities do not cancel one another.
d. Polar. In an angular molecule the bond polarities do not cancel one another.

5.79 a. Nonpolar. The molecule is symmetrical; effects of two identical polar bonds are cancelled.
b. Polar. In an angular molecule the bond polarities do not cancel one another; the electron distribution for the molecule is not symmetrical.
c. Polar. The two polar bonds do not cancel one another, both because of the angular molecular geometry and because the bonds have unequal polarity.
d. Polar. Although this molecule is linear, it is not symmetrical; its polar bonds do not cancel one another. They are two different bonds with differing polarities.

5.81 a. Polar. The molecule is not symmetrical; the polar N–Cl bonds do not cancel one another.
b. Polar. In an angular molecule the bond polarities do not cancel one another.
c. Nonpolar. The molecule is symmetrical; the effects of the two polar bonds are cancelled.
d. Polar. The tetrahedral molecule is not symmetrical. It has three C–Cl bonds and one C–H bond, so the electron distribution is not symmetrical.

5.83 The polarity of a molecule depends on both the polarity of the bonds and the arrangement of the bonds. A symmetrical molecule whose polar bonds have equal polarity may be polar or nonpolar, depending on the shape of the molecule.

a. There is insufficient information to determine the molecule's polarity. The shape of the molecule is unknown.
b. A molecule with two bonds, one polar and one nonpolar, would be polar. The molecule is not symmetrical.

5.85 Molecular polarity is a measure of the degree of inequality in the attraction of bonding electrons. The degree of molecular polarity depends on both the polarity and the arrangement of the bonds.

a. Each of the molecules has one bond, so the molecular polarity depends on the difference in electronegativity between the two atoms. For BrCl: 3.0 – 2.8 = 0.2; for BrI: 2.8 – 2.5 = 0.3 Therefore, BrI has the greater molecular polarity.

b. CO_2 is a linear molecule with two polar bonds; the molecule is nonpolar. SO_2 is an angular molecule with two polar bonds; the molecule is polar. Therefore, SO_2 has the greater molecular polarity.

c. SO_3 is a planar molecule with three polar bonds arranged symmetrically about the S atom; the molecule is nonpolar. NH_3 is a trigonal pyramidal molecule (See Chemistry at a glance, Section 5.8) with three polar bonds; the molecule is polar. Therefore, NH_3 has the greater molecular polarity.

d. CH_4 is a tetrahedral molecule with four equivalent (nonpolar) bonds; it is a nonpolar molecule. CH_3Cl is a tetrahedral molecule with three nonpolar bonds and one polar bond; it is a polar molecule. Therefore, CH_3Cl has the greater molecular polarity.

5.87 a. Draw the molecular structure with single covalent bonds between bonded atoms. Add electron dots nonbonding electron pairs to give each atom an octet of electrons.

```
     H            H           H            H             :F:
     ..           ..          ..           ..            ..
 H : C : H      H:C:F:      H:C:F:      :F:C:F:       :F:C:F:
     ..           ..          ..           ..            ..
     H            H          :F:          :F:           :F:

    CH4          CH3F        CH2F2        CHF3          CF4
```

b. The molecular geometry of all five molecules is tetrahedral; there are four single bonds around the central carbon atom and no nonbonding pairs of electrons on the central atom.

c. CH_4 and CF_4 are nonpolar; they are tetrahedral with electrons distributed symmetrically about the central atom. CH_3F, CH_2F_2, and CHF_3 are polar molecules; although the molecules are tetrahedral, there is unsymmetrical electron distribution in the molecule.

5.89 Binary molecular compounds are molecular compounds in which only two nonmetallic elements are present.

a. No. Na is a metal. b. No. Al is a metal. c. Yes d. Yes

5.91 Names for binary molecular compounds contain numerical prefixes that give the number of each type of atom present in addition to the names of the elements present. The nonmetal of lower electronegativity is named first followed by a separate word containing the stem of the name of the more electronegative nonmetal and the suffix *–ide*. A numerical prefix precedes the name of nonmetals.

a. N_2F_4 is named dinitrogen tetrafluoride. b. CCl_4 is named carbon tetrachloride.
c. OF_2 is named oxygen difluoride. d. CO is named carbon monoxide.

5.93 A numerical prefix precedes the name of each nonmetal; when only one atom of the first nonmetal is present, it is customary to omit the initial prefix *mono-*.

a. NO has one numerical prefix (nitrogen monoxide).
b. H_2S has zero numerical prefixes (hydrogen sulfide).
c. HI has zero numerical prefixes (hydrogen iodide).
d. N_2O_3 has two numerical prefixes (dinitrogen trioxide).

5.95 A few binary molecular compounds have common names that are unrelated to the systematic naming rules. These common names are given in Table 5.2.

a. H_2O_2 is hydrogen peroxide.
b. NO is nitric oxide.
c. CH_4 is methane.
d. NH_3 is ammonia.

5.97 a. diagram I b. diagram II c. diagram IV d. diagram III

5.99 In names for binary molecular compounds, numerical prefixes give the number of each type of atom present in addition to the names of the elements present. The nonmetal of lower electronegativity is named first followed by a separate word containing the stem of the name of the more electronegative nonmetal and the suffix *–ide*. A numerical prefix precedes the name of each nonmetal; when only one atom of the first nonmetal is present, it is customary to omit the initial prefix *mono-*.

a. ICl is iodine monochloride.
b. N_2O_2 is dinitrogen dioxide.
c. NCl_3 is nitrogen trichloride.
d. HBr is hydrogen bromide.

5.101 A few binary molecular compounds have common names that are unrelated to the systematic naming rules. These common names are given in Table 5.2.

a. H_2O_2 is hydrogen peroxide.
b. CH_4 is methane.
c. NH_3 is ammonia.
d. NO is nitric oxide.

5.103 The less electronegative element is named first.

5.105 Na_2CO_3 is an ionic compound. The name of an ionic compound is the name of the metal (followed by a Roman numeral if the metal is variably charged) and the name of the negative polyatomic ion. Na_2CO_3 is sodium carbonate.

5.107 a. The name of an ionic compound is the name of the metal followed by the stem name of the nonmetal with the suffix *–ide*. KCl is potassium chloride; HF is hydrogen fluoride.
A covalent compound ends in the stem name of the more electronegative nonmetal with the suffix *–ide*. The two compounds are N_2O_3 (dinitrogen trioxide) and CO_2 (carbon dioxide)

b. N_2O_3 (dinitrogen trioxide) has two numerical prefixes; prefixes are necessary because several different compounds may exist for the same pair of nonmetals.

c. None of the names of these five compounds contains the prefix *mono-*. The prefix *mono-* is used to modify the name of the second element in a molecular compound's name if only one atom of the element is present (which is not the case with any of these five molecules); *mono-* is never used to modify the name of the first element in a molecular compound's name.

d. The names of KCl (potassium chloride) and HF (hydrogen fluoride) do not contain numerical prefixes because they are ionic compounds. The name of CH_4 (methane) does not contain a numerical prefix because, for hydrocarbons, the naming system is different.

Chemical Calculations: Formula Masses, Moles, and Chemical Equations

Chapter 6

Solutions to Selected Problems

6.1 A formula mass is calculated by multiplying the atomic mass of each element by the number of atoms of that element in the chemical formula, and then summing all of the atomic masses of all the elements in the chemical formula.

a. [12(12.01) + 22(1.01) + 11(16.00)] amu = 342.34 amu
b. [7(12.01) + 16(1.01)] amu = 100.23 amu
c. [7(12.01) + 5(1.01) + 14.01 + 3(16.00) + 32.07] amu = 183.20 amu
d. [2(14.01) + 8(1.01) + 32.07 + 4(16.00)] amu = 132.17 amu

6.3 A formula mass is calculated by multiplying the atomic mass of each element by the number of atoms of that element in the chemical formula, and then summing all of the atomic masses of all the elements in the chemical formula (C_3H_yS):

$[3(12.01) + y(1.01) + 1(32.07)]$ amu = 76.18 amu Therefore: $y = 8$

6.5 The chemist's counting unit is the *mole*. A mole is 6.02×10^{23} objects.

a. 1.00 mole of apples = 6.02×10^{23} apples
b. 1.00 mole of elephants = 6.02×10^{23} elephants
c. 1.00 mole of Zn atoms = 6.02×10^{23} Zn atoms
d. 1.00 mole of CO_2 molecules = 6.02×10^{23} CO_2 molecules

6.7 Use a conversion factor derived from the definition of a mole. The equality is:
1 mole atoms = 6.02×10^{23} atoms

a. $0.500 \text{ moles Cu} \times \left(\dfrac{6.02 \times 10^{23} \text{ atoms Cu}}{1 \text{ mole Cu}}\right) = 3.01 \times 10^{23} \text{ atoms Cu}$

b. $0.500 \text{ moles Na} \times \left(\dfrac{6.02 \times 10^{23} \text{ atoms Na}}{1 \text{ mole Na}}\right) = 3.01 \times 10^{23} \text{ atoms Na}$

c. $0.500 \text{ moles } CO_2 \times \left(\dfrac{6.02 \times 10^{23} \text{ molecules } CO_2}{1 \text{ mole } CO_2}\right) = 3.01 \times 10^{23} \text{ molecules } CO_2$

d. $0.500 \text{ moles } SiF_4 \times \left(\dfrac{6.02 \times 10^{23} \text{ molecules } SiF_4}{1 \text{ mole } SiF_4}\right) = 3.01 \times 10^{23} \text{ molecules } SiF_4$

6.9 a. 0.200 mole Al atoms contains more moles (so more atoms) than 0.100 mole C atoms.
b. Avogadro's number (1.00 mole) of C atoms has more atoms than 0.750 mole Al atoms.
c. 1.50 moles Al has more atoms than 6.02×10^{23} atoms (1.00 mole) C atoms.
d. 6.50×10^{23} C atoms contains more atoms than Avogadro's number (6.02×10^{23}) of Al atoms.

6.11 The mass, in grams, of one mole of atoms is equal to the atomic mass of the element. The mass of one mole of molecules (formula mass) is equal to the sum of the atomic masses in one molecule.

a. One mole of Ca atoms has a mass of 40.08 g
b. One mole of Ag atoms has a mass of 107.9 g
c. One mole of NO molecules has a mass of 30.01 g (14.01 + 16.00)
d. One mole of HNO_3 molecules has a mass of 63.02 g (1.01 + 14.01 + 3 × 16.00)

6.13 To solve these problems use a conversion factor relating formula mass of the substance to moles of the substance. The equality will be:

Formula mass (g) substance = 1 mole substance

a. $2.00 \text{ moles C} \times \left(\frac{12.01 \text{ g C}}{1 \text{ mole C}}\right) = 24.0 \text{ g C}$

b. $3.00 \text{ moles CO} \times \left(\frac{28.01 \text{ g CO}}{1 \text{ mole CO}}\right) = 84.0 \text{ g CO}$

c. $5.00 \text{ moles F} \times \left(\frac{19.00 \text{ g F}}{1 \text{ mole F}}\right) = 95.0 \text{ g F}$

d. $5.00 \text{ moles } F_2 \times \left(\frac{38.00 \text{ g } F_2}{1 \text{ mole } F_2}\right) = 190. \text{ g } F_2$

6.15 Convert the given mass (5.00 g) to moles using the formula mass to form a conversion factor relating 1 mole to its formula mass. (For example, 1 mole CO = 28.01 g CO)

a. $5.00 \text{ g CO} \times \left(\frac{1 \text{ mole CO}}{28.01 \text{ g CO}}\right) = 0.179 \text{ mole CO}$

b. $5.00 \text{ g } CO_2 \times \left(\frac{1 \text{ mole } CO_2}{44.01 \text{ g } CO_2}\right) = 0.114 \text{ mole } CO_2$

c. $5.00 \text{ g } B_4H_{10} \times \left(\frac{1 \text{ mole } B_4H_{10}}{53.34 \text{ g } B_4H_{10}}\right) = 0.0937 \text{ mole } B_4H_{10}$

d. $5.00 \text{ g U} \times \left(\frac{1 \text{ mole U}}{238 \text{ g U}}\right) = 0.0210 \text{ mole U}$

6.17 Molar mass is the mass in grams of a substance that is numerically equal to the substance's formula mass.

formula mass (amu) = molar mass (g)
formula mass = 44.01 amu

6.19 Use the definition of formula mass (grams/mole) to solve the problem.

$$\text{formula mass} = \frac{\text{mass in grams}}{1.00 \text{ mole}} = \frac{140.07 \text{ g}}{7.00 \text{ moles}} = 20.0 \text{ amu}$$

6.21 From the chemical formula, find the number of moles of oxygen per formula unit and multiply this by the number of moles of compound.

a. 3 atoms of oxygen/1 formula unit means 3 moles oxygen/1 mole H_2SO_3
3 moles oxygen/1 mole H_2SO_3 × 2.00 moles H_2SO_3 = 6.00 moles oxygen

b. 7 moles oxygen/1 mole $H_2S_2O_7$ × 2.00 moles $H_2S_2O_7$ = 14.00 moles oxygen

c. 1 mole oxygen/1 mole $SOCl_2$ × 2.00 moles $SOCl_2$ = 2.00 moles oxygen

d. 2 moles oxygen/1 mole SO_2Cl_2 × 2.00 moles SO_2Cl_2 = 4.00 moles oxygen

6.23 total atoms/1 molecule of compound = total moles of atoms/1 mole of compound
moles of atoms/1 mole of compound × moles of atoms = total moles of atoms

a. 2 H + 1 S + 3 O = 6 atoms/1 H_2SO_3 molecule = 6 moles atoms/1 mole H_2SO_3
6 moles atoms/1 mole compound × 2.00 mole H_2SO_3 = 12.00 total moles of atoms

b. 2 H + 2 S + 7 O = 11 atoms/1 $H_2S_2O_7$ molecule = 11 moles atoms/1 mole $H_2S_2O_7$
11 moles atoms/1 mole compound × 2.00 mole $H_2S_2O_7$ = 22.00 total moles of atoms

c. 1 S + 1 O + 2 Cl = 4 atoms/1 $SOCl_2$ molecule = 4 moles atoms/1 mole $SOCl_2$
4 moles atoms/1 mole compound × 2.00 mole $SOCl_2$ = 8.00 total moles of atoms

d. 1 S + 2 O + 2 Cl = 5 atoms/1 SO_2Cl_2 molecule = 5 moles atoms/1 mole SO_2Cl_2
5 moles atoms/1 mole compound × 2.00 mole SO_2Cl_2 = 10.00 total moles of atoms

6.25 a. One mole H_2SO_4 contains 2 moles H atoms, 1 mole S atoms, and 4 moles O atoms. The conversion factors derived from this statement are:

$$\frac{2 \text{ moles H}}{1 \text{ mole H}_2\text{SO}_4}, \frac{1 \text{ mole H}_2\text{SO}_4}{2 \text{ moles H}}, \frac{1 \text{ mole S}}{1 \text{ mole H}_2\text{SO}_4}, \frac{1 \text{ mole H}_2\text{SO}_4}{1 \text{ mole S}},$$

$$\frac{4 \text{ moles O}}{1 \text{ mole H}_2\text{SO}_4}, \frac{1 \text{ mole H}_2\text{SO}_4}{4 \text{ moles O}}$$

b. One mole $POCl_3$ contains 1 mole P atoms, 1 mole O atoms, and 3 moles Cl atoms. The factors derived from this statement are:

$$\frac{1 \text{ mole P}}{1 \text{ mole POCl}_3}, \frac{1 \text{ mole POCl}_3}{1 \text{ mole P}}, \frac{1 \text{ mole O}}{1 \text{ mole POCl}_3}, \frac{1 \text{ mole POCl}_3}{1 \text{ mole O}},$$

$$\frac{3 \text{ moles Cl}}{1 \text{ mole POCl}_3}, \frac{1 \text{ mole POCl}_3}{3 \text{ moles Cl}}$$

6.27 One mole of H_3PO_4 contains 3 moles of H atoms, 1 mole of P atoms, 4 moles of O atoms, and 8 total moles of atoms (the 8 comes from the sum of the subscripts in the molecular formula).

a. $\frac{3 \text{ moles H}}{1 \text{ mole H}_3\text{PO}_4}$ b. $\frac{4 \text{ moles O}}{1 \text{ mole H}_3\text{PO}_4}$

c. $\frac{8 \text{ moles atoms}}{1 \text{ mole H}_3\text{PO}_4}$ d. $\frac{4 \text{ moles O}}{1 \text{ mole P}}$

6.29 Convert the given mass of the element to moles using the atomic mass to form a conversion factor relating 1 mole to its atomic mass. Use a second conversion factor based on the definition of Avogadro's number: 6.02×10^{23} atoms = 1 mole atoms

a. $20.0 \text{ g Be} \times \left(\dfrac{1 \text{ mole Be}}{9.01 \text{ g Be}}\right) \times \left(\dfrac{6.02 \text{ x } 10^{23} \text{ atoms Be}}{1 \text{ mole Be}}\right) = 1.34 \times 10^{24} \text{ atoms Be}$

b. $20.0 \text{ g Ar} \times \left(\dfrac{1 \text{ mole Ar}}{39.95 \text{ g Ar}}\right) \times \left(\dfrac{6.02 \text{ x } 10^{23} \text{ atoms Ar}}{1 \text{ mole Ar}}\right) = 3.01 \times 10^{23} \text{ atoms Ar}$

c. $20.0 \text{ g Ar} \times \left(\dfrac{1 \text{ mole Cr}}{52.00 \text{ g Cr}}\right) \times \left(\dfrac{6.02 \text{ x } 10^{23} \text{ atoms Cr}}{1 \text{ mole Cr}}\right) = 2.32 \times 10^{23} \text{ atoms Cr}$

d. $20.0 \text{ g Sn} \times \left(\dfrac{1 \text{ mole Sn}}{118.71 \text{ g Sn}}\right) \times \left(\dfrac{6.02 \text{ x } 10^{23} \text{ atoms Sn}}{1 \text{ mole Sn}}\right) = 1.01 \times 10^{23} \text{ atoms Sn}$

6.31 In these problems, first convert atoms to moles using the definition of Avogadro's number ($6.02 \text{ x } 10^{23}$ atoms = 1 mole atoms). Then multiply by a second conversion factor changing moles to grams of atoms (1 mole atoms = element's atomic mass in grams).

a. $6.02 \times 10^{23} \text{ atoms Cu} \times \left(\dfrac{1 \text{ mole Cu}}{6.02 \text{ x } 10^{23} \text{ atoms Cu}}\right) \times \left(\dfrac{63.55 \text{ g Cu}}{1 \text{ mole Cu}}\right) = 63.6 \text{ g Cu}$

b. $3.01 \times 10^{23} \text{ atoms Cu} \times \left(\dfrac{1 \text{ mole Cu}}{6.02 \text{ x } 10^{23} \text{ atoms Cu}}\right) \times \left(\dfrac{63.55 \text{ g Cu}}{1 \text{ mole Cu}}\right) = 31.8 \text{ g Cu}$

c. $557 \text{ atoms Cu} \times \left(\dfrac{1 \text{ mole Cu}}{6.02 \text{ x } 10^{23} \text{ atoms Cu}}\right) \times \left(\dfrac{63.55 \text{ g Cu}}{1 \text{ mole Cu}}\right) = 5.88 \times 10^{-20} \text{ g Cu}$

d. $1 \text{ atom Cu} \times \left(\dfrac{1 \text{ mole Cu}}{6.02 \text{ x } 10^{23} \text{ atoms Cu}}\right) \times \left(\dfrac{63.55 \text{ g Cu}}{1 \text{ mole Cu}}\right) = 1.06 \times 10^{-22} \text{ g Cu}$

6.33 To convert grams to moles (parts a. and b.), multiply by a conversion factor derived from the mass in grams of 1 mole (formula mass). To convert atoms to moles (parts c. and d.), use a conversion factor derived from the definition of Avogadro's number (1 mole = 6.02×10^{23}).

a. $10.0 \text{ g He} \times \left(\dfrac{1 \text{ mole He}}{4.00 \text{ g He}}\right) = 2.50 \text{ moles He}$

b. $10.0 \text{ g N}_2\text{O} \times \left(\dfrac{1 \text{ mole N}_2\text{O}}{44.02 \text{ g N}_2\text{O}}\right) = 0.227 \text{ mole N}_2\text{O}$

c. $4.0 \text{ x } 10^{10} \text{ atoms P} \times \left(\dfrac{1 \text{ mole P}}{6.02 \times 10^{23} \text{ atoms P}}\right) = 6.6 \times 10^{-14} \text{ moles P}$

d. $4.0 \times 10^{10} \text{ atoms Be} \times \left(\dfrac{1 \text{ mole Be}}{6.02 \times 10^{23} \text{ atoms Be}}\right) = 6.6 \times 10^{-14} \text{ moles Be}$

6.35 To change grams of molecules to atoms of S, we will use three conversion factors. 1) Change grams of molecules to moles of molecules using the definition of formula mass. 2) Change moles of molecules to moles of S in the molecule by determining the number of atoms of S in the molecule. 3) Change moles of S to atoms of S using Avogadro's number (1 mole $= 6.02 \times 10^{23}$). In part d., we are given the number of moles of molecules, so the first conversion factor is not needed.

a. $10.0 \text{ g } H_2SO_4 \times \left(\frac{1 \text{ mole } H_2SO_4}{98.09 \text{ g } H_2SO_4}\right) \times \left(\frac{1 \text{ mole S}}{1 \text{ mole } H_2SO_4}\right) \times \left(\frac{6.02 \times 10^{23} \text{ atoms S}}{1 \text{ mole S}}\right)$

$= 6.14 \times 10^{22}$ atoms S

b. $20.0 \text{ g } SO_3 \times \left(\frac{1 \text{ mole } SO_3}{80.07 \text{ g } SO_3}\right) \times \left(\frac{1 \text{ mole S}}{1 \text{ mole } SO_3}\right) \times \left(\frac{6.02 \times 10^{23} \text{ atoms S}}{1 \text{ mole S}}\right)$

$= 1.50 \times 10^{23}$ atoms S

c. $30.0 \text{ g } Al_2S_3 \times \left(\frac{1 \text{ mole } Al_2S_3}{150.17 \text{ g } Al_2S_3}\right) \times \left(\frac{3 \text{ moles S}}{1 \text{ mole } Al_2S_3}\right) \times \left(\frac{6.02 \times 10^{23} \text{ atoms S}}{1 \text{ mole S}}\right)$

$= 3.61 \times 10^{23}$ atoms S

d. $2 \text{ moles } S_2O \times \left(\frac{2 \text{ moles S}}{1 \text{ mole } S_2O}\right) \times \left(\frac{6.02 \times 10^{23} \text{ atoms S}}{1 \text{ mole S}}\right) = 2.41 \times 10^{24}$ atoms S

6.37 To calculate grams of S from molecules containing S atoms, use three conversion factors. 1) Change molecules of compound to moles of compound using Avogadro's number (1 mole $= 6.02 \times 10^{23}$). 2) Change moles of compound to moles S by determining the number of atoms of S in each molecule of the compound. 3) Convert moles S to grams of S using the atomic mass of sulfur. For parts c. and d., we are given moles of compound, so step 1 is not needed.

a. $3.01 \times 10^{23} \text{ molecules } S_2O \times \left(\frac{1 \text{ mole } S_2O}{6.02 \times 10^{23} \text{ molecules } S_2O}\right) \times \left(\frac{2 \text{ moles S}}{1 \text{ mole } S_2O}\right) \times \left(\frac{32.07 \text{ g S}}{1 \text{ mole S}}\right)$

$= 32.1$ g S

b. $3 \text{ molecules } S_4N_4 \times \left(\frac{1 \text{ mole } S_4N_4}{6.02 \text{ x } 10^{23} \text{ molecules } S_4N_4}\right) \times \left(\frac{4 \text{ moles S}}{1 \text{ mole } S_4N_4}\right) \times \left(\frac{32.07 \text{ g S}}{1 \text{ mole S}}\right)$

$= 6.39 \times 10^{-22}$ g S

c. $2.00 \text{ moles } SO_2 \times \left(\frac{1 \text{ mole S}}{1 \text{ mole } SO_2}\right) \times \left(\frac{32.07 \text{ g S}}{1 \text{ mole S}}\right) = 64.1$ g S

d. $4.50 \text{ moles } S_8 \times \left(\frac{8 \text{ mole S}}{1 \text{ mole } S_8}\right) \times \left(\frac{32.07 \text{ g S}}{1 \text{ mole S}}\right) = 1150$ g S

6.39 a. 1.00 mole S_8 contains a greater number of atoms than 1.00 mole S. The numerical subscripts in a chemical formula give the number of atoms of the various elements present in 1 formula unit of the substance. A molecule of S_8 contains 8 S atoms.

b. From the periodic table, it can be seen that 1.00 mole of Al has a mass of 26.98 g. Therefore, 28.00 g of Al would have more atoms than 1.00 mole (26.98 g) of Al.

c. From the periodic table, 28.09 g of Si is one mole of Si. 30.09 g of Mg has a greater mass than 1 mole (24.31 g) of Mg. Therefore, 30.09 g of Mg has a greater number of atoms than 28.09 g (1 mole) of Si.

d. 2.00 g of Na is less than 1 mole (22.99 g) of Na; 6.02×10^{23} atoms of He is 1 mole of He. Therefore, 6.02×10^{23} atoms of He is the greater number of atoms.

6.41 The chemical formula of a compound gives the number of atoms of each element in one formula unit (molecule) of the substance. Since the atoms of the molecule (H and O) are present in a 1:1 ratio, let x equal the number of atoms of each element in the molecule. Use the definition of formula mass to solve the problem.

formula mass = x(formula mass H) + x(formula mass O) = 34.02 g/mole (or amu/molecule)

$(x)(1.01) + (x)(16.0) = (x)(17.01) = 34.02$ amu/molecule

x = 34.02 amu/molecule ÷ 17.01 amu/molecule = 2 atoms of each element in the molecule.

The chemical formula is H_2O_2.

6.43 A balanced chemical equation has the same number of atoms of each element involved in the reaction on each side of the equation.

a. Balanced chemical equation

b. Balanced chemical equation

c. The chemical equation in part c. is not balanced; there are different numbers of both S atoms and O atoms on the two sides of the equation. The balanced chemical equation should be: $CS_2 + 3O_2 \rightarrow CO_2 + 2SO_2$

d. Balanced chemical equation

6.45 a. 2 SO_3: 1 S atom + 3 O atoms = 4 atoms/molecule SO_3
4 atoms/molecule SO_3 × 2 molecules SO_3 = 8 atoms

b. 4 N_2H_4: 2 N atoms + 4 H atoms = 6 atoms/molecule N_2H_4
6 atoms/molecule N_2H_4 × 4 molecules N_2H_4 = 24 atoms

c. 3 H_3PO_4: 3 H atoms + 1 P atom + 4 O atoms = 8 atoms/molecule H_3PO_4
8 atoms/molecule H_3PO_4 × 3 molecules H_3PO_4 = 24 atoms

d. 3 $Al(OH)_3$: 1 Al atom + 3 O atoms + 3 H atoms = 7 atoms/molecule $Al(OH)_3$
7 atoms/molecule $Al(OH)_3$ × 6 molecules $Al(OH)_3$ = 42 atoms

6.47 To determine the number of atoms of each element on each side of the chemical equation, multiply the number of atoms of the element in the molecule by the molecule's coefficient.

a. reactant side – two O atoms; products side – two O atoms

b. reactant side – four O atoms; product side – four O atoms

c. reactant side – 26 O atoms; product side – 26 O atoms

d. reactant side – 12 O atoms; product side – 12 O atoms

6.49 To balance a chemical equation, examine the equation and pick one element to balance first. Start with the compound that contains the greatest number of atoms, whether in the reactant or product. Add coefficients where necessary to balance this element, then continue adding coefficients to balance each of the other elements separately. As a final check, count the number of atoms of each element on each side of the equation to make sure they are equal.

a. $2Na + 2H_2O \rightarrow 2NaOH + H_2$
b. $2Na + ZnSO_4 \rightarrow Na_2SO_4 + Zn$
c. $2NaBr + Cl_2 \rightarrow 2NaCl + Br_2$
d. $2ZnS + 3O_2 \rightarrow 2ZnO + 2SO_2$

6.51 In the following chemical equations a carbon-containing compound is oxidized with molecular oxygen to form CO_2 and H_2O. It is usually convenient in this type of equation to begin by balancing the hydrogen atoms, then the carbon atoms, and finally the oxygen atoms.

a. $CH_4 + 2O_2 \rightarrow CO_2 + 2H_2O$
b. $2C_6H_6 + 15O_2 \rightarrow 12CO_2 + 6H_2O$
c. $C_4H_8O_2 + 5O_2 \rightarrow 4CO_2 + 4H_2O$
d. $C_5H_{10}O + 7O_2 \rightarrow 5CO_2 + 5H_2O$

6.53 The balanced chemical equation is:

2 butyne + $11O_2 \rightarrow 8CO_2 + 6H_2O$

On the right side of the chemical equation there are 22 O atoms, 12 H atoms and 8 C atoms. On the left side of the chemical equation there are 22 O atoms. Therefore, 2 molecules of butyne must contain 8 C atoms and 12 H atoms; 1 molecule of butyne contains 4 C atoms and 6 H atoms. The molecular formula of butyne is C_4H_6.

6.55 a. The reactant box contains $2A_2 + 6B_2 \rightarrow 4AB_3$. The product box contains 4 AB_3. The chemical equation for the reaction is $2A_2 + 6B_2 \rightarrow 4AB_3$. All the coefficients in this equation are divisible by 2, which simplifies the chemical equation to $A_2 + 3B_2 \rightarrow 2AB_3$.

b. The reactant box contains $6A_2 + 3B_2$. The product box contains $6AB + 3A_2$. Three of the $6A_2$ did not react since there are 3 A_2 in the product box. The chemical equation for the reaction, taking into account the unreacted molecules, is $3A_2 + 3B_2 \rightarrow 6AB$. All of the coefficients in this equation are divisible by 3, which simplifies the equation to $A_2 + B_2 \rightarrow 2AB$

6.57 Box I contains 6 orange atoms and 8 green atoms. Box II contains 7 orange atoms and 7 green atoms; box III, 6 orange atoms and 8 green atoms; and box IV, 6 orange atoms and 10 green atoms. The number of atoms of each kind must remain constant during a chemical reaction. Box III is, thus, the box consistent with box I.

6.59 The coefficients in the balanced chemical equation give the mole-to-mole ratios. The six mole-to-mole conversion factors that can be derived from the balanced equation are:

$$\left(\frac{2 \text{ moles } NH_3}{3 \text{ moles } H_2}\right) \left(\frac{3 \text{ moles } H_2}{2 \text{ moles } NH_3}\right) \left(\frac{2 \text{ moles } NH_3}{1 \text{ mole } N_2}\right) \left(\frac{1 \text{ mole } N_2}{2 \text{ moles } NH_3}\right) \left(\frac{3 \text{ moles } H_2}{1 \text{ mole } N_2}\right) \left(\frac{1 \text{ mole } N_2}{3 \text{ moles } H_2}\right)$$

6.61 The coefficients in the balanced chemical equation indicate that 1 mole of Sb_2S_3 reacts with 6 moles of HCl to produce 2 moles of $SbCl_3$ and 3 moles of H_2S. The coefficients of the various reactants and products are used in the conversion factors.

a. $\dfrac{3 \text{ moles } H_2S}{2 \text{ moles } SbCl_3}$ b. $\dfrac{6 \text{ moles HCl}}{1 \text{ mole } Sb_2S_3}$

c. $\dfrac{6 \text{ moles HCl}}{3 \text{ moles } H_2S}$ d. $\dfrac{2 \text{ moles } SbCl_3}{1 \text{ mole } Sb_2S_3}$

6.63 The coefficients from a balanced chemical equation can be used to form conversion factors to solve problems. In the problems below, the conversion factor is based on a mole-to-mole ratio using the coefficient of the first reactant and the coefficient of the CO_2 produced.

a. $5.00 \text{ moles } CO_2 \times \left(\dfrac{1 \text{ mole } C_7H_{16}}{7 \text{ moles } CO_2}\right) = 0.714 \text{ mole } C_7H_{16}$

b. $5.00 \text{ moles } CO_2 \times \left(\dfrac{1 \text{ mole } CS_2}{1 \text{ mole } CO_2}\right) = 5.00 \text{ moles } CS_2$

c. $5.00 \text{ moles } CO_2 \times \left(\dfrac{1 \text{ mole } Fe_2O_4}{1 \text{ mole } CO_2}\right) = 5.00 \text{ moles } Fe_2O_4$

d. $5.00 \text{ moles } CO_2 \times \left(\dfrac{2 \text{ moles HCl}}{1 \text{ mole } CO_2}\right) = 10.0 \text{ moles HCl}$

6.65 The coefficients in a balanced chemical equation give the numerical relationships among formula units consumed or produced in the chemical reaction. The balanced reaction is: $C_6H_{12}O_6 + 6O_2 \rightarrow 6CO_2 + 6H_2O$ The coefficients show that 1 molecule of $C_6H_{12}O_6$ gives 12 product molecules. Using this relationship as a conversion factor:

$$7 \text{ molecules } C_6H_{12}O_6 \times \frac{12 \text{ product molecules}}{1 \text{ molecule } C_6H_{12}O_6} = 84 \text{ product molecules}$$

6.67 The balanced chemical equation for the reaction, in which the coefficient for C_2H_4 is 8, is

$$8C_2H_4 + 8H_2O \rightarrow 8C_2H_5OH$$

Thus, 8 H_2O molecules are needed to react with 8 C_2H_4 molecules.

6.69 a. Red atoms are oxygen, black atoms carbon, and grayish blue atoms hydrogen. There are two different kinds of molecules in the product box: CO_2 and H_2O.

b. The balanced chemical equation for a reaction where CH_4 and O_2 are the reactants and CO_2 and H_2O are the products is $CH_4 + 2O_2 \rightarrow CO_2 + 2H_2O$. Thus, 1 mole of CO_2 and 2 moles of H_2O are produced for each mole of CH_4 that reacts. 6.0 moles of reacting CH_4 will produce 6.0 moles of CO_2 and 12 moles of H_2O.

6.71 First write the balanced equation: $3Be + N_2 \rightarrow Be_3N_2$ From the balanced equation, it can be seen that Be and N_2 react in a 3:1 ratio. Use this relationship as a conversion factor to calculate the moles Be used to react completely with 10.0 moles N_2.

$$10.0 \text{ moles } N_2 \times \frac{3.00 \text{ moles Be}}{1.00 \text{ mole } N_2} = 30.0 \text{ moles Be}$$

6.73 We can see from the road map in Figure 6.9 that the conversion of grams of CO_2 to grams of O_2 requires three conversion factors: 1) Use the molar mass of CO_2 to convert 3.50 g of CO_2 to moles of CO_2. 2) Use the coefficients from the balanced chemical equation to convert moles of CO_2 to moles of O_2. 3) Use the molar mass of O_2 to convert moles of O_2 to grams of O_2.

$$3.50 \text{ g } CO_2 \times \left(\frac{1 \text{ mole } CO_2}{44.01 \text{ g } CO_2}\right) \times \left(\frac{2 \text{ moles } O_2}{1 \text{ mole } CO_2}\right) \times \left(\frac{32.00 \text{ g } O_2}{1 \text{ mole } O_2}\right) = 5.09 \text{ g } O_2$$

6.75 Use the road map in Figure 6.9 to determine the conversion factors that will be needed. The conversion of grams of CO to grams of O_2 requires three conversion factors: 1) Use the molar mass of CO to convert 25.0 g of CO to moles of CO. 2) Use the coefficients from the balanced chemical equation to convert moles of CO to moles of O_2. 3) Use the molar mass of O_2 to convert moles of O_2 to grams of O_2.

$$25.0 \text{ g CO} \times \left(\frac{1 \text{ mole CO}}{28.01 \text{ g CO}}\right) \times \left(\frac{1 \text{ mole } O_2}{2 \text{ moles CO}}\right) \times \left(\frac{32.00 \text{ g } O_2}{1 \text{ mole } O_2}\right) = 14.3 \text{ g } O_2$$

6.77 Use the road map in Figure 6.9 to determine the conversion factors that will be needed. The conversion of grams of SO_2 to grams of H_2O requires three conversion factors: 1) Use the molar mass of SO_2 to convert 10.0 g of SO_2 to moles of SO_2. 2) Use the coefficients from the balanced chemical equation to convert moles of SO_2 to moles of H_2O. 3) Use the molar mass of H_2O to convert moles of H_2O to grams of H_2O.

$$10.0 \text{ g } SO_2 \times \left(\frac{1 \text{ mole } SO_2}{64.07 \text{ g } SO_2}\right) \times \left(\frac{2 \text{ moles } H_2O}{1 \text{ mole } SO_2}\right) \times \left(\frac{18.02 \text{ g } H_2O}{1 \text{ mole } H_2O}\right) = 5.63 \text{ g } H_2O$$

6.79 Write the balanced chemical equation: $3Be + N_2 \rightarrow Be_3N_2$ Use the road map in Figure 6.11 to determine the conversion factors for the problem: g $N_2 \rightarrow$ moles of $N_2 \rightarrow$ moles Be $\rightarrow$ g Be

$$45.0 \text{ g } N_2 \times \left(\frac{1 \text{ mole } N_2}{28.0 \text{ g } N_2}\right) \times \left(\frac{3.00 \text{ moles Be}}{1.00 \text{ mole } N_2}\right) \times \left(\frac{9.01 \text{ g Be}}{1 \text{ mole Be}}\right) = 43.4 \text{ g Be}$$

6.81 The theoretical yield for a chemical reaction is the maximum amount of a product that can be obtained from given amounts of reactants. The percent yield is the ratio of the actual (experimental) yield to the theoretical yield, multiplied by 100 (to give percent).

$$\text{percent yield} = \frac{\text{actual yield}}{\text{theoretical yield}} \times 100 = \frac{31.87 \text{ g}}{32.03 \text{ g}} \times 100 = 99.50\%$$

6.83 a. The theoretical yield for a chemical reaction is the maximum possible amount of product that can be obtained from given amounts of reactants. In this case the theoretical yield is 11.2 g of SO_2.

b. The actual yield is the experimental yield. In this case the actual yield is 9.75 g of SO_2.

c. The percent yield is the ratio of the actual (experimental) yield to the theoretical yield, multiplied by 100 (to give percent).

$$\text{percent yield} = \frac{\text{actual yield}}{\text{theoretical yield}} \times 100 = \frac{9.75 \text{ g } SO_2}{11.2 \text{ g } SO_2} \times 100 = 87.1\%$$

6.85 First, find the theoretical yield using the balanced chemical equation and the formula masses to determine the conversion factors.

$$29.0 \text{ g } N_2 \times \left(\frac{1 \text{ mole } N_2}{28.0 \text{ g } N_2}\right) \times \left(\frac{1.00 \text{ mole } Ca_3N_2}{1.00 \text{ mole } N_2}\right) \times \left(\frac{148 \text{ g } Ca_3N_2}{1 \text{ mole } Ca_3N_2}\right) = 153 \text{ g } Ca_3N_2$$

The percent yield is the ratio of the actual (experimental) yield to the theoretical yield, multiplied by 100 (to give percent).

$$\text{percent yield} = \frac{\text{actual yield}}{\text{theoretical yield}} \times 100 = \frac{126 \text{ g } Ca_3N_2}{153 \text{ g } Ca_3N_2} \times 100 = 82.0\%$$

Gases, Liquids, and Solids — Chapter 7

Solutions to Selected Problems

7.1
a. True. Both solids and liquids have a definite volume.
b. True. Thermal expansion for a liquid is generally greater than that of the corresponding solid.
c. False. The compressibility of a gas is more than that of the corresponding liquid.
d. False. The density of a solid is much greater than that of the corresponding gas.

7.3
a. Potential energy is related to cohesive forces.
b. The magnitude increases as temperature increases.
c. Cohesive forces cause order within the system.
d. Electrostatic attractions are particularly important.

7.5
a. liquid state
b. solid state
c. gaseous state
d. solid state

7.7
a. On a relative scale, density is described as high rather than low for both liquids and solids.
b. On a relative scale, the thermal expansion of solids is described as very small.
c. In general, the interactions between particles of a gas are very weak (almost zero).
d. In general, particles in a gas are far apart in a random arrangement.

7.9
a. amount
b. volume
c. pressure
d. temperature

7.11 Use the following relationships to set up conversion factors to complete the conversions:
1 atm = 760 mm Hg = 760 torr
1 atm = 14.7 psi

a. $555 \text{ mm of Hg} \times \left(\frac{1 \text{ atm}}{760 \text{ mm Hg}}\right) = 0.730 \text{ atm}$

b. $32.0 \text{ psi} \times \left(\frac{1 \text{ atm}}{14.7 \text{ psi}}\right) = 2.18 \text{ atm}$

c. $675 \text{ torr} \times \left(\frac{1 \text{ atm}}{760 \text{ torr}}\right) = 0.888 \text{ atm}$

d. $75.2 \text{ mm of Hg} \times \left(\frac{1 \text{ atm}}{760 \text{ mm Hg}}\right) = 0.0989 \text{ atm}$

7.13 Pressure readings in torr or mm of Hg have an uncertainty in the "ones place" even though no decimal point is specified.
a. 678 torr has 3 significant figures.
b. 680 torr has 3 significant figures.
c. 699 torr has 3 significant figures.
d. 700 torr has 3 significant figures.

7.15 Use the following relationships to set up conversion factors to complete the conversions:
1 atm = 760 mm Hg = 760 torr

a. $678 \text{ torr} \times \left(\dfrac{1 \text{ atm}}{760 \text{ torr}}\right) = 0.892 \text{ atm}$

b. $680 \text{ torr} \times \left(\dfrac{1 \text{ atm}}{760 \text{ torr}}\right) = 0.895 \text{ atm}$

c. $690 \text{ torr} \times \left(\dfrac{1 \text{ atm}}{760 \text{ torr}}\right) = 0.908 \text{ atm}$

d. $700 \text{ torr} \times \left(\dfrac{1 \text{ atm}}{760 \text{ torr}}\right) = 0.921 \text{ atm}$

7.17 According to Boyle's law, at a constant temperature, the volume of a gas is inversely proportional to the pressure applied to it: $P_1 \times V_1 = P_2 \times V_2$
Since P_1, V_1, and V_2 are given in the problem, we can solve the equation for P_2 and substitute the given values to find the value of P_2.

$$P_2 = \frac{P_1V_1}{V_2} = 3.0 \text{ atm} \times \left(\frac{6.0 \text{ L}}{2.5 \text{ L}}\right) = 7.2 \text{ atm}$$

(To check calculations: since volume decreases, P_2 should be larger than P_1, and this is true.)

7.19 The given quantities are P_1 (655 mm Hg), V_1 (3.00 L), and P_2 (725 mm Hg). We can solve Boyle's law for V_2 and substitute the given quantities.

$$V_2 = \frac{V_1P_1}{P_2} = 3.00 \text{ L} \times \left(\frac{655 \text{ mm Hg}}{725 \text{ mm Hg}}\right) = 2.71 \text{ L}$$

7.21 Use Boyle's Law ($P_1V_1 = P_2V_2$) to calculate the missing information.
a. $P_2 = 2.00$ atm b. $V_1 = 6.00$ L c. $V_2 = 1.00$ L d. $P_1 = 8.00$ atm

7.23 Diagram II. Doubling the pressure, at constant temperature and constant moles, will cut the volume in half (Boyle's law).

7.25 Since pressure is constant, we can use Charles's law, the relationship between temperature (using the Kelvin scale) and volume: $V_1/T_1 = V_2/T_2$
We are given V_1 (2.73 L), T_1 (27°C converted to 300 K), and T_2 (127°C = 400 K). We can solve Charles's law for V_2 and substitute the given quantities.

$$V_2 = \frac{V_1T_2}{T_1} = 2.73 \text{ L} \times \left(\frac{400 \text{ K}}{300 \text{ K}}\right) = 3.64 \text{ L}$$

7.27 The system is at constant pressure (2.0 atm), so we can use Charles's law to find the temperature (in degrees K) after a volume change. Given quantities are: T_1 (25°C = 298K), V_1 (375 mL), V_2 (525 mL). Solve Charles's law for T_2. Convert the Kelvin temperature to °C.

$$T_2 = \frac{T_1V_2}{V_1} = 298\text{ K} \times \left(\frac{525\text{ mL}}{375\text{ mL}}\right) = 417\text{ K}$$

$$417\text{ K} - 273 = 144^\circ\text{C}$$

7.29 Use Charles's law ($V_1/T_1 = V_2/T_2$) to calculate the missing information. Remember that in the gas laws temperature is expressed in degrees Kelvin. (K = °C + 273)

a. $V_1 = 0.974$ L b. $T_2 = 207$°C c. $T_1 = 127$°C d. $V_2 = 2.54$ L

7.31 Diagram II. Decreasing the Kelvin temperature by a factor of two, at constant pressure and constant moles, will cause the volume to decrease by a factor of two, that is, be cut in half (Charles's law).

7.33 The combined gas law ($P_1V_1/T_1 = P_2V_2/T_2$) can be rearranged to solve for any one of the variables.

a. $P_1 = P_2 \times \dfrac{V_2}{V_1} \times \dfrac{T_1}{T_2}$

b. $V_2 = V_1 \times \dfrac{P_1}{P_2} \times \dfrac{T_2}{T_1}$

7.35 There are changes in pressure, volume, and temperature in these problems, so the combined gas law is used to complete the calculations. The initial conditions for each part of the problem are: P_1 (1.35 atm), V_1 (15.2 L), and T_1 (33°C = 306 K)

a. T_2 (35°C = 308 K), P_2 (3.50 atm) are given. Rearrange the combined gas law to solve for V_2.

$$V_2 = \frac{V_1P_1T_2}{P_2T_1} = 15.2\text{ L} \times \left(\frac{1.35\text{ atm}}{3.50\text{ atm}}\right) \times \left(\frac{308\text{ K}}{306\text{ K}}\right) = 5.90\text{ L}$$

b. T_2 (42°C = 315 K), V_2 (10.0 L) are given. Rearrange the combined gas law to solve for P_2.

$$P_2 = \frac{P_1V_1T_2}{V_2T_1} = 1.35\text{ atm} \times \left(\frac{15.2\text{ L}}{10.0\text{ L}}\right) \times \left(\frac{315\text{ K}}{306\text{ K}}\right) = 2.11\text{ atm}$$

c. P_2 (7.00 atm), V_2 (0.973 L) are given. Rearrange the combined gas law to solve for T_2 and convert the Kelvin temperature to °C.

$$T_2 = \frac{T_1P_2V_2}{P_1V_1} = 306\text{ K} \times \left(\frac{7.00\text{ atm}}{1.35\text{ atm}}\right) \times \left(\frac{0.973\text{ L}}{15.2\text{ L}}\right) = 102\text{ K}$$

$$102\text{ K} - 273 = -171^\circ\text{C}$$

d. T_2 (97°C = 370 K), P_2 (6.70 atm) are given. Rearrange the combined gas law to solve for V_2. Convert V_1 (15.2 L) to mL in order to determine V_2 in mL.

$$V_2 = \frac{V_1P_1T_2}{P_2T_1} = 15{,}200 \text{ mL} \times \left(\frac{1.35 \text{ atm}}{6.70 \text{ atm}}\right) \times \left(\frac{370 \text{ K}}{306 \text{ K}}\right) = 3.70 \times 10^3 \text{ mL}$$

7.37 Diagram II. At constant moles, doubling the pressure will cut the volume in half and doubling the Kelvin temperature will double the volume. The halving and doubling effects cancel each other and the volume remains the same.

7.39 a. Since the temperature is constant, use Boyle's law ($P_1V_1 = P_2V_2$) to calculate the new pressure (P_2).

$$P_2 = \frac{P_1 \times V_1}{V_2} = \frac{1.25 \text{ atm} \times 575 \text{ mL}}{825 \text{ mL}} = 0.871 \text{ atm}$$

b. Since temperature, volume, and pressure all change, use the combined gas law to find P_2.

$$\frac{P_1 \times V_1}{T_1} = \frac{P_2 \times V_2}{T_2}$$

$$P_2 = \frac{P_1 \times V_1 \times T_2}{T_1 \times V_2} = \frac{1.25 \text{ atm} \times 575 \text{ mL} \times 418 \text{ K}}{825 \text{ mL} \times 398 \text{ K}} = 0.915 \text{ atm}$$

7.41 a. The number of moles is constant in Boyle's law, Charles's law, and the combined gas law.
b. The temperature is constant in Boyle's law.

7.43 When pressure is held constant, the combined gas law reverts to Charles's law.
When $P_1 = P_2$, $\frac{P_1 \times V_1}{T_1} = \frac{P_2 \times V_2}{T_2}$ reverts to $\frac{V_1}{T_1} = \frac{V_2}{T_2}$

7.45 Using the ideal gas law ($PV = nRT$) we can calculate any one of the four gas properties (P, V, T, or n) given the other three. R is the ideal gas constant (0.0821 atm · L/mole · K). Remember that in the gas laws temperature is expressed in degrees Kelvin. (K = °C + 273)

a. $P = nRT/V = 32.8$ atm
b. $V = nRT/P = 10.9$ L
c. $n = PV/RT = 0.459$ mole
d. $T = PV/nR = -214$°C

7.47 Using the ideal gas law ($PV = nRT$) we can calculate any one of the four gas properties (P, V, T, or n) given the other three. R is the ideal gas constant (0.0821 atm · L/mole · K). In this problem, n (5.23 moles), V (5.23 L), and P (5.23 atm) are given. Rearrange the equation to solve for T. Convert K to °C.

$$T = \frac{PV}{nR} = \frac{(5.23 \text{ atm})(5.23 \text{ L})}{(5.23 \text{ moles})\left(0.0821\frac{\text{atm L}}{\text{mole K}}\right)} = 63.7 \text{ K}$$

63.7 K − 273 = −209°C

7.49 In this problem, n (0.100 mole), T (0°C = 273 K), and P (2 atm) are given. Rearrange the ideal gas law to determine the volume in liters.

$$V = \frac{nRT}{P} = \frac{(0.100 \text{ mole})\left(0.0821 \frac{\text{atm L}}{\text{mole K}}\right)(273 \text{ K})}{(2.00 \text{ atm})} = 1.12 \text{ L}$$

7.51 Rearrange the ideal gas law in each problem below; use it to determine the unknown quantity.

a. $$V = \frac{nRT}{P} = \frac{(0.250 \text{ mole})\left(0.0821 \frac{\text{atm L}}{\text{mole K}}\right)(300 \text{ K})}{(1.50 \text{ atm})} = 4.11 \text{ L}$$

b. $$P = \frac{nRT}{V} = \frac{(0.250 \text{ mole})\left(0.0821 \frac{\text{atm L}}{\text{mole K}}\right)(308 \text{ K})}{(2.00 \text{ L})} = 3.16 \text{ atm}$$

c. $$T = \frac{PV}{nR} = \frac{(1.20 \text{ atm})(3.00 \text{ L})}{(0.250 \text{ mole})\left(0.0821 \frac{\text{atm L}}{\text{mole K}}\right)} = 175 \text{ K}$$

175 K – 273 = –98°C.

d. $$V = \frac{nRT}{P} = \frac{(0.250 \text{ mole})\left(0.0821 \frac{\text{atm L}}{\text{mole K}}\right)(398 \text{ K})}{(0.500 \text{ atm})} = 16.3 \text{ L} = 1.63 \times 10^4 \text{ mL}$$

7.53 To solve these problems, use the ideal gas law ($PV = nRT$). Pressure is expressed in atmospheres, n in moles, T in degrees Kelvin, and V in liters. R is equal to 0.0821 L atom/moles K.

a. $$P = \frac{nRT}{V} = \frac{(0.72 \text{ mole})\left(0.0821 \frac{\text{atm L}}{\text{mole K}}\right)(313 \text{ K})}{(4.00 \text{ L})} = 4.6 \text{ atm}$$

b. $$P = \frac{nRT}{V} = \frac{(4.5 \text{ mole})\left(0.0821 \frac{\text{atm L}}{\text{mole K}}\right)(313 \text{ K})}{(4.00 \text{ L})} = 29 \text{ atm}$$

c. In part c. and part d., change grams of O_2 to moles of O_2 by using the definition of formula mass as a conversion factor: formula mass = mass in grams/1 mole

$$P = \frac{nRT}{V} = \frac{\left[(0.72 \text{ g } O_2) \times \left(\frac{1.00 \text{ mole } O_2}{32.0 \text{ g } O_2}\right)\right]\left(0.0821 \frac{\text{atm L}}{\text{mole K}}\right)(313 \text{ K})}{(4.00 \text{ L})} = 0.14 \text{ atm}$$

d. $$P = \frac{nRT}{V} = \frac{\left[(4.5 \text{ g } O_2) \times \left(\frac{1.00 \text{ mole } O_2}{32.0 \text{ g } O_2}\right)\right]\left(0.0821 \frac{\text{atm L}}{\text{mole K}}\right)(313 \text{ K})}{(4.00 \text{ L})} = 0.90 \text{ atm}$$

7.55 Multiply the value of R (0.0821 L • atm/mole • K) by the conversion factor derived from the defined value of 1 atm (760 torr).

$$R = \frac{0.0821 \text{ L atm}}{\text{mole K}} \times \frac{760 \text{ torr}}{1.00 \text{ atm}} = 62.37 \frac{\text{L torr}}{\text{mole K}}$$

7.57 Dalton's law of partial pressures states that the total pressure exerted by a mixture of gases is the sum of the partial pressures of the individual gases present. For this mixture of three gases:

$P_{Total} = P_{N_2} + P_{He} + P_{O_2}$

The total pressure of the gas and the partial pressures of two of the gases are given:

$P_{Total} = 1.50 \text{ atm} = 0.75 \text{ atm} + 0.33 \text{ atm} + P_{O_2}$

Rearrange the equation to solve for the partial pressure of oxygen.

$P_{O_2} = P_{Total} - P_{N_2} - P_{He} = (1.50 - 0.75 - 0.33) \text{ atm} = 0.42 \text{ atm}$

7.59 Since the total pressure and three of the four partial pressures are given, rearrange Dalton's law to solve for the unknown partial pressure of the fourth gas:

$P_{CO_2} = P_{Total} - P_{O_2} - P_{N_2} - P_{Ar} = (623 - 125 - 175 - 225) \text{ mm Hg} = 98 \text{ mm Hg}$

7.61 a. 2.4 atm; 4 out of 10 molecules (0.40) are neon and 0.40 x 6.0 atm = 2.4 atm
b. 2.4 atm; 4 out of 10 molecules (0.40) are argon and 0.40 x 6.0 atm = 2.4 atm
c. 1.2 atm; 2 out of 10 molecules (0.20) are krypton and 0.20 x 6.0 atm = 1.2 atm

7.63 a. Since n (moles) is the same for each gas, the individual pressure (P) exerted by each gas is the same. $P_{He} + P_{Ne} + P_{Ar} = 3P = P_{Total} = 3.00 \text{ atm}$
$P = 1.00$ atm (partial pressure of each gas)

b. Since all three gases are monatomic, this is the same as the problem in a. An equal number of atoms means an equal number of moles: $P = 1.00$ atm (partial pressure of each gas)

c. The partial pressures of He, Ne, and Ar are in a 3:2:1 ratio, so $P_{He} = 3\,P_{Ar}$ and $P_{Ne} = 2P_{Ar}$,
$P_{Total} = P_{He} + P_{Ne} + P_{Ar} = 6\,P_{Ar} = 3.00 \text{ atm}$
$P_{Ar} = 0.50$ atm; $P_{He} = 3\,P_{Ar} = 1.50$ atm; and $P_{Ne} = 2P_{Ar} = 1.00$ atm
The partial pressures of He, Ne, and Ar are 1.50 atm, 1.00 atm, and 0.50 atm respectively.

d. $P_{He} = 1/2\,P_{Ne}$; $P_{He} = 1/3\,P_{Ar}$ and $P_{Total} = P_{He} + P_{Ne} + P_{Ar} = 3.00 \text{ atm}$
then $P_{Total} = P_{He} + 2P_{He} + 3P_{He} = 6P_{He} = 3.00$ atm and $P_{He} = 0.50$ atm
The partial pressures of He, Ne, and Ar are 0.50 atm, 1.00 atm, and 1.50 atm respectively.

7.65 a. No. Evaporation is endothermic, and freezing is exothermic.
b. No. Melting is endothermic, and deposition is exothermic.
c. Yes. Both freezing and condensation are exothermic.
d. Yes. Both sublimation and evaporation are endothermic.

7.67 a. No. Evaporation changes liquid to gas; freezing changes liquid to solid.
b. No. Melting changes solid to liquid; deposition changes gas to solid.
c. No. Freezing changes liquid to solid; condensation changes gas to liquid.
d. Yes. Sublimation changes solid to gas; evaporation changes liquid to gas.

7.69 a. No, they are not opposite. Evaporation changes liquid to gas; freezing changes liquid to solid.
b. No, they are not opposite. Melting changes solid to liquid; deposition changes gas to solid.
c. No, they are not opposite. Freezing changes liquid to solid; condensation changes gas to liquid.
d. No, they are not opposite. Sublimation changes solid to gas; evaporation changes liquid to gas.

7.71 Use the ideal gas law ($PV = nRT$) to find the volume of the CO_2 gas. Remember to change °C to K: K = °C + 273 = 296

$$V = \frac{nRT}{P} = \frac{(1.00 \text{ mole})\left(0.0821\frac{\text{atm L}}{\text{mole K}}\right)(296 \text{ K})}{(0.983 \text{ atm})} = 24.7 \text{ L}$$

7.73 The amount of liquid decreases; the temperature of the liquid decreases.

7.75 The rate increases.

7.77 At the same temperature, liquid B evaporates at a faster rate than liquid A because B has weaker intermolecular attraction forces between molecules than A does.

7.79 a. True b. True

7.81 The higher the temperature, the higher the vapor pressure.

7.83 The term is volatile.

7.85 a. True
b. True
c. False; "sea level" should be replaced by "external pressure of one atmosphere."
d. False; at 760 mm Hg, the boiling point is determined by the strength of intermolecular forces, which varies from substance to substance; thus, not all liquids have the same boiling point at 760 mm Hg pressure.

7.87 The higher the external pressure, the higher the boiling point.

7.89 A boiling point is the temperature at which the vapor pressure of a liquid becomes equal to the atmospheric pressure.
a. If vapor pressure and atmospheric pressure are equal, the liquid will boil.
b. If vapor pressure = 635 torr and atmospheric pressure = 735 torr, the liquid will not boil. The vapor pressure must be at least as high as the atmospheric pressure for the liquid to boil.

7.91 a. False. Hydrogen bonds are attractions between a hydrogen covalently bonded to a very electronegative atom (F, O, or N) and another electronegative atom (F, O, or N); London forces are the weakest type of intermolecular forces and occur between both polar and nonpolar molecules.
b. False. London forces are temporary weak intermolecular forces.
c. True.
d. False. Only molecules in which hydrogen is bonded to F, O, or N participate in hydrogen bonding.

7.93 Boiling point increases as intermolecular force strength increases. During boiling, molecules escape from the liquid state to the vapor state; for molecules with stronger intermolecular forces more heat energy is required for this escape.

7.95 Three types of intermolecular attractive forces act between a molecule and other molecules: dipole-dipole interactions (between polar molecules), hydrogen bonds (extra strong dipole-dipole interactions; a hydrogen is covalently bonded to a very electronegative atom (F, O, or N), and London forces (weakest type, between both polar and nonpolar molecules).

a. Since H_2 is a nonpolar molecule, the only intermolecular forces present are London forces (temporary weak intermolecular forces).

b. Since the bond in the HF molecule is between H and a very electronegative atom (F), there is hydrogen-bonding between molecules as well as London forces.

c. CO is a polar molecule; the intermolecular forces present are dipole-dipole interactions and London forces.

d. Since F_2 is a nonpolar molecule, the only intermolecular forces present are London forces (temporary weak intermolecular forces).

7.97 The dominant intermolecular force present in a liquid is the strongest intermolecular force present: hydrogen bonding > dipole-dipole interactions > London forces
a. London forces. H_2 is a nonpolar molecule.
b. Hydrogen bonding. H is covalently bonded to an electronegative atom, F.
c. Dipole-dipole interaction. CO is a polar molecule.
d. London forces. F_2 is a nonpolar molecule.

7.99 a. No hydrogen bonding. H is bonded to C, not to O.
b. Yes, there is hydrogen bonding. H is bonded to an electronegative atom, N.
c. Yes, there is hydrogen bonding. H is bonded to an electronegative atom, O.
d. No hydrogen bonding. H is bonded to I.

7.101 Four hydrogen bonds can form between a single water molecule and other water molecules. This is depicted in Figure 7.21.

7.103 The two types of hydrogen bonds that can form between this molecule and a water molecule are:

```
—Ö: ||||||| H—Ö:            —H ||||||| :Ö—H
               |                        |
               H                        H
```

7.105 A higher boiling point is due to stronger intermolecular forces.

a. Br_2 would be expected to have a higher boiling point than Cl_2 because it has greater mass and potentially greater London forces.

b. H_2O would be expected to have a higher boiling point than H_2S because it is a more polar molecule (O is more electronegative than S) and so has greater dipole-dipole interactions.

c. CO would be expected to have a higher boiling point than O_2 because CO is a polar molecule and O_2 is nonpolar. Dipole-dipole interactions are stronger intermolecular forces than London forces are.

d. HF would be expected to have a higher boiling point than HCl because F is more electronegative than Cl, so HF has stronger hydrogen-bonding than HCl has.

Solutions Chapter 8

Solutions to Selected Problems

8.1 a. True. It is possible to dissolve more than one substance in a given solvent.
b. True. Solutions are uniform throughout (homogeneous).
c. True. Since solutions are uniform throughout, every part is exactly the same.
d. False. A solution is, by definition, homogeneous; it does not separate into parts.

8.3 In a solution, a solute is a component of the solution that is present in a lesser amount relative to that of the solvent.
a. solute: sodium chloride; solvent: water
b. solute: sucrose; solvent: water
c. solute: water; solvent: ethyl alcohol
d. solute: ethyl alcohol; solvent: methyl alcohol

8.5 a. First solution. The solubility of a gas in water decreases with increasing temperature. Therefore, NH_3 gas is more soluble at 50°C than at 90°C.
b. First solution. The solubility of a gas in water increases as the pressure of the gas above the water increases. Therefore, CO_2 gas is more soluble at 2 atm than at 1 atm.
c. First solution. The solubility of NaCl in water increases with an increase in temperature (see Table 8.1).
d. Second solution. A change in pressure has little effect on the solubility of solids in water. However, the solubility of sugar in water increases with increasing temperature.

8.7 a. supersaturated b. saturated c. saturated d. unsaturated

8.9 Table 8.1 gives solubilities of various compounds (g solute/100 g H_2O) at 0°C, 50°C, and 100°C. A solution made by dissolving 34.0 g of NaCl in 100 g of H_2O at 0°C would be
a. unsaturated, because the solubility of NaCl at 0°C is 35.7 g/100 g H_2O.
b. concentrated, because it contains a large amount of solute relative to the amount that could dissolve.

8.11 If the solubility of a compound is 35 g/L at 0°C, a solution made with 25 g of the compound added to 0.5 L of water at 0°C results in a saturated solution. (The solubility of the compound in 0.5 L is: $35\text{g}/1\text{ L} \times 0.5\text{ L} = 17.5\text{ g}$)

8.13 a. A Na^+ ion surrounded by water molecules is a hydrated ion.
b. A Cl^- ion surrounded by water molecules is a hydrated ion.
c. The portion of a water molecule that is attracted to a Na^+ ion is the more electronegative atom of water, the oxygen atom.
d. The portion of a water molecule that is attracted to a Cl^- ion is the less electronegative atom of water, the hydrogen atom.

8.15 The rate of solution formation is affected by the state of subdivision of the solute (smaller subdivision: faster rate of solution), the degree of agitation during solution preparation (more agitation: faster rate of solution), and the temperature of the solution components (higher temperature: faster rate of solution).
a. Cooling the sugar cube-water mixture decreases the rate of solution.
b. Stirring the sugar cube-water mixture increases the rate of solution.
c. Breaking the sugar cube into smaller "chunks" increases the rate of solution.
d. Crushing the sugar cube to give a granulated form of sugar increases the rate of solution.

8.17 a. O_2 is a nonpolar molecule, and H_2O is a polar molecule. We predict that O_2 will be only slightly soluble in H_2O because of their differing polarities.
b. CH_3OH is a polar liquid, and H_2O is a polar liquid. We predict that CH_3OH will be very soluble in H_2O because of their similar polarities (like dissolves like).
c. CBr_4 is a nonpolar molecule, and H_2O is a polar molecule. We predict that CBr_4 will be only slightly soluble in H_2O because of their differing polarities.
d. AgCl, an ionic solid, is only slightly soluble in H_2O, a polar liquid. Most chlorides are soluble in water, but AgCl is one of the exceptions (see Table 8.2).

8.19 Like dissolves like; that is, polar solutes tend to be soluble in polar solvents, and nonpolar solutes tend to be soluble in nonpolar solvents.
a. ethanol; an ionic compound is a polar compound
b. carbon tetrachloride; a nonpolar solute and a nonpolar solvent
c. ethanol; a polar solute and a polar solvent
d. ethanol; a polar solute and a polar solvent

8.21 Locate each of the following types of compounds in Table 8.2.
a. Chlorides are soluble with exceptions.
b. Nitrates are soluble.
c. Carbonates are insoluble with exceptions.
d. Phosphates are insoluble with exceptions.

8.23 Solubility guidelines for ionic compounds in water are given in Table 8.2.
a. Calcium nitrate is soluble.
b. Sodium phosphate is soluble.
c. Silver chloride is insoluble.
d. Ammonium phosphate is soluble.

8.25 Solubility guidelines for ionic compounds in water are given in Table 8.2.
a. $CuSO_4$ is soluble.
b. $Cu(OH)_2$ is insoluble.
c. $Cu(NO_3)_2$ is soluble.
d. $CuCl_2$ is soluble.

8.27 Solubility guidelines for ionic compounds in water are given in Table 8.2.
a. Yes, both NH_4Cl and NH_4Br are soluble in water.
b. Yes, both KNO_3 and K_2SO_4 are soluble in sater.
c. Yes, both $CaCO_3$ and $Ca_3(PO_4)_2$ are insoluble in water.
d. Yes, both $Ni(OH)_2$ and $Ni_3(PO_4)_2$ are insoluble in water.

8.29 The compound would be classified as insoluble. The "insoluble" classification is used for compounds that have very limited solubility in water (less than 1 gram per liter of solution).

8.31 a. Diagram IV; concentration depends on amount and volume:
for diagram I, the concentration = 12 molecules/volume = 12/v
for diagram II, the concentration = 10 molecules/volume = 10/v
for diagram III, the concentration = 6 molecules/0.5 volume = 12/v
for diagram IV, the concentration = 5 molecules/0.25 volume = 20/v
b. Diagrams I and III; based on the information shown in part a, diagrams I and III have the same concentration (12/v).

8.33 Mass percent of solute is the mass of solute divided by the total mass of the solution, multiplied by 100 (to put the value in terms of percentage).

Mass of solution = mass of solute + mass of solvent.

a. $\frac{6.50 \text{ g}}{91.5 \text{ g}} \times 100 = 7.10\%(\text{m/m})$

b. $\frac{2.31 \text{ g}}{37.3 \text{ g}} \times 100 = 6.19\%(\text{m/m})$

c. $\frac{12.5 \text{ g}}{138 \text{ g}} \times 100 = 9.06\%(\text{m/m})$

d. $\frac{0.0032 \text{ g}}{1.2 \text{ g}} \times 100 = 0.27\%(\text{m/m})$

8.35 Mass percent of solute is the mass of solute divided by the total mass of the solution, multiplied by 100 (to put the value in terms of percentage).

Mass of solution = mass of solute + mass of solvent.

a. $\frac{10.0 \text{ g}}{50.0 \text{ g}} \times 100 = 20.0\% \text{ (m/m)}$

b. $\frac{5.00 \text{ g}}{360. \text{ g}} \times 100 = 1.38\% \text{ (m/m)}$

c. $\frac{20.0 \text{ g}}{585 \text{ g}} \times 100 = 3.42\% \text{ (m/m)}$

d. $\frac{0.520 \text{ g}}{2.52 \text{ g}} \times 100 = 20.6\% \text{ (m/m)}$

8.37 To convert grams of water to grams of glucose in each given solution, use a conversion factor based on percent-by-mass concentration: (g of glucose) ÷ (g of solution – grams of glucose)

a. $275 \text{ g } H_2O \times \left(\frac{1.30 \text{ g glucose}}{98.70 \text{ g } II_2O}\right) = 3.62 \text{ g glucose}$

b. $275 \text{ g } H_2O \times \left(\frac{5.00 \text{ g glucose}}{95.00 \text{ g } H_2O}\right) = 14.5 \text{ g glucose}$

c. $275 \text{ g } H_2O \times \left(\frac{20.0 \text{ g glucose}}{80.0 \text{ g } H_2O}\right) = 68.8 \text{ g glucose}$

d. $275 \text{ g } H_2O \times \left(\frac{31.0 \text{ g glucose}}{69.0 \text{ g } H_2O}\right) = 124 \text{ g glucose}$

8.39 The grams of K_2SO_4 needed can be determined by multiplying the grams of solution desired (32.00 g) by a conversion factor obtained from the definition of the percent by mass, 2.000%(m/m) K_2SO_4 [(2.000 g K_2SO_4 per 100.0 g solution) × 100].

$$\text{g } K_2SO_4 = 32.00 \text{ g solution} \times \left(\frac{2.000 \text{ g } K_2SO_4}{100.0 \text{ g solution}}\right) = 0.6400 \text{ g } K_2SO_4$$

8.41 For parts a, b, and c, use the definitions of deciliter and milliliter to form conversion factors. Part d: The mass-volume percent is the mass of solute in a solution (in grams) divided by the total volume of solution (in milliliters), multiplied by 100.

a. $\dfrac{0.55\text{ mg}}{\text{dL}} \times \dfrac{1.0\text{ dL}}{100\text{ mL}} = \dfrac{0.55\text{ mg}}{100\text{ mL}}$

b. $\dfrac{0.55\text{ mg}}{100\text{ mL}} \times \dfrac{1.0\text{ g}}{1000\text{ mg}} = \dfrac{0.00055\text{ g}}{100\text{ mL}}$

c. $\dfrac{0.00055\text{ g}}{100\text{ mL}} \times \dfrac{1000\text{ mL}}{1.0\text{ L}} = \dfrac{0.0055\text{ g}}{1.0\text{ L}}$

d. $\%(\text{m/v}) = \dfrac{0.00055\text{ mg}}{100\text{ mL}} \times 100 = 0.00055\%(\text{m/v})$

8.43 Percent by volume is the volume of solute in a solution divided by the total volume of the solution, multiplied by 100. Since the volumes of both the solute and the solution are given, we can use this definition to calculate the percent by volume.

a. $\%\text{ by volume} = \dfrac{20.0\text{ mL}}{475\text{ mL}} \times 100 = 4.21\%(\text{v/v})$

b. $\%\text{ by volume} = \dfrac{4.00\text{ mL}}{87.0\text{ mL}} \times 100 = 4.60\%(\text{v/v})$

8.45 Since the volumes of both the solute and the solution are given, we can use the definition of percent by volume (volume of solute in a solution divided by total volume of solution, multiplied by 100) to calculate the percent by volume.

$\%\text{ by volume} = \dfrac{22\text{ mL}}{125\text{ mL}} \times 100 = 18\%(\text{v/v})$

8.47 The mass-volume percent is the mass of solute in a solution (in grams) divided by the total volume of solution (in milliliters), multiplied by 100. Since both the mass of the solute and the volume of the solution are given, we can use this definition to calculate the percent by volume.

a. $\text{Mass-volume percent} = \dfrac{5.0\text{ g}}{250\text{ mL}} \times 100 = 2.0\%(\text{m/v})$

b. $\text{Mass-volume percent} = \dfrac{85\text{ g}}{580\text{ mL}} \times 100 = 15\%(\text{m/v})$

8.49 Since the volume of solution is given, we can convert this volume to mass of Na_2CO_3 needed by using the mass-volume percent as a conversion factor.

$\text{Mass of } Na_2CO_3 = 25.0\text{ mL solution} \times \left(\dfrac{2.00\text{ g } Na_2CO_3}{100.0\text{ mL solution}}\right) = 0.500\text{ g } Na_2CO_3$

8.51 Use the mass-volume percent (mass of solute in a solution in grams divided by the total volume of solution in milliliters) as a conversion factor to determine the number of grams of NaCl in 50.0 mL of solution.

$$\text{Mass of NaCl} = 50.0 \text{ mL solution} \times \left(\frac{7.50 \text{ g NaCl}}{100.0 \text{ mL solution}}\right) = 3.75 \text{ g NaCl}$$

8.53 Molarity is the moles of solute in a solution divided by liters of solution (moles/L). If both moles of solute and volume of solution are given, we can use the definition to find molarity. In parts c. and d., to convert grams to moles, use the formula mass as a conversion factor

a. $$\text{Molarity (M)} = \frac{\text{moles of solute}}{\text{liters of solution}} = \frac{1.40 \text{ moles}}{1.45 \text{ L}} = 0.966 \text{ M}$$

b. $$\text{Molarity (M)} = \frac{\text{moles of solute}}{\text{liters of solution}} = \frac{0.850 \text{ moles}}{0.867 \text{ L}} = 0.980 \text{ M}$$

c. $$\text{Moles of HCl} = 30.0 \text{ g HCl} \times \left(\frac{1 \text{ mole HCl}}{36.46 \text{ g HCl}}\right) = 0.823 \text{ mole HCl}$$

$$\text{Molarity (M)} = \frac{\text{moles of solute}}{\text{liters of solution}} = \frac{0.823 \text{ mole}}{1.45 \text{ L}} = 0.567 \text{ M}$$

d. $$\text{Moles of HCl} = 30.0 \text{ g HCl} \times \left(\frac{1 \text{ mole HCl}}{36.46 \text{ g HCl}}\right) = 0.823 \text{ mole HCl}$$

$$\text{Molarity (M)} = \frac{\text{moles of solute}}{\text{liters of solution}} = \frac{0.823 \text{ mole}}{0.875 \text{ L}} = 0.940 \text{ M}$$

8.55 Molarity is the moles of solute in a solution divided by liters of solution (moles/L). Use molarity as a conversion factor to convert liters of solution to moles of solute.

a. $$1.35 \text{ L solution} \times \left(\frac{0.300 \text{ moles NaOH}}{1.00 \text{ L solution}}\right) = 0.405 \text{ mole NaOH}$$

b. $$0.800 \text{ L solution} \times \left(\frac{0.100 \text{ moles NaOH}}{1.00 \text{ L solution}}\right) = 0.0800 \text{ mole NaOH}$$

c. $$0.875 \text{ L solution} \times \left(\frac{0.600 \text{ moles NaOH}}{1.00 \text{ L solution}}\right) = 0.525 \text{ mole NaOH}$$

d. $$0.125 \text{ L solution} \times \left(\frac{0.125 \text{ moles NaOH}}{1.00 \text{ L solution}}\right) = 0.0156 \text{ mole NaOH}$$

8.57 Molarity is the moles of solute in a solution divided by liters of solution (moles/L). Use molarity as a conversion factor to convert liters of solution to moles of solute. Then use formula mass as a conversion factor to convert moles of solute to grams of solute.

a. $1.00 \text{ L solution} \times \left(\frac{1.00 \text{ mole NaOH}}{1.00 \text{ L solution}}\right) \times \left(\frac{40.00 \text{ g NaOH}}{1 \text{ mole NaOH}}\right) = 40.0 \text{ g NaOH}$

b. $2.00 \text{ L solution} \times \left(\frac{2.00 \text{ moles NaOH}}{1.00 \text{ L solution}}\right) \times \left(\frac{40.00 \text{ g NaOH}}{1 \text{ mole NaOH}}\right) = 160.\text{ g NaOH}$

c. $0.100 \text{ L solution} \times \left(\frac{0.50 \text{ mole NaOH}}{1.00 \text{ L solution}}\right) \times \left(\frac{40.00 \text{ g NaOH}}{1 \text{ mole NaOH}}\right) = 2.0 \text{ g NaOH}$

d. $0.552 \text{ L solution} \times \left(\frac{0.333 \text{ mole NaOH}}{1.00 \text{ L solution}}\right) \times \left(\frac{40.00 \text{ g NaOH}}{1 \text{ mole NaOH}}\right) = 7.35 \text{ g NaOH}$

8.59 Use the definition of formula mass (grams/mole) as a conversion factor to convert grams of solute to moles of solute. Then use the definition of molarity to calculate volume of solution.

a. $\text{moles solute} = \frac{\text{g NaNO}_3}{\text{formula mass NaNO}_3} = \frac{60.0 \text{ g NaNO}_3}{85.0 \text{ g NaNO}_3} = 0.706 \text{ mole NaNO}_3$

$$M = \frac{\text{moles of solute}}{1.00 \text{ L solution}}$$

$$\text{L solution} = \frac{\text{moles solute}}{\text{molarity}} = \frac{0.706 \text{ mole NaNO}_3}{0.100 \text{ M}} = 7.06 \text{ L}$$

b. $\text{moles solute} = \frac{\text{g HNO}_3}{\text{formula mass HNO}_3} = \frac{60.0 \text{ g HNO}_3}{63.0 \text{ g HNO}_3} = 0.952 \text{ mole HNO}_3$

$$\text{L solution} = \frac{\text{moles solute}}{\text{molarity}} = \frac{0.952 \text{ mole HNO}_3}{0.100 \text{ M}} = 9.52 \text{ L}$$

c. $\text{moles solute} = \frac{\text{g KOH}}{\text{formula mass KOH}} = \frac{60.0 \text{ g KOH}}{56.1 \text{ g KOH}} = 1.07 \text{ moles KOH}$

$$\text{L solution} = \frac{\text{moles solute}}{\text{molarity}} = \frac{1.07 \text{ moles KOH}}{0.100 \text{ M}} = 10.7 \text{ L}$$

d. $\text{moles solute} = \frac{\text{g LiCl}}{\text{formula mass LiCl}} = \frac{60.0 \text{ g LiCl}}{42.4 \text{ g LiCl}} = 1.42 \text{ moles LiCl}$

$$\text{L solution} = \frac{\text{moles solute}}{\text{molarity}} = \frac{1.42 \text{ moles LiCl}}{0.100 \text{ M}} = 14.2 \text{ L}$$

8.61 Use the definition of mass-volume percent (grams of solute/100 mL solution) to calculate the grams of NaOH per liter of solution.

$$\text{number of g NaOH L solution} = \frac{10.0 \text{ g NaOH}}{100 \text{ mL}} \times \frac{1000 \text{ mL}}{1.00 \text{ L}} = \frac{100 \text{ g solute NaOH}}{1.00 \text{ L solution}}$$

Use the definition of formula mass (g/mole) to calculate the moles of NaOH.

$$\text{moles solute} = \frac{100 \text{ g solute NaOH}}{40.0 \text{ g NaOH}} = 2.50 \text{ moles NaOH}$$

Use the definition of molarity (molarity = moles/liter).

$$\text{M} = \frac{\text{moles NaOH}}{\text{L solution}} = \frac{2.50 \text{ moles NaOH}}{1.00 \text{ L}} = 2.50 \text{ M}$$

8.63 The original concentration of the $CaCl_2$ solution was 3.75 g in 10 mL of solution

a. To express this concentration in mass-volume percent, use the definition: mass volume percent = (grams solute)/(100 mL solution) × 100.

$$\text{number of g CsCl in 100 mL} = \frac{3.75 \text{ g CsCl}}{10.0 \text{ mL}} \times 100 \text{ mL} = 37.5 \text{ g CsCl}$$

$$\text{CsCl \% (m/v)} = \frac{\text{number of g CsCl}}{100 \text{ mL}} \times 100 = 37.5 \text{ \% (m/v)}$$

b. The definition of molarity is (moles of solute)/(L solution).

$$\text{moles of CsCl in 10.0 mL of solution} = \frac{3.75 \text{ g CsCl}}{168\text{g}} = 0.0223 \text{ moles CsCl}$$

$$\text{M} = \frac{\text{moles CsCl}}{\text{L solution}} = \frac{0.0223 \text{ mole CsCl}}{10.0 \text{ mL}} \times \frac{1000 \text{ mL}}{1.00 \text{ L}} = 2.23 \text{ M}$$

8.65 To determine the molarity of the diluted solution in each of the following problems, use the relationship between volumes and concentrations of diluted and stock solutions:

$$C_s \times V_s = C_d \times V_d$$

Rearrange this equation to solve for concentration of the diluted solution (C_d).

a. $C_d = C_s \times \dfrac{V_s}{V_d} = 0.220 \text{ M} \times \left(\dfrac{25.0 \text{ mL}}{30.0 \text{ mL}}\right) = 0.183 \text{ M}$

b. $C_d = C_s \times \dfrac{V_s}{V_d} = 0.220 \text{ M} \times \left(\dfrac{25.0 \text{ mL}}{75.0 \text{ mL}}\right) = 0.0733 \text{ M}$

c. $C_d = C_s \times \dfrac{V_s}{V_d} = 0.220 \text{ M} \times \left(\dfrac{25.0 \text{ mL}}{457 \text{ mL}}\right) = 0.0120 \text{ M}$

d. $C_d = C_s \times \dfrac{V_s}{V_d} = 0.220 \text{ M} \times \left(\dfrac{25.0 \text{ mL}}{2000 \text{ mL}}\right) = 0.00275 \text{ M}$

8.67 Use the relationship between volumes and concentrations of diluted and stock solutions:

$$C_s \times V_s = C_d \times V_d$$

Rearrange this equation to solve for the volume of the diluted solution (V_d). Since the problem asks for volume of water added, subtract the original volume from the final volume ($V_d - V_s$). We assume that the volumes of the two solutions are additive.

a. $V_d = V_s \times \dfrac{C_s}{C_d} = 50.0 \text{ mL} \times \left(\dfrac{3.00 \text{ M}}{0.100 \text{ M}}\right) = 1500 \text{ mL}$

Volume of water added = $V_d - V_s$ = 1500 mL – 50.0 mL = 1.45×10^3 mL

b. $V_d = V_s \times \dfrac{C_s}{C_d} = 2.00 \text{ mL} \times \left(\dfrac{1.00 \text{ M}}{0.100 \text{ M}}\right) = 20.0 \text{ mL}$

Volume of water added = $V_d - V_s$ = 20.0 mL – 2.00 mL = 18.0 mL

c. $V_d = V_s \times \dfrac{C_s}{C_d} = 1450 \text{ mL} \times \left(\dfrac{6.00 \text{ M}}{0.100 \text{ M}}\right) = 87000 \text{ mL}$

Volume of water added = $V_d - V_s$ = 87000 mL – 1450 mL = 8.56×10^4 mL

d. $V_d = V_s \times \dfrac{C_s}{C_d} = 75.0 \text{ mL} \times \left(\dfrac{0.110 \text{ M}}{0.100 \text{ M}}\right) = 82.5 \text{ mL}$

Volume of water added = $V_d - V_s$ = 82.5 mL – 75.0 mL = 7.5 mL

8.69 Solve the dilution equation ($C_s \times V_s = C_d \times V_d$) for C_d. Remember that the volume of the dilute solution is equal to the initial volume plus the water added ($V_d = V_s + 20.0$ mL).

a. $C_d = C_s \times \dfrac{V_s}{V_d} = 5.0 \text{ M} \times \left(\dfrac{30.0 \text{ mL}}{50.0 \text{ mL}}\right) = 3.0 \text{ M}$

b. $C_d = C_s \times \dfrac{V_s}{V_d} = 5.0 \text{ M} \times \left(\dfrac{30.0 \text{ mL}}{50.0 \text{ mL}}\right) = 3.0 \text{ M}$

c. $C_d = C_s \times \dfrac{V_s}{V_d} = 7.5 \text{ M} \times \left(\dfrac{30.0 \text{ mL}}{50.0 \text{ mL}}\right) = 4.5 \text{ M}$

d. $C_d = C_s \times \dfrac{V_s}{V_d} = 2.0 \text{ M} \times \left(\dfrac{60.0 \text{ mL}}{80.0 \text{ mL}}\right) = 1.5 \text{ M}$

8.71 Diagram II represents the new solution; concentration depends on amount and volume:

for diagram I, the concentration = 12 molecules/volume = 12/v
for diagram II, the concentration = 6 molecules/2 volume = 3/v
for diagram III, the concentration = 6 molecules/4 volume = 1.5/v
for diagram IV, the concentration = 12 molecules/4 volume = 3/v

One-half of the solution in diagram I will have 6 molecules/0.5 volume, which when diluted by a factor of 4 will have 6 molecules/2 volume (diagram II)

8.73 In this problem, the process is one of concentration, rather than dilution, but the dilution equation can still be used (C_d = final concentration, V_d = volume of concentrated solution). Solve the dilution equation for C_d, and substitute the given quantities:

$$C_s \times V_s = C_d \times V_d$$

$$C_d = \text{final concentration} = C_s \times \frac{V_s}{V_d} = 0.400\text{ M} \times \left(\frac{2212\text{ mL}}{853\text{ mL}}\right) = 1.04\text{ M}$$

8.75 a. The definition of %(m/v) is: $\%(m/v) = \frac{\text{mass solute}}{\text{volume solution}} \times 100$

Find mass solute: 2.00%(m/v) means $\frac{2.00\text{ g solute}}{100\text{ mL solution}}$; use %(m/v) as a conversion factor.

$$\text{mass solute} = \frac{2.00\text{ g solute}}{100\text{ mL solution}} \times 70.0\text{ mL} = 1.40\text{ g solute}$$

Then calculate the new mass volume percent:

$$\%(m/v) = \frac{1.40\text{ g solute}}{120\text{ mL solution}} \times 100 = 1.17\ \%(m/v)$$

b. The definition of %(v/v): $\%(v/v) = \frac{\text{volume solute}}{\text{volume solution}} \times 100$

Find solute volume: 2.00%(v/v) means $\frac{2.00\text{ mL solute}}{100\text{ mL solution}}$; use %(v/v) as a conversion factor.

$$\text{volume solute} = \frac{2.00\text{ mL solute}}{100\text{ mL solution}} \times 70.0\text{ mL solution} = 1.40\text{ mL solute}$$

Then calculate the new percent by volume:

$$\%(v/v) = \frac{1.40\text{ mL solute}}{120\text{ mL solution}} \times 100 = 1.17\ \%(v/v)$$

c. The definition of %(m/m) is: $\%(m/m) = \frac{\text{mass solute}}{\text{mass solution}} \times 100$

Find mass solute: 2.00%(m/m) means $\frac{2.00\text{ g solute}}{100\text{ g solution}}$; use %(m/m) as a conversion factor.

$$\text{mass solute} = \frac{2.00\text{ g solute}}{100\text{ g solution}} \times 70.0\text{ g solution} = 1.40\text{ g solute}$$

Then calculate the new percent by mass:

$$\%(m/m) = \frac{1.40\text{ g solute}}{120\text{ g solution}} \times 100 = 1.17\ \%(m/m)$$

d. First, calculate the moles of solute, using $M = \frac{\text{moles solute}}{1.00\text{ L solution}}$

$$\text{moles solute} = \frac{2.00\text{ moles solute}}{1.00\text{ L solution}} \times 0.070\text{ L solution} = 0.140\text{ mole solute}$$

Then calculate the new molarity:

$$M = \frac{\text{mole solute}}{1.00\text{ L solution}} = \frac{0.140\text{ mole solute}}{120\text{ mL}} \times \frac{1000\text{ mL}}{1.00\text{ L}} = 1.17\text{ M}$$

8.77 a. suspension b. suspension
b. true solution and colloidal dispersion d. true solution

8.79 a. False; the dispersed phase is not visible.
b. False; the dispersed phase particles are too small to be affected by gravity.
c. False; milk is a colloidal dispersion.
d. True.

8.81 The molecules of a nonvolatile solute take up space on the surface of the liquid. This means that there are fewer solvent molecules on the surface of the liquid so the molecules have fewer opportunities to escape than they do in the pure solvent.

8.83 Seawater (a more concentrated solution) has a lower vapor pressure than fresh water (a less concentrated solution) at the same temperature. This is because the solute molecules in seawater take up space on the surface of the liquid, giving water molecules fewer opportunities to escape.

8.85 Diagram III; the pure water will have a higher vapor pressure than the sugar solution so evaporation will occur faster from the pure water and its volume will decrease.

8.87 One mole of dissolved particles (molecular solutes or ions) raises the boiling point of one kilogram of water by 0.51°C. Use this as a conversion factor in the following problems.

a. Glucose is a molecular solute; one mole of dissolved glucose is equal to 1 mole of dissolved particles.

$$\text{boiling point increase} = 3 \text{ moles glucose} \times \frac{1 \text{ mole particles}}{1 \text{ mole glucose}} \times \frac{0.51^\circ\text{C}}{1 \text{ mole particles}} = 1.53^\circ\text{C}$$

Therefore, the boiling point of the solution = 100 °C + 1.53 °C = 101.53 °C

b. Lactose is a molecular solute; one mole of dissolved lactose is equal to 1 mole of dissolved particles.

$$\text{boiling point increase} = 3 \text{ moles lactose} \times \frac{1 \text{ mole particles}}{1 \text{ mole lactose}} \times \frac{0.51^\circ\text{C}}{1 \text{ mole particles}} = 1.53^\circ\text{C}$$

Therefore, the boiling point of the solution = 100 °C + 1.53 °C = 101.53 °C

c. NaCl is an ionic compound that produces two ions per formula unit in solution.

$$\text{boiling point increase} = 3 \text{ moles NaCl} \times \frac{2 \text{ moles particles}}{1 \text{ mole NaCl}} \times \frac{0.51^\circ\text{C}}{1 \text{ mole particles}} = 3.06^\circ\text{C}$$

Therefore, the boiling point of the solution = 100°C + 3.06°C = 103.06°C

d. One formula unit of Na_3PO_4 in solution produces four ions: $Na_3PO_4 \rightarrow 3Na^+ + PO_4^{2-}$

$$\text{boiling point increase} = 3 \text{ moles Na}_3\text{PO}_4 \times \frac{4 \text{ moles particles}}{1 \text{ mole Na}_3\text{PO}_4} \times \frac{0.51^\circ\text{C}}{1 \text{ mole particles}} = 6.12^\circ\text{C}$$

Therefore, the boiling point of the solution = 100°C + 6.12°C = 106.12°C

8.89 One mole of solute particles lowers the freezing point of one kilogram of water by 1.86°C.

a. Glucose is a molecular solute; one mole of dissolved glucose is equal to 1 mole of dissolved particles.

$$\text{freezing point decrease} = 3 \text{ moles glucose} \times \frac{1 \text{ mole particles}}{1 \text{ mole glucose}} \times \frac{1.86^\circ\text{C}}{1 \text{ mole particles}} = 5.58^\circ\text{C}$$

Therefore, the freezing point of the solution = 0°C – 5.58°C = –5.58°C

b. Lactose is a molecular solute; one mole of dissolved lactose is equal to 1 mole of dissolved particles.

$$\text{freezing point decrease} = 3 \text{ moles lactose} \times \frac{1 \text{ mole particles}}{1 \text{ mole lactose}} \times \frac{1.86^\circ\text{C}}{1 \text{ mole particles}} = 5.58^\circ\text{C}$$

Therefore, the freezing point of the solution = 0°C – 5.58°C = –5.58°C

c. NaCl is an ionic compound that produces two ions per formula unit in solution.

$$\text{freezing point decrease} = 3 \text{ moles NaCl} \times \frac{2 \text{ moles particles}}{1 \text{ mole NaCl}} \times \frac{1.86^\circ\text{C}}{1 \text{ mole particles}} = 11.16^\circ\text{C}$$

Therefore, the freezing point of the solution = 0°C – 11.16°C = –11.16°C

d. One formula unit of Na_3PO_4 in solution produces four ions: $Na_3PO_4 \rightarrow 3Na^+ + PO_4^{2-}$

$$\text{freezing point decrease} = 3 \text{ moles } Na_3PO_4 \times \frac{4 \text{ moles particles}}{1 \text{ mole } Na_3PO_4} \times \frac{1.86^\circ\text{C}}{1 \text{ mole particles}} = 22.32^\circ\text{C}$$

Therefore, the freezing point of the solution = 0°C – 22.32°C = –22.32°C

8.91 Dissolved particles raise the boiling point of a solution, so the solute concentration that gives the higher particle molarity will have the higher boiling point.

a. 1.0 M glucose produces 1 mole of particles per liter. 1.0 M NaCl produces 2 moles of particles per liter. Therefore, 1.0 M NaCl will have the higher boiling point.
b. 1.0 M NaCl produces 2 moles of particles per liter. 1.0 M Na_2SO_4 produces 3 moles of particles per liter. Therefore, 1.0 M Na_2SO_4 will have the higher boiling point.
c. 1.0 M KBr produces 2 moles of particles per liter. 2.0 M KBr produces 4 moles of particles per liter. Therefore, 2.0 M KBr will have the higher boiling point.
d. 2.0 M glucose produces 2 moles of particles per liter. 1.0 M Na_2SO_4 produces 3 moles of particles per liter. Therefore, 1.0 M Na_2SO_4 will have the higher boiling point.

8.93 Glucose is a molecular solid; one mole of dissolved glucose produces one mole of dissolved particles. One mole of solute particles lowers the freezing point of one kilogram of water by 1.86°C. Calculate the moles per kilogram in each problem and multiply by –1.86°C to find the freezing point decrease.

a. $$\frac{1.0 \text{ mole glucose}}{250 \text{ g}} \times \frac{1000 \text{ g}}{1.00 \text{ kg}} = \frac{4.0 \text{ moles glucose}}{1.00 \text{ kg}}$$

(4.0 moles glucose) × (–1.86°C) = –7.4°C Therefore: f.p. = 0 °C – 7.4°C = –7.4°C

b. $$\frac{1.0 \text{ mole glucose}}{250 \text{ g}} \times \frac{1000 \text{ g}}{1.00 \text{ kg}} = \frac{2.0 \text{ moles glucose}}{1.00 \text{ kg}}$$

(2.0 moles glucose) × (–1.86°C) = –3.7°C Therefore: f.p. = 0 °C – 3.7°C = –3.7°C

c. The concentration is expressed in mole/kg; (1.0 mole glucose) × (1.86°C) = −1.9°C
Therefore: f.p. = 0 °C − 1.9°C = −1.9°C

d. $\frac{1.0 \text{ mole glucose}}{2000 \text{ g}} \times \frac{1000 \text{ g}}{1.00 \text{ kg}} = \frac{0.50 \text{ mole glucose}}{1.00 \text{ kg}}$

(0.50 mole glucose) × (−1.86°C) = −0.93°C Therefore: f.p. = 0 °C − 0.93°C = −0.93°C

8.95 Osmotic pressure (the pressure needed to prevent the net flow of solvent across a membrane) depends on the concentration of particles in solution (osmolarity). The higher the osmolarity, the higher the osmotic pressure.

a. Both NaCl and NaBr dissociate into two moles particles per mole of solute, so osmolarity is the same for both solutions (0.2 osmol); they have the same osmotic pressure.

b. NaCl dissociates into two moles of particles per mole of solute, so the osmolarity for NaCl is 2 × 0.1 M = 0.2 osmol. $MgCl_2$ dissociates into three particles per mole, so 0.050 M $MgCl_2$ is 3 × 0.050 M = 0.15 osmol. Since the NaCl solution has a higher osmolarity than the $MgCl_2$ solution, it has an osmotic pressure greater than that of $MgCl_2$.

c. Since 0.1 M NaCl (two particles per mole) has an osmolarity of 0.2 osmol, and 0.1 M $MgCl_2$ (three particles per mole) has an osmolarity of 0.3 osmol, the NaCl solution has a lower osmolarity than the $MgCl_2$ solution and an osmotic pressure that is less than that of the $MgCl_2$ solution.

d. Glucose does not dissociate; the osmolarity of the glucose solution is the same as its molarity (0.1 osmol). The NaCl solution, with an osmolarity of 0.2 osmol, will have an osmotic pressure that is greater than that of the glucose solution.

8.97 The ratio of osmolarities for the NaCl solution (0.30 M × 2 particles/mole = 0.60 osmol) and the $CaCl_2$ solution (0.10 M × 3 particles/mole = 0.30 osmol) is 2:1, so the ratio of osmotic pressures for the two solutions is also 2:1.

8.99 Osmolarity = molarity × i, where i is the number of particles per formula unit of the solute.

a. Osmolarity of KNO_3 solution = 2.0 M × 2 = 4.0 OsM

b. Osmolarity of Na_2SO_4 solution = 2.0 M × 3 = 6.0 OsM

c. Osmolarity of KNO_3 solution = 2.0 M × 2 = 4.0 OsM;
osmolarity of NaCl solution = 2.0 M × 2 = 4.0 OsM.
The total osmolarity is 8.0 OsM.

d. Osmolarity of KNO_3 solution = 2.0 M × 2 = 4.0 OsM;
osmolarity of glucose solution = 2.0 M × 1 = 2.0 OsM.
The total osmolarity is 6.0 OsM.

8.101 A 5.0%(m/v) glucose solution and a 0.92(m/v) NaCl solution are isotonic solutions for red blood cells. A hypertonic solution has a higher m/v concentration than that within the cell; a hypotonic solution has a lower m/v concentration than that within the cell.

a. A 0.92%(m/v) glucose solution is hypotonic.
b. 0.92%(m/v) NaCl solution is isotonic.
c. 2.3%(m/v) glucose solution is hypotonic.
d. 5.0%(m/v) NaCl solution is hypertonic.

8.103 Red blood cells in a hypertonic solution shrink in size, a process called crenation. If the red blood cells are in a hypotonic solution they enlarge in size and finally burst, a process called hemolysis. In an isotonic solution red blood cells remain unaffected.

a. A 0.92%(m/v) glucose solution is hypotonic. Red blood cells will swell.
b. 0.92%(m/v) NaCl solution is isotonic. Red blood cells remain the same size.
c. 2.3%(m/v) glucose solution is hypotonic. Red blood cells will swell.
d. 5.0%(m/v) NaCl solution is hypertonic. Red blood cells will shrink.

8.105 Red blood cells in a hypertonic solution shrink in size, a process called crenation. If the red blood cells are in a hypotonic solution they enlarge in size and finally burst, a process called hemolysis. In an isotonic solution red blood cells remain unaffected.

a. A 0.92%(m/v) glucose solution is hypotonic. Red blood cells will hemolyze.
b. 0.92%(m/v) NaCl solution is isotonic. Red blood cells remain unaffected.
c. 2.3%(m/v) glucose solution is hypotonic. Red blood cells will hemolyze.
d. 5.0%(m/v) NaCl solution is hypertonic. Red blood cells will crenate.

8.107 All solutions whose total solute concentration, taking into account the formation of ions by some solutes, is 0.28 M are isotonic with red blood cells.

a. A 0.28 M glucose solution is isotonic relative to red blood cells.
b. A solution that is 0.28 M in both glucose and sucrose would be a 0.56 M solution. The solution is hypertonic relative to red blood cells.
c. A solution that is 0.14 M in both glucose and sucrose would be a 0.28 M solution. The solution is isotonic relative to red blood cells.
d. A solution that is 0.28 M NaCl would be 0.56 M in particles. The solution is hypertonic relative to red blood cells.

8.109 Red blood cells in a hypertonic solution shrink in size, a process called crenation. If the red blood cells are in a hypotonic solution they enlarge in size and finally burst, a process called hemolysis. In an isotonic solution red blood cells remain unaffected.

a. A 0.28 M glucose solution is isotonic relative to red blood cells. The red blood cells will remain the same size.
b. A solution that is 0.28 M in both glucose and sucrose would be a 0.56 M solution. The solution is hypertonic relative to red blood cells. The red blood cells will shrink.
c. A solution that is 0.14 M in both glucose and sucrose would be a 0.28 M solution. The solution is isotonic relative to red blood cells. The red blood cells will remain the same size.
d. A solution that is 0.28 M NaCl would be 0.56 M in particles. The solution is hypertonic relative to red blood cells. The red blood cells will shrink.

8.111 Red blood cells in a hypertonic solution shrink in size, a process called crenation. If the red blood cells are in a hypotonic solution they enlarge in size and finally burst, a process called hemolysis. In an isotonic solution red blood cells remain unaffected.

a. A 0.28 M glucose solution is isotonic relative to red blood cells. The red blood cells will remain unaffected.
b. A solution that is 0.28 M in both glucose and sucrose would be a 0.56 M solution. The solution is hypertonic relative to red blood cells. The red blood cells will crenate.
c. A solution that is 0.14 M in both glucose and sucrose would be a 0.28 M solution. The solution is isotonic relative to red blood cells. The red blood cells will remain unaffected.
d. A solution that is 0.28 M NaCl would be 0.56 M in particles. The solution is hypertonic relative to red blood cells. The red blood cells will crenate.

8.113 The net transfer of solvent is to the more concentrated solution. The level in A will

a. decrease; B is the more concentrated solution.
b. increase; A is the more concentrated solution.
c. not change; A and B have the same osmolarity, since $i = 2$ in each case.
d. decrease; A is 1.0 osmol ($i = 1$) and B is 2.0 osmol ($i = 2$).

8.115 Each of the properties listed is a colligative property, a physical property of a solution that depends only on the number (concentration) of particles (molecules or ions) present in the solution. Vapor pressure of a solvent is lowered by dissolved solute particles. Boiling point is raised by dissolved solute particles. Freezing point is lowered by dissolved solute particles. Osmotic pressure is raised by dissolved solute particles.

a. 1.0 M glucose has a lower particle concentration than 1.0 M NaCl, so 1.0 M glucose has a higher vapor pressure.
b. 1.0 M NaCl has a higher particle concentration than 1.0 M glucose, so 1.0 M NaCl has a higher boiling point.
c. 1.0 M glucose has a lower particle concentration than 1.0 M NaCl, so 1.0 M glucose has a higher freezing point.
d 1.0 M NaCl has a higher particle concentration than 1.0 M glucose, so 1.0 M NaCl has a higher osmotic pressure.

8.117 The solution giving rise to the higher osmotic pressure is the one with the greater osmolarity. To determine which of the solutions has the greater osmotic pressure, first calculate the molarity of the solution and then calculate its osmolarity.

a. Use the definition of molarity (molarity = moles/liter). Osmolarity = molarity × i, where i is the number of particles per formula unit of the solute.

$$\text{M} = \frac{\left(\dfrac{8.00\text{ g NaCl}}{58.45\text{ g NaCl}}\right)}{375\text{ mL}} \times \frac{1000\text{ mL}}{1.00\text{ L}} = 0.365\text{ M} \qquad \text{M} = \frac{\left(\dfrac{4.00\text{ g NaBr}}{102.9\text{ g NaBr}}\right)}{155\text{ mL}} \times \frac{1000\text{ mL}}{1.00\text{ L}} = 0.251\text{ M}$$

Osmolarity of NaCl solution = 0.365 M × 2 = 0.730 osmol
Osmolarity of NaBr solution = 0.251 M × 2 = 0.502 osmol
The NaCl solution has the greater osmotic pressure.

b. Use the definition of molarity (molarity = moles/liter). Osmolarity = molarity × i, where i is the number of particles per formula unit of the solute.

$$\text{M} = \frac{\left(\dfrac{7.00\text{ g NaCl}}{58.5\text{ g NaCl}}\right)}{775\text{ mL}} \times \frac{1000\text{ mL}}{1.00\text{ L}} = 0.154\text{ M} \qquad \text{M} = \frac{\left(\dfrac{6.00\text{ g NaBr}}{103\text{ g NaBr}}\right)}{375\text{ mL}} \times \frac{1000\text{ mL}}{1.00\text{ L}} = 0.155\text{ M}$$

Osmolarity of NaCl solution = 0.154 M × 2 = 0.308 osmol
Osmolarity of NaBr solution = 0.155 M × 2 = 0.310 osmol
The NaBr solution has the greater osmotic pressure.

Chemical Reactions — Chapter 9

Solutions to Selected Problems

9.1 a. $X + YZ \rightarrow Y + XZ$ b. $X + Y \rightarrow XY$

9.3 a. The reaction is a displacement reaction; one atom (Al) replaces another atom (Cu) in a compound ($CuSO_4$).
b. The reaction is a decomposition reaction; a single reactant (K_2CO_3) is converted into two simpler substances (K_2O and CO_2).
c. The reaction is an exchange reaction; two substances ($AgNO_3$ and K_2SO_4) exchange parts with one another and form two different substances (Ag_2SO_4 and KNO_3).
d. The reaction is a combination reaction; a single product (PH_3) is produced from two reactants (P and H_2).

9.5 In hydrocarbon combustion, the carbon of the hydrocarbon combines with the oxygen of the air to produce carbon dioxide (CO_2) and the hydrogen of the hydrocarbon reacts with oxygen to give water (H_2O).

a. CO_2, H_2O b. CO_2, H_2O c. CO_2, H_2O d. CO_2, H_2O

9.7 A combustion reaction is a reaction between a substance and oxygen.

a. Yes, it is a combustion reaction. b. Yes, it is a combustion reaction.
c. Yes, it is a combustion reaction. d. No, it is not a combustion reaction.

9.9 There are five general types of chemical reactions: combination (a single product is produced from reactants), decomposition (a reactant is converted to simpler substances), displacement (an atom or molecule replaces an atom or group of atoms from a compound), exchange (two substances exchange parts with one another), combustion (reaction between a substance and oxygen).
a. An element may be a reactant in the following types of reactions: combination, displacement, combustion.
b. An element may be a product in the following types of reactions: decomposition, displacement.
c. A compound may be a reactant in the following types of reactions: combination, decomposition, displacement, exchange, combustion.
d. A compound may be a product in the following types of reactions: combination, decomposition, displacement, exchange, combustion.

9.11 An oxidation number represents the charge that an atom bonded to another atom appears to have when its electrons are assigned to the more electronegative of the two atoms in the bond. For a compound, the sum of the individual oxidation numbers is equal to zero.
a. The oxidation number of S in S_2 is 0; the oxidation number of an element in its elemental state is zero.
b. The oxidation number of S in S_8 is 0; the oxidation number of an element in its elemental state is zero.
c. The oxidation number of S in H_2S is –2; the oxidation number of hydrogen is +1 because sulfur is more electronegative than hydrogen, which has one electron.
d. The oxidation number of S in SO_3 is +6, because the oxidation number of oxygen is –2 (except in peroxides, and this compound is not a peroxide).

9.13 The oxidation number of a monatomic ion is the same as the charge on the ion; for a polyatomic ion, the sum of the oxidation numbers is equal to the charge on the ion.

a. Al in Al^{3+} has an oxidation number of +3.
b. S in S^{2-} has an oxidation number of –2.
c. N in NO_3^- has an oxidation number of +5 because the oxidation number of O is –2.
d. The oxidation number of P in PO_4^{3-} is +5 because the oxidation number of O is –2.

9.15 For a compound, the sum of the individual oxidation numbers is equal to zero.

a. The oxidation number of Cl in NaClO is +1, because Na = +1 and, O = –2
b. The oxidation number of Cl in $NaClO_3$ is +5, because Na = +1 and, O = –2
c. The oxidation number of Cl in $HClO_4$ is +7, because H = +1 and, O = –2
d. The oxidation number of Cl in PCl_3 is –1, because chlorine is more electronegative than phosphorus; the P atom contributes one electron to each P–Cl bond.

9.17 An oxidation number represents the charge that an atom bonded to another atom appears to have when its electrons are assigned to the more electronegative of the two atoms in the bond. For a compound, the sum of the individual oxidation numbers is equal to zero; for a polyatomic ion, the sum of the oxidation numbers is equal to the charge on the ion.

a. SiF_4: Si = +4, F = –1
b. H_2SO_4: H = +1, S = +6, O = –2
b. Na_3PO_4: Na = +1, P = +5, O = –2
d. $Cr_2O_7^{2-}$: Cr = +6, O = –2

9.19 In a redox reaction there is a transfer of electrons from one reactant to another; in a nonredox reaction there is no electron transfer. Oxidation numbers are used as a "bookkeeping system" to identify electron transfer in a redox reaction.

a. This a redox reaction; there is a transfer of electrons from Cu to O. The oxidation number of Cu goes from 0 to +2; the oxidation number of O goes from 0 to – 2.
b. This is a nonredox reaction; no electrons are transferred, no oxidation numbers change.
c. This is a redox reaction; electrons are transferred from O to Cl. The oxidation number of O goes from –2 to 0; the oxidation number of Cl goes from +5 to –1.
d. This is a redox reaction; electrons are transferred from C to O. The oxidation number of C goes from –4 to +4; the oxidation number of oxygen goes from 0 to –2.

9.21 An oxidation-reduction (redox) reaction is a chemical reaction in which there is a transfer of electrons from one reactant to another reactant.

a. A combination reaction in which one reactant is an element is a redox reaction. The oxidation number of the reactant element is zero; when it combines with another substance, its oxidation number will change (it loses or gains electrons).
b. A decomposition reaction in which the products are all elements is a redox reaction. Elements have an oxidation number of zero. The reactant in a decompositions reaction is a compound; the elements in the compound lose or gain electrons to reach a zero oxidation number.
c. A decomposition reaction in which one of the products is an elements is a redox reaction. Elements have an oxidation number of zero. The reactant in the decomposition reaction is a compound; one of the elements in the compound loses or gains electrons to reach a zero oxidation number.
d. The description "a displacement reaction in which both of the reactants are compounds" cannot be classified as redox or nonredox because it does not contain enough information. We don't know whether the elements in the products have changed in oxidation number.

9.23 An oxidation-reduction (redox) reaction is a chemical reaction in which there is a transfer of electrons from one reactant to another reactant.

a. This is a redox reaction because both zinc and copper change in oxidation number (gain or lose electrons). It is a displacement reaction because Zn displaces Cu in $Cu(NO_3)_2$.
b. This is a combustion reaction (reaction with O_2). It is a redox reaction because O_2 changes oxidation number. (All combustion reactions are redox reactions.)
c. This is a decomposition reaction because CuO is converted into two simpler substances, Cu and O_2. It is a redox reaction because Cu gains electrons and O loses electrons.
d. This is an exchange reaction; two substances exchange parts with one another and form two different substances. It is a nonredox reaction; none of the atoms loses or gains electrons.

9.25 Oxidation is a loss of one or more electrons (increase in oxidation number); reduction is a gain of one or more electrons (a decrease in oxidation number).

a. $SO_2 \rightarrow SO_3$; this is an oxidation because the oxidation number of S changes from +4 to +6.
b. $N_2O \rightarrow NO$; this is an oxidation because the oxidation number of N changes from +1 to +2.
c. $Cr^{3+} \rightarrow Cr^{2+}$; this a reduction (gain of electrons).
d. $S^{2-} \rightarrow S$; this is an oxidation (loss of electrons). S oxidation number changes from –2 to 0.

9.27 In a redox reaction, both oxidation and reduction take place. During oxidation, a reactant loses one or more electrons and increases in oxidation number. During reduction, a reactant gains one or more electrons and decreases in oxidation number.

a. H is oxidized (oxid. no. goes from 0 to +1); N is reduced (oxid. no. goes from 0 to –3).
b. I is oxidized (oxid. no. goes from –1 to 0); Cl is reduced (oxid. no. goes from 0 to –1).
c. Fe is oxidized (oxid. no. goes from 0 to +2); Sb is reduced (oxid. no. goes from +3 to 0).
d. S is oxidized (oxid. no. goes from +4 to +6); N is reduced (oxid. no. goes from +5 to +2).

9.29 In a redox reaction, an oxidizing agent causes oxidation of another reactant by accepting electrons from it. A reducing agent causes reduction of another reactant by giving up electrons for the other reactant to accept. An oxiding agent is reduced; a reducing agent is oxidized.

a. N_2 is the oxidizing agent; H_2 is the reducing agent.
b. Cl_2 is the oxidizing agent; KI is the reducing agent.
c. Sb_2O_3 is the oxidizing agent; Fe is the reducing agent.
d. HNO_3 is the oxidizing agent; H_2SO_3 is the reducing agent.

9.31 Oxidation is a loss of one or more electrons; reduction is a gain of one or more electrons. In a redox reaction, an oxidizing agent is reduced and a reducing agent is oxidized.

a. The oxidizing agent gains electrons.
b. The reducing agent loses electrons.
c. The substance undergoing oxidation loses electrons.
d. The substance undergoing reduction gains electrons.

9.33
a. Incorrect. A reducing agent is oxidized (loses electrons).
b. Correct. Oxidation is a loss of electrons.
c. Incorrect. An oxidizing agent is reduced (decreases in oxidation number).
d. Incorrect. An oxidizing agent is reduced.

9.35 The three central concepts associated with collision theory are: molecular collisions, activation energy, and collision orientation.

9.37 The two factors are total kinetic energy of colliding reactants and collision orientation.

9.39 An exothermic reaction is a reaction in which energy is released as the reaction occurs; an endothermic reaction is one in which a continuous input of energy is needed for the reaction to occur.

a. This reaction is exothermic because heat is given off.
b. This reaction is endothermic because heat must be provided for the reaction to occur.
c. This reaction is endothermic because heat must be provided for the reaction to occur.
d. This reaction is exothermic because heat is given off.

9.41 In an exothermic reaction, energy is released as the reaction occurs (heat is produced). In an endothermic reaction, energy is needed for the reaction to occur (heat is added).

a. The reaction is exothermic, so heat is part of the product.
b. The reaction is endothermic, so heat is added as a reactant.
c. The reaction is endothermic, so heat is added as a reactant.
d. The reaction is exothermic, so heat is part of the product.

9.43 a. The reaction is exothermic; for the product energies to be lower than the reactant energies, energy must be released.
b. Energy is released.

9.45

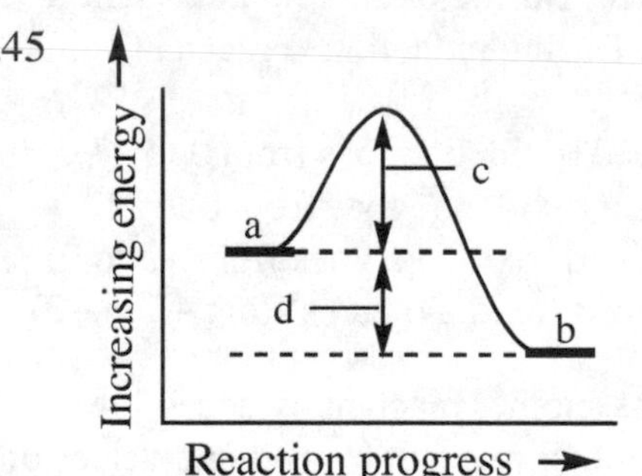

In this energy diagram:
a = average energy of reactants,
b = average energy of products,
c = activation energy,
d = amount of energy liberated during the reaction.

9.47 a. An increase in the temperature of a system results in an increase in the average kinetic energy (the average speed) of the reacting molecules. As the average kinetic energy of the molecules increases, the number of collisions per second increases, and a larger fraction of the collisions have enough kinetic energy to reach the activation energy.
b. A catalyst lowers the activation energy (thus increasing the rate of reaction) by providing an alternate reaction pathway that has a lower activation energy than the original pathway.

9.49 The concentration of O_2 in air is 21%; in pure oxygen the concentration is 100%. During oxidation, a substance reacts with oxygen molecules. If more molecules of oxygen are present, collisions take place more frequently, and the rate of reaction is increased.

9.51 a. The rate will increase; with more reactant molecules present, more collisions will occur.
b. The rate will decrease; lowering the temperature decreases molecular energies, and fewer molecules will have the required energy for an effective collision.
c. The rate will increase; the catalyst lowers the activation energy, and more molecules will now have the required activation energy.
d. The rate will decrease; with fewer reactant molecules present, fewer collisions will occur.

9.53 a. If the concentration of H_2SO_4 (a reactant) is increased, more collisions will occur. The rate of reaction will increase.
b. If the copper (a reactant) is ground into a powder, surface area increases and more collisions will occur. The rate of reaction will increase.
c. If the mixture is stirred rapidly, more collisions will occur. The rate of reaction will increase.
d. If the temperature is increased, more molecules will have the required energy for an effective collision. The rate of reaction will increase.

9.55

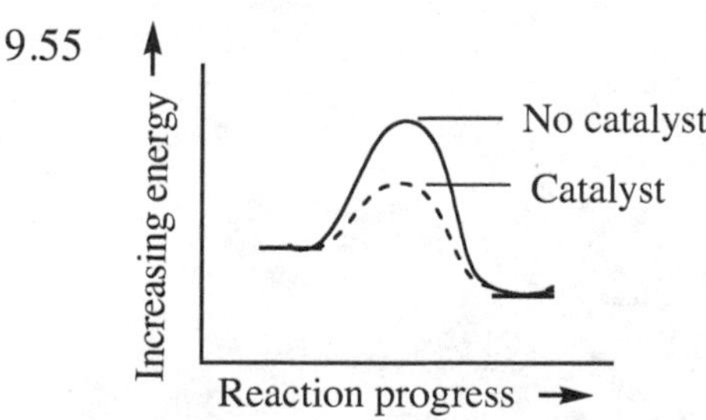

As can be seen from the diagram, activation energy is lower when a catalyst is present. The diagrams are similar in that the average energy of the reactants and the average energy of the products remain the same.

9.57 a. The activation energy for reaction 1 is lower than that for reaction 2, so the reaction rate is greater for reaction 1. More collisions have enough kinetic energy to reach the activation energy.
b. The temperature for reaction 3 is higher than that for reaction 1, so the reaction rate is higher for reaction 3. More molecules have enough kinetic energy to reach the activation energy when they collide.
c. The concentration of reactants in reaction 4 is greater than that in reaction 1, so the reaction rate for reaction 4 is greater. At higher reactant concentrations, collisions take place more frequently.
d. Reaction 3 has both a lower activation energy and a higher temperature than reaction 2 does. Both of these conditions favor a faster reaction rate for reaction 3.

9.59 In a system at chemical equilibrium (a system in which the concentrations of all reactants and all products remain constant), the rate of the forward reaction is equal to rate of the reverse reaction.

9.61 The reversible reaction is a chemical reaction in which the conversion of reactants to products (the forward reaction) and the conversion of products to reactants (the reverse reaction) occur simultaneously.

9.63 a. $N_2(g) + O_2(g) \rightarrow 2NO(g)$ b. $2NO(g) \rightarrow N_2(g) + O_2(g)$

9.65 The concentration of the reactants decreases during the course of the chemical reaction (reactants are used up) and then remains constant when equilibrium is reached. The concentration of the products increases during the course of the reaction and then remains constant when equilibrium is reached.

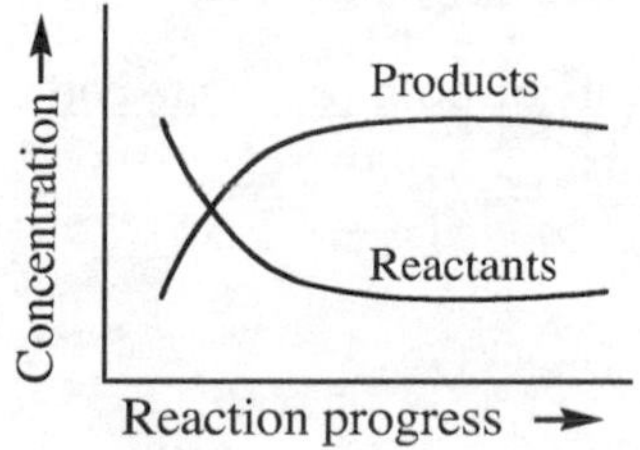

9.67 Yes, the concentrations of molecules in diagrams III and IV are the same.

9.69 Diagrams II and IV; diagram III cannot be produced from the original mixture since there are more green atoms than in the starting mixture.

9.71 For a chemical equation of the general form: $wA + xB \rightleftharpoons yC + zD$,

the equilibrium constant expression is: $K_{eq} = \dfrac{[C]^y[D]^z}{[A]^w[B]^x}$

a. $\dfrac{[Z]^2}{[X]^2[Y]^3}$

b. $\dfrac{[Z]^2}{[X]^2}$

c. $[Z]^2$

d. $\dfrac{1}{[X]^2[Y]^3}$

9.73 For a chemical equation of the general form: $wA + xB \rightleftharpoons yC + zD$,

the equilibrium constant expression is: $K_{eq} = \dfrac{[C]^y[D]^z}{[A]^w[B]^x}$

a. $\dfrac{[NO]^2[Cl_2]}{[NOCl]^2}$

b. $\dfrac{[NO]^4[H_2O]^6}{[NH_3]^4[O_2]^5}$

c. $\dfrac{1}{[Cl_2]}$

d. $\dfrac{[NaNO_3]}{[AgNO_3][NaCl]}$

9.75 First, write the equilibrium constant expression; then substitute the given concentrations and calculate the value of the equilibrium constant for the chemical reaction.

$$K_{eq} = \frac{[NO_2]^2}{[N_2O_2]} = \frac{[0.0032]^2}{[0.213]} = 4.8 \times 10^{-5}$$

9.77 Table 9.2 describes the extent to which a chemical reaction takes place, given various values of the equilibrium constant.

a. The value of K_{eq} is large so more products than reactants are formed.

b. The value of K_{eq} is very small so the reaction mixture is essentially all reactants.

c. The value of K_{eq} is near unity so there are significant amounts of both reactants and products in the reaction mixture at equilibrium.

d. The value of K_{eq} is near unity so there are significant amounts of both reactants and products in the reaction mixture at equilibrium.

9.79 In an equilibrium constant expression, product concentrations are placed in the numerator and reactant concentrations are placed in the denominator. The powers to which the concentrations are raised indicate the coefficients in the balanced chemical equation for the equilibrium system.

$$K_{eq} = \frac{[N_2]^2[H_2O]^6}{[NH_3]^4[O_2]^3}$$

Using the information given by the equilibrium constant expression above, the equation for the reversible reaction can be written.

$$4\ NH_3(g) + 3O_2(g) \rightleftharpoons 2N_2(g) + 6H_2O(g)$$

9.81 Diagram IV; this diagram has the greatest number of product molecules and the more product molecules present, the greater the value of the equilibrium constant.

9.83 Diagram IV; here there are 8 product molecules and 1 molecule each of the two reactants.

$$K_{eq} = \frac{[8]^2}{[1][1]} = 64$$

9.85 According to Le Chatelier's principle, if a stress (change of conditions) is applied to a system in equilibrium, the system will readjust (change the equilibrium position) in the direction that reduces the stress imposed on the system.

a. No, the change will not affect the value of the equilibrium constant. Removal of a reactant from the equilibrium mixture will shift the equilibrium to the left, but the system will readjust to establish a new equilibrium condition with the same equilibrium constant.

b. No, the change will not affect the value of the equilibrium constant. A decrease in the system's total pressure may shift the equilibrium (if gases are present in reactants or products), but the system will readjust to establish a new equilibrium condition with the same equilibrium constant.

c. Yes, a decrease in the system's temperature will affect the value of the equilibrium constant. In an exothermic or an endothermic reaction, heat may be treated as a reactant or product in the chemical reaction, but it is not considered in the equilibrium expression. Therefore, a change in temperature will affect the equilibrium and result in a new equilibrium constant.

d. No, the addition of a catalyst will not affect the value of the equilibrium constant. Addition of a catalyst affects only the rate of reaction, not the position of the equilibrium.

9.87 According to Le Chatelier's principle, if a stress (change of conditions) is applied to a system in equilibrium, the system will readjust (change the equilibrium position) in the direction that reduces the stress imposed on the system.

a. If the concentration of B is increased, the reaction will shift to the right and use up more A. The concentration of A will decrease.

b. If the concentration of C is decreased, the reaction will shift to the right to produce more C. The concentration of A will decrease.

c. If the concentration of D is increased, the reaction will shift to the left to use up more D. The concentration of A will increase.

d. If the concentration of D is decreased, the reaction will shift to the right to produce more D. The concentration of A will decrease.

9.89 We can use Le Chatelier's principle to explain the changes in the direction of the equilibrium.

a. Increasing the concentration of a product (C_6H_{12}) shifts the equilibrium to the left, thus using up some of the added product.

b. Decreasing the concentration of a reactant (C_6H_6) shifts the equilibrium to the left, thus producing more of the substance that was removed.

c. Increasing the temperature of the reaction mixture favors the endothermic reaction; the equilibrium shifts to the left and heat is absorbed.

d. Decreasing the pressure by increasing the volume of the container decreases the concentration of all of the gases. However, there are four moles of reactant gases to one mole of product, so the reaction shifts to the left (the direction that will give more moles of gases).

9.91 a. Since this is an endothermic reaction, refrigerating the equilibrium mixture (decreasing the temperature) produces a shift to the left (to generate more heat).

b. A catalyst has no effect on equilibrium position, only on rate of reaction.

c. Increasing the concentration of a reactant (CO) shifts the equilibrium to the right, thus using up some of the added reactant.

d. Increasing the size of the reaction container has no effect on the equilibrium position. When the number of moles of reactant gases equals the number of moles of product gases, a change in volume affects reactants and products equally; stresses on the equilibrium system are equal.

9.93 It is an endothermic reaction; there are more product molecules in the second diagram than in the first, meaning the reaction shifts to the right as the temperature is increased, the characteristic for an endothermic reaction.

9.95 For exothermic reactions, heat can be treated as one of the products; for endothermic reactions, heat can be treated as one of the reactants. In an exothermic reaction, equilibrium shifts to the left as temperature is increased; in an endothermic reaction, equilibrium shifts to the right as temperature is increased.

a. Yes, there is an increase in product formation; it is an endothermic reaction.

b. Yes, there is an increase in product formation; it is an endothermic reaction.

c. No, there is not an increase in product formation; it is an exothermic reaction.

d. Yes, there is an increase in product formation; it is an endothermic reaction.

9.97 An increase in volume would decrease the pressure on the reaction mixture. A decrease in pressure shifts the equilibrium position in the direction that increases the number of moles of gases in the system.

a. No, the product formation does not increase; it decreases because there are more moles of gas on the reactant side than on the product side.

b. No, the product formation does not increase; it decreases because there are more moles of gas on the reactant side than on the product side.

c. No, there would be no shift in the equilibrium position; the number of moles of gases is the same for reactants and products.

d. Yes, the product formation does increase, because there are more moles of gas on the product side than on the reactant side.

Acids, Bases, and Salts — Chapter 10

Solutions to Selected Problems

10.1 a. In Arrhenius acid-base theory, H^+ ion is responsible for properties of acidic solutions.
b. In Arrhenius acid-base theory, OH^- ion is responsible for properties of basic solutions.

10.3 a. A sour taste is a property of an Arrhenius acid.
b. A bitter taste is a property of an Arrhenius base.

10.5 In water, Arrhenius acids ionize to form H^+ ions and Arrhenius bases ionize to form OH^- ions.

a. $HI \xrightarrow{H_2O} H^+ + I^-$ b. $HClO \xrightarrow{H_2O} H^+ + ClO^-$

c. $LiOH \xrightarrow{H_2O} Li^+ + OH^-$ d. $CsOH \xrightarrow{H_2O} Cs^+ + OH^-$

10.7 A Brønsted-Lowry acid donates protons to a Brønsted-Lowry base; a Brønsted-Lowry base accepts protons from a Brønsted-Lowry acid.

a. HF is a Brønsted-Lowry acid because it donates a proton to H_2O.
b. CN^- is Brønsted-Lowry base because it accepts a proton from H_2O.
c. HCN is a Brønsted-Lowry acid because it donates a proton to NO_2^-.
d. NH_3 is Brønsted-Lowry base because it accepts a proton from HNO_3.

10.9 A Brønsted-Lowry acid donates protons to a Brønsted-Lowry base; a Brønsted-Lowry base accepts protons from a Brønsted-Lowry acid.

a. $HOCl + H_2O \rightarrow H_3O^+ + OCl^-$ b. $NH_3 + H_2O \rightarrow NH_4^+ + OH^-$
c. $H_2PO_4^- + H_2O \rightarrow H_3O^+ + HPO_4^{2-}$ d. $CO_3^{2-} + H_2O \rightarrow HCO_3^- + OH^-$

10.11 A conjugate acid-base pair is two species, one an acid and one a base, that differ from each other through the loss or gain of a proton (H^+ ion).

a. Yes, HCl and Cl^- are a conjugate acid-base pair. HCl, the acid, loses a proton to become Cl^-, the conjugate base.
b. Yes, NH_4^+ and NH_3 are a conjugate acid-base pair. NH_4^+, the acid, loses a proton to become NH_3, the conjugate base.
c. No, H_2CO_3 and CO_3^{2-} are not a conjugate acid-base pair. H_2CO_3 is an acid but its conjugate base is HCO_3^-, not CO_3^{2-}.
d. Yes, $H_2PO_4^-$ and HPO_4^{2-} are a conjugate acid-base pair. $H_2PO_4^-$, the acid, loses a proton to become HPO_4^{2-}, the conjugate base.

10.13 A conjugate acid-base pair is two species, one an acid and one a base, that differ from each other through the loss or gain of a proton (H^+ ion).

a. The conjugate base of H_2SO_3 is HSO_3^-. (The acid loses a proton.)
b. The conjugate acid of CN^- is HCN. (The base gains a proton.)
c. The conjugate base of $HC_2O_4^-$ is $C_2O_4^{2-}$. (The acid loses a proton.)
d. The conjugate acid of HPO_4^{2-} is $H_2PO_4^-$. (The base gains a proton.)

10.15 A conjugate acid-base pair is two species, one an acid and one a base, that differ from each other through the loss or gain of a proton (H^+ ion). The two conjugate acid-base pairs are: $HC_2H_3O_2$/ $C_2H_3O_2^-$ and H_3O^+/H_2O

10.17 An amphiprotic substance is a substance that can either lose or accept a proton and thus can function as either a Brønsted-Lowry acid or a Brønsted-Lowry base.

a. The ion HCO_3^- accepts a proton and acts as a Brønsted-Lowry base.
$HCO_3^- + H_2O \rightarrow H_2CO_3 + OH^-$

b. The ion HCO_3^- loses a proton and acts as a Brønsted-Lowry acid.
$HCO_3^- + H_2O \rightarrow H_3O^+ + CO_3^{2-}$

10.19 Acids can be classified according to the number of protons (H^+ ions) they can transfer per molecule during an acid-base reaction: a monoprotic acid supplies one proton, a diprotic acid supplies two protons, and a triprotic acid supplies three protons.

a. $HClO_3$ is a monoprotic acid.
b. $HC_3H_5O_4$ is a monoprotic acid
c. $H_3C_6H_5O_7$ is a triprotic acid
d. H_3PO_4 is a triprotic acid

10.21 a. $HClO_3$ has one acidic hydrogen atom and zero nonacidic hydrogen atoms.
b. $HC_3H_5O_4$ has one acidic hydrogen atom and five nonacidic hydrogen atoms.
c. $H_3C_6H_5O_7$ has three acidic hydrogen atoms and five nonacidic hydrogen atoms.
d. H_3PO_4 has three acidic hydrogen atoms and zero nonacidic hydrogen atoms.

10.23 Diprotic acids in aqueous solution transfer protons in two steps.

a. $H_2CO_3 + H_2O \rightarrow H_3O^+ + HCO_3^-$
$HCO_3^- + H_2O \rightarrow H_3O^+ + CO_3^{2-}$

b. $H_2C_3H_2O_4 + H_2O \rightarrow H_3O^+ + HC_3H_2O_4^-$
$HC_3H_2O_4^- + H_2O \rightarrow H_3O^+ + C_3H_2O_4^{2-}$

10.25 Acidic hydrogen atoms are written first in the formula of an acid, thus separating them from the other hydrogen atoms in the formula. There is one hydrogen atom at the beginning of the chemical formula ($HC_3H_5O_3$) indicating that this is a monoprotic acid.

10.27 From the structure of pyruvic acid we can see that there is only one H atom that is involved in a polar bond, making this a monoprotic acid.

10.29 In an aqueous solution, a strong acid transfers nearly 100% of its protons to water, and a weak acid transfers a small percentage (usually less than 5%) of its protons to water. Table 10.1 lists commonly encountered strong acids. Referring to the acids in Problem 10.19:

a. $HClO_3$ is a strong acid.
b. $HC_3H_5O_4$ is a weak acid.
c. $H_3C_6H_5O_7$ is a weak acid.
d. H_3PO_4 is a weak acid.

10.31 In aqueous solution a strong acid transfers very nearly 100% of its protons (H^+ ions) to water, while a weak acid transfers only a small percentage of its protons to water. See Table 10.1 for a list of some commonly encountered strong acids.

a. HNO_3 is stronger than HNO_2.
b. HF is weaker than HBr.
c. H_2CO_3 is weaker than $HClO_3$.
d. HCN is weaker than HCl.

10.33 See Table 10.1 for a list of some commonly encountered strong acids and 10.2 for some commonly encountered strong hydroxide bases.

a. Yes, both are strong.
b. Yes, both are strong.
c. Yes, both are strong.
d. No. HF is a weak acid.

10.35 The molar concentrations are 0.10 M in both H_3O^+ and Cl^- ions and zero in HCl.

10.37 The strongest acid is the one in diagram IV; it has the greatest relative amount of HA molecules ionized.

10.39 An acid is monoprotic if it supplies one proton per molecule in an acid-base reaction, diprotic is it supplies two protons, and triprotic if it supplies three protons. See Table 10.1 for a list of some commonly encountered strong acids.

a. H_3PO_4 is a weak triprotic acid.
b. H_3PO_3 is a weak triprotic acid.
c. HBr is a strong monoprotic acid.
d. $HC_2H_3O_2$ is a weak monoprotic acid. (The other three hydrogen atoms are not acidic.)

10.41 The acid ionization constant for a monoprotic weak acid is obtained by writing the equilibrium constant for the reaction of the weak acid with water.

a. $K_a = \dfrac{[H^+][F^-]}{[HF]}$
b. $K_a = \dfrac{[H^+][C_2H_3O_2^-]}{[HC_2H_3O_2]}$

10.43 The base ionization constant for a weak base is obtained by writing the equilibrium constant for the reaction of the weak base with water.

a. $K_b = \dfrac{[NH_4^+][OH^-]}{[NH_3]}$
b. $K_b = \dfrac{[C_6H_5NH_3^+][OH^-]}{[C_6H_5NH_2]}$

10.45 The strength of an acid is indicated by the magnitude of its K_a; the larger the K_a, the stronger the acid. Table 10.3 gives the ionization constant values for selected weak acids.

a. H_3PO_4 is a stronger acid than HNO_2.
b. HF is a stronger acid than HCN.
c. H_2CO_3 is a stronger acid than HCO_3^-.
d. HNO_2 is a stronger acid than HCN.

10.47 Since the acid, HA, is 12% ionized, the concentration of H_3O^+ is 12% of the molarity of HA. The ionization of HA produces 1 H_3O^+ ion and 1 A^- ion per molecule, so the concentration of the two ions will be the same.

$[H_3O^+] = [A^-] = (0.12)(0.00300\text{ M}) = 0.00036\text{ M}$

The concentration of HA is equal to the original concentration minus the amount that ionizes.

$[HA] = (0.00300 - 0.00036)\text{ M} = 0.00264\text{ M}$

Substitute these values in the equilibrium expression to calculate the value of K_a.

$$K_a = \frac{[0.00036][0.00036]}{[0.00264]} = 4.9 \times 10^{-5}$$

10.49 A strong acid dissociates almost completely; a weak acid dissociates to a small degree.

a. Since Y dissociates to a greater extent than Z in water, Y transfers more protons to water; Y is a stronger acid than Z.
b. Acid strength increases as K_a increases. The K_a for Z is larger than the K_a for Y; Z is a stronger acid than Y.
c. In the equilibrium equation for an acid, the equilibrium lies further to the right if the acid produces more protons. Y produces more protons than Z does, so Y is a stronger acid.
d. Y is a stronger acid because proton transfer occurs to a greater extent than it does for Z.

10.51 An Arrhenius acid must contain hydrogen, written first in the molecular formula. An Arrhenius base must have OH^- present. A salt is an ionic compound containing a metal or a polyatomic ion as the positive ion and a nonmetal or polyatomic ion (except hydroxide ion) as the negative ion.

a. acid b. salt c. salt d. base

10.53 A salt is an ionic compound containing a metal or a polyatomic ion as the positive ion and a nonmetal or polyatomic ion (except hydroxide ion) as the negative ion.

a. Yes, both are salts.
b. No. HCl is an acid.
c. No. NaOH is a base
d. Yes, both are salts.

10.55 All common soluble salts are completely dissociated into ions in solution.

a. $Ba(NO_3)_2 \xrightarrow{H_2O} Ba^{2+} + 2NO_3^-$

b. $Na_2SO_4 \xrightarrow{H_2O} 2Na^+ + SO_4^{2-}$

c. $CaBr_2 \xrightarrow{H_2O} Ca^{2+} + 2Br^-$

d. $K_2CO_3 \xrightarrow{H_2O} 2K^+ + CO_3^{2-}$

10.57 A neutralization reaction is the chemical reaction between an acid and a base in which a salt and water are the products.

a. No, this reaction is not a neutralization reaction because only salts are present.
b. Yes, this reaction is a neutralization reaction.
c. Yes, this reaction is a neutralization reaction.
d. No, this reaction is not a neutralization reaction because there is no hydroxide base present.

10.59 The molecular ratio to which these acid-base pairs will react is the inverse of the ratio of the number of H atoms in the chemical formula to the number of OH groups in the chemical formula.

a. 1 HNO_3 molecule to 1 NaOH molecule
b. 1 H_2SO_4 molecule to 2 NaOH molecules
c. 1 H_2SO_4 molecule to 1 $Ba(OH)_2$ molecule
d. 2 HNO_3 molecules to 1 $Ba(OH)_2$ molecule

10.61 In an acid-base neutralization reaction, an acid reacts with a base to produce a salt and water. The chemical equations for the neutralization reactions between the given acid-base pairs can be balanced by using the ratio between acidic hydrogen atoms and hydroxide groups.

a. $HCl + NaOH \rightarrow NaCl + H_2O$

b. $HNO_3 + KOH \rightarrow KNO_3 + H_2O$

c. $H_2SO_4 + 2LiOH \rightarrow Li_2SO_4 + 2H_2O$

d. $2H_3PO_4 + 3Ba(OH)_2 \rightarrow Ba_3(PO_4)_2 + 6H_2O$

10.63 Write these balanced chemical equations by working backwards from the salt: The positive metal ion in the salt comes from the base; the negative ion in the salt comes from the acid. 1) Write the formula for the acid, using the number of H^+ ions needed to balance the charge on the negative ion. 2) Write the chemical formula for the base, using the number of OH^- ions needed to balance the charge on the positive ion. 3) Balance the equation using the ratio between the acidic hydrogen ions and the hydroxide groups.

a. $H_2SO_4 + 2LiOH \rightarrow Li_2SO_4 + 2H_2O$

b. $HCl + NaOH \rightarrow NaCl + H_2O$

c. $HNO_3 + KOH \rightarrow KNO_3 + H_2O$

d. $2H_3PO_4 + 3Ba(OH)_2 \rightarrow Ba_3(PO_4)_2 + 6H_2O$

10.65 The ion product constant for water (1.00×10^{-14}) is obtained by multiplying the molar concentrations of H_3O^+ ion and OH^- ion present in pure water. If the OH^- ion concentration of an aqueous solution is known, the H_3O^+ ion concentration can be calculated by rearranging the ion product expression: $[H_3O^+] \times [OH^-] = 1.00 \times 10^{-14}$

a. $$\left[H_3O^+\right] = \frac{1.00 \times 10^{-14}\ M}{\left[OH^-\right]} = \frac{1.00 \times 10^{-14}\ M}{3.5 \times 10^{-3}\ M} = 2.9 \times 10^{-12}\ M$$

b. $$\left[H_3O^+\right] = \frac{1.00 \times 10^{-14}\ M}{\left[OH^-\right]} = \frac{1.00 \times 10^{-14}\ M}{4.7 \times 10^{-6}\ M} = 2.1 \times 10^{-9}\ M$$

c. $$\left[H_3O^+\right] = \frac{1.00 \times 10^{-14}\ M}{\left[OH^-\right]} = \frac{1.00 \times 10^{-14}\ M}{1.1 \times 10^{-8}\ M} = 9.1 \times 10^{-7}\ M$$

d. $$\left[H_3O^+\right] = \frac{1.00 \times 10^{-14}\ M}{\left[OH^-\right]} = \frac{1.00 \times 10^{-14}\ M}{8.7 \times 10^{-10}\ M} = 1.1 \times 10^{-5}\ M$$

10.67 In an acidic solution, $[H_3O^+]$ is higher than 1.00×10^{-7}M; in a basic solution, $[OH^-]$ is higher than 1.00×10^{-7} M.

a. A solution whose $[H_3O^+]$ is 2.9×10^{-12} M is basic.

b. A solution whose $[H_3O^+]$ is 2.1×10^{-9} M is basic.

c. A solution whose $[H_3O^+]$ is 9.1×10^{-7} M is acidic.

d. A solution whose $[H_3O^+]$ is 1.1×10^{-5} M is acidic.

10.69 The ion product constant for water (1.00×10^{-14}) is obtained by multiplying the molar concentrations of H_3O^+ ion and OH^- ion present in pure water. If the H_3O^+ ion concentration of an aqueous solution is known, the OH^- ion concentration can be calculated by rearranging the ion product expression: $[H_3O^+] \times [OH^-] = 1.00 \times 10^{-14}$

a. $\left[OH^-\right] = \dfrac{1.00 \times 10^{-14}\text{ M}}{\left[H_3O^+\right]} = \dfrac{1.00 \times 10^{-14}\text{ M}}{5.5 \times 10^{-2}\text{ M}} = 1.8 \times 10^{-13}\text{ M}$

b. $\left[OH^-\right] = \dfrac{1.00 \times 10^{-14}\text{ M}}{\left[H_3O^+\right]} = \dfrac{1.00 \times 10^{-14}\text{ M}}{9.4 \times 10^{-5}\text{ M}} = 1.1 \times 10^{-10}\text{ M}$

c. $\left[OH^-\right] = \dfrac{1.00 \times 10^{-14}\text{ M}}{\left[H_3O^+\right]} = \dfrac{1.00 \times 10^{-14}\text{ M}}{2.3 \times 10^{-7}\text{ M}} = 4.3 \times 10^{-8}\text{ M}$

d. $\left[OH^-\right] = \dfrac{1.00 \times 10^{-14}\text{ M}}{\left[H_3O^+\right]} = \dfrac{1.00 \times 10^{-14}\text{ M}}{6.6 \times 10^{-12}\text{ M}} = 1.5 \times 10^{-3}\text{ M}$

10.71 In an acidic solution, $[H_3O^+]$ is higher than 1.00×10^{-7}M; in a basic solution, $[OH^-]$ is higher than 1.00×10^{-7} M.

a. A solution whose $[OH^-]$ is 1.8×10^{-13} M is acidic.

b. A solution whose $[OH^-]$ is 1.1×10^{-10} M is acidic.

c. A solution whose $[OH^-]$ is 4.3×10^{-8} M is acidic.

d. A solution whose $[OH^-]$ is 1.5×10^{-3} M is basic.

10.73 Use the following relationship to complete the table: $[H_3O^+] \times [OH^-] = 1.00 \times 10^{-14}$ If $[H_3O^+]$ is larger than $[OH^-]$, the solution is acidic; if $[OH^-]$ is larger than $[H_3O^+]$, the solution is basic.

	$[H_3O^+]$	$[OH^-]$	Acidic or Basic
	2.2×10^{-2}	4.5×10^{-13}	acidic
a.	3.0×10^{-12}	3.3×10^{-3}	basic
b.	6.8×10^{-8}	1.5×10^{-7}	basic
c.	1.4×10^{-7}	7.2×10^{-8}	acidic
d.	4.7×10^{-5}	2.12×10^{-10}	acidic

10.75 The definition of pH is $-\log[H_3O^+]$. If the $[H_3O^+]$ is given, and its coefficient in the exponential expression is 1.0, the pH can easily be obtained from the relationship: $[H_3O^+] = 1.0 \times 10^{-x}$ (where pH = x). If the coefficient is not 1.0, use an electronic calculator to find pH from the above expression.

a. $[H_3O^+] = 1.00 \times 10^{-5}$; pH = 5.000

b. $[H_3O^+] = 1.00 \times 10^{-8}$; pH = 8.000

c. $[H_3O^+] = 4.75 \times 10^{-6}$; pH = 5.323

d. $[H_3O^+] = 8.88 \times 10^{-8}$; pH = 7.052

10.77 If the $[OH^-]$ is given, first calculate the $[H_3O^+]$ using the ion product constant for water. Then determine pH from the relationship: $[H_3O^+] = 1.0 \times 10^{-x}$ (where pH = x).

a. $[OH^-] = 1.00 \times 10^{-4}$; $[H_3O^+] = 1.00 \times 10^{-10}$; pH = 10.000

b. $[OH^-] = 1.00 \times 10^{-10}$; $[H_3O^+] = 1.00 \times 10^{-4}$; pH = 4.000

c. $[OH^-] = 1.11 \times 10^{-3}$; $[H_3O^+] = 9.01 \times 10^{-10}$; pH = 11.045

d. $[OH^-] = 6.05 \times 10^{-7}$; $[H_3O^+] = 1.65 \times 10^{-8}$; pH = 7.782

10.79 If pH is given, use the relationship pH = $-\log[H_3O^+]$. The antilog of the –pH is the $[H_3O^+]$.

a. pH = 5.00; $[H_3O^+]$ = antilog (–5) = 1.0×10^{-5} M

b. pH = 7.00; $[H_3O^+]$ = antilog (–7) = 1.0×10^{-7} M

c. pH = 3.45; $[H_3O^+]$ = antilog (–3.45) = 3.5×10^{-4} M

d. pH = 7.15; $[H_3O^+]$ = antilog (–7.15) = 7.1×10^{-8} M

10.81 Molar hydronium ion concentration may be written $[H_3O^+]$. If pH is given, use the relationship pH = $-\log[H_3O^+]$. The antilog of the –pH is the $[H_3O^+]$.

a. pH = 3.13; $[H_3O^+]$ = antilog (–3.13) = 7.4×10^{-4} M

b. pH = 3.25; $[H_3O^+]$ = antilog (–3.25) = 5.6×10^{-4} M

c. pH = 3.50; $[H_3O^+]$ = antilog (–3.50) = 3.2×10^{-4} M

d. pH = 3.75; $[H_3O^+]$ = antilog (–3.75) = 1.8×10^{-4} M

10.83 Acidic solutions have a pH below 7. Basic solutions have a pH above 7. If the pH = 7, the solution is neutral.

a. Milk (pH = 6.4) is acidic.

b. Sea water (pH = 8.5) is basic.

c. Drinking water (pH = 7.2) is basic

d. A strawberry (pH = 3.4) is acidic.

10.85 Use the following relationships to complete the table: $[H_3O^+] \times [OH^-] = 1.00 \times 10^{-14}$ If $[H_3O^+]$ is larger than $[OH^-]$, the solution is acidic; if $[OH^-]$ is larger than $[H_3O^+]$, the solution is basic. pH = $-\log[H_3O^+]$

	$[H_3O^+]$	$[OH^-]$	pH	Acidic or Basic
	6.2×10^{-8}	1.6×10^{-7}	7.21	basic
a.	7.2×10^{-10}	1.4×10^{-5}	9.14	basic
b.	5.0×10^{-6}	2.0×10^{-9}	5.30	acidic
c.	1.4×10^{-5}	7.2×10^{-8}	4.85	acidic
d.	5.8×10^{-9}	1.7×10^{-6}	8.23	basic

10.87 a. Solution 3, the carbonated beverage, is the solution with the lowest pH and highest $[H_3O^+]$.

b. Solution 4, drinking water, is the solution with the highest pH and therefore the highest $[OH^-]$.

c. The solutions in order of increasing acidity are: 4, 1, 2, 3.

d. The solutions in order of decreasing basicity are: 4, 1, 2, 3.

10.89 a. 1.0 M NaOH has a lower $[H_3O^+]$ than the 1.0 M HCl; 1.0 M NaOH has a higher pH.

b. 1.0 M HNO_3 has a higher $[H_3O^+]$ than 0.10 M HNO_3; 1.0 M HNO_3 has a lower pH.

c. 0.10 M $HClO_4$ has a higher $[H_3O^+]$ than 0.10 M HCN because HCN is a weak acid ($K_a = 4.9 \times 10^{-10}$); 0.10 M $HClO_4$ has a lower pH.

d. $[H_3O^+] = 3.3 \times 10^{-3}$ is lower than $[H_3O^+] = 9.3 \times 10^{-3}$; $[H_3O^+] = 3.3 \times 10^{-3}$ has a higher pH.

10.91 K_a, the acid ionization constant, is a measure of acid strength. Another method for expressing the strengths of acids is in terms of pK_a units (p$K_a = -\log K_a$). The pK_a for an acid is calculated from K_a in the same way that pH is calculated from $[H_3O^+]$.

a. $K_a = 4.5 \times 10^{-4}$; p$K_a = 3.35$

b. $K_a = 4.3 \times 10^{-7}$; p$K_a = 6.37$

c. $K_a = 6.2 \times 10^{-8}$; p$K_a = 7.21$

d. $K_a = 1.5 \times 10^{-2}$; p$K_a = 1.82$

10.93 The acid with the larger K_a (more ionization) is the stronger acid. Since p$K_a = -\log K_a$, we can see that an acid with a larger K_a has a smaller pK_a. Acid B has a smaller pK_a than Acid A does; therefore, Acid B is the stronger acid.

10.95 Use the definition p$K_a = -\log K_a$. Since p$K_a = 8.73$, use your calculator to find the antilog of -8.73. The answer is 1.9×10^{-9}.

10.97 Analyze each salt to determine which acid contributed the negative ion and which base contributed the positive ion to the salt.

a. NaCl is the salt of a strong acid (HCl) and a strong base (NaOH).

b. $KC_2H_3O_2$ is the salt of a weak acid ($HC_2H_3O_2$) and a strong base (KOH).

c. NH_4Br is the salt of a strong acid (HBr) and a weak base (NH_4OH).

d. $Ba(NO_3)_2$ is the salt of a strong acid (HNO_3) and a strong base ($Ba(OH)_2$).

10.99 In water, the negative ion of a weak acid or the positive ion of a weak base will undergo hydrolysis (reaction with water).

a. Neither of the ions of NaCl undergoes hydrolysis because NaCl is the salt of a strong acid and a strong base.

b. The $C_2H_3O_2^-$ ion undergoes hydrolysis; $HC_2H_3O_2$ is a weak acid.

c. The NH_4^+ ion undergoes hydrolysis; NH_4OH is a weak base.

d. Neither of the ions of $Ba(NO_3)_2$ undergoes hydrolysis because $Ba(NO_3)_2$ is the salt of a strong acid and a strong base.

10.101 Guidelines for determining if a salt solution will be acidic, basic, or neutral: Table 10.7.

a. A NaCl solution will be neutral; the salt of a strong acid and a strong base does not hydrolyze, so the solution is neutral.

b. A $KC_2H_3O_2$ solution will be basic; the salt of a weak acid and a strong base hydrolyzes to produce a basic solution.

c. An NH_4Br solution will be acidic; the salt of a strong acid and a weak base hydrolyzes to produce an acidic solution.

d. A $Ba(NO_3)_2$ solution will be neutral; the salt of a strong acid and a strong base does not hydrolyze, so the solution is neutral.

10.103 Table 10.8 gives the pH of a 0.1 M NH_4CN as 9.3. The hydrolysis equations for two ions are:

$NH_4^+ + H_2O \rightarrow NH_3 + H_3O^+$ and $CN^- + H_2O \rightarrow HCN + OH^-$

The strength of NH_3 as a base is greater than the strength of HCN as an acid. Therefore, the CN^- ion hydrolyzes to a greater extent than the NH_4^+ ion, and more OH^- ion is produced than H_3O^+ ion.

10.105 The pHs of the four aqueous solutions in order of increasing pH are:

HCl (acidic), HCN (weakly acidic), NaCl (neutral), and NaCN (weakly basic).

10.107 A buffer is an aqueous solution containing substances that prevent major changes in solution pH. Buffer solutions contain either a weak acid and a salt of that weak acid or a weak base and a salt of that weak base.

a. No. HNO_3 is a strong acid, and $NaNO_3$ is a salt of a strong acid.

b. Yes. HF is a weak acid, and NaF is a salt of that weak acid.

c. No. Both KCl and KCN are salts. KCN is the salt of a weak acid, but no weak acid is present.

d. Yes. H_2CO_3 is a weak acid, and $NaHCO_3$ is the salt of a weak acid.

10.109 The active species in a buffered system are the substance that reacts with and removes added base and the substance that reacts with and removes added acid.

a. HCN reacts with added base; CN^- reacts with added acid.

b. H_3PO_4 reacts with added base; $H_2PO_4^-$ reacts with added acid.

c. H_2CO_3 reacts with added base; HCO_3^- reacts with added acid.

d. HCO_3^- reacts with added base; CO_3^{2-} reacts with added acid.

10.111 Buffering actions are the reactions that take place in the buffer system with the addition of a small amount of acid or base.

a. Addition of acid to HF/F^- buffer: $F^- + H_3O^+ \rightarrow HF + H_2O$

b. Addition of base to H_2CO_3/HCO_3^- buffer: $H_2CO_3 + OH^- \rightarrow HCO_3^- + H_2O$

c. Addition of acid to HCO_3^-/CO_3^{2-} buffer: $CO_3^{2-} + H_3O^+ \rightarrow HCO_3^- + H_2O$

d. Addition of a base to $H_3PO_4/H_2PO_4^-$ buffer: $H_3PO_4 + OH^- \rightarrow H_2PO_4^- + H_2O$

10.113 The buffer equation is $NO_2^- + H_3O^+ \rightarrow HNO_2 + H_2O$
When acid is added, it reacts with NO_2^- using some of it up and producing more HNO_2; therefore, $[NO_2^-]$ decreases and $[HNO_2]$ increases.

10.115 The addition of a small amount of acid to a buffer system lowers the pH of the buffer slightly; the addition of a small amount of base to a buffer system raises the pH of the buffer slightly.

a. The most likely value of the buffer system when a small amount of acid is added is 8.1.

b. The most likely value of the buffer system when a small amount of base is added is 8.3.

10.117 All four diagrams; HA and A^- are present in each case.

10.119 Buffer solutions contain a conjugate acid-base pair, that is, either a weak acid and a salt of that weak acid or a weak base and a salt of that weak base.

a. The acid HCN and the salt KCN are a weak acid and the salt of that weak acid; the buffer system is HCN and CN^-.

b. The acid HNO_2 and the salt $NaNO_2$ are a weak acid and the salt of that weak acid; the buffer system is HNO_2 and NO_2^-. (H_2CO_3 is also a weak acid, but removal of one proton gives HCO_3^-, not CO_3^{2-}.)

10.121 The four given species are: Cl^-, HPO_4^{2-}, CO_3^{2-}, and H_2SO_3.

a. An amphiprotic substance can either lose or accept a proton. Of the four given species, only HPO_4^{2-} is amphiprotic.

b. The salt of a strong acid does not hydrolyze in solution. Cl^- is the negative ion of a salt of HCl and so does not hydrolyze in solution.

c. The equation for hydrolysis to give a basic solution is: $A^- + H_2O \quad HA + OH^-$ where A is the negative ion of a weak acid. This description fits two of the given species, HPO_4^{2-} and CO_3^{2-}.

d. A buffer sytem consists of either a weak acid and a salt of that weak acid or a weak base and a salt of that weak base. H_2SO_3 (a weak acid), CO_3^{2-} (the negative ion of a weak acid), and HPO_4^{2-} (a weak acid and also the negative ion of a weak acid) can all function as buffer system components. Only Cl^- (the salt of a strong acid) cannot.

10.123 To calculate the pH of a buffer solution, we can use the Henderson-Hasselbalch equation. In this equation, HA is the weak acid and A^- is the acid's conjugate base.

$$\text{pH} = \text{p}Ka + \log\frac{[A^-]}{[HA]} = 6.72 + \log\left[\frac{0.500\text{ M}}{0.230\text{ M}}\right] = 7.06$$

10.125 First change the K_a to pK_a (p$K_a = -\log K_a$). Then use the Henderson-Hasselbalch equation to calculate the pH of the buffer solution.

$$\text{pH} = \text{p}Ka + \log\frac{[A^-]}{[HA]} = -\log\left[6.8 \times 10^{-6}\right] + \log\left[\frac{0.150\text{ M}}{0.150\text{ M}}\right] = 5.17$$

10.127 In water solution, a strong electrolyte completely dissociates into ions; a weak electrolyte ionizes only slightly.

a. H_2CO_3 is a weak acid and is therefore a weak electrolyte.

b. KOH is a strong base and is therefore a strong electrolyte.

c. NaCl is a soluble salt and is therefore a strong electrolyte.

d. H_2SO_4 is a strong acid and is therefore a strong electrolyte.

10.129 a. both; a weak acid does not dissociate 100%

b. molecules; a nonelectrolyte does not dissociate at all

c. ions; soluble salts dissociate 100%

d. both; a weak electrolyte does not dissociate 100%

10.131 a. 2 (Na^+ and Cl^-) b. 3 (Mg^{2+} and two NO_3^-)
c. 3 (two K^+ and S^{2-}) d. 2 (NH_4^+ and CN^-)

10.133 a. $NaCl \rightarrow Na^+ + Cl^-$ b. $Mg(NO_3)_2 \rightarrow Mg^{2+} + 2NO_3^-$
c. $K_2S \rightarrow 2K^+ + S^{2-}$ d. $NH_4CN \rightarrow NH_4^+ + CN^-$

10.135 Diagram 3; this solution contains the greatest number of ions.

10.137 One equivalent is the amount of ions needed to supply one mole of charge.
a. 1 Eq b. 1 Eq c. 2 Eq d. 1 Eq

10.139 a. $2 \text{ moles } K^+ \times \left(\frac{1 \text{ Eq } K^+}{1 \text{ mole } K^+}\right) = 2 \text{ Eq } K^+$

b. $3 \text{ moles } H_2PO_4^- \times \left(\frac{1 \text{ Eq } H_2PO_4^-}{1 \text{ mole } H_2PO_4^-}\right) = 3 \text{ Eq } H_2PO_4^-$

c. $2 \text{ moles } HPO_4^{2-} \times \left(\frac{2 \text{ Eq } HPO_4^{2-}}{1 \text{ mole } HPO_4^{2-}}\right) = 4 \text{ Eq } HPO_4^{2-}$

d. $7 \text{ moles } Ca^{2+} \times \left(\frac{2 \text{ Eq } Ca^{2+}}{1 \text{ mole } Ca^{2+}}\right) = 14 \text{ Eq } Ca^{2+}$

10.141 1 Eq = 1000 mEq; Eq × moles of charge/Eq = moles

a. $48 \text{ mEq } Ca^{2+} \times \frac{1 \text{ Eq } Ca^{2+}}{10^3 \text{ mEq } Ca^{2+}} = 0.048 \text{ Eq } Ca^{2+}$

b. $0.048 \text{ Eq } Ca^{2+} \times \frac{1 \text{ mole } Ca^{2+}}{2 \text{ Eq } Ca^{2+}} = 0.024 \text{ mole } Ca^{2+}$

10.143 $\frac{0.0030 \text{ M } HPO_4^{2-}}{L} \times \left(\frac{2 \text{ Eq } HPO_4^{2-}}{1 \text{ mole } HPO_4^{2-}}\right) \times \left(\frac{1 \text{ mEq } HPO_4^{2-}}{10^{-3} \text{ Eq } HPO_4^{2-}}\right) = \frac{6.0 \text{ mEq } HPO_4^{2-}}{L}$

10.145 A charge balance must exist among the ions present in an electrolyte solution. This means that the Eq/L of positive charges and the Eq/L of negative charges must be equal.

a. 75 mEq Na^+ + 25 mEq K^+ = 100 mEq positive charges;
95 mEq Cl^- + 5 mEq NO_3^- = 100 mEq of negative charges. Yes, it is possible to prepare this solution.

b. 73 Eq K^+ = 73 Eq positive charges; 55 Eq Cl^- + 25 Eq $C_2H_3O_2^-$ = 80 Eq of negative charges. No, it is not possible to prepare this solution.

c. 750 mEq Na^+ = 750 mEq positive charges; 0.750 Eq Cl^-= 0.750 Eq = 750 mEq of negative charges. Yes, it is possible to prepare this solution.

d. 0.025 mole Na^+ + 0.025 mole Ca^{2+} = 0.025 mEq Na^+ + 0.050 mEq Ca^{2+}
= 0.075 mEq positive charges; 0.075 mole Cl^-= 0.075 mEq of negative charges. Yes, it is possible to prepare this solution.

10.147 A charge balance must exist among the ions present in an electrolyte solution. This means that the Eq/L of positive charges and the Eq/L of negative charges must be equal.
Summing the positive charges:
75 mEq Na^+ + 83 mEq K^+ + 10 mEq Ca^{2+} = 168 mEq of positive charge
Summing the negative charges:
153 mEq of Cl^- + x mEq HCO_3^- = 168 mEq of negative charge
x = 15 mEq HCO_3^-

10.149 In the following problems, an acid neutralizes 25.0 mL of a NaOH solution of unknown molarity. To determine the molarity of the NaOH solution:

1) Convert mL of acid solution to moles of acid using the molarity of the acid solution as a conversion factor.
2) Write the balanced equation between the acid and the base (neutralization). Use the ratio of the coefficients of the acid and base as a conversion factor to change moles of acid to moles of NaOH.
3) Use the definition of molarity (moles/L) to calculate the molarity of the NaOH solution.

a. $$5.00\text{ mL HNO}_3 \times \left(\frac{0.250\text{ mole HNO}_3}{1000\text{ mL HNO}_3}\right) \times \left(\frac{1\text{ mole NaOH}}{1\text{ mole HNO}_3}\right) = 0.00125\text{ mole NaOH}$$

$$\frac{0.00125\text{ mole NaOH}}{0.0250\text{ L solution}} = 0.0500\text{ M}$$

b. $$20.00\text{ mL H}_2\text{SO}_4 \times \left(\frac{0.500\text{ mole H}_2\text{SO}_4}{1000\text{ mL H}_2\text{SO}_4}\right) \times \left(\frac{2\text{ moles NaOH}}{1\text{ mole H}_2\text{SO}_4}\right) = 0.0200\text{ mole NaOH}$$

$$\frac{0.0200\text{ mole NaOH}}{0.0250\text{ L solution}} = 0.800\text{ M}$$

c. $$23.76\text{ mL HCl} \times \left(\frac{1.00\text{ mole HCl}}{1000\text{ mL HCl}}\right) \times \left(\frac{1\text{ mole NaOH}}{1\text{ mole HCl}}\right) = 0.02376\text{ mole NaOH}$$

$$\frac{0.02376\text{ mole NaOH}}{0.0250\text{ L solution}} = 0.950\text{ M}$$

d. $$10.00\text{ mL H}_3\text{PO}_4 \times \left(\frac{0.100\text{ mole H}_3\text{PO}_4}{1000\text{ mL H}_3\text{PO}_4}\right) \times \left(\frac{3\text{ moles NaOH}}{1\text{ mole H}_3\text{PO}_4}\right) = 0.00300\text{ mole NaOH}$$

$$\frac{0.00300\text{ mole NaOH}}{0.0250\text{ L solution}} = 0.120\text{ M}$$

Nuclear Chemistry — Chapter 11

Solutions to Selected Problems

11.1 The complete symbol $^{A}_{Z}X$ is a notation used to identify a nucleus or atom uniquely. The superscript *A* before the symbol for the element is the mass number, and the subscript *Z* is the atomic number. Z gives the number of protons in the nucleus; $A - Z$ gives the number of neutrons in the nucleus.

a. 4, 6 b. 11, 12 c. 47, 61 d. 16, 18

11.3 a. $^{14}_{7}N$ and nitrogen-14 indicate the same nuclide. 14 is the mass number in each case.
b. The two notations have a different mass number (27 and 28), so are not the same nuclide.
c. The mass number is the same in the two notations; they are the same nuclide.
d. The two notations have a different mass number (18 and 36) and a different atomic number (8 and 18), so they are not the same nuclide.

11.5 Spontaneous emission of radiation from the nucleus of an atom indicates an unstable nucleus.

11.7 The neutron-to-proton ratios are: approximately a 1-to-1 ratio for low-atomic-numbered stable nuclei and a 3-to-2 ratio for higher-atomic-numbered stable nuclei.

11.9 a. $^{4}_{2}\alpha$ An alpha particle consists of two protons and two neutrons. The superscript indicates a mass of 4 amu (two protons + two neutrons); the subscript indicates that the charge is a +2 (from two protons).

b. $^{0}_{-1}\beta$ A beta particle has a mass (very close to zero amu) and charge (–1) identical to those of an electron.

c. $^{0}_{0}\gamma$ A gamma ray is a form of high-energy radiation without mass or charge.

11.11 An alpha particle consists of two protons and two neutrons.

11.13 Alpha particle decay is the emission of an alpha particle from a nucleus. The product nucleus has a mass number that is four less than the original nucleus, and an atomic number that is two less than the original nucleus. In writing nuclear equations, make sure that the mass numbers and the atomic numbers balance on the two sides of the equation.

a. $^{200}_{84}Po \rightarrow ^{4}_{2}\alpha + ^{196}_{82}Pb$ b. $^{240}_{96}Cm \rightarrow ^{4}_{2}\alpha + ^{236}_{94}Pu$

c. $^{244}_{96}Cm \rightarrow ^{4}_{2}\alpha + ^{240}_{94}Pu$ d. $^{238}_{92}U \rightarrow ^{4}_{2}\alpha + ^{234}_{90}Th$

11.15 In beta particle decay, the mass number of the new nuclide is the same as that of the original nuclide; however, the atomic number has increased by one unit. Balance the superscripts and subscripts on both sides of the equation.

a. $^{10}_{4}Be \rightarrow ^{0}_{-1}\beta + ^{10}_{5}B$ b. $^{14}_{6}C \rightarrow ^{0}_{-1}\beta + ^{14}_{7}N$

c. $^{21}_{9}F \rightarrow ^{0}_{-1}\beta + ^{21}_{10}Ne$ d. $^{25}_{11}Na \rightarrow ^{0}_{-1}\beta + ^{25}_{12}Mg$

11.17 In alpha particle decay, the radionuclide loses an alpha particle ($^{4}_{2}\alpha$), which has two protons, two neutrons, and a mass number of four.

a. The atomic number decreases by two. b. The mass number decreases by four.

11.19 a. Polonium-210 (84 p, 126 n) decays to lead-206 (82 p, 124 n). The loss of 2 p and 2 n is alpha particle decay.

b. Thorium-225 (90 p, 135 n) decays to protactinium-225 (91 p, 134 n). The gain of 1 p and loss of 1 n is beta particle decay.

c. Pt-190 (78 p, 112 n) decays to Os-186 (76 p, 110 n). The loss of 2 p and 2 n is alpha particle decay.

d. O-19 (8 p, 11 n) decays to F-19 (9 p, 10 n). Gain of 1 p, loss of 1 n is beta particle decay.

11.21 In a balanced nuclear equation the sum of the superscripts (mass numbers) and the sum of the subscripts (atomic numbers or particle charges) on the two sides of the equation must be equal.

a. $^{210}_{84}\text{Po} \rightarrow {}^{206}_{82}\text{Pb} + {}^{4}_{2}\alpha$ b. $^{225}_{90}\text{Th} \rightarrow {}^{225}_{91}\text{Pa} + {}^{0}_{-1}\beta$

c. $^{190}_{78}\text{Pt} \rightarrow {}^{186}_{76}\text{Os} + {}^{4}_{2}\alpha$ d. $^{19}_{8}\text{O} \rightarrow {}^{19}_{9}\text{F} + {}^{0}_{-1}\beta$

11.23 In a balanced nuclear equation both the sums of the subscripts and the sums of the superscripts on the two sides of the equation are equal.

a. Write complete symbols $^{A}_{Z}\text{X}$ for both the product and the emitted beta particle on the right side of the equation, and, on the left, write the complete symbol for the radionuclide whose decay will balance the sums of *A* and *Z* on the two sides of the equation.

$$^{199}_{79}\text{Au} \rightarrow {}^{199}_{80}\text{Hg} + {}^{0}_{-1}\beta$$

b. Write the complete symbols for palladium-109 on the left side of the equation and for the beta particle on the right side. Also on the right side, write the complete symbol for the product that will balance the values of *A* and *Z* on the two sides of the equation.

$$^{109}_{46}\text{Pd} \rightarrow {}^{109}_{47}\text{Ag} + {}^{0}_{-1}\beta$$

c. Write complete symbols for terbium-148 and an alpha particle on the right side and, on the left side, write the complete symbol for the radionuclide whose decay will balance the sums of *A* and *Z* on the two sides of the equation.

$$^{152}_{67}\text{Ho} \rightarrow {}^{148}_{65}\text{Tb} + {}^{4}_{2}\alpha$$

d. Write the complete symbol for fermium-249 on the left side of the equation and, on the right side, the complete symbols for an alpha particle and for the product that will balance *A* and *Z* on the two sides of the equation.

$$^{249}_{100}\text{Fm} \rightarrow {}^{245}_{98}\text{Cm} + {}^{4}_{2}\alpha$$

11.25 The half-life ($t_{1/2}$) is the time required for one-half of a given quantity of a radioactive substance to undergo decay. To calculate the fraction of nuclide remaining, determine the number of half-lives elapsed, and use the equation below (*n* is number of half-lives):

$$\left(\begin{array}{c}\text{Amount of radionuclide}\\ \text{undecayed after } n \text{ half-lives}\end{array}\right) = \left(\begin{array}{c}\text{original amount}\\ \text{of radionuclide}\end{array}\right) \times \left(\frac{1}{2^n}\right)$$

a. $\frac{1}{2^2} = \frac{1}{4}$ of original is undecayed. b. $\frac{1}{2^6} = \frac{1}{64}$ of original is undecayed.

c. $\frac{1}{2^3} = \frac{1}{8}$ of original is undecayed. d. $\frac{1}{2^6} = \frac{1}{64}$ of original is undecayed.

11.27 First, determine the number of half-lives that have elapsed (using the equation in Problem 11.25). Then divide the time that has elapsed by the number of half-lives to find the length of one half-life.

a. $\frac{1}{16} = \frac{1}{2^4}$; $\frac{5.4 \text{ days}}{4} = 1.4$ days b. $\frac{1}{64} = \frac{1}{2^6}$; $\frac{5.4 \text{ days}}{6} = 0.90$ day

c. $\frac{1}{256} = \frac{1}{2^8}$; $\frac{5.4 \text{ days}}{8} = 0.68$ day d. $\frac{1}{1024} = \frac{1}{2^{10}}$; $\frac{5.4 \text{ days}}{10} = 0.54$ day

11.29 A half-life ($t_{1/2}$) is the time required for one-half of a given quantity of a radioactive substance to undergo decay. Use the following relationship to find the values for the four different radionuclides below that will fill in the blanks in the table.

$$\begin{pmatrix}\text{Amount of radionuclide}\\ \text{undecayed after } n \text{ half-lives}\end{pmatrix} = \begin{pmatrix}\text{original amount}\\ \text{of radionuclide}\end{pmatrix} \times \left(\frac{1}{2^n}\right)$$

	Time elapsed	Half-life	Fraction undecayed	Fraction decayed
	3.0 hr	1.0 hr	1/8	7/8
a.	4.0 hr	2.0 hr	1/4	3/4
b.	6.0 hr	3.0 hr	1/4	3/4
c.	4.0 hr	1.0 hr	1/16	15/16
d.	5.0 hr	5.0 hr	1/2	1/2

11.31 a. It takes two half-lifes for the iodine-125 to decay to ¼ its original level (½ × ½). 2 × 60 days = 120 days
b. It takes four half-lifes for the iodine-125 to decay to 1/16 its original level (½ × ½ × ½ × ½). 4 × 60 days = 240 days

11.33 a. First, determine the number of half-lives by dividing the total time elapsed by the length of one half-life: 60.0 hr ÷ 15.0 hr = 4 half lives. The fraction of undecayed nuclide remaining is ½ × ½ × ½ × ½ or 1/16. Therefore, the fraction of decayed nuclide is 15/16. 15/16 × 4.00 = 3.75 g of the nuclide has decayed.
b. To go from 28 mg to 7 mg (½ × ½ × 28) requires 2 half-lives (½ × ½ × 28); 2(15.0 hr) = 30.0 hr

11.35 Three half-lives have elapsed; 2 out of 16 atoms remain undecayed; 2/16 = 1/8 and 1/8 = $1/2^3$, which means three half-lives have elapsed.

11.37 Over 2000 bombardment-produced radionuclides that do not occur naturally are now known.

11.39 Uranium, element 92, has the highest atomic number of any naturally occurring element.

11.41 To write each of these equations for bombardment reactions, put the parent nuclide and the bombardment particle on the reactant side and the daughter nuclides and decay particles on the product side. Balance the superscripts (mass numbers) and the subscripts (atomic numbers or charges).

a. ${}^{4}_{2}\alpha$ Superscripts: 24 + x (reactant side) = 27 + 1 (product side); $x = 4$
Subscripts: 12 + y (reactant side) = 14 + 0 (product side); $y = 2$

b. ${}^{25}_{12}\text{Mg}$ Superscripts: 27 + 2 (reactant side) = x + 4 (product side); $x = 25$
Subscripts: 13 + 1 (reactant side) = y + 2 (product side); $y = 12$

c. ${}^{4}_{2}\alpha$ Superscripts: 9 + x (reactant side) = 12 + 1 (product side); x = 4
Subscripts: 4 + y (reactant side) = 6 + 0 (product side); y = 2

d. ${}^{1}_{1}\mathrm{p}$ Superscripts: 6 + x (reactant side) = 4 + 3 (product side); x = 1
Subscripts: 3 + y (reactant side) = 2 + 2 (product side); y = 1

11.43 In a bombardment reaction, the bombarding particle is written on the reactant side of the nuclear equation. Write the given reactants and/or products in complete symbols on the correct side of the equation. Balance the mass numbers and the atomic numbers to identify the missing parts of the equations.

a. ${}^{9}_{4}\mathrm{Be} + {}^{4}_{2}\alpha \rightarrow {}^{12}_{6}\mathrm{C} + {}^{1}_{0}n$

b. ${}^{23}_{11}\mathrm{Na} + {}^{2}_{1}\mathrm{H} \rightarrow {}^{21}_{10}\mathrm{Ne} + {}^{4}_{2}\alpha$

c. ${}^{113}_{48}\mathrm{Cd} + {}^{1}_{0}n \rightarrow {}^{114}_{48}\mathrm{Cd} + {}^{0}_{0}\gamma$

d. ${}^{27}_{13}\mathrm{Al} + {}^{4}_{2}\alpha \rightarrow {}^{30}_{15}\mathrm{P} + {}^{1}_{0}n$

11.45 a. From Table 11.2, it can be determined that there are 7 transuranium elements that have a half-life greater than one year.
b. From Table 11.2, it can be determined that there are 17 transuranium elements that have a half-life less than one day.

11.47 We would expect lead-207 to be a stable nuclide, since it terminates the uranium-235 decay series; termination of a decay series requires a stable nuclide.

11.49 The two equations are: ${}^{222}_{86}\mathrm{Rn} \rightarrow {}^{218}_{84}\mathrm{A} + {}^{4}_{2}\alpha$ and ${}^{218}_{84}\mathrm{A} \rightarrow {}^{214}_{82}\mathrm{B} + {}^{4}_{2}\alpha$
The identity of B: ${}^{214}_{82}\mathrm{Pb}$

11.51 In each equation in this decay series, a parent nuclide produces a daughter nuclide and an alpha or beta particle. For each equation (after the first), the parent nuclide is the daughter nuclide from the previous equation. Balance the superscripts and the subscripts.

${}^{232}_{90}\mathrm{Th} \rightarrow {}^{4}_{2}\alpha + {}^{228}_{88}\mathrm{Ra}$ Superscripts: 232 (reactants) = 4 + x (products); x = 228
Subscripts: 90 (reactant) = 2 + y (product); y = 88

${}^{228}_{88}\mathrm{Ra} \rightarrow {}^{0}_{-1}\beta + {}^{228}_{89}\mathrm{Ac}$ Superscripts: 228 (reactants) = 0 + x (products); x = 228
Subscripts: 88 (reactant) = –1 + y (product); y = 89

${}^{228}_{89}\mathrm{Ac} \rightarrow {}^{0}_{-1}\beta + {}^{228}_{90}\mathrm{Th}$ Superscripts: 228 (reactants) = 0 + x (products); x = 228
Subscripts: 89 (reactant) = –1 + y (product); y = 90

${}^{228}_{90}\mathrm{Th} \rightarrow {}^{4}_{2}\alpha + {}^{224}_{88}\mathrm{Ra}$ Superscripts: 228 (reactants) = 4 + x (products); x = 224
Subscripts: 90 (reactant) = 2 + y (product); y = 88

11.53 Technicians who work around radiation usually wear film badges to detect the extent of their exposure to radiation. Radiation ionizes atoms and molecules in the badges, and the resulting charged particles can be detected.

11.55 An ion pair is the electron and the positive ion that are produced during an ionizing interaction between a molecule (or an atom) and radiation.

11.57 Draw the Lewis structure for each of the species; from the Lewis structure, determine whether there is an unpaired electron (a free radical).

a. 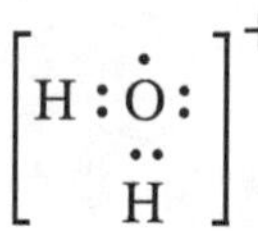

Yes, there is an unpaired electron; H_2O^+ is a free radical.

b. $\left[\mathrm{H:\ddot{O}:H} \atop \mathrm{H}\right]^+$

No, there is no unpaired electron; H_3O^+ is not a free radical.

c. $\mathrm{H:\dot{O}:}$

Yes, there is an unpaired electron; OH is a free radical.

d. $\left[\mathrm{H:\ddot{O}:}\right]^-$

No, there is no unpaired electron; OH^- is not a free radical.

11.59 The fate of a radiation particle involved in an ion pair formation is to continue on, interacting with other atoms and forming more ion pairs.

11.61 a. Beta radiation and gamma radiation can "pass through" a thick sheet of paper.
b. Gamma radiation can "pass through" one centimeter thick aluminum foil.
c. Gamma radiation can "pass through" five centimeter thick concrete.
d. Gamma radiation can "pass through" outer layers of human skin.

11.63 α-Particles can travel 6 cm in air before energy is dissipated; β-particles can travel 1000 cm.

11.65 Table 11.3 gives the effects of short-term whole-body radiation exposure on humans. Whole-body radiation exposure of:
a. 10 rems has no detectable effects.
b. 150 rems can cause nausea, fatigue, and lowered blood cell count.

11.67 Background radiation is naturally-occurring ionizing radiation.

11.69 The major sources are radon seepage, cosmic radiation, rocks and minerals, food and drink.

11.71 Nearly all diagnostic radionuclides are gamma emitters because the penetrating power of alpha and beta particles is too low; it is necessary to be able to detect the radiation externally (outside the body).

11.73 a. Gallium-67 is used to locate sites of infection.
b. Potassium-42 is used in the determination of intercellular spaces in fluids.
c. Thallium-201 is used in the assessment of blood flow in heart muscle.
d. Barium-131 is used in the detection of bone tumors.

11.75 Radionuclides used for therapeutic purposes are usually α or β emitters instead of γ emitters. Therapeutic radionuclides are used to selectively destroy abnormal (usually cancerous) cells; an intense dose of radiation in a small localized area is needed.

11.77 a. The fact that a large nucleus splits into two intermediate-sized nuclei indicates nuclear fission.
b. Hydrogen nuclei are important reactants in nuclear fusion processes.

11.79 The two sides of the nuclear equation for a nuclear fission reaction must have an equal number of protons and an equal number of neutrons. Don't forget to add one neutron on the reactant side to initiate the fission reaction.

a. $X + {}^{1}_{0}n \rightarrow {}^{143}_{56}Ba + {}^{94}_{37}Rb + 3{}^{1}_{0}n$ Particle X is ${}^{239}_{93}Np$

b. ${}^{235}_{92}U + {}^{1}_{0}n \rightarrow {}^{142}_{56}Ba + {}^{91}_{36}X + 3{}^{1}_{0}n$ Particle X is ${}^{91}_{36}Kr$

11.81 a. Balance the mass numbers: products [2(3)] = reactants [2(1) + x] = 6; x = 4
Balance the atomic numbers: products (2 + x) = reactants [2(2)] = 4; x = 2
The particle having a mass number of 4 and an atomic number of 2 is:
${}^{4}_{2}\alpha$

b. Balance the mass numbers: products [2(4) + 1] = reactants [7 + x] = 9; x = 2
Balance the atomic numbers: products [2(2) + 0] = reactants [3 + x] = 4; x = 1
The particle having a mass number of 2 and an atomic number of 1 is:
${}^{2}_{1}H$

11.83 a. Nuclear fusion is a nuclear reaction in which two small nuclei are put together to make a larger one. For fusion to occur, an extremely high temperature is required.
b. The process of nuclear fusion occurs on the sun.
c. During transmutation a nuclide of one element is changed into another nuclide of another element. Transmutation occurs in both fission and fusion reactions.
d. Bombardment of a nuclide with neutrons may cause it to split into two fragments; this process is called nuclear fission.

11.85 During a nuclear fission reaction, a large nucleus splits into two medium-sized nuclei. During a nuclear fusion reaction, two small nuclei are put together to make a larger one.
a. This is a fusion reaction because two small nuclei come together to form a larger one.
b. This is a fission reaction because one large nucleus splits into two medium sized nuclei.
c. This reaction is neither fission nor fusion because the large nucleus does not split into two medium-sized nuclei.
d. This reaction is neither fission nor fusion because the large nucleus does not split into two medium-sized nuclei.

11.87 Different isotopes of an element have the same chemical properties but different nuclear properties.

11.89 Temperature, pressure, and catalysts affect chemical reaction rates but do not affect nuclear reaction rates.

Saturated Hydrocarbons — Chapter 12

Solutions to Selected Problems

12.1 a. False. There are approximately seven million organic compounds known and 1.5 million inorganic compounds.
b. False. Chemists have successfully synthesized man organic compounds, starting with Wöhler's 1820 synthesis of urea; the "vital force" theory has been completely abandoned.
c. True. The *org-* of the term *organic* came from the term *living organism*.
d. True. Most compounds found in living organisms are organic compounds, but there are also inorganic salts and inorganic carbon compounds, such as CO_2.

12.3 Organic compounds are hydrocarbons (compounds of carbon and hydrogen) and their derivatives. Inorganic compounds are all substances other than hydrocarbons and their derivatives.
a. HNO_3 is an inorganic compound.
b. C_4H_{10} is an organic compound.
c. C_2H_6O is an organic compound.
d. CO_2 is an inorganic compound.

12.5 The bonding requirement for a carbon atom is that each carbon atom shares its four valence electrons in four covalent bonds.
a. Two single bonds and a double bond (equivalent to two single bonds) are a total of four covalent bonds; this meets the bonding requirement.
b. A single bond and two double bonds are equivalent to five covalent bonds; this does not meet the bonding requirement.
c. Three single bonds and a triple bond (equivalent to three single bonds) are six covalent bonds; this does not meet the bonding requirement.
d. A double bond and a triple bond are equivalent to five covalent bonds; this does not meet the bonding requirement.

12.7 A hydrocarbon contains only the elements carbon and hydrogen; a hydrocarbon derivative contains at least one additional element besides carbon and hydrogen.

12.9 All bonds are single bonds in a saturated hydrocarbon; at least one carbon–carbon multiple bond is present in an unsaturated hydrocarbon.

12.11 In a saturated hydrocarbon all carbon–carbon bonds are single bonds; in an unsaturated hydrocarbon carbon-carbon multiple bonds are present.
a. Saturated. All of the carbon bonds are single bonds.
b. Unsaturated. The molecule contains a double bond.
c. Unsaturated. The molecule contains a double bond.
d. Unsaturated. The molecule contains a triple bond.

12.13 The general formula for alkanes is C_nH_{2n+2}, where n is the number of carbon atoms present.
a. C_8H_{18} contains 18 hydrogen atoms ($n = 8$; $2n + 2 = 18$)
b. C_4H_{10} contains 4 carbon atoms.
c. $n + (2n + 2) = 41$; $n = 13$. The formula for the alkane contains 13 carbon atoms. It is $C_{13}H_{28}$.
d. The total number of covalent bonds for C_7H_{16} is 22 (16 carbon–hydrogen covalent bonds and 6 carbon–carbon covalent single bonds).

12.15 A condensed structural formula uses groupings of atoms, in which central atoms and the atoms connected to them are written as a group.

a. $CH_3-CH_2-CH_2-CH_3$ b. $CH_3-CH_2-CH_2-CH_2-CH_2-CH_2-CH_3$

12.17 A condensed structural formula uses groupings of atoms, in which central atoms and the atoms connected to them are written as a group.

a. $CH_3-(CH_2)_2-CH_3$ b. $CH_3-(CH_2)_5-CH_3$

12.19 A skeletal structural formula shows the arrangement and bonding of carbon atoms present in an organic molecule, but does not show the hydrogen atoms bonded to the carbon atoms.

a. C–C–C–C b. C–C–C–C–C–C–C

12.21 a.

$$\begin{array}{ccccccccccc} & & H & & H & & H & & H & & H & & \\ & & | & & | & & | & & | & & | & & \\ H & - & C & - & C & - & C & - & C & - & C & - & H \\ & & | & & | & & | & & | & & | & & \\ & & H & & H & & H & & H & & H & & \end{array}$$

The molecular formula shows that the compound has 5 carbon atoms and 12 hydrogen atoms. Connect carbon atoms by single bonds, and attach hydrogen atoms to fulfill each carbon atom's bonding requirements.

b.

$$\begin{array}{ccccccccccccccccc} & & H & & H & & H & & H & & H & & H & & H & & H & & \\ & & | & & | & & | & & | & & | & & | & & | & & | & & \\ H & - & C & - & C & - & C & - & C & - & C & - & C & - & C & - & C & - & H \\ & & | & & | & & | & & | & & | & & | & & | & & | & & \\ & & H & & H & & H & & H & & H & & H & & H & & H & & \end{array}$$

Expand the condensed structural formula so that all covalent bonds (C–C and C–H) are shown.

c. $CH_3 \text{—}(CH_2)_8\text{—} CH_3$

The compound has 10 carbon atoms and 22 hydrogen atoms. The formula contains 10 carbon-centered groups. The groups on both ends contain 3 hydrogen atoms; the rest contain two.

d. C_6H_{14}

Count the number of carbon atoms and hydrogen atoms in the given condensed structural formula and write a molecular formula (no bonds are shown).

12.23 The skeletal formula of an alkane shows the bonds between carbon atoms, but not the bonds between carbon and hydrogen atoms. Because each carbon atom is bonded to four other atoms (either carbon or hydrogen) a carbon atom shown with one bond to another carbon atom is also bonded to three hydrogen atoms, a carbon with two bonds to two carbon atoms is also bonded to two hydrogen atoms.

a. There are 14 hydrogen atoms present.
b. There are 5 carbon-carbon bonds.
c. There are 4 CH_2 groups.
d. The total number of covalent bonds in the molecule is 19 (5 carbon-carbon bonds and 14 carbon-hydrogen bonds).

12.25 a. The last carbon in the carbon chain should have three hydrogen atoms.
b. The third carbon in the carbon chain should have only one hydrogen atom.

12.27 a. the same; constitutional isomers have the same molecular formula
b. different; constitutional isomers have the same molecular formula but different structural formulas
c. different; constitutional isomers have different physical properties
d. different; constitutional isomers have different physical properties

12.29 Constitutional isomers must have the same molecular formula.

12.31 a. 2 b. 5 c. 18 d. 75

12.33 There is one; for any set of constitutional isomers there is only one continuous chain isomer.

12.35 Isomers have the same molecular formulas. Constitutional isomers differ in the order in which atoms are connected to each other within molecules. A conformation is the specific three-dimensional arrangement of atoms in an organic molecule that results from rotation about carbon-carbon single bonds.
a. The two compounds have different molecular formulas; they are not constitutional isomers.
b. The two compounds are different compounds that are constitutional isomers.
c. The two structural formulas show different conformations of the same molecule.
d. The two compounds are different compounds that are constitutional isomers.

12.37 a. $CH_3-CH_2-CH-CH_2-CH_3$ (with CH_3 on the middle CH)

b. $CH_3-CH-CH_2-CH-CH_3$ (with CH_3 on each CH)

c. $CH_3-CH-CH_3$ (with CH_3 on the CH)

d. $CH_3-CH_2-CH-CH_2-CH_3$ (with CH_2-CH_3 on the CH)

12.39 A condensed structural formula without parentheses conveys molecular structural information in a less abbreviated form.

a. $CH_3-CH_2-CH_2-CH_2-CH_3$

b. $CH_3-CH-CH_2-CH_3$ (with CH_3 on the CH)

c. CH_3-C-CH_3 (with CH_3 above and CH_3 below the C)

d. $CH_3-CH-CH-CH_3$ (with CH_3 on each CH)

12.41 The IUPAC prefix associated with
a. two carbons is eth–
b. four carbons is but–
c. six carbons is hex–
d. eight carbons is oct–

12.43 The IUPAC name for
a. $-CH_2-CH_3$ is ethyl.
b. $-CH_2-CH_2-CH_2-CH_2-CH_3$ is pentyl.

12.45 The chemical formula for
a. ethane is C_2H_6
b. the ethyl group is C_2H_5
c. pentane is C_5H_{12}
d. the pentyl group C_5H_{11}

12.47 The longest continuous carbon chain (the parent chain) may or may not be shown in a straight line. In the given skeletal carbon arrangements, the longest continuous carbon chain contains:
a. 7 carbon atoms
b. 8 carbon atoms
c. 8 carbon atoms
d. 7 carbon atoms

12.49 a. 3-methylpentane
b. 2-methylhexane
c. 2-methylhexane
d. 2,4-dimethylhexane

12.51 a. 2,3,5-trimethylhexane
b. 2,2,4-trimethylpentane
c. 3-ethyl-3-methylpentane
d. 3-ethyl-3-methylhexane

12.53 In the condensed structural formula for an alkane, a central carbon atom and the hydrogen atoms connected to it are written as a group. In the problems below, draw the parent chain, choose one end to number from, attach the alkyl groups to the correct carbons, and complete each carbon-centered group by adding enough hydrogen atoms to give four bonds to each carbon atom.

```
a. CH3-CH2-CH—CH—CH2-CH3
           |   |
          CH3 CH3

          CH3
           |
b. CH3-CH2-C—CH2-CH3
           |
          CH2
           |
          CH3

c. CH3-CH2-CH—CH2-CH—CH2-CH2-CH3
           |       |
          CH2     CH2
           |       |
          CH3     CH3

d. CH3-CH2-CH2-CH—CH2-CH2-CH2-CH2-CH3
               |
              CH2
               |
              CH2
               |
              CH3
```

12.55 The only substituents on these alkanes are alkyl groups.
a. There are two alkyl groups (two methyl groups); there are two substituents.
b. There are two alkyl groups (one ethyl and one methyl); there are two substituents.
c. There are two alkyl groups (two ethyl groups); there are two substituents.
d. There is one alkyl group (one propyl group); there is one substituent.

12.57 a. The name is not based on the longest carbon chain; the correct name is 2,2-dimethylbutane.
b. The carbon chain is numbered from the wrong end; the correct name is 2,2,3-trimethylbutane.
c. The carbon chain is numbered from the wrong end, and the alkyl groups are not listed alphabetically; the correct name is 3-ethyl-4-methylhexane
d. Like alkyl groups are listed separately; the correct name is 2,4-dimethylhexane.

12.59 a. No. Butane and pentane do not have the same molecular formula.
b. Yes. The two compounds, 2-methylhexane and 3-methylhexane, have the same molecular formula, but a different connectivity of atoms.
c. Yes. Butane and 2-methylpropane have the same molecular formula, but a different connectivity of atoms.
d. Yes. 2-Methylhexane and 3,3-dimethylpentane have the same molecular formula, but a different connectivity of atoms.

12.61 The base-chain name for an alkane is the name of the longest continuous carbon chain.

a. There is only one unbranched C_8 alkane, so only one C_8 alkane is named as an octane.
b. Three constitutional isomers of the C_8 alkane have a C_7 parent chain and are named as heptanes. These three isomers each have a methyl substituent.
c. Seven constitutional isomers of the C_8 alkane have a C_6 (hexane) chain; there are five isomers with two methyl substituents each and two isomers with one ethyl substituent.
d. Six constitutional isomers of the C_8 alkane have a C_5 (pentane) chain with methyl and ethyl substituents.

12.63 In a line-angle structural formula, a carbon atom is present at every point where two lines meet and at the ends of the lines. In a skeletal structural formula, carbon atoms are shown but hydrogen atoms are not.

```
a.  C—C—C—C—C—C—C—C          b.  C—C—C—C—C—C
      |                                |  |
      C                                C  C

c.  C—C—C—C—C                d.  C—C—C—C—C—C—C—C
        |                            |        |
        C                            C        C—C
                                              |
                                              C
```

12.65 In a line-angle structural formula, a carbon atom is present at every point where two lines meet and at the ends of the lines. In a condensed structural formula, bonds between carbon atoms are shown, and hydrogen atoms are shown as part of a carbon-centered group.

```
a.  CH3—CH—CH—CH—CH3
        |   |   |
       CH3 CH3 CH3

b.  CH3—CH—CH—CH2—CH2—CH3
        |   |
       CH3 CH2
            |
           CH3

c.  CH3-CH2-CH—CH-CH2-CH2-CH3
            |   |
           CH3 CH2
                |
               CH3

d.  CH3—CH—CH2-CH2-CH—CH2—CH2—CH3
        |           |
       CH3         CH2
                    |
                   CH3
```

12.67 Name the compounds being compared (from their line-angle structural formulas), and determine their molecular formulas. If the name of the two line-angle structural formulas is the same, the compounds are the same. If the names are different, but the two compounds have the same molecular formula, they are constitutional isomers. If neither the names nor the molecular formulas are the same, they are two different compounds that are not constitutional isomers.

a. The two structures are constitutional isomers. They have the same molecular formula (C_6H_{14}) and different names (2-methylpentane and 2,4-dimethylbutane).
b. The two structures are the same compound (C_7H_{16}; 2,3-dimethylpentane).

12.69 In a line-angle structural formula, a carbon atom is represented by every point where two lines meet and by the end of each line.

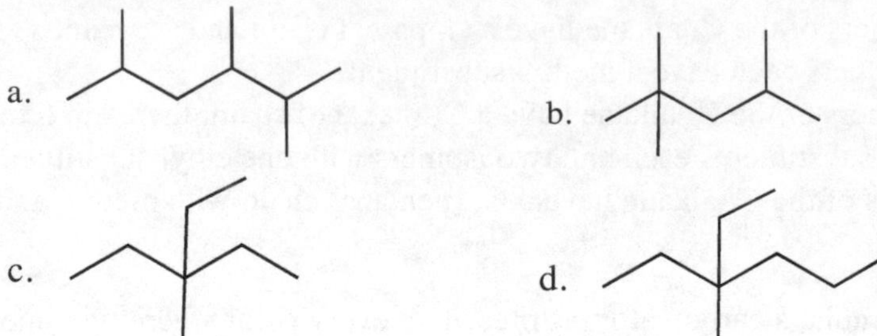

12.71 a. 2-methyloctane. Number the carbon atoms in the parent chain from the end nearest the alkyl group.
b. 2,3-dimethylhexane. Two or more alkyl groups of the same kind are combined and a prefix indicates how many there are. Their positions are indicated together, with a comma separating them.
c. 3-methylpentane. Number the carbon atoms in the parent chain. In this compound, numbering from either end gives the same number for the alkyl group.
d. 5-isopropyl-2-methyloctane. The longest carbon chain has eight carbons. Number the chain from the end that gives the alkyl group nearest the end the lowest possible number. Give the names of the substituents in alphabetical order.

12.73 In a line-angle structural formula, a carbon atom is present at every point where two lines meet and at the ends of the lines. The molecular formula for an alkane is C_nH_{2n+2}, where n is the number of carbon atoms present.

a. C_8H_{18} b. C_9H_{20} c. $C_{10}H_{22}$ d. $C_{11}H_{24}$

12.75 A primary carbon atom is bonded to only one other carbon atom, a secondary carbon atom to two other carbon atoms, a tertiary carbon atom to three other carbon atoms, and a quaternary carbon atom to four other carbon atoms.

a. primary – 5, secondary – 1, tertiary – 3, quaternary – 0.
b. primary – 5, secondary – 1, tertiary – 1, quaternary – 1.
c. primary – 4, secondary – 3, tertiary – 0, quaternary – 1.
d. primary – 4, secondary – 4, tertiary – 0, quaternary – 1.

12.77 A secondary carbon atom is bonded to two other carbon atoms, a tertiary carbon atom is bonded to three other carbon atoms.

a. Yes, both secondary and tertiary carbon atoms are present in the first structure. No, only tertiary carbon atoms are present in the second structure.
b. Yes, both secondary and tertiary carbon atoms are present in the first structure. Yes, both secondary and tertiary carbon atoms are present in the second structure.

12.79 A primary carbon atom is bonded to only one other carbon atom, a secondary carbon atom to two other carbon atoms, a tertiary carbon atom to three other carbon atoms, and a quaternary carbon atom to four other carbon atoms.

a. A C_7 alkane (molecular formula C_7H_{16}) with only 1° and 2° carbon atoms has no branching; it is a straight chain heptane.

$$CH_3-CH_2-CH_2-CH_2-CH_2-CH_2-CH_3$$

b. A C_7 alkane (molecular formula C_7H_{16}) with a 3° and a 4° carbon has one carbon atom with three carbon-carbon bonds and one carbon with 4 carbon-carbon bonds. The only possibility is shown below.

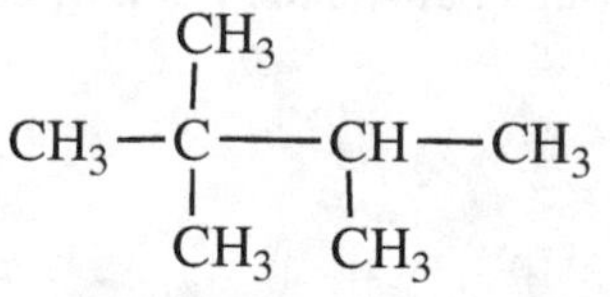

```
       CH3
       |
  CH3—C——CH—CH3
       |    |
      CH3  CH3
```

12.81 Figure 12.5 gives the IUPAC names of the four most common branched-chain alkyl groups.

a. isopropyl b. isobutyl c. isopropyl d. *sec*-butyl

12.83 Find the longest continuous carbon chain, and identify the names and the positions of the alkyl groups attached to the chain. (Figure 12.5 gives the IUPAC names and structures of the four most common branched-chain alkyl groups).

Part d. contains a complex branched alkyl group that has been named in a somewhat different manner. Parentheses are used to set off the name of the complex alkyl group, and it has been numbered, beginning with the carbon atom attached to the main carbon chain. The substituents (two methyl groups) on the base alkyl group (ethyl) are listed with appropriate numbers.

```
a. CH3-CH2-CH2-CH2-CH—CH2-CH2-CH2-CH2-CH3
                   |
                   CH—CH3
                   |
                   CH2
                   |
                   CH3
```

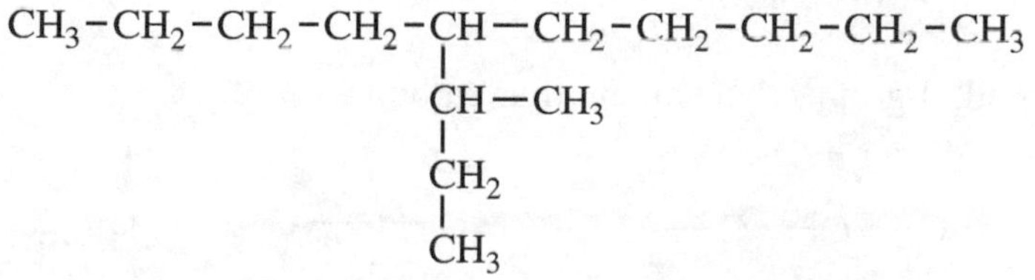

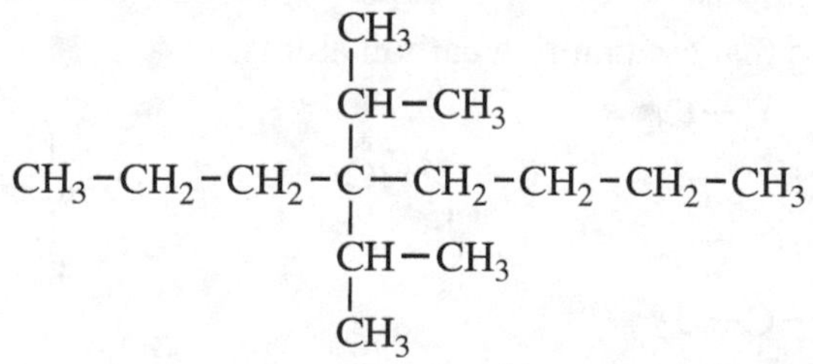

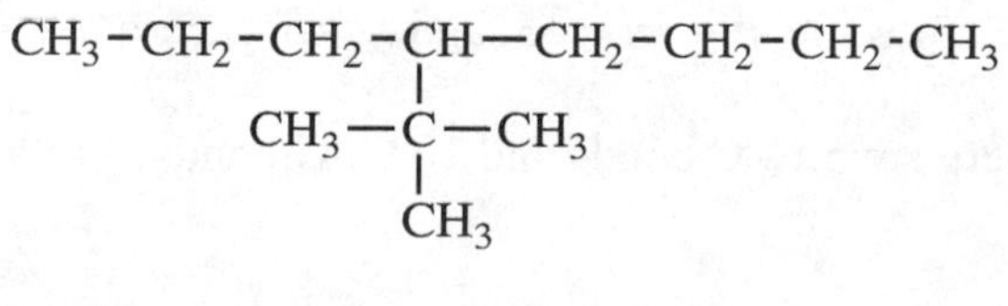

```
               CH3
               |
               CH-CH3
               |
b. CH3-CH2-CH2-C—CH2-CH2-CH2-CH3
               |
               CH-CH3
               |
               CH3
```

```
c. CH3-CH—CH—CH2-CH—CH2-CH2-CH2-CH3
       |   |      |
      CH3 CH3     CH2
                  |
                  CH—CH3
                  |
                  CH3
```

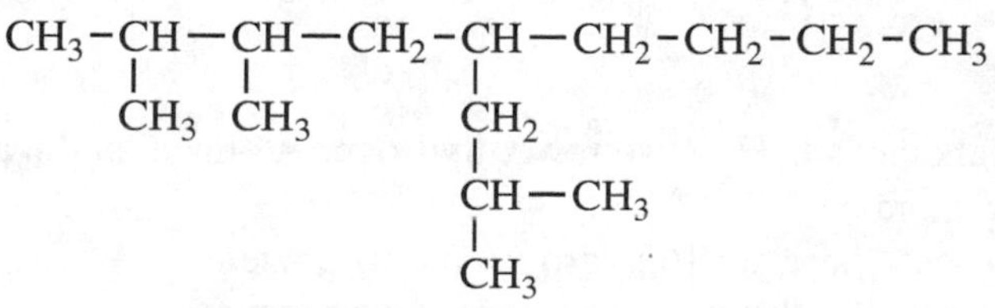

```
d. CH3-CH2-CH2-CH—CH2-CH2-CH2-CH3
               |
          CH3—C—CH3
               |
               CH3
```

12.85 a. carbons 3 or 4; placing the ethyl group on any other carbon lengthens the carbon chain
b. carbons 3 or 4; placing the isopropyl group on any other carbon lengthens the carbon chain
c. none of them; placing the isobutyl group on any carbon atom lengthens the carbon chain
d. carbons 3 or 4; placing the *tert*-butyl group on any other carbon lengthens the carbon chain

12.87 Numbering of the attached carbon chain begins with the carbon atom that is attached to the parent carbon chain.
a. (2-methylbutyl) group
b. (1,1-dimethylpropyl) group

12.89 In the skeletal structures for alkyl groups below, the point of attachment to the parent chain is shown with a heavier line.

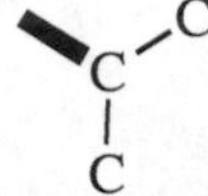

a. An isopropyl group can also be called 1-methylethyl.

b. A *tert*-butyl group can also be called 1,1-dimethylethyl.

c. A 1-methylpropyl group can also be called *sec*-butyl.

d. A 2-methylpropyl group can also be called isobutyl.

12.91 In the skeletal structures for alkyl groups below, the point of attachment to the parent chain is shown with a heavier line.

a. There are four different alkyl groups that contain four carbon atoms.

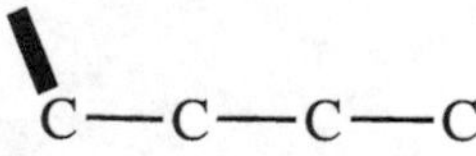

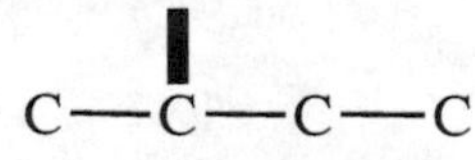

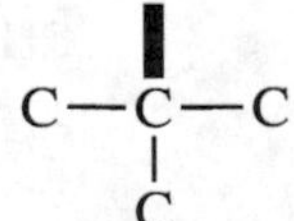

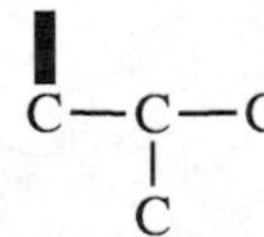

b. There are eight different alkyl groups that contain five carbon atoms.

12.93 The general formula for a cycloalkane is C_nH_{2n} (two fewer hydrogen atoms than the alkane with the same number of carbon atoms).
a. When 8 carbon atoms are present, there are 16 hydrogen atoms present.
b. When 12 hydrogen atoms are present, there are 6 carbon atoms present.
c. When there are fifteen atoms in the molecule, $n + 2n = 15$, and $n = 5$ (5 carbon atoms present).
d. Since this is a cyclic compound, there are 5 C–C bonds and 10 C–H bonds, a total of 15 covalent bonds.

12.95 The molecular formula of a compound tells how many atoms of each kind are in a molecule of the compound. The molecular formulas for the cycloalkane molecules are:
a. C_4H_8 b. C_8H_{16} c. C_9H_{18} d. C_8H_{16}

12.97 a. There are zero alkyl groups present in this cycloalkane.
b. There are two alkyl groups (two methyl groups) present in this cycloalkane.
c. There is one alkyl group (isopropyl) present in this cycloalkane.
d. There are three alkyl groups (three methyl groups) present in this cycloalkane.

12.99 A secondary carbon atom is bonded to two other carbon atoms.
a. There are four secondary carbon atoms present.
b. There are five secondary carbon atoms present.
c. There are five secondary carbon atoms present.
d. There are two secondary carbon atoms present.

12.101 The IUPAC names for the cycloalkanes are
a. cyclopentane b. ethylcyclopentane
c. 1,2-dimethylcyclopentane d. 1-isopropyl-2,4-dimethylcyclopentane

12.103 a. When there are two methyl groups, you must locate methyl groups with numbers.
b. This is the wrong numbering system for a ring; when there are two substituents, the first is given the number 1.
c. No number is needed; if there is just one ring substituent, it is not necessary to locate it by number.
d. This is the wrong numbering system for alkyl groups on a ring; the alkyl groups should be numbered alphabetically (ethyl should be 1, methyl should be 2).

12.105 In a line-angle structural formula, a carbon atom is present at every point where two lines meet and at the ends of lines. Two or more substituents on a ring are numbered in relation to the first substituent.

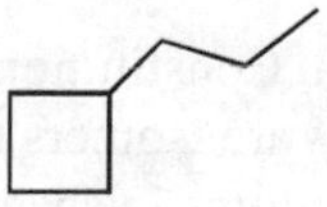

a. propylcyclobutane 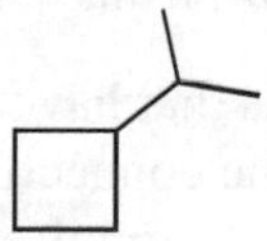 b. isopropylcyclobutane

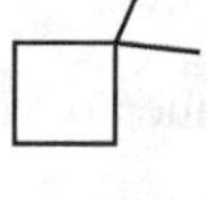

c. 1,1-dimethylcyclobutane 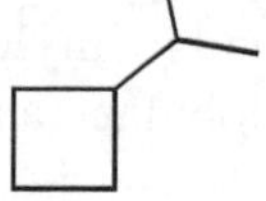 d. (1-methylethyl)cyclobutane

12.107 The molecular formula of a compound tells how many atoms of each kind are in a molecule of the compound. The molecular formulas of the compounds are:
a. C_8H_{16} b. C_8H_{16} c. C_8H_{18} d. C_8H_{18}

12.109 A saturated hydrocarbon is hydrocarbon in which all carbon-carbon bonds are single bonds.
a. saturated b. saturated c. saturated d. saturated

12.111 a. two (cyclobutane, methylcyclopropane)
b. three (1,1-dimethylcyclopropane, 1,2-dimethylcyclopropane, ethylcyclopropane
c. one (methylcyclopentane)
d. four (1,1-dimethylcyclopentane, 1,2-dimethylcyclopentane, 1,3-dimethylcyclopentane, ethylcyclopentane)

12.113 *Cis-trans* isomerism can exist for disubstituted cycloalkanes. *Cis-trans* isomers have the same molecular and structural formulas, but different arrangement of atoms in space because of restricted rotation about bonds.

a. *Cis-trans* isomerism is not possible for isopropylcyclobutane because it has only one substituent on the ring.

b. H_3C-CH_2 CH_2-CH_3 H H

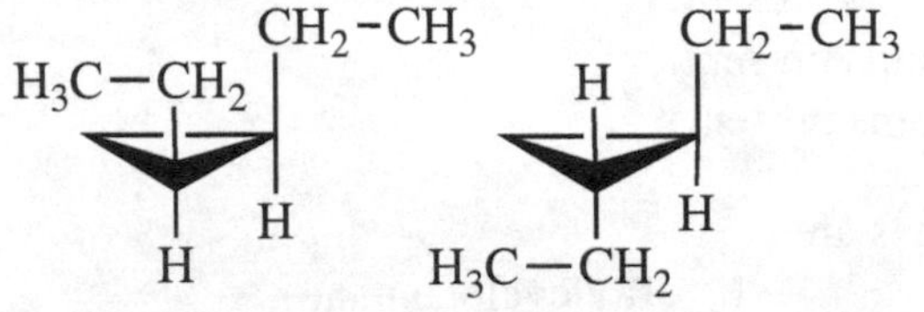

cis

CH_2-CH_3 H H H_3C-CH_2

trans

c. *Cis-trans* isomerism is not possible because both substituents are attached to the same carbon atom in the ring.

d.

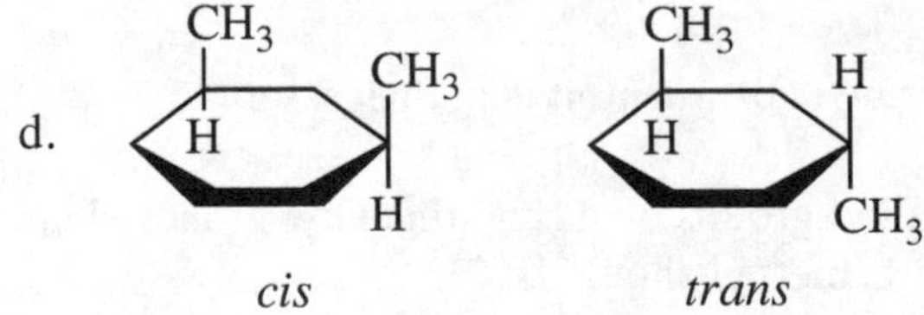

cis *trans*

12.115 In cyclic compounds, two alkyl groups in the *cis*- position are on the same side of the ring; two in the *trans*- position are on opposite sides of the ring.

a. *cis*- b. *trans*- c. *trans*- d. *cis*-

12.117 Isomers are compounds that have the same molecular formula. Constitutional isomers are isomers that differ in the connectivity of atoms. Stereoisomers are isomers that have the same molecular and structural formulas but different orientation of atoms in space.

a. Hexane and cyclohexane are not isomers.
b. Hexane and methylcyclopentane are not isomers.
c. Methylcyclobutane and methylcyclopentane are not isomers.
d. *Cis*-1,2-dimethylcyclobutane and *trans*-1,2-dimethylcyclobutane are stereoisomers.

12.119 50–90% methane, 1–10% ethane, up to 8% propane and butanes

12.121 Hydrocarbons can be separated by fractional distillation because of their differing boiling points.

12.123 a. Octane has the higher boiling point; the boiling point increases with an increase in carbon chain length.
b. Cyclopentane has a higher boiling point; boiling point increases with increase in ring size.
c. Pentane has a higher boiling point; branching on a carbon chain lowers the boiling point of an alkane.
d. Cyclopentane has a higher boiling point; cycloalkanes have higher boiling points than their noncyclic counterparts.

12.125 Figure 12.12 summarizes the physical-states of unbranched alkanes and unsubstituted cycloalkanes.

a. Different states. Ethane is a gas, and hexane is a liquid.
b. Same state. Cyclopropane and butane are both gases.
c. Same state. Octane and 3-methyloctane are both liquids.
d. Same state. Pentane and decane are both liquids.

12.127 The unbranched alkane that contains 6 carbon atoms is hexane. Facts about hexane and other unbranched alkanes can be found in Sections 12.16 and 12.17 of your textbook.

a. It is a liquid.
b. It is less dense than water.
c. It is insoluble in water.
d. It is flammable.

12.129 The complete combustion of alkanes and cycloalkanes produces carbon dioxide and water.

a. CO_2 and H_2O b. CO_2 and H_2O c. CO_2 and H_2O d. CO_2 and H_2O

12.131 Halogenation of an alkane usually results in the formation of a mixture of products because more than one hydrogen atom can be replaced with halogen atoms. In this case, each of the four hydrogen atoms in methane can be replaced by a bromine atom. The four products are:

$$CH_3Br, CH_2Br_2, CHBr_3, CBr_4$$

12.133 The structural formula of a monochlorinated alkane may depend on which hydrogen the chlorine is substituted for.

a. CH_3-CH_2
$\quad\quad\quad|$
$\quad\quad\;Cl$

Only one product is possible because all of the hydrogen atoms are equivalent.

b. $CH_2-CH_2-CH_2-CH_3$, $CH_3-CH-CH_2-CH_3$
$\;|\quad\quad\quad\quad\quad\quad\quad\quad\quad\quad\quad\;|$
$Cl\quad\quad\quad\quad\quad\quad\quad\quad\quad\quad\;Cl$

Two different kinds of hydrogen atoms are present and can be replaced by a chlorine atom, so there are two possible products.

c. $Cl-CH_2-CH-CH_3$, $CH_3-C(Cl)-CH_3$
$\quad\quad\quad\quad\quad|\quad\quad\quad\quad\quad\quad\quad|$
$\quad\quad\quad\quad CH_3\quad\quad\quad\quad\quad\;CH_3$

Two different kinds of hydrogen atoms are present and can be replaced by a chlorine atom, so there are two possible products.

d. —Cl

Only one product is possible, because all of the hydrogen atoms on the ring of a cycloalkane are equivalent.

12.135 The IUPAC names for the given compounds are:
a. iodomethane b. 1-chloropropane
c. 2-fluorobutane d. chlorocyclobutane

12.137 a. iodomethane is methyl iodide b. 1-chloropropane is propyl chloride
c. 2-fluorobutane is *sec*-butyl fluoride d. chlorocyclobutane is cyclobutyl chloride.

12.139 Structural formulas for halogenated alkanes can be written in the same way as those for alkyl-substitued alkanes. The halogen atom takes the place of a hydrogen atom. Remember, a halogen atom forms one bond, and each carbon participates in a total of four bonds.

```
       Cl                F   F         c. CH3−CH−Br
       |                 |   |                |
a.  H−C−Cl        b. Cl−C−C−Cl               CH3
       |                 |   |
       Cl                F   F
```

d. 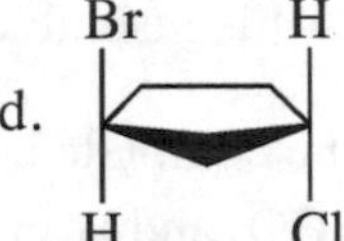

12.141 The general formula for alkanes is C_nH_{2n+2}; the general formula for cycloalkanes is C_nH_{2n}. Add one hydrogen atom for each halogen atom in the formula of the halogenated hydrocarbon to obtain the formula for the corresponding alkane or cycloalkane.

a. $C_4H_8Br_2$ corresponds to C_4H_{10}, an alkane.
b. $C_5H_9Cl_3$ corresponds to C_5H_{12}, an alkane.
c. $C_6H_{11}F$ corresponds to C_6H_{12}, a cycloalkane.
d. $C_6H_{12}F_2$ corresponds to C_6H_{14}, an alkane.

12.143 For similar compounds, the one with the greater mass and/or polarity usually has the higher boiling point.

a. CH_3I has a greater mass than CH_3Br, and so would be expected to have a higher boiling point.
b. CH_3CH_2Cl has a greater mass than CH_3Cl, and so would be expected to have a higher boiling point.
c. CH_3CH_2I has a greater mass than CH_3Br, and so would be expected to have a higher boiling point.
d. CH_3Br has a greater mass than CH_4, and so would be expected to have a higher boiling point.

12.145 a. The name is incorrect. CH_3Cl is methyl chloride, not chloroform.
b. The name is incorrect. CCl_4 is tetrachloromethane or carbon tetrachloride, not chloromethane.
c. The name is incorrect. $CHCl_3$ is trichloromethane or chloroform, not methylene chloride.
d. The name is correct. CH_2Cl_2 is dichloromethane or methylene chloride.

12.147 The IUPAC names for the eight isomeric halogenated hydrocarbons having the molecular formula $C_5H_{11}Cl$ are: 1-chloropentane, 2-chloropentane, 3-chloropentane, 1-chloro-2-methylbutane, 2-chloro-2-methylbutane, 2-chloro-3-methylbutane, 1-chloro-3-methylbutane, and 1-chloro-2,2-trimethylpropane

Unsaturated Hydrocarbons — Chapter 13

Solutions to Selected Problems

13.1 All bonds in saturated hydrocarbons are single bonds. An unsaturated hydrocarbon is a hydrocarbon in which one or more carbon-carbon multiple bonds (double bonds, triple bonds, or both) are present.

a. The hydrocarbon is saturated.
b. The hydrocarbon is unsaturated, one carbon-to-carbon double bond.
c. The hydrocarbon is unsaturated, one carbon-to-carbon triple bond.
d. The hydrocarbon is unsaturated, one carbon-to-carbon double bond.

13.3 An alkene has a carbon-to-carbon double bond; an alkyne has a carbon-to-carbon triple bond.

a. an alkane b. an alkene c. an alkyne d. an alkene

13.5 An alkene has a carbon-to-carbon double bond.

13.7 The physical properties of saturated and unsaturated hydrocarbons are similar.

13.9 The molecular formula of a compound tells how many atoms of each kind are in a molecule of the compound.

a. The molecular formula for a 4-carbon alkene with one double bond is C_4H_8.
b. The molecular formula for a 4-carbon alkene with two double bonds is C_4H_6.
c. The molecular formula for a 4-carbon cycloalkene with one double bond is C_4H_6.
d. The molecular formula for a 4-carbon cycloalkene with two double bonds is C_4H_4.

13.11 a. C_6H_{12} is an alkene with one double bond.
b. C_5H_{10} is an alkene with one double bond.
c. C_6H_{10} is a diene (with two double bonds).
d. C_5H_8 is a diene (with two double bonds).

13.13 a. There is nothing wrong with the condensed structural formula.
b. The third and fourth carbon atoms have one too many hydrogen atoms.

13.15 The spatial arrangement for bonds about a carbon atom that participates in one carbon-carbon double bond is trigonal planar.

13.17 To name alkenes, choose the longest continuous chain of carbon atoms containing the double bond(s) as the parent chain, number the chain from the end that will give the lowest number to the double bond, and use the suffix *–ene*. Multiple double bonds are denoted by the prefixes *di-* and *tri-*.

a. 2-butene b. 2,4-dimethyl-2-pentene
c. 3-methylcyclohexene d. 1,3-cyclopentadiene

13.19 Condensed structural formulas for unsaturated hydrocarbons are drawn in the same way as those for saturated hydrocarbons. The number denoting the double bond is assigned to the first carbon atom of the double bond.

a. $H_2C{=}CH{-}CH{-}CH_2{-}CH_3$
$\qquad\qquad\quad |$
$\qquad\qquad\; CH_3$

b. 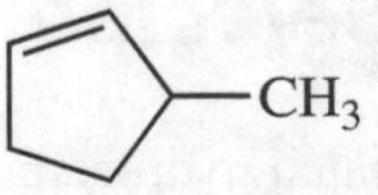 $-CH_3$

c. $H_2C{=}CH{-}CH{=}CH_2$

d. $H_2C{=}CH{-}CH{-}CH{=}CH_2$
$\qquad\qquad\quad |$
$\qquad\qquad\; CH_2$
$\qquad\qquad\quad |$
$\qquad\qquad\; CH_3$

13.21 a. The correct name is 3-methyl-3-hexene; the parent carbon chain is the longest continuous chain of carbon atoms that contains both carbon atoms of the double bond.
b. The correct name is 2,3-dimethyl-2-hexene; the parent carbon chain is numbered at the end nearest the double bond.
c. The correct name is 1,3-cyclopentadiene; in cycloalkenes with more than one double bond within the ring, one double bond is assigned the numbers 1 and 2 and the other double bonds the lowest possible number.
d. The correct name is 4,5-dimethylcyclohexene; in substituted cycloalkenes with only one double bond, no number is needed to locate the double bond.

13.23 a. $H_2C{=}CH_2$

b. 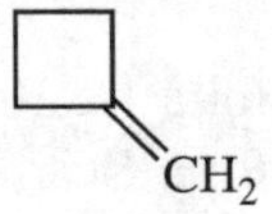 CH_2

c. $H_2C{=}CH{-}Br$

d. $H_2C{=}CH{-}CH_2{-}I$

13.25 An unsaturated hydrocarbon contains one or more carbon-carbon multiple bonds. Alkenes and cycloalkenes contain at least one double bond and have a name ending in *–ene*.
a. Ethylcyclopentane is saturated; it is a cycloalkane.
b. Ethylcyclopentene is unsaturated; the ring contains a double bond.
c. 1,3-Butadiene is unsaturated. It contains two double bonds.
d. 2-Methyl-2-pentene is unsaturated. It contains one double bond.

13.27 a. The general molecular formula for a cycloalkane is C_nH_{2n}. Ethylcyclopentane has the molecular formula C_7H_{14}; it has 14 hydrogen atoms.
b. A cycloalkene has the general formula C_nH_{2n-2}. Ethylcyclopentene has the molecular formula C_7H_{12}; it has 12 hydrogen atoms.
c. A diene has the general formula C_nH_{2n-2}. 1,3-Butadiene has the molecular formula C_4H_6; it has 6 hydrogen atoms.
d. An alkene has the general formula C_nH_{2n}. 2-Methyl-2-pentene has the molecular formula C_6H_{12}; it has 12 hydrogen atoms.

13.29 a.

b.

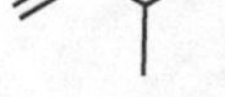

c.

d.

13.31 In line-angle structural formulas, a line represents a carbon-carbon bond and a carbon atom is understood to be present at every point where two lines meet and at the ends of lines.

a.

b.

c.

d. 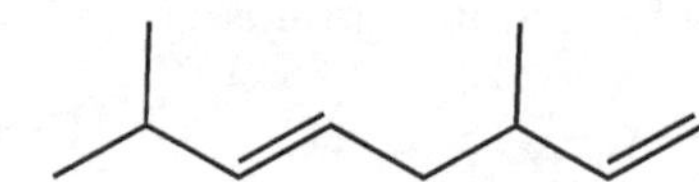

Each of the molecules represented by structures a., b., and c. contains 8 carbon atoms. The molecule represented by structure d. contains 10 carbon atoms.

13.33 The molecular formula formula for an alkene is C_nH_{2n}. For a diene, the molecular formula is C_nH_{2n-2}.

a. The molecular formula is C_8H_{16}.
b. The molecular formula is C_8H_{16}.
c. The molecular formula is C_8H_{14}.
d. The molecular formula is $C_{10}H_{18}$.

13.35 To name alkenes, choose the longest continuous chain of carbon atoms containing the double bond(s) as the parent chain, number the chain from the end that will give the lowest number to the double bond, and use the suffix *–ene*. Multiple double bonds are denoted by the prefixes *di-* and *tri-*.

a. The compound is named 3-octene.
b. The compound is named 3-octene.
c. The compound is named 1,3-octadiene.
d. The compound is named 3,7-dimethyl-1,5-octadiene.

13.37 a. The left-most carbon has four single bonds; it is tetrahedral.
b. The left-most carbon has four single bonds; it is tetrahedral.
c. The left-most carbon has two single bonds and one double bond; it is trigonal planar.
d. The left-most carbon has four single bonds; it is tetrahedral.

13.39 a. Positional; the carbon skeletons are identical with the difference being the position of the double bond.
b. Skeletal; the carbon skeletons are different.
c. Skeletal; the carbon skeletons are different.
d. Positional; the carbon skeletons are identical with the difference being the position of the double bond.

13.41 a. two (1-pentene and 2-pentene)
b. four (1,2-pentadiene, 1,3-pentadiene, 1,4-pentadiene, and 2,3-pentadiene)
c. three (2-methyl-1-butene, 3-methyl-1-butene, and 2-methyl-2-butene)
d. zero (dimethylpropenes do not exist, as the middle carbon atom would have 5 bonds)

13.43 Constitutional isomers have the same molecular formula but differ in the order in which atoms are connected to each other. Alkenes have more isomers than alkanes because in addition to skeletal isomers (different carbon-chain and hydrogen atom arrangements), they can also form positional isomers (the same carbon-chain arrangement, but different locations of hydrogen atoms and functional groups).

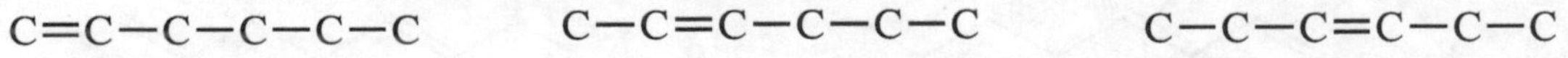

1-hexene 2-hexene 3-hexene

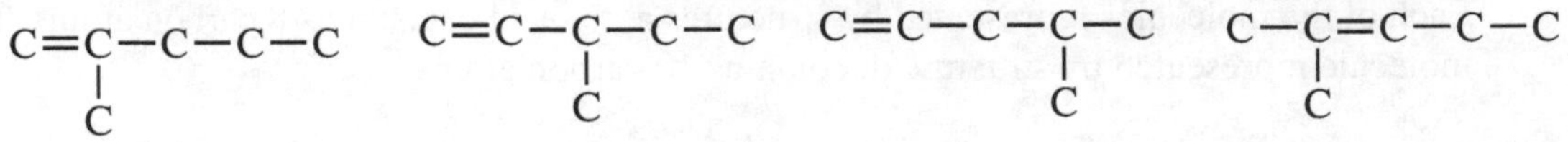

2-methyl-1-pentene 3-methyl-1-pentene 4-methyl-1-pentene 2-methyl-2-pentene

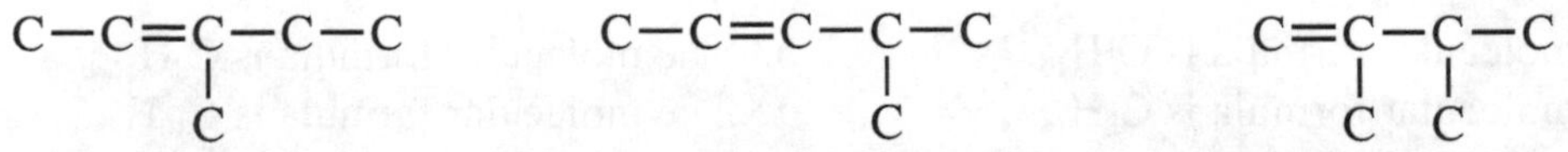

3-methyl-2-pentene 4-methyl-2-pentene 2,3-dimethyl-1-butene

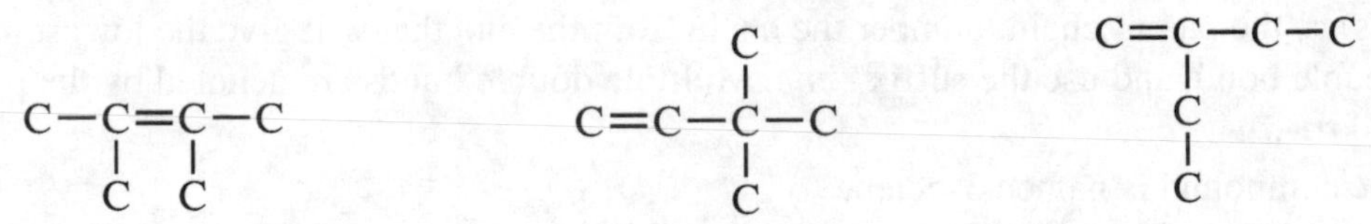

2,3-dimethyl-2-butene 3,3-dimethyl-1-butene 2-ethyl-1-butene

13.45 The general molecular formula for both alkenes and cycloalkanes is C_nH_{2n}. The skeletal structural formulas for the four alkenes or cycloalkanes with the formula C_4H_8 are shown below.

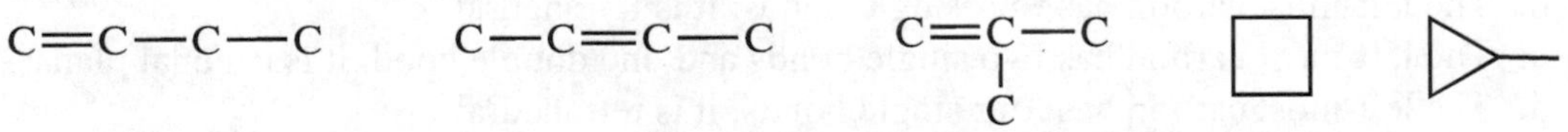

13.47 The molecules in parts a. and b. do not have *cis-trans* isomers. For *cis-trans* isomers to exist, each of the two carbons of the double bond must have two different groups attached to it.

The molecules in parts c. and d. do have *cis-trans* isomers, because each of the two carbons of each double bond has two different groups attached to it.

c.

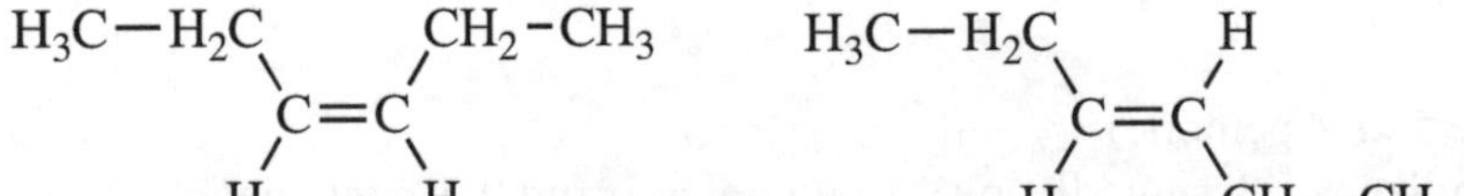

cis *trans*

d.

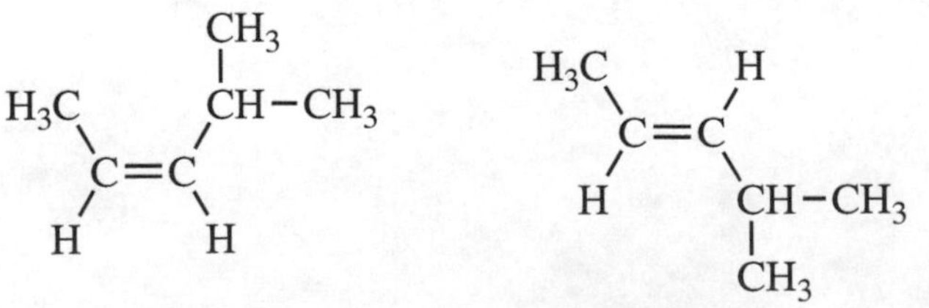

cis *trans*

13.49 When naming alkenes, choose the longest continuous chain of carbon atoms containing the double bond(s) as the parent chain, number the chain from the end that will give the lowest number to the double bond, and use the suffix *–ene*. If both carbon atoms of the double bond bear two different attachments, *cis-trans* isomerism is possible; use *cis-* (on the same side) or *trans* (across) as a prefix at the beginning of the name.

a. *cis*-2-pentene
b. *trans*-1-bromo-2-iodoethene
c. tetrafluoroethene
d. 2-methyl-2-butene

13.51 In a *cis*-isomer, both of the double bond substituents are on the same side of the double bond; in a *trans*-isomer, the two double bond substituents are on opposite sides of the double bond.

a. $H_3C{-}H_2C$ and H_3C on one carbon, H and $CH_2{-}CH_3$ on the other: $(H_3C{-}H_2C)(H_3C)C{=}C(H)(CH_2{-}CH_3)$

b. H_3C and H on one carbon, $CH_2{-}CH_3$ and H on the other: $(H_3C)(H)C{=}C(CH_2{-}CH_3)(H)$

c. H_3C and H on one carbon, H and $CH_2{-}CH(CH_3){-}CH_2{-}CH_3$ on the other: $(H_3C)(H)C{=}C(H)(CH_2{-}CH(CH_3){-}CH_2{-}CH_3)$

d. $H_2C{=}CH$ and H on one carbon, H and CH_3 on the other: $(H_2C{=}CH)(H)C{=}C(H)(CH_3)$

13.53 If each of the two carbons of the double bond has two different groups attached to it, *cis* and *trans* isomer exist. To determine whether an alkene has *cis* and *trans* isomers, draw the alkene structure in a way that emphasizes the four attachments to the double-bonded carbon atoms.

a. $CH_3{-}CH_2{-}CH_2{-}C(CH_3){=}CH_2$ No, *cis-trans* isomerism is not possible.

b. $CH_3{-}CH_2{-}CH_2{-}CH_2{-}CH{=}CH_2$ No, *cis-trans* isomerism is not possible.

c. Cyclohexane with H_3C and H on the same ring carbon. No, *cis-trans* isomerism is not possible.

d. Cyclopentane with $CH_3{-}CH_2$ and $CH_2{-}CH_3$ on adjacent ring carbons (cis, drawn in perspective); cyclopentane with $CH_2{-}CH_3$ above and $CH_2{-}CH_3$ below on adjacent ring carbons (trans). Yes, *cis-trans* isomerism is possible.

13.55 A pheromone is a chemical substance produced by a species that triggers a response in other members of the same species.

13.57 The condensed structural formula for isoprene is: $CH_2{=}C(CH_3){-}CH{=}CH_2$

13.59 The number of carbon atoms in a terpene is always a multiple of the number 5 because the structures of terpenes are formed from C_5 isoprene units.

13.61 a. True.
b. False. 2-Butene is soluble in nonpolar solvents.
c. True.
d. True.

13.63 Use Figure 13.7 in your textbook to answer this question.
a. Ethene is a gas.
b. 1-Pentene is a liquid.
c. Cyclobutane is a gas.
d. Cyclohexane is a liquid.

13.65 During a dehydrogenation reaction a hydrogen atom is lost from each of two adjacent carbon atoms, and a carbon-carbon double bond is formed.

13.67 When ethane is subjected to dehydrogenation, the two products are ethene and H_2.

13.69 An addition reaction is a reaction in which atoms or groups of atoms are added to each carbon atom of a carbon-carbon multiple bond.
a. Yes, this reaction is an addition reaction; a chlorine atom is added to each carbon atom of the double bond.
b. No, this reaction is not an addition reaction; one chlorine atom is substituted for one of the hydrocarbon's hydrogen atoms.
c. Yes, this is an addition reaction; a hydrogen atom and a chlorine atom are added, one to each of the two carbon atoms of the double bond.
d. No, this is not an addition reaction; four atoms of hydrogen are eliminated.

13.71 The addition of H_2 to a double bond requires the presence of a catalyst, Ni.

a. $CH_2{=}CH_2 + Cl_2 \longrightarrow \underset{\displaystyle Cl}{CH_2}-\underset{\displaystyle Cl}{CH_2}$

b. $CH_2{=}CH_2 + HCl \longrightarrow CH_3-\underset{\displaystyle Cl}{CH_2}$

c. $CH_2{=}CH_2 + H_2 \xrightarrow{Ni} CH_3-CH_3$

d. $CH_2{=}CH_2 + HBr \longrightarrow CH_3-\underset{\displaystyle Br}{CH_2}$

13.73 Markovnikov's rule states that when an unsymmetrical molecule of the form HQ adds to an unsymmetrical alkene, the hydrogen atom from the HQ becomes attached to the unsaturated carbon atom that already has the most hydrogen atoms.
a. Hydration of propene produces two products.
b. Hydration of 3-hexene produces one product.
c. Hydration of cyclopropene produces one product.
d. Hydration of cyclopentene produces one product.

13.75 Markovnikov's rule states that when an unsymmetrical molecule of the form HQ adds to an unsymmetrical alkene, the hydrogen atom from the HQ becomes attached to the unsaturated carbon atom that already has the most hydrogen atoms. apply this rule in parts b. and d.

a. 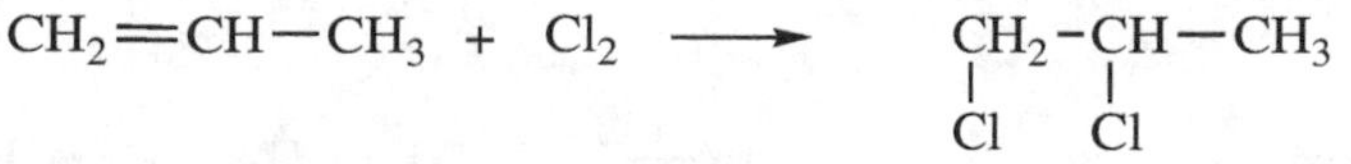 $CH_2{=}CH{-}CH_3 + Cl_2 \longrightarrow \underset{\mathrm{Cl}}{\underset{|}{CH_2}}{-}\underset{\mathrm{Cl}}{\underset{|}{CH}}{-}CH_3$

b. 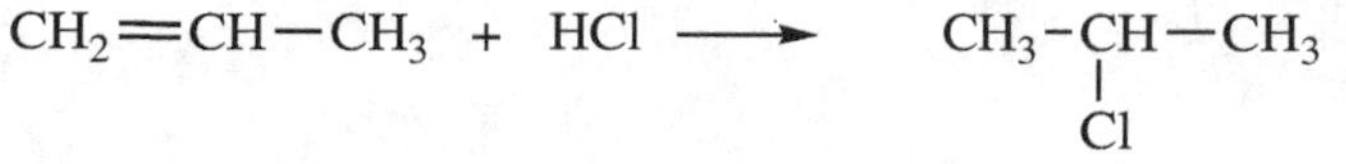 $CH_2{=}CH{-}CH_3 + HCl \longrightarrow CH_3{-}\underset{\mathrm{Cl}}{\underset{|}{CH}}{-}CH_3$

c. $CH_2{=}CH{-}CH_3 + H_2 \xrightarrow{\mathrm{Ni}} CH_3{-}CH_2{-}CH_3$

d. 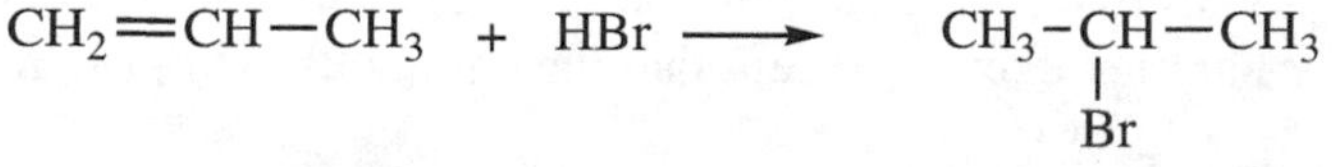 $CH_2{=}CH{-}CH_3 + HBr \longrightarrow CH_3{-}\underset{\mathrm{Br}}{\underset{|}{CH}}{-}CH_3$

13.77 In these reactions, the two atoms of the molecule being added are attached to the two carbons of the double bond. In part d., the molecule being added is H_2O; a hydrogen atom adds to one carbon atom and a hydroxyl group to the other. Apply Markovnikov's rule in part b.

a. 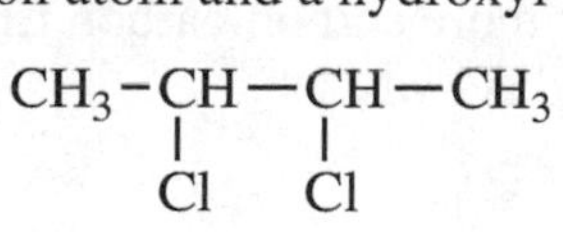 $CH_3{-}\underset{\mathrm{Cl}}{\underset{|}{CH}}{-}\underset{\mathrm{Cl}}{\underset{|}{CH}}{-}CH_3$

b. 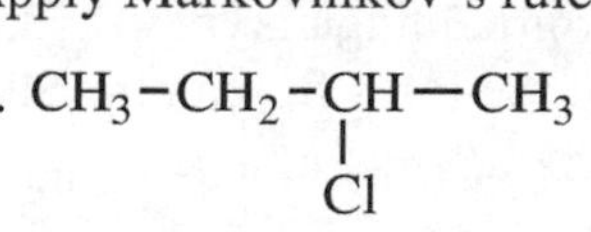 $CH_3{-}CH_2{-}\underset{\mathrm{Cl}}{\underset{|}{CH}}{-}CH_3$

c.

d. 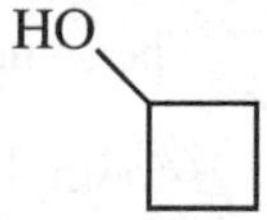 HO

13.79 All of the compounds can be prepared by addition reactions to cyclohexene. Analyze the compounds to see which atoms have been added to the double bond. Additions of H_2 and H_2O require the presence of a catalyst.

a. Br_2

b. H_2 + Ni catalyst

c. HCl

d. $H_2O + H_2SO_4$ catalyst

13.81 One molecule of H_2 gas will react with each double bond in the molecule.

a. Two molecules of H_2 gas react with the two double bonds per molecule.
b. Two molecules of H_2 gas react with the two double bonds per molecule.
c. Two molecules of H_2 gas react with the two double bonds per molecule.
d. Three molecules of H_2 gas react with the three double bonds per molecule.

13.83 The addition polymers are made by addition of monomers to themselves to form a long chain. The double bond of the monomer changes to a single bond in the polymer. Analyze each polymer to identify the unsaturated monomer that could form the polymer by addition to itself. The polymer in part b. has one double bond per unit, which means that the monomer must have two double bonds.

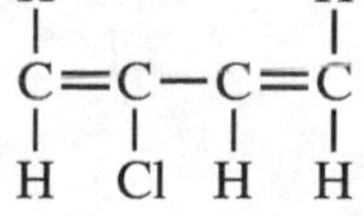

a. 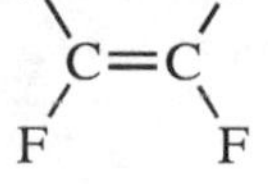 F F / C=C / F F

b. H H / C=C−C=C / H Cl H H

c. H H / C=C / H Cl

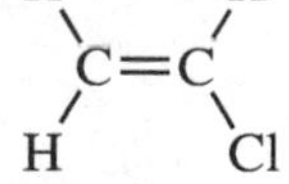

d. 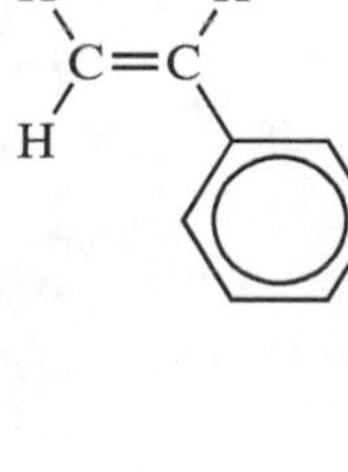 H H / C=C / H

13.85 In the formation of an addition polymer the double bond of the monomer changes to a single bond in the polymer as the monomer adds to itself.

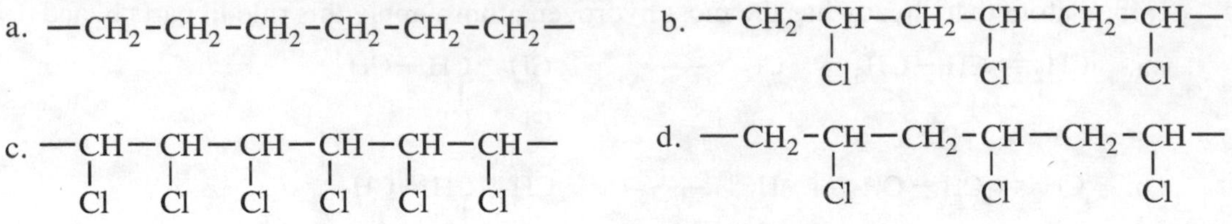

a. $-CH_2-CH_2-CH_2-CH_2-CH_2-CH_2-$

b. $-CH_2-CH(Cl)-CH_2-CH(Cl)-CH_2-CH(Cl)-$

c. $-CH(Cl)-CH(Cl)-CH(Cl)-CH(Cl)-CH(Cl)-CH(Cl)-$

d. $-CH_2-CH(Cl)-CH_2-CH(Cl)-CH_2-CH(Cl)-$

13.87 C_nH_{2n-6}; four hydrogens are lost for each triple bond, so $2n + 2 - 8 = 2n - 6$, where $2n + 2$ is the maximum hydrogen count possible (an alkane).

13.89 IUPAC rules for naming alkynes (hydrocarbons with one or more carbon–carbon triple bonds) are identical to the rules for naming alkenes, except that the ending used is *–yne* rather than *–ene*.

a. 1-hexyne
b. 4-methyl-2-pentyne
c. 2,2-dimethyl-3-heptyne
d. 1-butyne

13.91 IUPAC rules for naming alkynes (hydrocarbons with one or more carbon-carbon triple bonds) are identical to the rules for naming alkenes; the ending used is *–yne*.

$C\equiv C-C-C-C$ 1-pentyne

$C-C\equiv C-C-C$ 2-pentyne

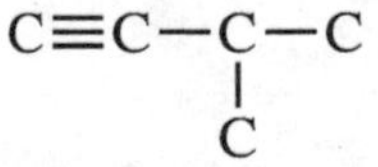

$C\equiv C-C(C)-C$ 3-methyl-1-butyne

13.93 *Cis-trans* isomerism is not possible for an alkyne because of the linearity (180° angles) about an alkyne's carbon–carbon triple bond.

13.95 Their physical properties are very similar.

13.97 Addition reactions in alkynes are similar to those in alkenes, except that two molecules of a specific reagent can add to the triple bond. In the reactions below, four atoms of two molecules are attached to the two carbons of the triple bond. In part d. only one molecule is added. Apply Markovnikov's rule in parts c. and d.; molecules with the form HQ are added in those parts.

a. CH_3-CH_3

b. 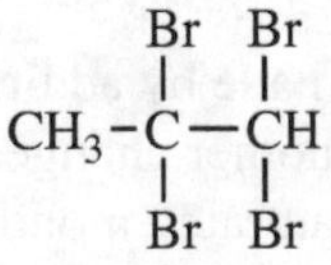 $CH_3-CBr_2-CHBr_2$

c. 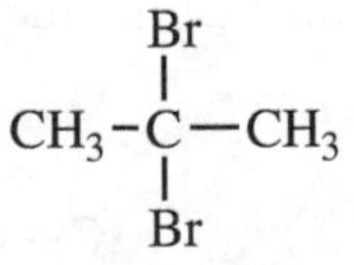 $CH_3-CBr_2-CH_3$

d. 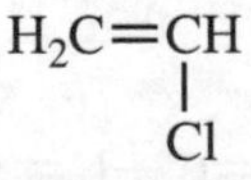 $H_2C=CH(Cl)$

13.99 An alkene contains a carbon-carbon double bond. An alkyne contains a carbon-carbon triple bond. The prefix *di* means two. The condensed structural formulas for a. 5-methyl-2-hexyne, b. 2-methyl-2-butene, c. 1,6-heptadiyne, and d. 3-penten-1-yne are shown below.

a. $CH_3-C\equiv C-CH_2-\underset{\displaystyle CH_3}{\underset{|}{CH}}-CH_3$

b. $CH_3-\underset{\displaystyle CH_3}{\underset{|}{C}}=CH-CH_3$

c. $CH\equiv C-CH_2-CH_2-CH_2-C\equiv CH$

d. $CH\equiv C-CH=CH-CH_3$

13.101 a. The alkenyl group with the common name allyl is a 2-propenyl group. The compound has three carbon atoms.
b. Acetylene is the common name for ethyne; it has two carbon atoms.
c. Dimethylacetylene has 2 methyl substituents on acetylene; it has four carbon atoms.
d. The parent chain (1-hexyne) has six carbon atoms and the cyclobutyl group has four carbon atoms; the compound has 10 carbon atoms.

13.103 

13.105 That representation implies that there are two types of carbon-carbon bonds present, which is not the case.

13.107 The IUPAC system of naming monosubstituted benzene derivatives uses the name of the substituent as a prefix to the name benzene. When two substituents are attached to a benzene ring, we specify the positions of the substituents, in alphabetical order, by using numbers as prefixes; the first substituent is 1, the second is numbered relative to the first.

A few monosubstituted benzenes have particular names: methylbenzene is toluene, and the substituents are numbered with relation to the methyl group.

a. 1,3-dibromobenzene
b. 1-chloro-2-fluorobenzene
c. 1-chloro-4-fluorobenzene
d. 3-chlorotoluene

13.109 A second IUPAC system of naming is the prefix naming system which uses *ortho-* (1,2 disubstitution), *meta-* (1,3 disubstitution), and *para-* (1,4 disubstitution). These may be abbreviated in the name of the compound as *o-*, *m-*, and *p-*.

a. *m*-dibromobenzene
b. *o*-chlorofluorobenzene
c. *p*-chlorofluorobenzene
d. *m*-chlorotoluene

13.111 The positions of more than two substituents on a benzene ring are indicated with numbers; the ring is numbered to obtain the lowest possible numbers for the carbon atoms that have substituents.

a. 2,4-dibromo-1-chlorobenzene
b. 3-bromo-5-chlorotoluene
c. 1-bromo-3-chloro-2-fluorobenzene
d. 1,4-dibromo-2,5-dichlorobenzene

13.113 Sometimes a benzene ring is treated as a substituent on a carbon chain. In this case the substituent is called *phenyl-*.

a. 2-phenylbutane
b. 3-phenyl-1-butene
c. 3-methyl-1-phenylbutane
d. 2,4-diphenylpentane

13.115 In writing structural formulas, a six-membered ring with a circle inside denotes benzene or a phenyl- group.

a. CH_2-CH_3 / CH_2-CH_3 b. CH_3 / CH_3 c. CH_3 / CH_2-CH_3 d.

13.117 The molecular formula for benzene is C_6H_6. For substituted benzenes that have the molecular formula C_9H_{12}, the substituents have a total of 3 carbon atoms. The substituents on the ring therefore will be: (1) 3 methyl groups (2) 1 methyl and 1 ethyl group or (3) 1 propyl group. The IUPAC names for the eight substituted benzenes are: 1,2,3-trimethylbenzene, 1,2,4-trimethylbenzene, 1,3,5-trimethylbenzene, 2-ethyltoluene, 3-ethyltoluene, 4-ethyltoluene, propylbenzene. and isopropylbenzene.

13.119 Of the four compounds cyclohexane, cyclohexene, 1,3-cyclohexadiene, and benzene, those listed below have the given characteristics.

a. All four of the compounds contain a 6-membered ring.
b. Cyclohexene has the generalized formula C_nH_{2n-2}.
c. Cyclohexane and benzene undergo substitution reactions.
d. Delocalized bonding is present in benzene.

13.121 They are in the liquid state.

13.123 Petroleum is the primary source for aromatic hydrocarbons.

13.125 Alkanes undergo substitution and combustion reactions. Alkenes undergo addition reactions to the double bonds. Aromatic compounds undergo substitution reactions rather than addition reactions.

a. Alkanes undergo substitution reactions.
b. Dienes (two double bonds) undergo addition reactions.
c. Alkylbenzenes undergo substitution reactions.
d. Cycloalkenes undergo addition reactions.

13.127 Each of these three reactions involves substitution on an aromatic ring.

a. The reagent used for bromination of an aromatic ring is Br_2 (in the presence of a $FeBr_3$ catalyst).

b. $-CH-CH_3$ (with CH_3 below) Benzene and an alkyl chloride, in the presence of an aluminum chloride catalyst, undergo a substitution reaction, alkylation.

c. An alkyl group on a benzene ring can be produced by the a substitution reaction on benzene with an alkyl bromide and $AlBr_3$ as a catalyst; the alkyl halide in this case is CH_3-CH_2-Br.

13.129 Carbon atoms are shared between rings.

Alcohols, Phenols, and Ethers — Chapter 14

Solutions to Selected Problems

14.1 a. Oxygen has 6 valence electrons and so forms 2 covalent bonds.
b. Hydrogen has 1 valence electron and forms 1 covalent bond to complete its "octet" of 2.
c. Carbon has 4 valence electrons and so forms 4 covalent bonds.
d. A halogen atom has 7 valence electrons and so forms 1 covalent bond.

14.3 Alcohols may be viewed as being alkyl derivatives of water in which a hydrogen atom has been replaced by an alkyl group.

14.5 The generalized formula for an alcohol with a saturated carbon attached to the OH is ROH. The molecular formula for an alcohol having
a. 3 carbon atoms and one –OH group is C_3H_8O.
b. 5 carbon atoms and one –OH group is $C_5H_{12}O$.

14.7 To name an alcohol by the IUPAC rules, find the longest carbon chain to which the hydroxyl group is attached, number the chain starting at the end nearest the hydroxyl group, and name and locate any other substituents present. Use the suffix *-ol*.
a. 2-pentanol
b. 3-methyl-2-butanol
c. 2-ethyl-1-pentanol
d. 2-butanol

14.9 To name an alcohol by the IUPAC rules, find the longest carbon chain to which the hydroxyl group is attached, number the chain starting at the end nearest the hydroxyl group, and name and locate any other substituents present. Use the suffix *-ol*.
a. 2-hexanol
b. 3-methyl-1-butanol
c. 3-methyl-1-pentanol
d. 4-heptanol

14.11 In an alcohol name, the number before the parent chain designates the position of the –OH group. The positions of the substituents are numbered relative to the –OH group.

a. $CH_2-CH-CH_3$ (OH on CH_2; CH_3 on CH)

b. $CH_3-CH-CH_2-CH-CH_3$ (OH on C-2; CH_3 on C-4)

c. CH_3-C-CH_3 (OH on C; phenyl ring on C)

d. cyclobutane ring with OH and adjacent CH_3

14.13 Common names exist for alcohols with simple alkyl groups. The word alcohol, as a separate word, is placed after the name of the alkyl group.

a. $CH_3-CH_2-CH_2-CH_2-CH_2-OH$

1-pentanol

b. $CH_3-CH_2-CH_2-OH$

1-propanol

c. 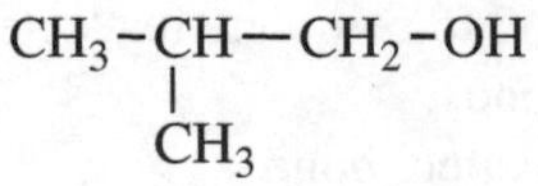 $CH_3-CH(CH_3)-CH_2-OH$

2-methyl-1-propanol

d. 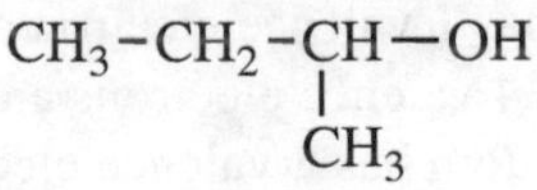 $CH_3-CH_2-CH(CH_3)-OH$

2-butanol

14.15 Polyhydroxy alcohols (more than one hydroxyl group) can be named with a slight modification: a compound with two hydroxyl groups is named a diol, one with three hydroxyl groups is a triol.

a. 1,2-butanediol
b. 1,5-hexanediol
c. 2,4-hexanediol
d. 3-methyl-1,2,5-pentanetriol

14.17 In the naming of alcohols with unsaturated carbon chains, the longest chain must contain both the carbon atom to which the hydroxyl group is attached and also the carbon atoms which are unsaturated. The chain is numbered from the end that gives the lowest number to the carbon to which the hydroxyl group is attached. Two endings are needed: one for the double or triple bond and one for the –OH group. Unsaturated alcohols are named as *alkenols* or *alkynols*.

a. 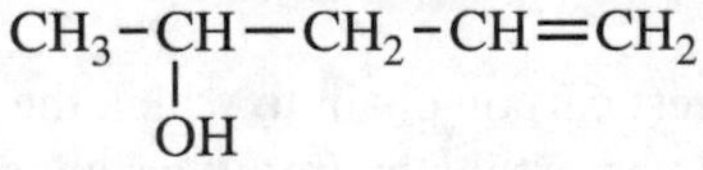 $CH_3-CH(OH)-CH_2-CH=CH_2$

b. 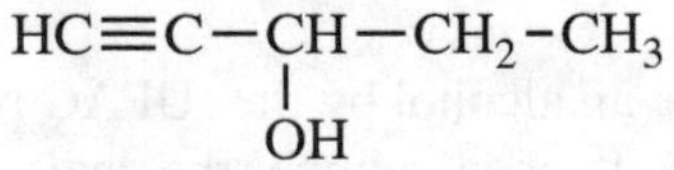 $HC\equiv C-CH(OH)-CH_2-CH_3$

c. $CH_3-CH(OH)-C(CH_3)=CH_2$

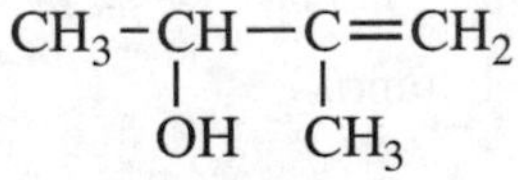

d. 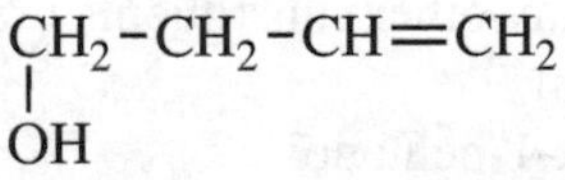 $CH_2(OH)-CH_2-CH=CH_2$

14.19 a. 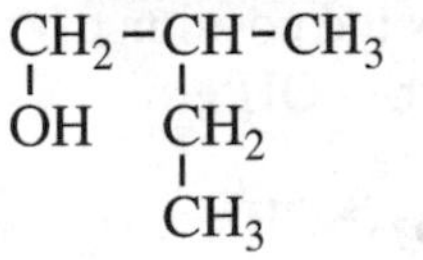 $CH_2(OH)-CH(CH_2CH_3)-CH_3$

For the incorrect name, the wrong parent chain was chosen. The correct name is 2-methyl-1-butanol.

b. 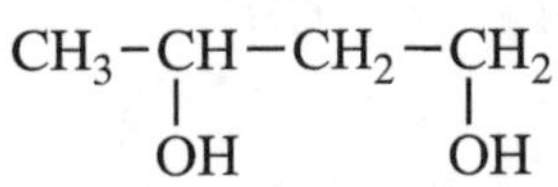 $CH_3-CH(OH)-CH_2-CH_2(OH)$

For the incorrect name, the parent chain was numbered from the wrong end. The correct name is 1,3-butanediol.

c. 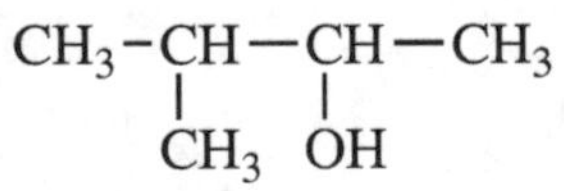 $CH_3-CH(CH_3)-CH(OH)-CH_3$

For the incorrect name, the parent chain was numbered from the wrong end. The correct name is 3-methyl-2-butanol.

d. 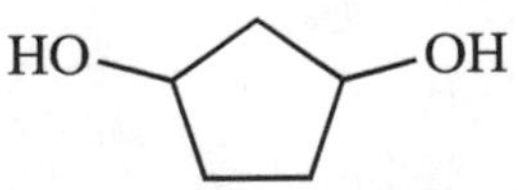 HO–(cyclopentane ring)–OH

When there are two possible ways to number the substituents on a ring, choose the way that gives the lowest possible total number. The correct name is 1,3-cyclopentanediol.

14.21 a. No, this is not a constitutional isomer of 1-hexanol; it has a different molecular formula. (This is 1-pentanol.)
b. Yes, this is a constitutional isomer of 1-hexanol; the position of the functional group (OH) has changed. (This is 3-hexanol.)
c. Yes, this is a constitutional isomer of 1-hexanol; the carbon-chain arrangement has changed. (This is 4-methyl-2-pentanol.)
d. Yes, this is a constitutional isomer of 1-hexanol; the position of the functional group (OH) has changed. (This is the same compound as the one in part b., 3-hexanol.)

14.23 a. There are four C_7 alcohols that are named as heptanols.
b. There are eight C_7 alcohols that are named as dimethylcyclopentanols.
c. There are eight alcohols that have the molecular formula $C_5H_{12}O$.
d. There are four saturated alcohols that have the molecular formula C_4H_8O.

14.25 *Cis-trans* isomers have the same molecular and structural formulas, but different arrangements of atoms in space because of restricted rotation about bonds.
a. a *trans*-isomer b. a *cis*-isomer c. a *cis*-isomer d. a *trans*-isomer

14.27 a. *trans*-2-buten-1-ol b. *cis*-3-penten-1-ol
c. *cis*-2-methylcyclohexanol d. *trans*-3-chlorocyclohexanol

14.29 a. Absolute alcohol is 100% ethyl alcohol, with all traces of water removed.
b. Grain alcohol is ethyl alcohol; ethyl alcohol can be synthesized from grains such as corn, rice, and barley.
c. Rubbing alcohol is 70% isopropyl alcohol; because of its high evaporation rate, it is used for alcohol rubs combating high body temperatures.
d. Drinking alcohol is ethyl alcohol; it is the alcohol produced by yeast fermentation of sugars, and it is present in all alcoholic beverages.

14.31 Use information about production and uses of alcohols in Section 14.5 of your textbook.
a. 1,2,3-propanetriol (also called glycerol) is a moistening agent in many cosmetics.
b. 1,2-propanediol (propylene glycol) is a major ingredient in "environmentally friendly" antifreeze formulations.
c. Methanol is industrially produced from CO and H_2.
d. Ethanol is often produced via a fermentation process.

14.33 The fuel of choice for Indianapolis Speedway racecars changed in 2005 from gasoline to methanol because methanol fires, unlike gasoline fires, can be put out with water.

14.35 Alcohols can form hydrogen bonds with one another (see Figure 14.10). Alkane molecules do not form hydrogen bonds.

14.37 a. 1-Heptanol has a higher boiling point than 1-butanol because boiling point increases as the length of the carbon chain increases.
b. 1-Propanol has a higher boiling point than butane; 1-propanol forms hydrogen bonds between molecules.
c. 1,2-Ethanediol has a higher boiling point than ethanol; because of increased hydrogen bonding, alcohols with multiple –OH groups have higher boiling points than their monohydroxy counterparts.

14.39 a. 1-Butanol is more soluble than butane because alcohol molecules can form hydrogen bonds with water molecules.

b. 1-Pentanol is more water-soluble than 1-octanol; as the carbon-chain (nonpolar) increases in length, solubility in water (polar) decreases.

c. 1,2-Butanediol is more water-soluble than 1-butanol; increased hydrogen bonding makes an alcohol with two –OH groups more soluble than its counterpart with one –OH group.

14.41 a. Three hydrogen bonds can form between ethanol molecules (see Figure 14.10).

b. Three hydrogen bonds can form between ethanol molecules and water molecules (see Figure 14.11).

c. Three hydrogen bonds can form between methanol molecules (see Figure 14.10).

d. Three hydrogen bonds can form between 1-propanol molecules (see Figure 14.10).

14.43 Two general methods of preparing alcohols are: 1) hydration of alkenes, in which a molecule of water is added to a double bond in the presence of a catalyst (sulfuric acid), and 2) addition of H_2 to a carbon-oxygen double bond (a carbonyl group) in the presence of a catalyst.

a. CH_3-CH_2 (with OH on CH_2)

b. $CH_3-CH_2-CH_2$ (with OH on terminal CH_2)

c. $CH_3-CH_2-C-CH_3$ (with OH and CH_3 on C)

d. $CH_3-CH_2-CH-CH_2-CH_3$ (with OH on CH)

14.45 Alcohols are classified by the number of carbons bonded to the hydroxyl-bearing carbon atom: in a primary alcohol the hydroxyl-bearing carbon atom is bonded to one other carbon atom, in a secondary alcohol it is bonded to two other carbon atoms, and in a tertiary alcohol it is bonded to three other carbon atoms.

a. 2-Pentanol is secondary alcohol.
b. 3-Methyl-2-butanol is a secondary alcohol.
c. 2-Ethyl-1-pentanol is a primary alcohol.
d. 2-Butanol is a secondary alcohol.

14.47 In a primary alcohol the hydroxyl-bearing carbon atom is bonded to one other carbon atom, in a secondary alcohol it is bonded to two other carbon atoms, and in a tertiary alcohol it is bonded to three other carbon atoms.

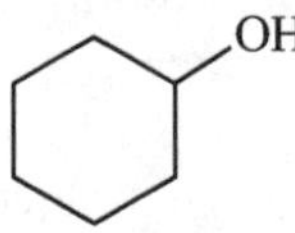

a. secondary

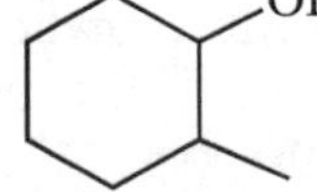

b. secondary

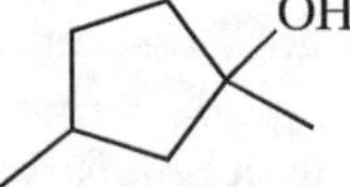

c. tertiary

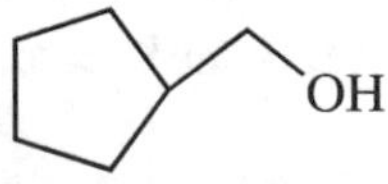

d. primary

14.49 a. 1-Pentanol is a primary alcohol; the carbon atom to which the –OH group is attached is bonded to only one other carbon atom.

b. 2-Pentanol is a secondary carbon; the carbon atom to which the –OH group is attached is bonded to two other carbon atoms.

c. 2-Methyl-1-pentanol is a primary alcohol; the carbon atom to which the –OH group is attached is bonded to only one other carbon atom.

d. 2-Methyl-2-pentanol is a tertiary carbon; the carbon atom to which the –OH group is attached is bonded to three other carbon atoms.

14.51 The hydroxyl-bearing carbon atom in a secondary alcohol is bonded to two other carbon atoms; in a tertiary alcohol it is bonded to three other carbon atoms.

a. The simplest 3° alcohol with an acyclic R group is: $C-C(C)(C)-OH$

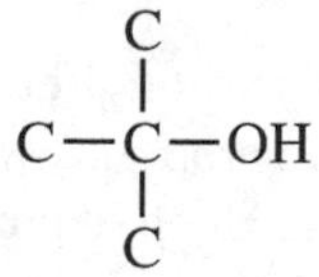

b. The simplest 2° alcohol with a cyclic R group is: cyclopropyl$-OH$

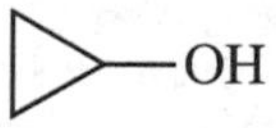

14.53 Considering the alcohol constitutional isomers having the formula $C_5H_{12}O$:

a. four isomers that are 1° alcohols
b. three isomers that are 2° alcohols
c. one isomer that is a 3° alcohol
d. zero isomeric alcohols with a carbon ring

14.55 In the dehydration of an alcohol the components of a water molecule (H and OH) are removed from a single molecule or from two molecules. Sulfuric acid is the catalyst. Notice that in parts a. and c. the starting material is the same, but the temperature differs. The same product is formed at both 140°C as at 180°C.

a. $CH_2=CH-CH_3$

b. 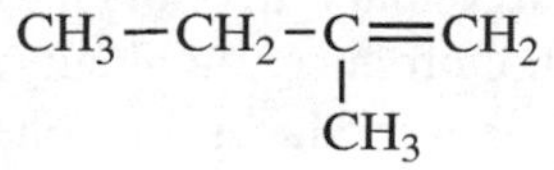 $CH_3-CH_2-C(CH_3)=CH_2$

c. $CH_2=CH-CH_3$

d. $CH_3-CH_2-CH_2-O-CH_2-CH_2-CH_3$

14.57 In the dehydration of an alcohol, the components of a water molecule (H and OH) are removed from a single molecule or from two molecules. Apply Zaitsev's rule in parts a. and b.: the major product in an intramolecular alcohol dehydration reaction is the alkene that has the greatest number of alkyl groups attached to the carbon atoms of the double bond.

a. 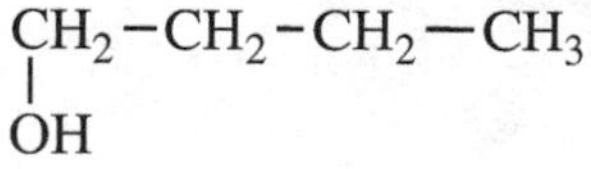 $CH_2(OH)-CH_2-CH_2-CH_3$

b. 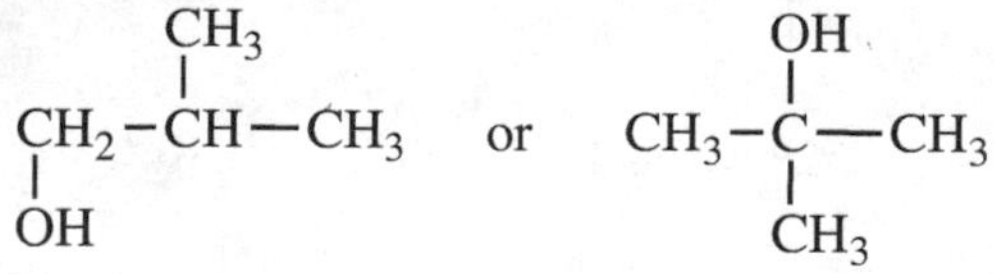 $CH_2(OH)-CH(CH_3)-CH_3$ or $CH_3-C(OH)(CH_3)-CH_3$

c. CH_3-CH_2-OH

d. 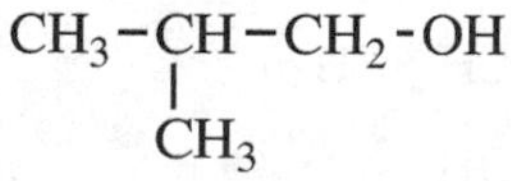 $CH_3-CH(CH_3)-CH_2-OH$

14.59 When 3-methyl-3-hexanol is dehydrated, the structures of the possible products are:

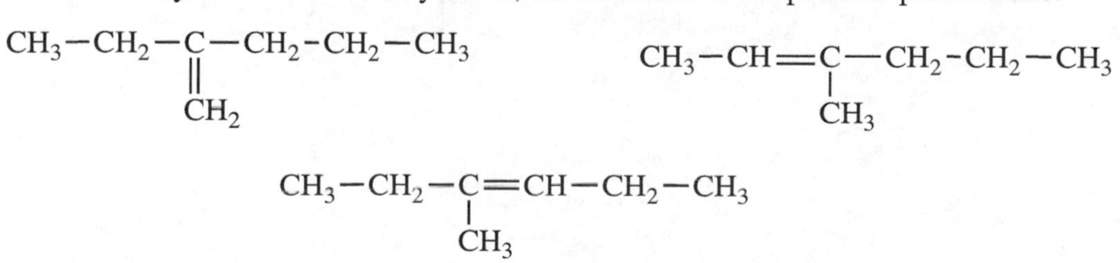

$CH_3-CH_2-C(=CH_2)-CH_2-CH_2-CH_3$

$CH_3-CH=C(CH_3)-CH_2-CH_2-CH_3$

$CH_3-CH_2-C(CH_3)=CH-CH_2-CH_3$

14.61 The alcohol 2,2-dimethyl-1-butanol cannot be dehydrated because the carbon to which the hydroxyl-bearing carbon is attached has no hydrogen atoms.

14.63 In an aldehyde, the carbonyl-carbon atom is bonded to at least one hydrogen atom; in a ketone, the carbonyl-carbon atom has two other carbon atoms directly attached to it. Oxidation of a 1° alcohol produces an aldehyde; oxidation of a 2° alcohol produces a ketone.

a. aldehyde b. ketone c. aldehyde d. ketone

14.65 The structures of the aldehydes and ketones produced in Problem 14.63 are

a. 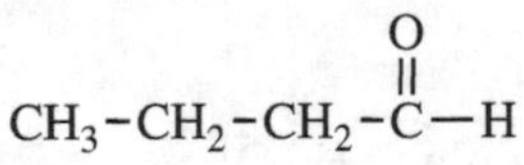 $CH_3-CH_2-CH_2-\overset{\overset{\displaystyle O}{\|}}{C}-H$

b. $CH_3-\overset{\overset{\displaystyle O}{\|}}{C}-CH_2-CH_3$

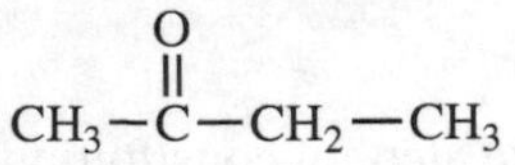

c. $CH_3-CH_2-\underset{\underset{\displaystyle CH_3}{|}}{CH}-\overset{\overset{\displaystyle O}{\|}}{C}-H$

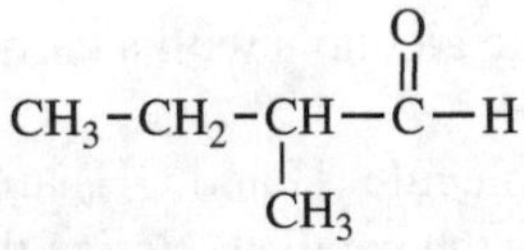

d. 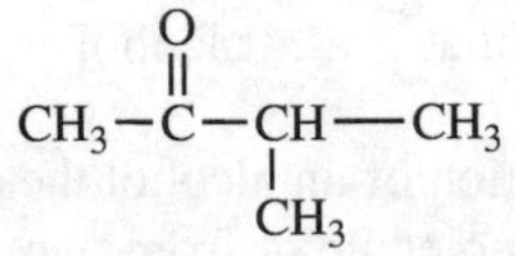 $CH_3-\overset{\overset{\displaystyle O}{\|}}{C}-\underset{\underset{\displaystyle CH_3}{|}}{CH}-CH_3$

14.67 Primary and secondary alcohols may be oxidized in the presence of a mild oxidizing agent. A primary alcohol produces an aldehyde that is often further oxidized to a carboxylic acid. A secondary alcohol produces a ketone.

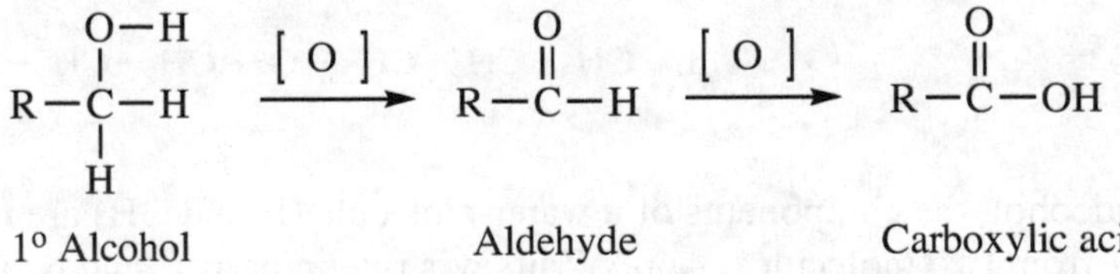

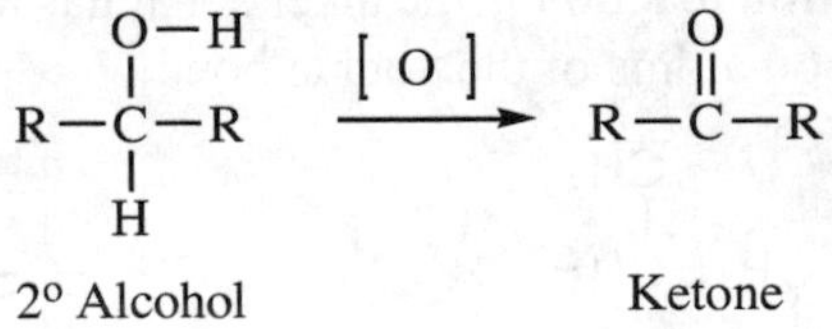

$$\underset{\text{1° Alcohol}}{R-\overset{\overset{\displaystyle O-H}{|}}{\underset{\underset{\displaystyle H}{|}}{C}}-H} \xrightarrow{[O]} \underset{\text{Aldehyde}}{R-\overset{\overset{\displaystyle O}{\|}}{C}-H} \xrightarrow{[O]} \underset{\text{Carboxylic acid}}{R-\overset{\overset{\displaystyle O}{\|}}{C}-OH}$$

$$\underset{\text{2° Alcohol}}{R-\overset{\overset{\displaystyle O-H}{|}}{\underset{\underset{\displaystyle H}{|}}{C}}-R} \xrightarrow{[O]} \underset{\text{Ketone}}{R-\overset{\overset{\displaystyle O}{\|}}{C}-R}$$

a. 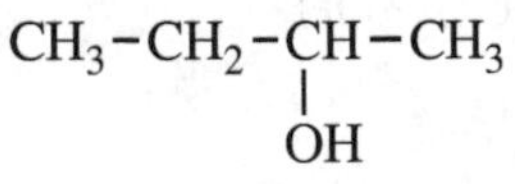 $CH_3-CH_2-\underset{\underset{\displaystyle OH}{|}}{CH}-CH_3$

b. $CH_3-CH_2-CH_2-OH$

c. $CH_3-CH_2-CH_2-OH$

d. 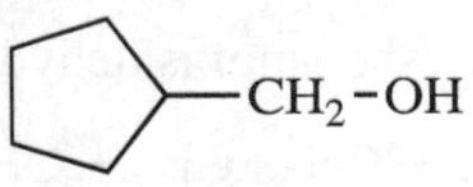 (cyclopentyl)$-CH_2-OH$

14.69 Alcohols undergo several types of reactions.

a. This is a halogenation reaction; a halogen atom is substituted for the hydroxyl group.

b. A water molecule is removed (dehydration reaction) within the molecule (180°C). Remember to use Zaitsev's rule.

c. This is the mild oxidation of a secondary alcohol.

d. This is also a dehydration reaction but at a lower temperature (140°C); a water molecule is removed from two alcohol molecules to produce an ether.

a. $CH_3-CH_2-CH_2-Cl$

b. CH_3 (on cyclopentene ring)

c. $CH_3-\overset{O}{\overset{\|}{C}}-CH_2-CH_3$

d. $CH_3-CH_2-O-CH_2-CH_3$

14.71 The three isomeric pentanols with unbranched carbon chains are 1-pentanol, 2-pentanol, and 3-pentanol. 1-Pentanol, upon dehydration at 180 °C, loses –OH from carbon 1 and –H from carbon 2, and so yields only 1-pentene. 2-Pentanol yields a mixture of two dehydration products, 1-pentene and 2-pentene, because –H can come from either carbon 1 or carbon 3. 3-Pentanol yields only 3-pentene.

14.73 In a phenol the –OH group is attached to a carbon atom that is part of an aromatic ring.

a. No, it is not a phenol; it does not contain an aromatic ring.
b. No, it is not a phenol; it does not contain an aromatic ring.
c. Yes, it is a phenol.
d. No, it is not a phenol; the –OH group is not attached to an aromatic ring.

14.75 In naming phenols, the parent name is phenol; substituents are numbered beginning with the –OH group and proceeds in the direction that gives the lower number to the next carbon atom bearing a substituent. The –OH group is not specified in the name because it is 1 by definition.

a. 3-ethylphenol
b. 2-chlorophenol
c. *o*-cresol
d. hydroquinone

14.77 The positions of the substituents are relative to the –OH group (carbon 1). Methylphenols are called cresols. Each of the three hydroxyphenols has a different name: resorcinol is the *meta*-hydroxyphenol.

a. 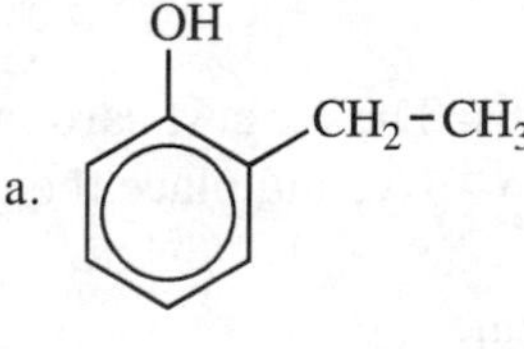

b. 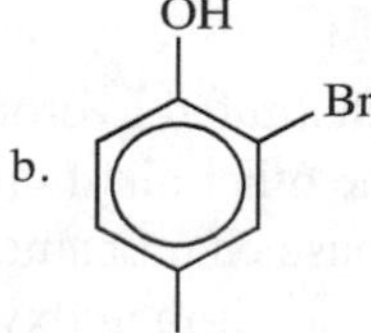

c. 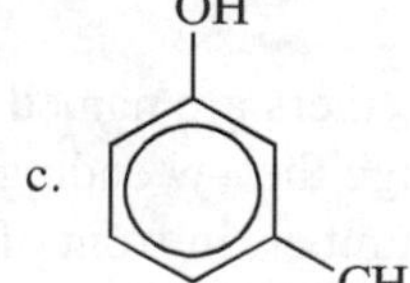

d. 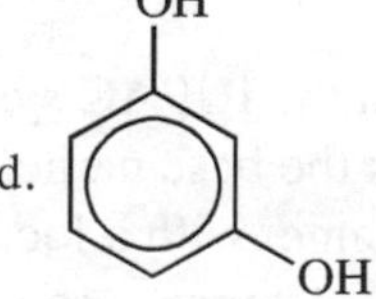

14.79 Phenols are low-melting solids or oily liquids.

14.81 a. Both are flammable. b. Both undergo halogenation.

14.83 Phenols are weak acids in water solution, and like other weak acids, they ionize in water to form the hydronium ion and a negative ion (the phenoxide ion).

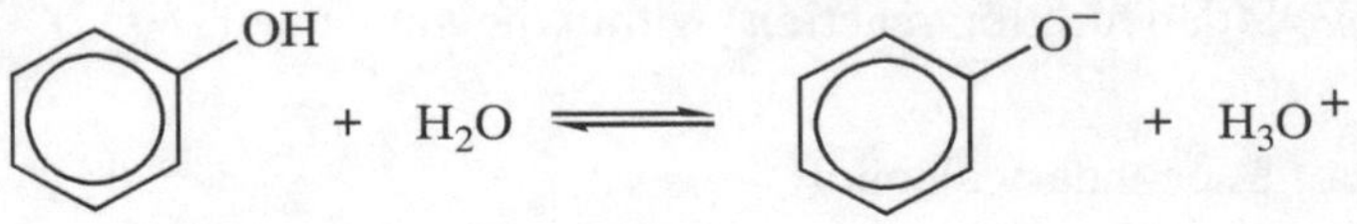

$$C_6H_5OH + H_2O \rightleftharpoons C_6H_5O^- + H_3O^+$$

14.85 An antiseptic kills microorganisms on living tissue; a disinfectant kills microorganisms on inanimate objects.

14.87 BHA has methoxy and *tert*-butyl groups; BHT has a methyl group and two *tert*-butyl groups.

14.89 In an ether, an oxygen atom is bonded to two carbon atoms by single bonds.
a. Yes, this is an ether.
b. No, this is not an ether; it is an alcohol.
c. Yes, this is an ether.
d. Yes, this is an ether.

14.91 In an ether molecule an oxygen atom is bonded to two carbon atoms by single bonds. In an alcohol molecule an –OH group is bonded to a saturated carbon atom.
a. The compound is an ether.
b. The compound is an alcohol.
c. The compound is an alcohol.
d. The compound is an ether.

14.93 The simplest ether in which both R groups are identical is dimethyl ether.

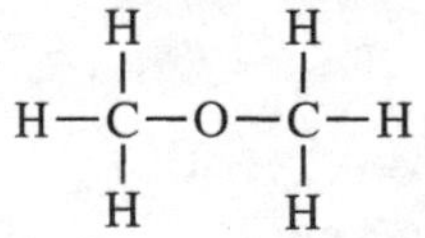

```
    H     H
    |     |
H—C—O—C—H
    |     |
    H     H
```

The expanded structural formula shows all atoms and all bonds in the molecule.

$CH_3—O—CH_3$

The condensed structural formula uses groupings of atoms in which central atoms and the atoms connected to them are written as a group.

C—O—C

The skeletal structural formula shows carbon and oxygen atoms and the bonds between them, but no hydrogen atoms or bonds to them.

C_2H_6O

The molecular formula gives the number of each type of atoms in the molecule.

14.95 In the IUPAC system, ethers are named as substituted hydrocarbons. The longest carbon chain is the base name. Change the *–yl* ending of the other alkyl group to *–oxy*, and place the alkoxy name, with a locator number, in front of the base chain name.
a. 1-ethoxypropane
b. 2-methoxypropane
c. methoxybenzene
d. cyclohexoxycyclohexane

14.97 The common names for ethers use the form: alkyl alkyl ether or dialkyl ether. Two different alkyl groups are written in alphabetical order.
a. ethyl propyl ether
b. isopropyl methyl ether
c. methyl phenyl ether
d. dicyclohexyl ether

14.99 In the IUPAC system, ethers are named as substituted hydrocarbons. The longest carbon chain is the base name. Change the *–yl* ending of the other alkyl group to *–oxy*, and place the alkoxy name, with a locator number, in front of the base chain name.

a. 1-methoxypentane
b. 1-ethoxy-2-methylpropane
c. 2-ethoxybutane
d. 2-methoxybutane

14.101 The common names for ethers use the form: alkyl alkyl ether or dialkyl ether.

a. 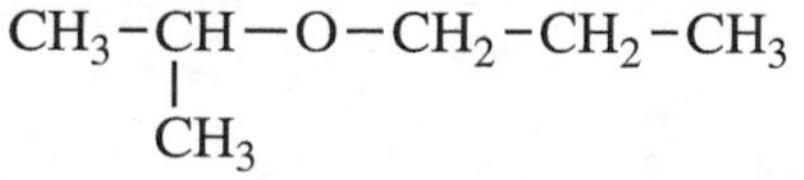 $CH_3-\underset{\displaystyle CH_3}{\underset{|}{CH}}-O-CH_2-CH_2-CH_3$

b. 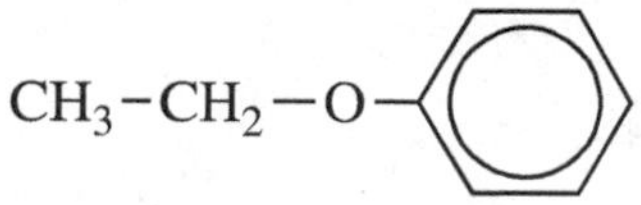 CH_3-CH_2-O- (benzene ring)

c. 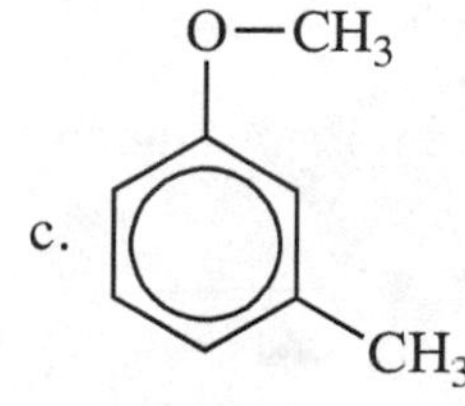

d. 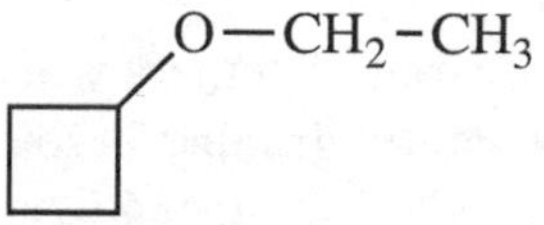 (cyclobutane ring)$-O-CH_2-CH_3$

14.103 In a line-angle structural formula, a line represents a carbon-carbon bond and a carbon atom is understood to be present at every point where lines meet and at the ends of lines. The names of the compounds are:

a. ethoxyethane
b. 2-methoxy-2-methylpropane
c. methoxymethanol
d. 2-methylanisole

14.105 The letters in the name MTBE stand for methyl *tert*-butyl ether.

14.107 The use of MTBE has been discontinued in the United States because spillage has been found in water supplies.

14.109 a. Eugenol has the odor of cloves
b. Isoeugenol has the odor of nutmeg.
c. Vanillin has the odor of vanilla
d. Allyl anisole has the odor of anise and fennel.

14.111 Constitutional isomers have the same molecular formulas, but different bonding arrangements between atoms. Ethyl propyl ether has a molecular formula of $C_5H_{12}O$.

a. No, this is not a constitutional isomer of ethyl propyl ether; the molecular formula is $C_6H_{14}O$.
b. No, this is not a constitutional isomer of ethyl propyl ether; the molecular formula is $C_6H_{14}O$.
c. Yes, this is a constitutional isomer of ethyl propyl ether; the molecular formula is $C_5H_{12}O$, and the name is *sec*-butyl methyl ether.
d. No, this is not a constitutional isomer of ethyl propyl ether; the molecular formula is $C_6H_{14}O$.

14.113 The easiest way to find the common names for the five ethers that are constitutional isomers of ethyl propyl ether is to draw the isomers and then name them.

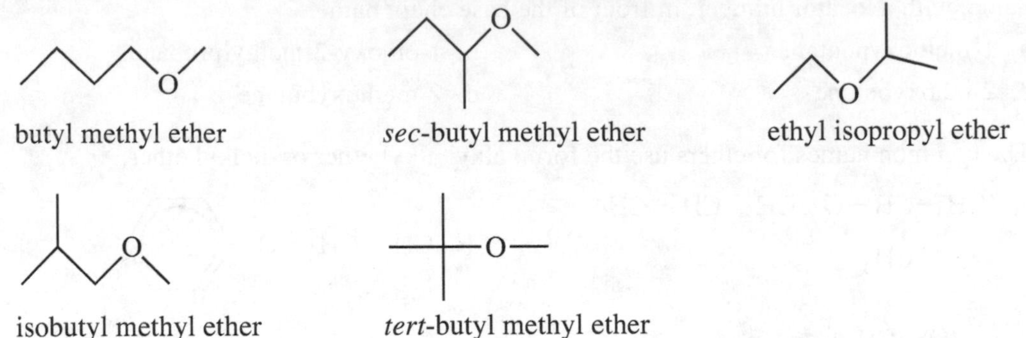

14.115 The number of isomeric ethers when the following R groups are present in the ether are:

a. for C_1 and C_3 alkyl groups, 2 isomeric ethers
b. for C_2 and C_4 alkyl groups, 4 isomeric ethers
c. for C_3 and C_3 alkyl groups, 3 isomeric ethers
d. for C_3 and C_4 cycloalkyl groups, 3 isomeric ethers

14.117 Functional group isomers are constitutional isomers that contain different functional groups.

a. These three ethers are functional group isomers of 1-butanol.

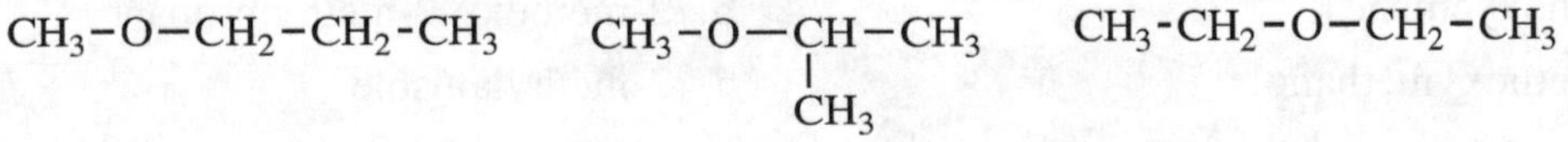

$CH_3-O-CH_2-CH_2-CH_3$

$CH_3-O-\underset{\displaystyle CH_3}{\underset{|}{CH}}-CH_3$

$CH_3-CH_2-O-CH_2-CH_3$

b. These four alcohols are functional group isomers of 1-methoxypropane.

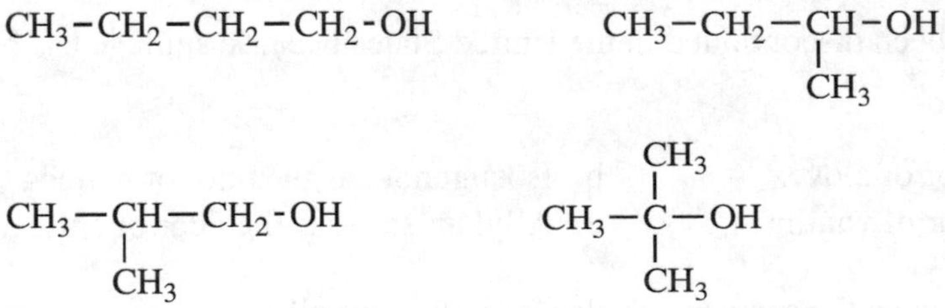

$CH_3-CH_2-CH_2-CH_2-OH$

$CH_3-CH_2-\underset{\displaystyle CH_3}{\underset{|}{CH}}-OH$

$CH_3-\underset{\displaystyle CH_3}{\underset{|}{CH}}-CH_2-OH$

$CH_3-\overset{\displaystyle CH_3}{\overset{|}{\underset{\displaystyle CH_3}{\underset{|}{C}}}}-OH$

14.119 Condensed structural formulas to the eight isomeric alcohols and six isomeric ethers that have the molecular formula $C_5H_{12}O$ are shown below.

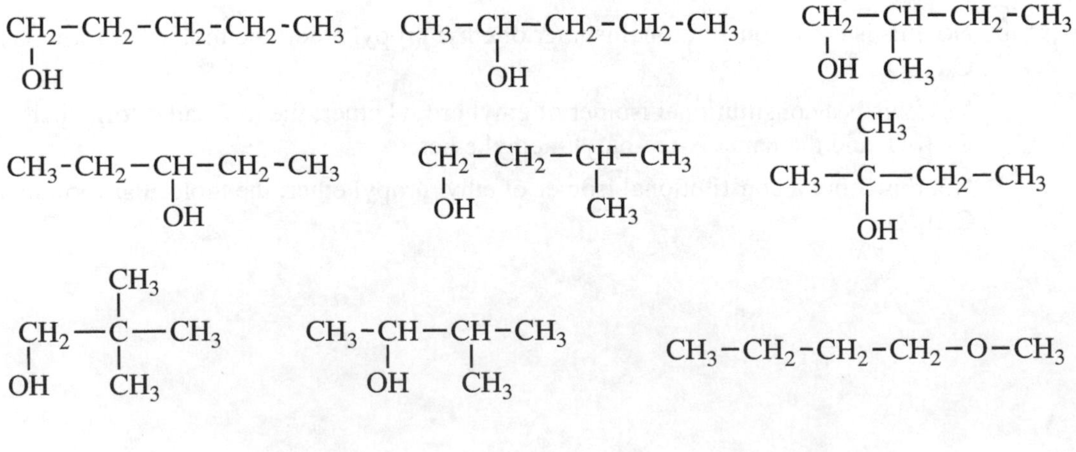

$\underset{\displaystyle OH}{\underset{|}{CH_2}}-CH_2-CH_2-CH_2-CH_3$

$CH_3-\underset{\displaystyle OH}{\underset{|}{CH}}-CH_2-CH_2-CH_3$

$\underset{\displaystyle OH}{\underset{|}{CH_2}}-\underset{\displaystyle CH_3}{\underset{|}{CH}}-CH_2-CH_3$

$CH_3-CH_2-\underset{\displaystyle OH}{\underset{|}{CH}}-CH_2-CH_3$

$\underset{\displaystyle OH}{\underset{|}{CH_2}}-CH_2-\underset{\displaystyle CH_3}{\underset{|}{CH}}-CH_3$

$CH_3-\overset{\displaystyle CH_3}{\overset{|}{\underset{\displaystyle OH}{\underset{|}{C}}}}-CH_2-CH_3$

$\underset{\displaystyle OH}{\underset{|}{CH_2}}-\overset{\displaystyle CH_3}{\overset{|}{\underset{\displaystyle CH_3}{\underset{|}{C}}}}-CH_3$

$CH_3-\underset{\displaystyle OH}{\underset{|}{CH}}-\underset{\displaystyle CH_3}{\underset{|}{CH}}-CH_3$

$CH_3-CH_2-CH_2-CH_2-O-CH_3$

$CH_3-\underset{\underset{\displaystyle CH_3}{|}}{CH}-CH_2-O-CH_3$ $CH_3-CH_2-\underset{\underset{\displaystyle CH_3}{|}}{CH}-O-CH_3$ $CH_3-\overset{\overset{\displaystyle CH_3}{|}}{\underset{\underset{\displaystyle CH_3}{|}}{C}}-O-CH_3$

$CH_3-CH_2-CH_2-O-CH_2-CH_3$ $CH_3-\underset{\underset{\displaystyle CH_3}{|}}{CH}-O-CH_2-CH_3$

14.121 There is no hydrogen bonding between molecules of dimethyl ether (disruptive forces are greater than cohesive forces); there is hydrogen bonding between molecules of ethyl alcohol (cohesive forces are of about the same magnitude as disruptive forces).

14.123 The two chemical hazards associated with ether use are flammability and peroxide formation.

14.125 Ether molecules cannot form hydrogen bonds with one another because there are no oxygen-hydrogen bonds in ethers.

14.127 In cyclic ethers, the ether functional group is part of a ring system.
a. This is a noncyclic ether; the ether functional group is not included in the ring.
b. This is a noncyclic ether; the ether functional group is not included in the ring.
c. This is a cyclic ether; the functional group is part of the ring system.
d. This is a nonether; it is an alcohol.

14.129 A thioalcohol has the general formula R–S–H where R is an alkyl group; an alcohol has the general formula R–O–H.

14.131 The IUPAC names of thiols are similar to those of alcohols; *–ol* is replaced by *–thiol*.

a. $\underset{\underset{\displaystyle SH}{|}}{CH_2}-CH_2-CH_2-CH_3$

b. $\underset{\underset{\displaystyle SH}{|}}{CH_2}-CH_2-\underset{\underset{\displaystyle CH_3}{|}}{CH}-CH_2-CH_3$

c. (cyclopentane ring)—SH

d. $\underset{\underset{\displaystyle SH}{|}}{CH_2}-\underset{\underset{\displaystyle SH}{|}}{CH_2}$

14.133 In writing common names for thiols, the name of the alkyl group (as a separate word) precedes the word *mercaptan*. The structures of thiols are similar to those of alcohols, except the oxygen atom is replaced by a sulfur atom.
a. methyl mercaptan
b. propyl mercaptan
c. *sec*-butyl mercaptan
d. isobutyl mercaptan

14.135 The oxidation of an alcohol produces aldehydes (which may be further oxidized to carboxylic acids) and ketones; the oxidation of a thiol produces disulfides.

14.137 a. $CH_3-CH_2-CH_2-OH$ The solubility of alcohols in water is greater than that of thiols because of hydrogen bonding between alcohols and water.

b. $CH_3-CH_2-CH_2-SH$ The solubility of thiols in water is greater than that of alkanes because thiols are more polar (like dissolves like).

14.139 Alcohols are named for the longest carbon chain to which the hydroxyl group is attached; the *–e* ending of the alkane becomes *–ol*. Thiols are named for the longest chain to which the sulfhydryl group is attached, with the addition of *–thiol* to the alkane name. The names of the compounds are:

a. 1-pentanethiol
b. 1-pentanol
c. 3-methyl-2-butanethiol
d. 3-methyl-2-butanol

14.141 R–S–R versus R–O–R

14.143 For common name, thioethers are named in the same way as ethers, with *sulfide* used in place of *ether*, and *alkylthio* used in place of *alkoxy* in IUPAC names.

a. methylthioethane, ethyl methyl sulfide
b. 2-methylthiopropane, isopropyl methyl sulfide
c. methylthiocyclohexane, cyclohexyl methyl sulfide
d. 3-(methylthio)-1-propene, allyl methyl sulfide

14.145 a. This is the structural formula of a disulfide. A disulfide consists of two R groups connected by two sulfur atoms bonded together.

b. This compound is a peroxide. A peroxide consists of two R groups connected by two oxygen atoms bonded together.

c. This compound is a diol, a compound having two hydroxyl groups on different carbon atoms.

d. The compound having this structural formula is characterized as an ether, a thioether, and a sulfide. The molecule contains two different functional groups; the sulfur-containing group can be called a sulfide or a thioether.

Aldehydes and Ketones — Chapter 15

Solutions to Selected Problems

15.1 A carbonyl group is a carbon atom double bonded to an oxygen atom.
a. Yes, this molecule contains a carbonyl group.
b. No, this molecule contains an hydroxyl group but not a carbonyl group.
c. No, this molecule contains an ether group but not a carbonyl group.
d. Yes, this molecule contains a carbonyl group.

15.3 A carbon–oxygen double bond is polar; a carbon–carbon double bond is nonpolar.

15.5 Bond angles of 120° are associated with a trigonal planar arrangement of bonds.

15.7 A carbonyl group is a carbon atom double bonded to an oxygen atom.
a. Yes, an aldehyde contains a carbonyl group.
b. Yes, a ketone contains a carbonyl group.
c. Yes, an amide contains a carbonyl group.
d. Yes, a carboxylic acid contains a carbonyl group.

15.9 a. No. An aldehyde does not contain a carbon-oxygen single bond. It contains a carbon-oxygen double bond.
b. No. A ketone does not contain a carbon-oxygen single bond. It contains a carbon-oxygen double bond.
c. Yes. An ester contains both a carbon-oxygen single bond and a carbon-oxygen double bond.
d. Yes. A carboxylic acid contains both a carbon-oxygen single bond and a carbon-oxygen double bond.

15.11 An aldehyde is a carbonyl-containing compound in which the carbonyl carbon has at least one hydrogen atom attached directly to it. A ketone is a carbonyl-containing compound in which the carbonyl carbon atom has two other carbon atoms directly attached to it.
a. The compound is neither an aldehyde nor a ketone; it is a carboxylic acid.
b. The compound is a ketone.
c. The compound is neither an aldehyde nor a ketone; it is an ether.
d. The compound is an aldehyde.

15.13
a. $H-\overset{\overset{\displaystyle O}{\|}}{C}-H$

b. Structure cannot exist. A ketone must have at least three carbon atoms.

c. $CH_3-CH_2-\overset{\overset{\displaystyle O}{\|}}{C}-H$

d. $CH_3-\overset{\overset{\displaystyle O}{\|}}{C}-CH_2-CH_3$

15.15 In an aldehyde the carbonyl carbon has at least one hydrogen atom attached directly to it. In a ketone the carbonyl carbon atom has two other carbon atoms directly attached to it. The compounds below are:
a. an aldehyde
b. neither an aldehyde nor a ketone
c. a ketone
d. a ketone

15.17 In naming aldehydes using the IUPAC system, select the parent carbon chain that includes the carbon of the carbonyl group. The carbonyl carbon has the number 1 on the chain, but this is not specified in the name. Identify and locate any substituents. Use the suffix *–al*.

a. 2-methylbutanal b. 4-methylheptanal
c. 3-phenylpropanal d. propanal

15.19 In the IUPAC naming system, aldehydes have the suffix *–al*. The carbonyl carbon is number 1 on the chain, but this is not specified in the name. Identify and locate any substituents.

a. pentanal b. 3-methylbutanal
c. 3-methylpentanal d. 2-ethyl-3-methylpentanal

15.21 An aldehyde parent chain is numbered with the carbonyl carbon as number 1.

a. $CH_3-CH_2-CH(CH_3)-CH_2-C(=O)-H$

b. $CH_3-CH_2-CH_2-CH_2-CH(CH_2CH_3)-C(=O)-H$

c. $CH_3-CCl_2-C(=O)-H$

d. $CH_3-CH_2-CH_2-CH_2-CH(OH)-CH_2-CH(CH_3)-C(=O)-H$

15.23 The correct IUPAC name for:

a. 3-methylpropanal is butanal b. 2-ethylpropanal is 2-methylbutanal
c. 3,4-dimethylbutanal is 3-methylpentanal d. dimethylethanal is 2-methylpropanal

15.25 There are common names for the first four straight-chain aldehydes: the prefixes *form-*, *acet-*, *propion-* and *butyr-* are used with the suffix *–aldehyde*. The common name of the compound in which an aldehyde group is attached to a benzene ring is benzaldehyde.

a. $H-C(=O)-H$

b. $CH_3-CH_2-C(=O)-H$

c. benzene ring bearing $-C(=O)-H$ with Cl on the adjacent (ortho) carbon

d. benzene ring bearing $-C(=O)-H$, with CH_3 on the adjacent (ortho) carbon and H_3C on the carbon para to that CH_3

15.27 Common names for the first four straight-chain aldehydes use the prefixes *form-*, *acet-*, *propion-* and *butyr-* and the suffix *–aldehyde*. The common name of the compound in which an aldehyde group is attached to a benzene ring is benzaldehyde.

a. propionaldehyde b. propionaldehyde
c. dichloroacetaldehyde d. 2-chlorobenzaldehyde

15.29 a. An *–al* ending indicates that propanal is an aldehyde.
b. In the name propenal, the *–en–* indicates an alkene (a double bond) and the *–al* indicates an aldehyde.
c. In the name 2-propanol, the *–ol* ending indicates an alcohol.
d. In the name 3-hydroxypropanal, there are two functional groups that are indicated: 3-hydroxy indicates an alcohol and *–al* indicates an aldehyde.

15.31 In naming ketones with the IUPAC system, select the parent chain that includes the carbon of the carbonyl group and number in the way that will give the carbonyl group the lowest number. Determine and identify the locations of substituents. The ketone name has the suffix *–one*.

a. 2-butanone
b. 2,4,5-trimethyl-3-hexanone
c. 6-methyl-3-heptanone
d. 1,1-dichloro-2-butanone

15.33 In naming ketones using the IUPAC system, select the parent chain that includes the carbon of the carbonyl group, number it in the way that will give the carbonyl group the lowest number, and determine and identify the locations of substituents. The ketone name has the suffix *–one*.

a. 2-hexanone
b. 5-methyl-3-hexanone
c. 2-pentanone
d. 4-ethyl-3-methyl-2-hexanone

15.35 Cyclic ketones in which the carbonyl group is part of a five- or six-membered ring are named as cyclopentanones and cyclohexanones. The carbonyl carbon has the number 1, and the ring is numbered to give the lowest possible numbers to the substituents.

a. cyclohexanone
b. 3-methylcyclohexanone
c. 2-methylcyclohexanone
d. 3-chlorocyclopentanone

15.37 Ketone names use the parent chain that includes the carbon of the carbonyl group. The chain is numbered in the way that will give the carbonyl group the lowest number; the substituents are numbered relative to the carbonyl group.

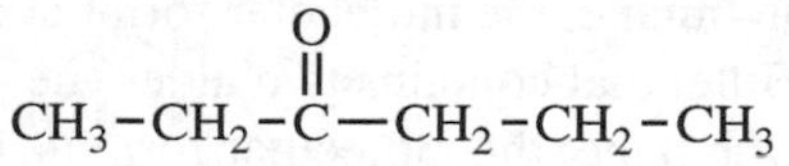

a. 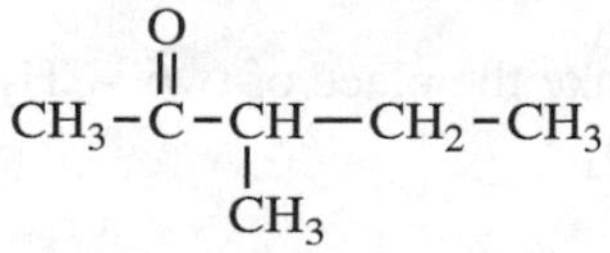 $CH_3-\overset{O}{\overset{\|}{C}}-\underset{CH_3}{\underset{|}{CH}}-CH_2-CH_3$

b. $CH_3-CH_2-\overset{O}{\overset{\|}{C}}-CH_2-CH_2-CH_3$

c. O

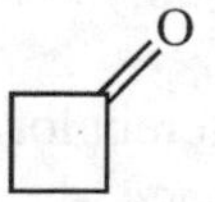

d. 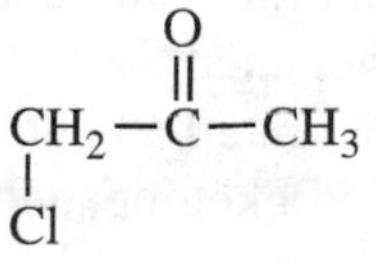 $\underset{Cl}{\underset{|}{CH_2}}-\overset{O}{\overset{\|}{C}}-CH_3$

15.39 The correct IUPAC name for:

a. 3-butanone is 2-butanone
b. 2-methyl-3-butanone is 3-methyl-2-butanone
c. 3-ethyl-2-butanone is 3-methyl-2-pentanone
d. 1,1-dimethyl-2-butanone is 2-methyl-3-pentanone

15.41 Common names for ketones are similar to the common names used for ethers with the carbonyl group taking the place of the oxygen atom. The alkyl groups are named, and the word ketone is added as a separate word: alkyl alkyl ketone or dialkyl ketone. Three ketones have additional common names; acetophenone (methyl phenyl ketone) is one of these.

a. 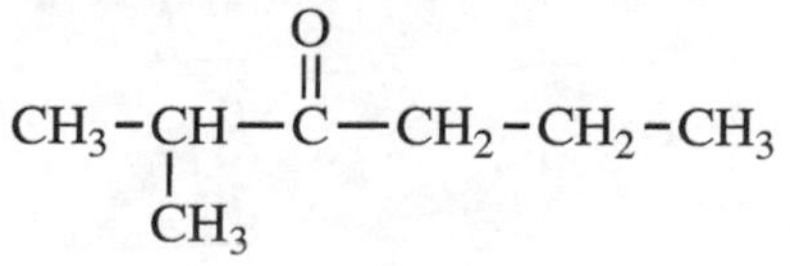 $CH_3-\underset{CH_3}{\underset{|}{CH}}-\overset{O}{\overset{\|}{C}}-CH_2-CH_2-CH_3$

b. 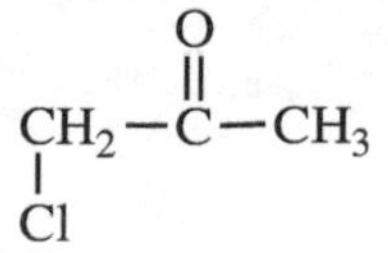 $\underset{Cl}{\underset{|}{CH_2}}-\overset{O}{\overset{\|}{C}}-CH_3$

c. $CH_3-\overset{O}{\overset{\|}{C}}-$

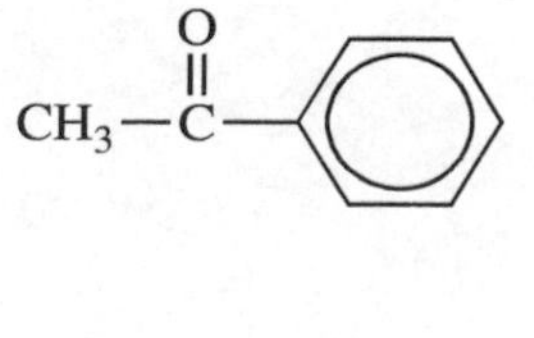

d. 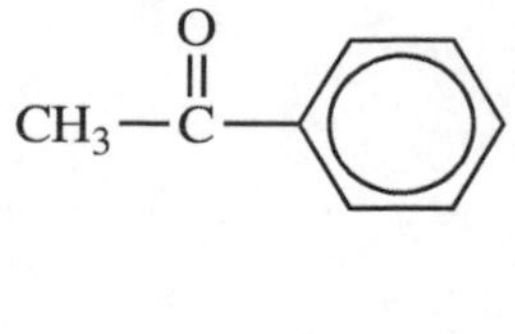 $CH_3-\overset{O}{\overset{\|}{C}}-$

15.43 a. In the name propanone, the ending *–one* indicates a ketone.
b. In the name propanal, the ending *–al* indicates an aldehyde.
c. In the name 4-oxohexanal, *–oxo–* indicates a ketone and the ending *–al* indicates an aldehyde. When both an aldehyde and a ketone are present, the compound is named as an aldehyde.
d. In 4-octen-2-one, *–en–* indicates an alkene (double bond) and *–one* indicates a ketone.

15.45 In naming a compound containing two or more of the following – aldehyde, ketone, alcohol – the aldehyde name takes precedence over the other two and the ketone takes precedence over the alcohol.

a. The compound contains both an alcohol and a ketone; it is named as a ketone.
b. The compound contains both a ketone and an aldehyde; it is named as an aldehyde.
c. The compound contains an alcohol and an aldehyde; it is named as an aldehyde.
d. The compound contains an aldehyde, an alcohol, and a ketone; it is named as an aldehyde.

15.47 The four compounds are based on the C_4 alkane, butane (C_4H_{10}).
a. 2-Butanol contains a hydroxyl group, –OH, that takes the place of one H atom in butane; the molecular formula of 2-butanol is $C_4H_{10}O$.
b. Butanal contains an aldehyde group, –HC=O, that takes the place of a $–CH_3$ group of butane; the molecular formula for butanal is C_4H_8O.
c. Butanone contains a ketone (carbonyl) group, C=O, that takes the place of a $–CH_2–$ group in butane; the molecular formula of butanone is C_4H_8O.
d. Butendial contains two aldehyde groups, –HC=O, that take the place of two $–CH_3$ groups of butane; the molecular formula of butanedial is $C_4H_6O_2$.

15.49 a. Since the chain is saturated and unbranched, and aldehyde groups occur only at the end of a chain, there is one heptanal.
b. Since the chain is saturated and unbranched, there are three possible different locations for the carbonyl group of a ketone, and so three possible ketones: 2-heptanone, 3-heptanone, and 4-heptanone.

15.51 a. CH_2O – only one aldehyde is possible; no ketones are possible (ketones must have two carbon atoms attached to the carbonyl group).
b. C_3H_6O – one aldehyde is possible; one ketone is possible.

15.53 The name *x*-methyl-3-hexanone is a correct name for $x = 2$, 4, or 5. The values $x = 1$ or 6 would be different parent compounds (and improper designations for a methyl substituent). The value $x = 3$ is impossible (the carbonyl carbon already has four bonds).

15.55 One carbon atom of $C_5H_{10}O$ belongs to the carbonyl group. This leaves four carbon atoms for the alkyl group(s). For the aldehydes, there are four possible R— groups; for the ketones, there are three possible combinations of the two ketone R— groups.

```
            O                          O
            ‖                          ‖
C—C—C—C—C—H              C—C—C—C—H
                                   |
                                   C

        O                        C  O
        ‖                        |  ‖
C—C—C—C—H                C—C—C—H
  |                              |
  C                              C

   O               O                  O
   ‖               ‖                  ‖
C—C—C—C—C       C—C—C—C         C—C—C—C—C
                     |
                     C
```

15.57 Formaldehyde is the simplest (C_1) aldehyde and is an irritating gas at room temperature; formalin is an aqueous solution containing 37% formaldehyde by mass.

15.59 Acetone is a colorless, volatile liquid that is miscible with both water and nonpolar solvents; its main use is as a solvent.

15.61 The structures for these four compounds may be found in Section 15.7 in your textbook. Count the aldehyde and ketone groups present in each structure.

a. Vanillin contains one aldehyde group and no ketone groups.
b. Carvone contains no aldehyde groups and one ketone group.
c. Cortisone contains no aldehyde groups and three ketone groups.
d. Avobenzene contains no aldehyde groups and two ketone groups

15.63 Classify the aldehydes and ketones as liquids or gases using information from Fig. 15.6

a. Methanal (C_1) is a gas.
b. Butanal (C_4) is a liquid.
c. Propanone (C_3) is a liquid.
d. 2-Hexanone (C_6) is a liquid.

15.65 a. Pentanal has a greater solubility in water than pentane because pentanal has polar carbonyl group; hydrogen bonding can occur.

b. Pentanal has a greater solubility in water than heptanal because pentanal has a shorter nonpolar carbon chain.

15.67 a. 2-Pentanol has a higher boiling point than 2-pentanone because there is hydrogen bonding between the hydroxyl groups of the 2-pentanol molecules.

b. 2-Pentanone has a higher boiling point than 2-methylpentane because the polar carbonyl groups on the 2-pentanone molecules are attracted to one another.

15.69 a. Zero hydrogen bonds can form between two ethanal molecules; there are no hydrogen atoms attached to O, N, or F atoms.

b. Two hydrogen bonds can form between an ethanal molecule and water molecules. See Figure 15.7 in your textbook.

15.71 Primary and secondary alcohols can be oxidized, using mild oxidizing agents, to produce aldehydes and ketones.

a. 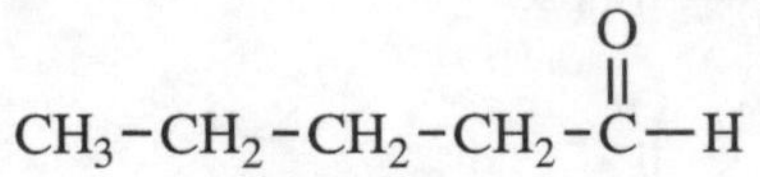 $CH_3-CH_2-CH_2-CH_2-\overset{\overset{\displaystyle O}{\|}}{C}-H$

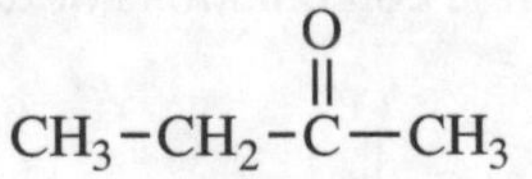

b. $CH_3-CH_2-\overset{\overset{\displaystyle O}{\|}}{C}-CH_3$

c. $CH_3-\underset{\underset{\displaystyle CH_3}{|}}{\overset{\overset{\displaystyle CH_3}{|}}{C}}-CH_2-\overset{\overset{\displaystyle O}{\|}}{C}-H$

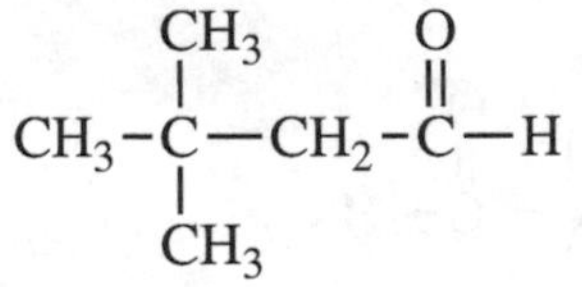

d. 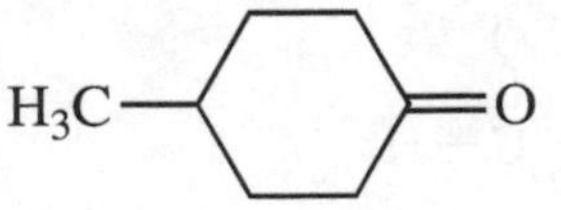 H_3C-(cyclohexane ring)$=O$

15.73 Aldehydes and ketones can be produced by the oxidation of primary and secondary alcohols, respectively, using mild oxidizing agents.

a. 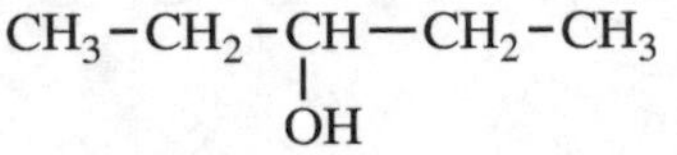 $CH_3-CH_2-\underset{\underset{\displaystyle OH}{|}}{CH}-CH_2-CH_3$

b. 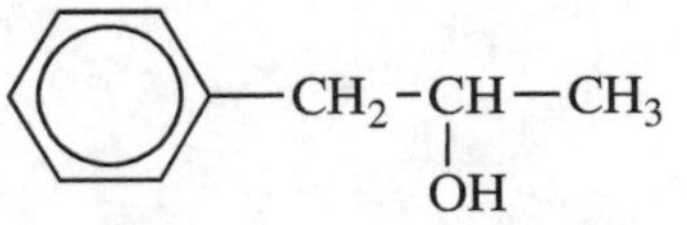 (benzene ring)$-CH_2-\underset{\underset{\displaystyle OH}{|}}{CH}-CH_3$

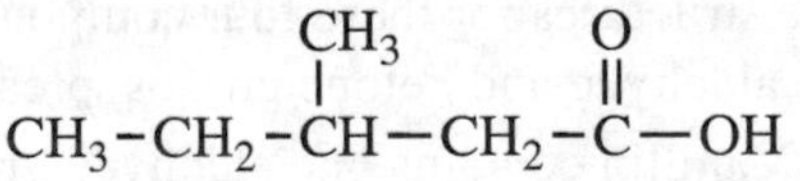

c. CH_3-CH_2-OH

d. 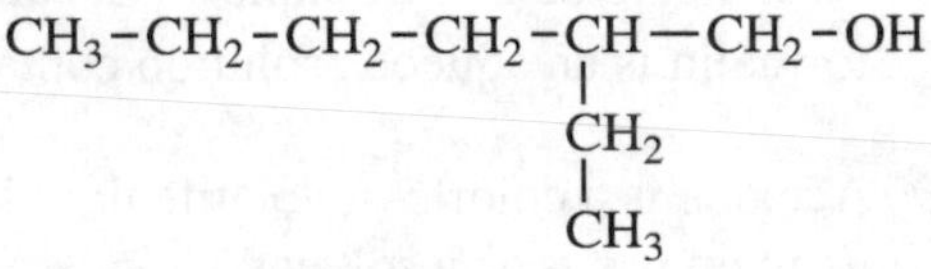 $CH_3-CH_2-CH_2-CH_2-\underset{\underset{\displaystyle \underset{\underset{\displaystyle CH_3}{|}}{CH_2}}{|}}{CH}-CH_2-OH$

15.75 Aldehydes readily undergo oxidation to carboxylic acids.

a. 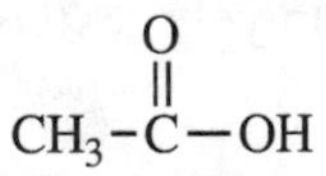 $CH_3-\overset{\overset{\displaystyle O}{\|}}{C}-OH$

b. $CH_3-CH_2-\overset{\overset{\displaystyle CH_3}{|}}{CH}-CH_2-\overset{\overset{\displaystyle O}{\|}}{C}-OH$

c. $H-\overset{\overset{\displaystyle O}{\|}}{C}-OH$

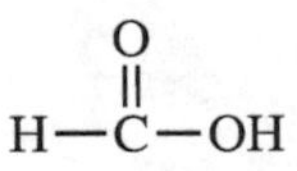

d. 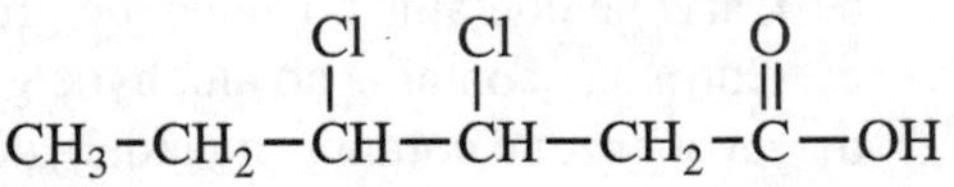 $CH_3-CH_2-\overset{\overset{\displaystyle Cl}{|}}{CH}-\overset{\overset{\displaystyle Cl}{|}}{CH}-CH_2-\overset{\overset{\displaystyle O}{\|}}{C}-OH$

15.77 Tollens and Benedict's reagents react with aldehydes because they are easily oxidized to carboxylic acids. Ketones are not oxidized and so do not give a positive test.

a. 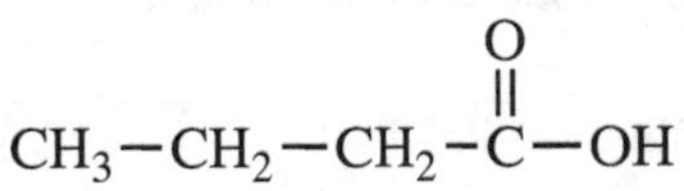 $CH_3-CH_2-CH_2-\overset{\overset{\displaystyle O}{\|}}{C}-OH$

b. No reaction occurs.

c. 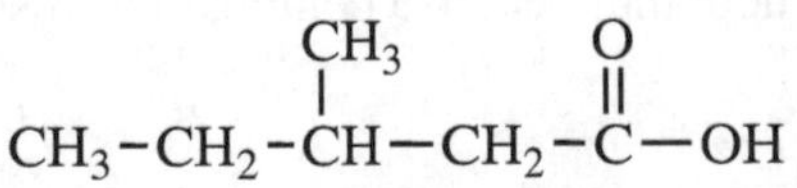 $CH_3-CH_2-\overset{\overset{\displaystyle CH_3}{|}}{CH}-CH_2-\overset{\overset{\displaystyle O}{\|}}{C}-OH$

d. No reaction occurs.

15.79 a. When Tollens solution is added to an aldehyde, Ag^+ ion acts as an oxidizing agent to oxidize the aldehyde to a carboxylic acid. In the process, Ag^+ ion is reduced to Ag.

b. No reaction occurs.

c. The oxidizing agent in Benedict's solution is Cu^{2+} ion; when an aldehyde is oxidized the Cu^{2+} ion becomes Cu^+ ion, which precipitates from solution as Cu_2O (a brick red solid).

d. No reaction occurs.

15.81 Tollens reagent reacts with aldehydes because they are easily oxidized to carboxylic acids. Ketones are not oxidized with Tollens reagent and so do not give a positive test.

a. No, it does not react; it is a ketone.
b. Yes, it does react; it is an aldehyde.
c. Yes, it does react; it is an aldehyde.
d. No, it does not react; it is a ketone.

15.83 Aldehydes are reduced by H_2 in the presence of a catalyst to form primary alcohols. The reduction of ketones by H_2 in the presence of a catalyst produces secondary alcohols.

a. $\begin{array}{l} CH_3-CH_2-CH_2-CH_2 \\ \qquad\qquad\qquad\qquad | \\ \qquad\qquad\qquad\quad OH \end{array}$

b. $\begin{array}{l} CH_3-CH_2-CH-CH_2-CH_3 \\ \qquad\qquad\quad | \\ \qquad\qquad\; OH \end{array}$

c. $\begin{array}{l} CH_3-CH-CH_2-CH_2 \\ \qquad\;\; | \qquad\qquad\; | \\ \qquad CH_3 \qquad\;\; OH \end{array}$

d. $\begin{array}{l} CH_3-CH-CH-CH_2-CH_2-CH_3 \\ \qquad\;\; | \qquad | \\ \qquad CH_3 \; OH \end{array}$

15.85 Aldehydes undergo oxidation to carboxylic acids with $K_2Cr_2O_7$, Tollens solution, or Benedict's solution. Ketones are not easily oxidized. Alcohols can be oxidized by a strong oxidizing agent such as $K_2Cr_2O_7$, but not by weak oxidizing agents such as Tollens or Benedict's solutions.

a. Pentanal and 2-pentanol can be oxidized by $K_2Cr_2O_7$.
b. Pentanal can be oxidized by Tollens solution.
c. Pentanal can be oxidized by Benedict's solution.
d. Pentanal and 2-pentanone can be reduced by H_2 in the presence of a Ni catalyst.

15.87 a. Ethanal can be reduced to ethyl alcohol with H_2 in the presence of a Ni catalyst.
b. Butanone can be reduced to 2-butyl alcohol with H_2 in the presence of a Ni catalyst.
c. Propanal can be oxidized to propanoic acid with a weak or a strong oxidizing agent.
d. Benzaldehyde can be oxidized to benzoic acid with a weak or a strong oxidizing agent.

15.89 When an alcohol molecule (R–O–H) adds across a carbon–oxygen double bond, the fragments of the alcohol are R–O– and H–.

15.91 In a hemiacetal, a carbon atom is bonded to both a hydroxyl group (–OH) and an alkoxy group (–OR).

a. No, this compound is an ether; there is no hydroxyl group.
b. Yes, this is a hemiacetal; a carbon atom is bonded to both –OH and –OR.
c. Yes, this is a cyclic hemiacetal; a carbon atom in a ring is bonded to both –OH and –OR.
d. No, this is not a hemiacetal; there is no –OH.

15.93 In a hemiacetal, a carbon atom is bonded to both a hydroxyl group (–OH) and an alkoxy group (–OR).

a. The hemiacetal carbon atom is C_2.
b. The hemiacetal carbon atom is C_1.

15.95 Hemiacetal formation is an addition reaction in which a molecule of alcohol adds to the carbonyl group of an aldehyde or ketone. The H portion of the alcohol adds to O atom of the carbonyl group; the R–O– portion of the alcohol adds to the C atom of the carbonyl group.

a. 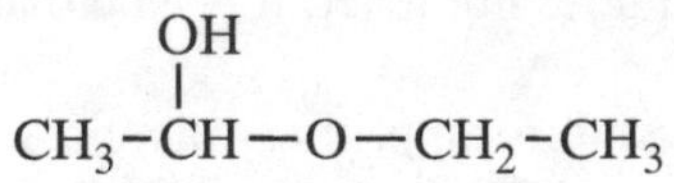 $CH_3-CH(OH)-O-CH_2-CH_3$

b. 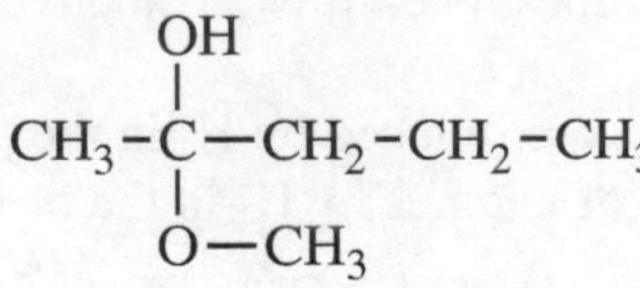 $CH_3-C(OH)(O-CH_3)-CH_2-CH_2-CH_3$

c. $CH_3-CH_2-CH_2-CH(OH)-O-CH_2-CH_3$

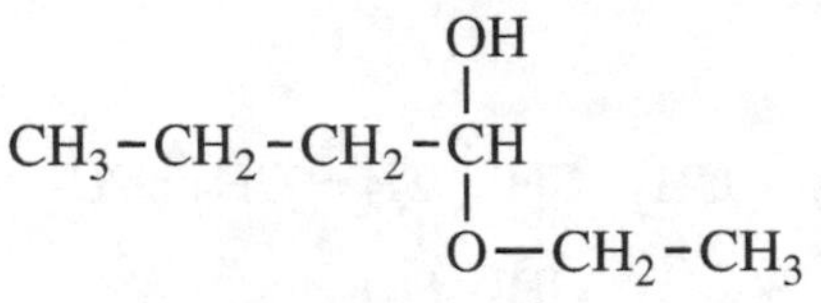

d. 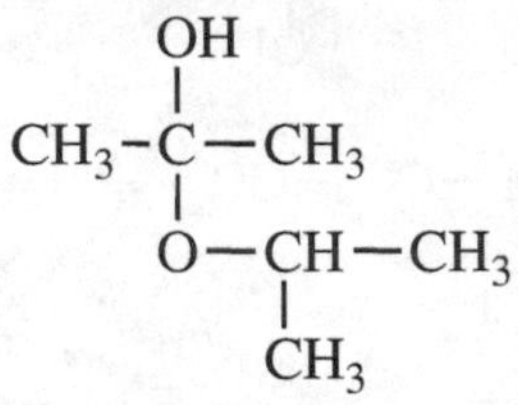 $CH_3-C(OH)(CH_3)-O-CH(CH_3)-CH_3$

15.97 Hemiacetal formation is an addition reaction in which a molecule of alcohol adds to the carbonyl group of an aldehyde or ketone. The H portion of the alcohol adds to O atom of the carbonyl group; the R–O– portion of the alcohol adds to the C atom of the carbonyl group.

a. 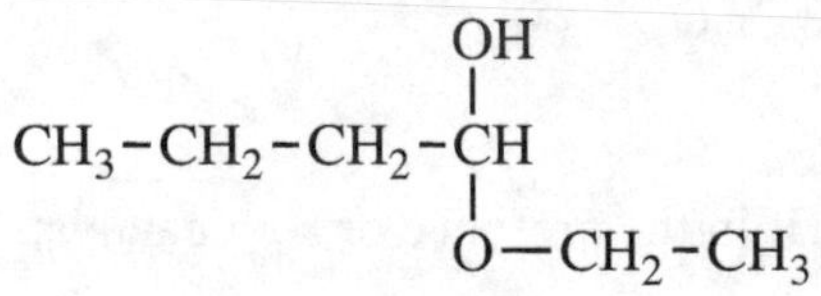 $CH_3-CH_2-CH_2-CH(OH)-O-CH_2-CH_3$

b. 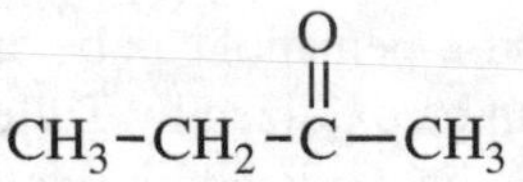 $CH_3-CH_2-C(=O)-CH_3$

c. $CH_3-CH_2-C(OH)(O-CH_3)-CH_3$

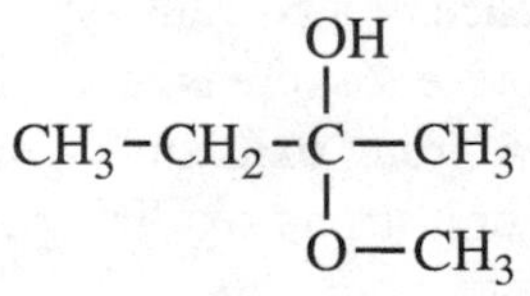

d. 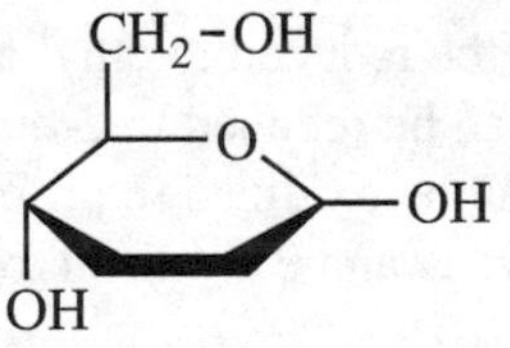 CH_2-OH, O, OH, OH (cyclic hemiacetal ring structure)

15.99 An acetal contains a carbon atom bonded to two alkoxy groups (–OR).

a. Yes, this is an acetal; two identical alkoxy groups are bonded to the same carbon atom.
b. Yes, this is an acetal; two different alkoxy groups are bonded to the same carbon atom.
c. No, this is not an acetal. One alkoxy group and one hydroxyl group are bonded to the same carbon atom; it is a hemiacetal.
d. Yes, this is an acetal; two identical alkoxy groups are bonded to the same carbon atom.

15.101 If a small amount of acid catalyst is added to a hemiacetal reaction mixture, the hemiacetal reacts with a second alcohol molecule in a condensation reaction to form an acetal and water.

a. The acetal carbon is C_2.
b. The acetal carbon is C_1.

15.103 If a small amount of acid catalyst is added to a hemiacetal reaction mixture, the hemiacetal reacts with a second alcohol molecule in a condensation reaction to form an acetal and water. The structures of the compounds missing from the equations are:

a. CH_3-OH

b. $CH_3-\underset{\underset{\displaystyle OH}{|}}{CH}-O-CH_3$

c. $CH_3-CH_2-\underset{\underset{\displaystyle O-\underset{\underset{\displaystyle CH_3}{|}}{CH}-CH_3}{|}}{CH}-O-CH_3$

d. $CH_3-\underset{\underset{\displaystyle OH}{|}}{CH}-O-CH_3$, CH_3-OH

15.105 In the hydrolysis reaction of an acetal, the acetal splits into three fragments as the elements of water (H– and –OH) are added to the compound. The products of the acetal hydrolysis are the aldehyde or ketone and the alcohols that originally reacted to form the acetal.

a. $CH_3-\overset{\overset{\displaystyle O}{\|}}{C}-H$, $2\ CH_3-OH$

b. $CH_3-\overset{\overset{\displaystyle O}{\|}}{C}-CH_3$, $2\ CH_3-OH$

c. $CH_3-CH_2-\overset{\overset{\displaystyle O}{\|}}{C}-CH_2-CH_3$, CH_3-OH , CH_3-CH_2-OH

d. $CH_3-CH_2-CH_2-CH_2-\overset{\overset{\displaystyle O}{\|}}{C}-H$, $2\ CH_3-OH$

15.107 The names of acetals are based on the word *acetal* and the name of the aldehyde or ketone that formed the acetal, preceded by the names of the alkoxy groups in the acetal.

a. dimethyl acetal of ethanal
b. dimethyl acetal of propanone
c. ethyl methyl acetal of 3-pentanone
d. dimethyl acetal of pentanal

15.109 The products of acetal hydrolysis are the aldehyde or ketone and alcohols that originally reacted to form the acetal.

a. Hydrolysis of the given hemiacetal produces methanol and propanal.
b. Hydrolysis of one mole acetal produces two moles methanol and one mole propanal.
c. Hydrolysis of the hemiacetal produces methanol and 2-methylpropanal.
d. Hydrolysis of the acetal produces methanol, ethanol, and propanal.

15.111 The reactants and products for the given reactions with pentanal are shown below.

a. 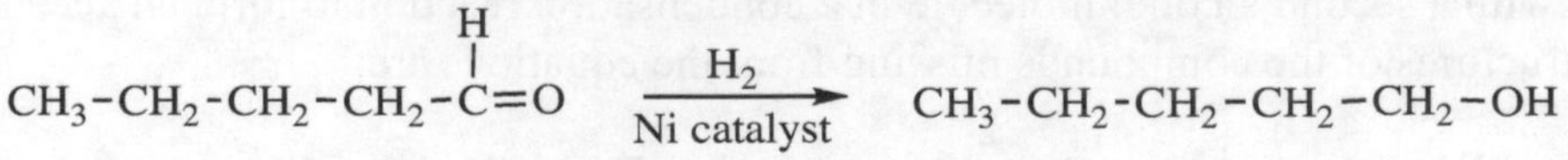 $CH_3-CH_2-CH_2-CH_2-\overset{H}{\overset{|}{C}}=O \xrightarrow[\text{Ni catalyst}]{H_2} CH_3-CH_2-CH_2-CH_2-CH_2-OH$

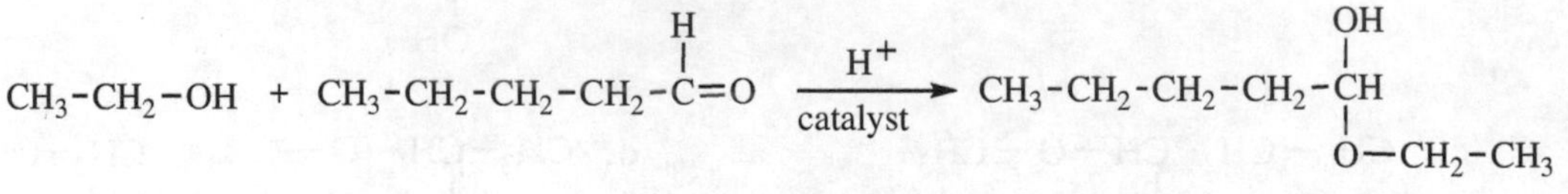

b. $CH_3-CH_2-OH + CH_3-CH_2-CH_2-CH_2-\overset{H}{\overset{|}{C}}=O \xrightarrow[\text{catalyst}]{H^+} CH_3-CH_2-CH_2-CH_2-\underset{O-CH_2-CH_3}{\underset{|}{\overset{OH}{\overset{|}{CH}}}}$

c. $2\,CH_3-OH + CH_3-CH_2-CH_2-CH_2-\overset{H}{\overset{|}{C}}=O \xrightarrow[\text{catalyst}]{H^+} CH_3-CH_2-CH_2-CH_2-\underset{O-CH_3}{\underset{|}{\overset{O-CH_3}{\overset{|}{CH}}}}$

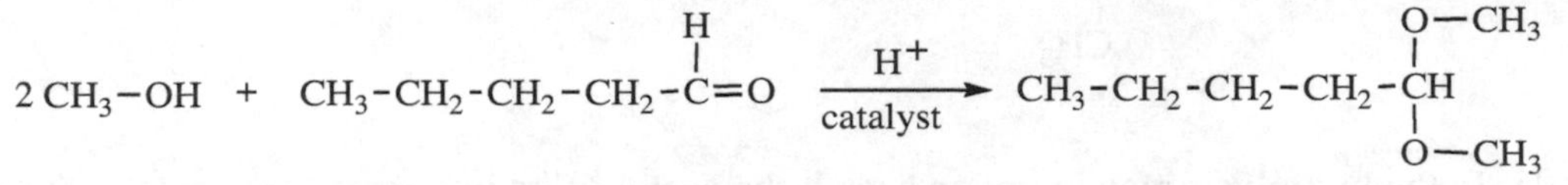

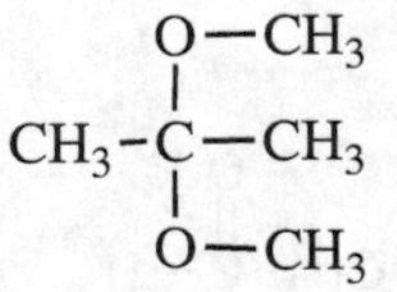

d. 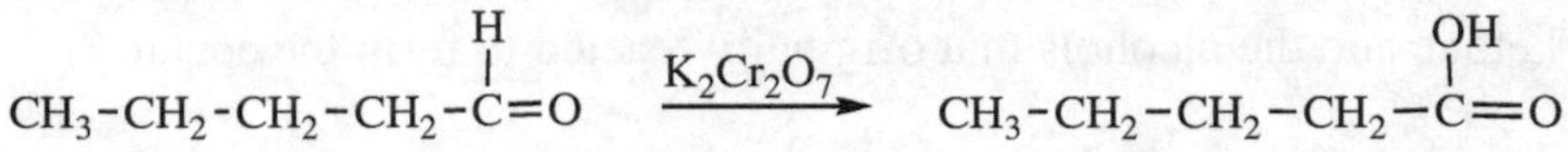 $CH_3-CH_2-CH_2-CH_2-\overset{H}{\overset{|}{C}}=O \xrightarrow{K_2Cr_2O_7} CH_3-CH_2-CH_2-CH_2-\overset{OH}{\overset{|}{C}}=O$

15.113 a. 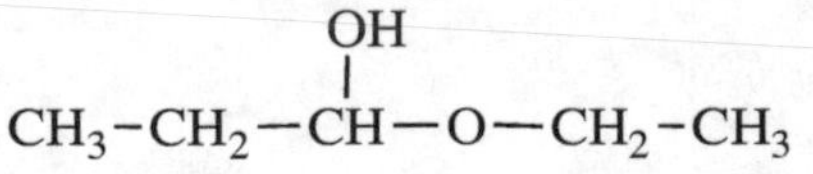 $CH_3-CH_2-\overset{OH}{\overset{|}{CH}}-O-CH_2-CH_3$ b. $CH_3-\underset{O-CH_2-CH_3}{\underset{|}{\overset{OH}{\overset{|}{C}}}}-CH_3$

Hemiacetal of propanal $C_5H_{12}O_2$ Hemiacetal of propanone $C_5H_{12}O_2$

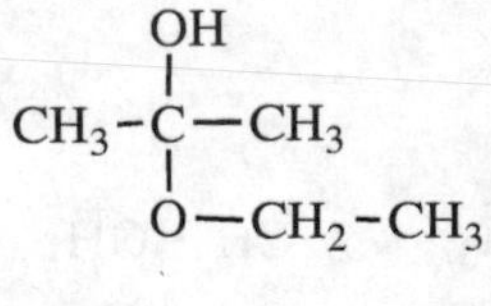

c. 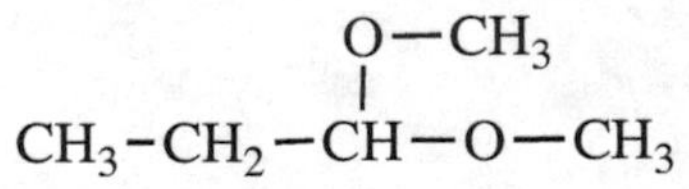 $CH_3-CH_2-\overset{O-CH_3}{\overset{|}{CH}}-O-CH_3$ d. $CH_3-\underset{O-CH_3}{\underset{|}{\overset{O-CH_3}{\overset{|}{C}}}}-CH_3$

Acetal of propanal $C_5H_{12}O_2$ Acetal of propanone $C_5H_{12}O_2$

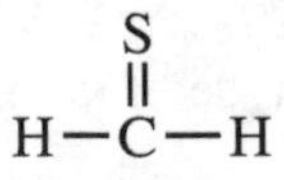

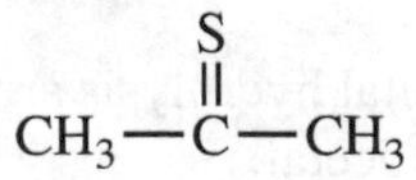

15.115 The compound formed by replacement of the carbonyl oxygen atom with a sulfur atom is a thiocarbonyl compound

15.117 a. 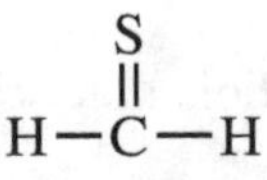 $H-\overset{S}{\overset{\|}{C}}-H$ b. $H-\overset{S}{\overset{\|}{C}}-H$ c. $CH_3-\overset{S}{\overset{\|}{C}}-CH_3$ d. 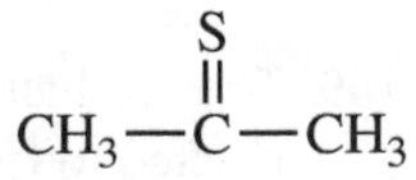 $CH_3-\overset{S}{\overset{\|}{C}}-CH_3$

Carboxylic Acids, Esters, and Other Acid Derivatives — Chapter 16

Solutions to Selected Problems

16.1 A carboxyl group is a carbonyl group (C=O) with a hydroxyl group (–OH) bonded to the carbonyl carbon atom.

a. Yes, this compound does contain a carboxyl group.
b. No, this compound is an ester; it does not have a hydroxyl group (–OH) bonded to the carbonyl carbon atom.
c. No, this compound is a ketone; it has both a carbonyl group (C=O) and a hydroxyl group (–OH), but they are bonded to different carbon atoms.
d. Yes, this compound does contain a carboxyl group (the carboxyl group notation is the linear abbreviated form).

16.3 A carboxylic acid contains a carboxyl group, a carbonyl group (C=O) with a hydroxyl group (—OH) bonded to the carbonyl carbon atom.

a. Yes, it is a carboxylic acid.
b. No, it is not a carboxylic acid.
c. No, it is not a carboxylic acid.
d. Yes, it is a carboxylic acid.

16.5 The general structural formulas for the carboxylic acid derivatives are given below.

$R-\overset{O}{\overset{\|\|}{C}}-O-R$	$R-\overset{O}{\overset{\|\|}{C}}-Cl$	$R-\overset{O}{\overset{\|\|}{C}}-NH_2$	$R-\overset{O}{\overset{\|\|}{C}}-O-\overset{O}{\overset{\|\|}{C}}-R$
a. an ester	b. an acid chloride	c. an amide	d. an acid anhydride

16.7 An acyl group contains a carbonyl carbon bonded to an R group (R–C=O). If the carbonyl carbon of the acyl group is attached (C—Z) to an electronegative atom (oxygen, nitrogen, or halogen), the bond is polar. If the carbonyl carbon of the acyl group is attached to an atom of similar electronegativity (carbon or hydrogen), the bond is nonpolar.

a. In an amide, the C—N bond is polar.
b. In a ketone, the C—C bond is nonpolar.
c. In an acid anhydride, the C—O bond is polar.
d. In an ester, the C—O bond is polar.

16.9 A carbonyl group is a carbon atom double-bonded to an oxygen atom (C=O). An acyl group contains a carbonyl group bonded to an R group. A compound in which the acyl group is attached (C—Z) to an oxygen, nitrogen, or halogen atom is called an acyl compound. A compound in which the acyl group is bonded to carbon or hydrogen is called a carbonyl compound.

$$R-\overset{O}{\overset{||}{C}}-Z$$

acyl compound: C—Z = C—O, C—N, or
carbonyl compound: C—Z = C—H, C—C

a. For an amide the C—Z bond is C—N; an amide is an acyl compound.
b. For a ketone the C—Z bond is C—O; a ketone is a carbonyl compound.
c. For an acid anhydride the C—Z bond is C—O; an acid anhydride is an acyl compound.
d. For an ester the C—Z bond is C—O; an ester is an acyl compound.

16.11 A carbonyl group is a carbon atom double-bonded to an oxygen atom (C=O). A carboxyl group is a carbonyl group (C=O) with a hydroxyl group (—OH) bonded to the carbonyl atom. An acyl group is a carbonyl group bonded to an R group.

a. An amide contains a carbonyl group and an acyl group.
b. A carboxylic acid contains a carbonyl group, a carboxyl group, and an acyl group.
c. An aldehyde contains a carbonyl group and an acyl group.
d. An ester contains a carbonyl group and an acyl group.

16.13 When naming carboxylic acids using the IUPAC system, select as the parent chain the longest carbon chain that includes the carbon atom of the carboxyl group, number the parent chain by assigning the number 1 to the carboxyl carbon atom, determine the identity and location of any substituents. The suffix used is *–oic acid.*

a. butanoic acid
b. 2-methylpropanoic acid
c. 2-ethylbutanoic acid
d. propanoic acid

16.15 In the IUPAC system, carboxylic acids have ending *–oic acid*. The parent chain is the longest carbon chain that includes the carbon atom of the carboxyl group; the number 1 is assigned to the carboxyl carbon atom, with the locations of substituents numbered relative to it.

a. pentanoic acid
b. 3-methylbutanoic acid
c. 2-methylbutanoic acid
d. 4-methylhexanoic acid

16.17 The parent chain of a carboxylic acid includes the carboxyl group carbon atom (number 1). The locations for the substituents are relative to the carboxyl carbon atom.

a. $CH_3-CH_2-\underset{\substack{|\\ CH_2\\ |\\ CH_3}}{CH}-\overset{\substack{O\\ \|}}{C}-OH$

b. $CH_3-\underset{\substack{|\\ CH_3}}{CH}-CH_2-CH_2-\underset{\substack{|\\ CH_3}}{CH}-\overset{\substack{O\\ \|}}{C}-OH$

c. $CH_3-\overset{\substack{CH_3\\ |}}{CH}-\overset{\substack{O\\ \|}}{C}-OH$

d. $Cl-\overset{\substack{Cl\\ |}}{CH}-\overset{\substack{O\\ \|}}{C}-OH$

16.19 A dicarboxylic acid has two carboxyl groups, one at each end of the parent carbon chain; the suffix used is *dioic acid*. The simplest aromatic carboxylic acid is called benzoic acid. Other simple aromatic acids are named as derivatives of benzoic acid.

a. propanedioic acid
b. 3-methylpentanedioic acid
c. 2-chlorobenzoic acid
d. 2-chlorocyclopentanecarboxylic acid

16.21 Dicarboxylic acids have carboxyl groups at both ends of the parent carbon chains. Substituents on the ring of a cyclic carboxylic acid are given the lowest possible numbers relative to the carboxyl group.

a. $CH_3-CH_2-\overset{\substack{CH_3\\ |}}{\underset{\substack{|\\ CH_3}}{C}}-\overset{\substack{O\\ \|}}{C}-OH$

b. $HO-\overset{\substack{O\\ \|}}{C}-\overset{\substack{CH_3\\ |}}{\underset{\substack{|\\ CH_3}}{C}}-CH_2-\overset{\substack{O\\ \|}}{C}-OH$

c. $HO-\overset{\substack{O\\ \|}}{C}-\overset{\substack{CH_3\\ |}}{\underset{\substack{|\\ CH_3}}{C}}-CH_2-CH_2-\overset{\substack{O\\ \|}}{C}-OH$

d. COOH (cyclohexane ring with COOH and two methyl groups on the adjacent carbon)

16.23 Table 16.1 gives common names for the first six monocarboxylic acids. Table 16.2 gives the common names for the first six dicarboxylic acids.

a. The common name for the five-carbon monocarboxylic acid is valeric acid.
b. The common name for the two-carbon monocarboxylic acid is acetic acid.
c. The common name for the five-carbon dicarboxylic acid is glutaric acid.
d. The common name for the two-carbon dicarboxylic acid is oxalic acid.

16.25 Table 16.1 gives common names for the first six monocarboxylic acids. Table 16.2 gives the common names for the first six dicarboxylic acids.

a. Formic acid has one carbon atom.
b. Caproic acid has six carbon atoms.
c. Succinic acid has four carbon atoms.
d. Malonic acid has three carbon atoms.

16.27 The condensed structural formulas for the carboxylic acids in Problem 16.23 are:

a. $CH_3-(CH_2)_3-\overset{\displaystyle O}{\overset{\|}{C}}-OH$

b. $CH_3-\overset{\displaystyle O}{\overset{\|}{C}}-OH$

c. $HO-\overset{\displaystyle O}{\overset{\|}{C}}-(CH_2)_3-\overset{\displaystyle O}{\overset{\|}{C}}-OH$

d. $HO-\overset{\displaystyle O}{\overset{\|}{C}}-\overset{\displaystyle O}{\overset{\|}{C}}-OH$

16.29 The condensed structural formulas for the given carboxylic acids are:

a. $CH_3-CH_2-\overset{\displaystyle Cl}{\overset{|}{C}}H-COOH$

b. $CH_3-\overset{\displaystyle Cl}{\overset{|}{C}}H-CH_2-COOH$

c. $HOOC-\overset{\displaystyle Cl}{\overset{|}{C}}H-CH_2-COOH$

d. $HOOC-\overset{\displaystyle Cl}{\overset{|}{C}}H-\overset{\displaystyle Cl}{\overset{|}{C}}H-COOH$

16.31 a. Acrylic acid contains a carbon–carbon double bond.
b. Lactic acid contains a hydroxyl group.
c. Maleic acid contains a carbon–carbon double bond.
d. Glycolic acid contains a hydroxyl group.

16.33 a. Acrylic acid contains three carbon atoms (*prop-*) and a double bond between carbon atoms 2 and 3 (the only possible position, so no locant is specified); the IUPAC name is propenoic acid.
b. Lactic acid has three carbons with a hydroxyl group on carbon atom 2; the IUPAC name is 2-hydroxypropanoic acid.
c. Maleic acid is a four-carbon dicarboxylic acid with a carbon–carbon double bond whose hydrogen atoms are *cis-* to one another; the IUPAC name is *cis*-butenedioic acid.
d. Glycolic acid is a two-carbon acid with a hydroxyl group on carbon 2; the IUPAC name is 2-hydroxyethanoic acid.

16.35 When both a carboxyl group and a carbonyl group are present in the same molecule, the carboxyl group has precedence and the carbonyl group is named with the prefix *oxo*-. The name in part d. is a common name: the substituents are located using α amd β, relative to one of the carboxyl groups.

a. 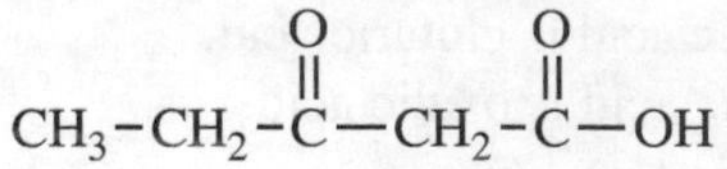 $CH_3-CH_2-C(=O)-CH_2-C(=O)-OH$

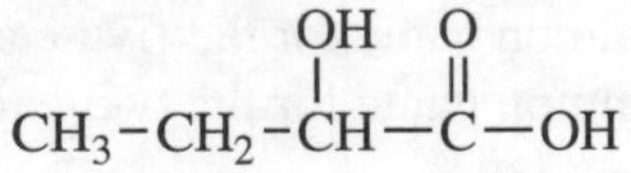

b. $CH_3-CH_2-CH(OH)-C(=O)-OH$

c. $(H_3C)(H)C=C(H)(CH_2-CH_2-C(=O)-OH)$

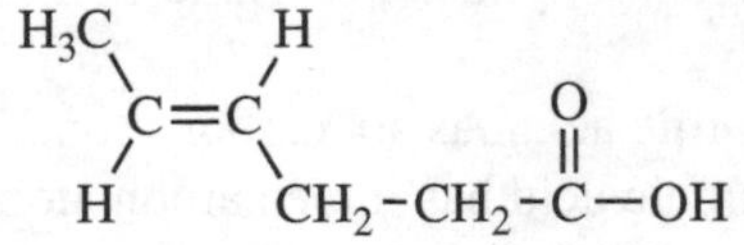

d. 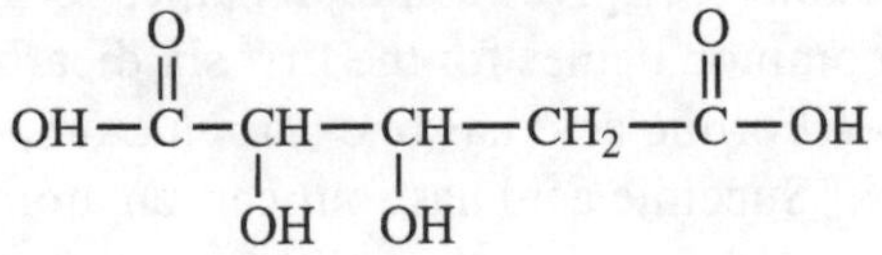 $OH-C(=O)-CH(OH)-CH(OH)-CH_2-C(=O)-OH$

16.37 a. Acrylic acid has one double bond and one carboxyl group; it is an unsaturated monocarboxylic acid.
b. Maleic acid (one double bond, two carboxyl groups) is an unsaturated dicarboxylic acid.
c. Succinic acid has two carboxyl groups; it is a dicarboxylic acid.
d. Caproic acid has one carboxyl group; it is a monocarboxylic acid.

16.39 Use Figure 16.8 in your textbook to obtain information about the physical states of the following compounds at room temperature.
a. The C_4 monocarboxylic acid is a liquid at room temperature.
b. The C_4 dicarboxylic acid is a solid at room temperature.
c. The C_5 monocarboxylic acid is a liquid at room temperature.
d. The C_5 dicarboxylic acid is a solid at room temperature.

16.41 a. A given carboxylic acid molecule can form two hydrogen bonds to another carboxylic acid molecule, producing a dimer (Figure 16.9).
b. The maximum number of hydrogen bonds that can form between an acetic acid molecule and water molecules is five, two hydrogen bonds between C=O and water molecules (Figure 15.7) and three hydrogen bonds between OH and water molecules (Figure 14.11).

16.43 Figure 16.8 gives the physical state of unbranched monocarboxylic and unbranched dicarboxylic acids at room temperature and pressure. A unique hydrogen bonding arrangement between two molecules of a carboxylic acid produces a dimer that has twice the mass of a single molecule, leading to high boiling and melting points.
a. Oxalic acid (a dicarboxylic acid) is a solid.
b. Decanoic acid (a 10-carbon acid) is a solid.
c. Hexanoic acid (a 6-carbon acid) is a liquid.
d. Benzoic acid (an aromatic carboxylic acid) is a solid.

16.45 Oxidation of primary alcohols or aldehydes using an oxidizing agent, such as CrO_3 or $K_2Cr_2O_7$, produces carboxylic acids. Aromatic acids can be prepared by oxidizing a carbon-containing side chain (an alkyl group) on a benzene derivative.

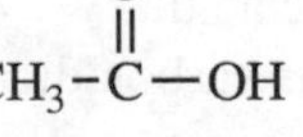

a. 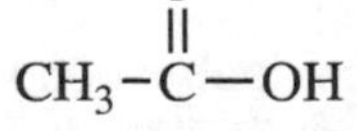 $CH_3-C(=O)-OH$

b. $CH_3-C(=O)-OH$

c. $CH_3-CH_2-CH(CH_3)-CH_2-C(=O)-OH$

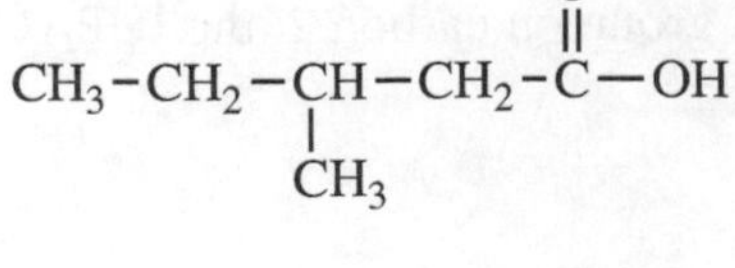

d. 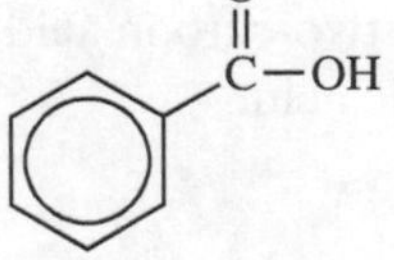 $C_6H_5-C(=O)-OH$

16.47 There is one acidic hydrogen for each carboxyl group in a carboxylic acid. Common names and structures for the acids in this problem can be found in Section 16.3.

a. Pentanoic acid is a monocarboxylic acid; it has one acidic hydrogen atom.
b. Citric acid has three carboxyl groups; it has three acidic hydrogen atoms.
c. Succinic acid is a dicarboxylic acid; it has two acidic hydrogen atoms.
d. Oxalic acid is a dicarboxylic acid; it has two acidic hydrogen atoms.

16.49 A carboxylate ion is the negative ion (–1) produced when a carboxylic acid loses its acidic hydrogen atom.

a. Pentanoic acid is a monocarboxylic acid; it loses one acidic hydrogen atom and forms one carboxylate ion, and so has a –1 charge.
b. Citric acid has three carboxyl groups; it loses three acidic hydrogen atoms and forms three carboxylate ions on the same molecule, so it has a –3 charge.
c. Succinic acid is a dicarboxylic acid; it loses two acidic hydrogen atoms, and so has a –2 charge.
d. Oxalic acid is a dicarboxylic acid; it loses two acidic hydrogen atoms, and has a –2 charge.

16.51 Carboxylate ions are named by dropping the *–ic acid* ending from the name of the parent acid and replacing it with *–ate*.

a. Pentanoic acid forms a pentanoate ion.
b. Citric acid forms a citrate ion.
c. Succinic acid forms a succinate ion.
d. Oxalic acid forms an oxalate ion.

16.53 When a carboxylic acid is placed in water, hydrogen ion transfer occurs to produce hydronium ion and carboxylate ion.

a. $CH_3-\overset{\displaystyle O}{\overset{\|}{C}}-OH + H_2O \longrightarrow H_3O^+ + CH_3-\overset{\displaystyle O}{\overset{\|}{C}}-O^-$

b. $HO-\overset{\displaystyle O}{\overset{\|}{C}}-CH_2-\underset{\underset{\displaystyle OH}{|}}{\underset{\underset{\displaystyle C=O}{|}}{\overset{\displaystyle OH}{\overset{|}{C}}}}-CH_2-\overset{\displaystyle O}{\overset{\|}{C}}-OH + 3H_2O \longrightarrow 3H_3O^+ + {}^-O-\overset{\displaystyle O}{\overset{\|}{C}}-CH_2-\underset{\underset{\displaystyle O^-}{|}}{\underset{\underset{\displaystyle C=O}{|}}{\overset{\displaystyle OH}{\overset{|}{C}}}}-CH_2-\overset{\displaystyle O}{\overset{\|}{C}}-O^-$

c. $CH_3-\overset{\displaystyle O}{\overset{\|}{C}}-OH + H_2O \longrightarrow H_3O^+ + CH_3-\overset{\displaystyle O}{\overset{\|}{C}}-O^-$

d. $CH_3-CH_2-\underset{\underset{\displaystyle CH_3}{|}}{CH}-\overset{\displaystyle O}{\overset{\|}{C}}-OH + H_2O \longrightarrow H_3O^+ + CH_3-CH_2-\underset{\underset{\displaystyle CH_3}{|}}{CH}-\overset{\displaystyle O}{\overset{\|}{C}}-O^-$

16.55 The structural formulas for the given ions are:

a. $CH_3-CH_2-\overset{\displaystyle O}{\overset{\|}{C}}-O^-$

b. $CH_3-CH_2-\overset{\displaystyle O}{\overset{\|}{C}}-O^-\ Na^+$

c. $CH_3-\overset{\displaystyle O}{\overset{\|}{C}}-O^-$

d. $CH_3-\overset{\displaystyle O}{\overset{\|}{C}}-O^-\ Na^+$

16.57 Carboxylate salts are named similarly to other ionic compounds. The positive ion is named first; it is followed by a separate word giving the negative ion (named by dropping the *–ic acid* ending from the name of the parent acid and replacing it with *–ate*).

a. potassium ethanoate
b. calcium propanoate
c. potassium butanedioate
d. sodium pentanoate

16.59 a. The common name for the potassium salt of a C_2 carboxylic acid is potassium acetate.
b. The common name for the calcium salt of a C_3 carboxylic acid is calcium propionate.
c. The common name for the potassium salt of a C_4 dicarboxylic acid is potassium succinate.
d. The common name for the sodium salt of a C_5 carboxylic acid is sodium valerate.

16.61 Like inorganic acids, carboxylic acids react with strong bases to produce water and a salt; the salt formed is a carboxylic acid salt (a carboxylate).

a. $CH_3-\overset{O}{\overset{\|}{C}}-OH + KOH \rightarrow CH_3-\overset{O}{\overset{\|}{C}}-O^- K^+ + H_2O$

b. $2\ CH_3-CH_2-\overset{O}{\overset{\|}{C}}-OH + Ca(OH)_2 \rightarrow \left(CH_3-CH_2-\overset{O}{\overset{\|}{C}}-O^-\right)_2 Ca^{2+} + 2H_2O$

c. $HO-\overset{O}{\overset{\|}{C}}-CH_2-CH_2-\overset{O}{\overset{\|}{C}}-OH + 2\ KOH \rightarrow K^+\ {}^-O-\overset{O}{\overset{\|}{C}}-CH_2-CH_2-\overset{O}{\overset{\|}{C}}-O^- K^+ + 2H_2O$

d. $CH_3-CH_2-CH_2-CH_2-\overset{O}{\overset{\|}{C}}-OH + NaOH \rightarrow CH_3-CH_2-CH_2-CH_2-\overset{O}{\overset{\|}{C}}-O^- Na^+ + H_2O$

16.63 Converting a carboxylic acid salt back to its carboxylic acid is very simple. The salt reacts with a solution of a strong acid (in this case HCl) to give the carboxylic acid and the inorganic salt.

a. $CH_3-CH_2-CH_2-\overset{O}{\overset{\|}{C}}-O^- Na^+ + HCl \rightarrow CH_3-CH_2-CH_2-\overset{O}{\overset{\|}{C}}-OH + NaCl$

b. $K^+\ {}^-O-\overset{O}{\overset{\|}{C}}-\overset{O}{\overset{\|}{C}}-O^- K^+ + 2HCl \rightarrow HO-\overset{O}{\overset{\|}{C}}-\overset{O}{\overset{\|}{C}}-OH + 2KCl$

c. $\left({}^-O-\overset{O}{\overset{\|}{C}}-CH_2-\overset{O}{\overset{\|}{C}}-O^-\right) Ca^{2+} + 2HCl \rightarrow HO-\overset{O}{\overset{\|}{C}}-CH_2-\overset{O}{\overset{\|}{C}}-OH + CaCl_2$

d. $C_6H_5-\overset{O}{\overset{\|}{C}}-O^- Na^+ + HCl \rightarrow C_6H_5-\overset{O}{\overset{\|}{C}}-OH + NaCl$

16.65 An antimicrobial is a compound used as a food preservative.

16.67 a. benzoic acid b. sorbic acid c. sorbic acid d. propionic acid

16.69 In the IUPAC sytem, dicarboxylic acid salts are named in the same way as monocarboxylic acid salts, except that the prefix *di-* precedes the *–ate* ending.
a. Sodium formate is a monocarboxylic acid salt. A formate is the salt of formic acid, the common name for the C_1 monocarboxylic acid.
b. Potassium pentanedioate is a dicarboxylic acid salt.
c. Calcium oxalate is a dicarboxylic acid salt. Oxalic acid is the common name for the C_2 diacid.
d. Magnesium propanoate is a monocarboxylic acid salt.

16.71 a. There are two ions in one formula unit of sodium formate: one sodium ion, one formate ion.
b. There are three ions in one formula unit of potassium pentanedioate, two potassium ions and one pentanedioate ion.
c. There are two ions in a formula unit of calcium oxalate: one calcium ion, one oxalate ion.
d. There are three ions in one formula unit of magnesium propanoate, one magnesium ion and two propanoate ions.

16.73 The small inorganic molecule that is always a product of a decarboxylation reaction is CO_2.

16.75 The structures of the organic compounds formed by decarboxylation of each of these carboxylic acids:

a. $CH_3-CH_2-CH_2-CH_3$ b. $CH_3-CH_2-\overset{\overset{\displaystyle O}{\|}}{C}-CH_3$

16.77 a. two oxygen atoms b. two carbon atoms

16.79 An ester is a carboxylic acid derivative in which the –OH portion of the carboxyl group has been replaced with a –OR group.
a. Yes, this compound is an ester.
b. Yes, this compound is an ester
c. No, not an ester. The –OR group is not attached to the carbonyl carbon atom.
d. Yes, this compound is an ester (a cyclic ester); the oxygen atom in the ring is attached to the carbonyl carbon atom (also in the ring).

16.81 a. Acetic acid contains a carboxyl group; it is a carboxylic acid.
b. Methyl acetate contains a methyl group bonded to an acetate group; it is an ester.
c. Sodium acetate contains a positive metal ion and an acetate group; it is a carboxylic acid salt.
d. Acetic acid contains a carboxyl group; it is a carboxylic acid.

16.83 An ester is a carboxylic acid derivative in which the –OH portion of the carboxyl group has been replaced with a –OR group. The structures of the esters formed by the condensation reactions of the given acid with the four given alcohols are:

a. $CH_3-CH_2-\overset{\overset{\displaystyle O}{\|}}{C}-O-CH_2-CH_3$ b. $CH_3-CH_2-\overset{\overset{\displaystyle O}{\|}}{C}-O-CH_3$

c. $CH_3-CH_2-\overset{\overset{\displaystyle O}{\|}}{C}-O-CH_2-\overset{\overset{\displaystyle CH_3}{|}}{CH}-CH_3$ d. $CH_3-CH_2-\overset{\overset{\displaystyle O}{\|}}{C}-O-\overset{\overset{\displaystyle CH_2-CH_3}{|}}{CH}-CH_2-CH_3$

16.85 The reaction of a carboxylic acid with an alcohol produces an ester. To determine the parent acid and parent alcohol of the ester, split the ester molecule between the carbonyl group and the alkoxy group; the carbonyl portion adds –OH to become the carboxylic acid, and the alkoxy group adds a –H atom to become the alcohol.

a. $CH_3-CH_2-\overset{\overset{\Large O}{\|}}{C}-OH + HO-CH_2-CH_3$ (acid; alcohol)

b. $CH_3-CH_2-CH_2-\overset{\overset{\Large O}{\|}}{C}-OH + CH_3-OH$ (acid; alcohol)

c. $CH_3-\overset{\overset{\Large O}{\|}}{C}-OH$ + phenol (C_6H_5-OH) (acid; alcohol)

d. benzoic acid ($C_6H_5-\overset{\overset{\Large O}{\|}}{C}-OH$) + CH_3-OH (acid; alcohol)

16.87 Lactones are cyclic esters; they are produced via an intramolecular esterification reaction.

16.89 To name an ester using the IUPAC system, visualize it as having an acid part and an alcohol part. The name of the alcohol appears first, followed by a separate word giving the acid name with the suffix *–ate*.

a. ethyl propanoate
b. propyl ethanoate
c. ethyl propanoate
d. butyl methanoate

16.91 Common names for esters are similar to IUPAC names. The common name for the acid is used.

a. ethyl propionate b. propyl acetate c. ethyl propionate d. butyl formate

16.93 To name an ester using the IUPAC system, visualize it as having an acid part and an alcohol part. The name of the alcohol appears first, followed by a separate word giving the acid name with the suffix *–ate*.

a. ethyl butanoate
b. propyl pentanoate
c. methyl 3-methylpentanoate
d. ethyl propanoate

16.95 Common names for esters are similar to IUPAC names, except that the common name for the acid is used. Visualize the ester as having an acid part and an alcohol part; the carbonyl portion of the ester is contributed by the acid and the alkoxy portion is contributed by the alcohol.

a. $H-\overset{\overset{\Large O}{\|}}{C}-O-CH_3$

b. $C_6H_5-CH_2-\overset{\overset{\Large O}{\|}}{C}-O-CH_2-CH_3$

c. $CH_3-\overset{\overset{\Large O}{\|}}{C}-O-\underset{\underset{\Large CH_3}{|}}{CH}-CH_3$

d. $CH_3-\overset{\overset{\Large O}{\|}}{C}-O-CH_2-\underset{\underset{\Large Br}{|}}{CH}-CH_3$

16.97 The reaction of a carboxylic acid with an alcohol produces an ester. To name an ester using the IUPAC system, visualize it as having an acid part and an alcohol part. The name of the alcohol appears first, followed by a separate word giving the acid name with the suffix *–ate*.

a. ethyl ethanoate
b. methyl ethanoate
c. ethyl butanoate
d. 1-methylpropyl hexanoate (or *sec*-butyl hexanoate)

16.99 A carboxylic acid is named by changing the *–e* ending of the alkane to *–oic acid*. In an ester name, the name of the alcohol appears first, followed by a separate word giving the acid name with the suffix *–ate*. The name of a carboxylic acid salt is the name of the positive metal ion followed by the acid name with the suffix *–ate*.

a. Methyl succinate is an ester.
b. Sodium succinate is a carboxylic acid salt.
c. Methyl butanoate is an ester.
d. Methyl 2-methylbutanoate is an ester.

16.101 a. Methyl succinate is the ester of methyl alcohol (C_1) and succinic acid (C_4); there are five carbon atoms.

b. Sodium succinate is the salt of succinic acid (C_4); there are four carbon atoms.
c. Methyl butanoate is the ester of methyl alcohol (C_1) and butanoic acid (C_4); there are five carbon atoms.
d. Methyl 2-methylbutanoate is the ester of methyl alcohol (C_1) and 2-methylbutanoic acid (C_5); there are six carbon atoms.

16.103 See Table 16.4, which gives the odor of some esters.

a. Isobutyl methanoate has the odor of raspberry.
b. Pentyl ethanoate has the odor of banana.
c. Methyl butanoate has the odor of apple.

16.105 The acid part is the same; the alcohol part is methyl (apple) versus ethyl (pineapple).

16.107 The –OH group of salicylic acid has been esterified (methyl ester) in aspirin.

16.109 Nepetalactone, a lactone present in the catnip plant, is an attractant for cats but is not considered to be a pheromone because different species are involved (plant and animal).

16.111 In a C_5 monocarboxylic acid, the first carbon atom belongs to the carboxyl group. This leaves four carbon atoms for the saturated alkyl portion of the molecule. There are four possible four-carbon alkyl groups. The IUPAC names for the four acids are: pentanoic acid, 2-methylbutanoic acid, 3-methylbutanoic acid, and 2,2-dimethylpropanoic acid.

16.113 In a methyl ester containing six carbon atoms, one carbon atom belongs to the carbonyl group and one to the methoxy group. This leaves four carbon atoms for the saturated alkyl side chain; there are four possible four-carbon alkyl groups. The IUPAC names for the four esters are: methyl pentanoate, methyl 2-methylbutanoate, methyl 3-methylbutanoate, and methyl 2,2-dimethylpropanoate.

16.115 One way to do this problem is to draw skeletal structures of esters isomeric with 2-methylbutanoic acid (a C_5 acid). Ester isomers will have one carbon atom in the carbonyl group and four carbon atoms distributed between the alcohol and acid portions of the ester.

$$\mathrm{C-C-C-\overset{\overset{\large O}{\|}}{C}-O-C}\ ,\quad \mathrm{C-\underset{\underset{\large C}{|}}{C}-\overset{\overset{\large O}{\|}}{C}-O-C}$$

Consider the possible methyl esters first: there are two possible C_3 side chains for the acid. The two esters are: methyl butanoate and methyl 2-methylpropanoate.

$$\mathrm{C-C-\overset{\overset{\large O}{\|}}{C}-O-C-C}$$

Next consider the ethyl esters: there is only one possible C_2 side chain for the acid (propanoic acid). The name of the ester is ethyl propanoate.

$$\mathrm{C-\overset{\overset{\large O}{\|}}{C}-O-C-C-C}\ ,\quad \mathrm{C-\overset{\overset{\large O}{\|}}{C}-O-\overset{\overset{\large C}{|}}{C}-C}$$

Consider the propyl esters: there is a C_2 acid (ethanoic) and two possible C_3 alcohols (propyl and isopropyl). The names of the esters are: propyl ethanoate and isopropyl ethanoate.

Finally, consider the C_4 alcohol groups: there are four alkyl side chains having four carbons. The names of the esters are: butyl methanoate, *sec*-butyl methanoate, isobutyl methanoate, and *tert*-butyl methanoate.

$$\mathrm{\overset{\overset{\large O}{\|}}{C}-O-C-C-C-C}\ ,\ \mathrm{\overset{\overset{\large O}{\|}}{C}-O-\underset{\underset{\large C}{|}}{C}-C-C}\ ,\ \mathrm{\overset{\overset{\large O}{\|}}{C}-O-C-\underset{\underset{\large C}{|}}{C}-C}\ ,\ \mathrm{\overset{\overset{\large O}{\|}}{C}-O-\overset{\overset{\large C}{|}}{\underset{\underset{\large C}{|}}{C}}-C}$$

There are a total of nine ester isomers.

16.117 A C_3 carboxylic acid has one carbon atom in the carboxyl group and two carbon atoms in the side chain. There are two possibilities for a C_3 ester: one has one carbon atom in the carbonyl group, one in the alcohol portion and one in the side chain of the acid portion, and the other has one carbon atom in the carbonyl group and two in the alcohol portion.

$$\mathrm{CH_3-CH_2-\overset{\overset{\large O}{\|}}{C}-OH}\qquad \mathrm{CH_3-\overset{\overset{\large O}{\|}}{C}-O-CH_3}\qquad \mathrm{H-\overset{\overset{\large O}{\|}}{C}-O-CH_2-CH_3}$$

16.119 Use the simplest ester, methyl formate, as an example. The condensed structural formula is $HCOOCH_3$ and the molecular formula is $C_2H_4O_2$. Since the ester functional group always has two oxygen atoms, the general molecular formula for an ester is $C_nH_{2n}O_2$.

16.121 The maximum number of hydrogen bonds that can form between this ester molecule and
a. another ester molecule is zero.
b. a water molecule is four.

16.123 Figure 16.15 gives the physical states of methyl and ethyl esters at room temperature.
a. Methyl ethanoate is a liquid.
b. Methyl propanoate is a liquid.
c. Ethyl ethanoate is a liquid.
d. Ethyl butanoate is a liquid.

16.125 Hydrolysis of an ester with a strong acid yields a carboxylic acid and an alcohol.

a. $CH_3-CH_2-\overset{O}{\overset{\|}{C}}-OH + CH_3-OH$

b. $CH_3-\overset{CH_3}{\overset{|}{C}H}-\overset{O}{\overset{\|}{C}}-OH + CH_3-CH_2-CH_2-OH$

c. $CH_3-CH_2-CH_2-\overset{O}{\overset{\|}{C}}-OH + C_6H_5-OH$

d. $C_6H_5-\overset{O}{\overset{\|}{C}}-OH + CH_3-OH$

16.127 Saponification of an ester with a strong base yields the carboxylate salt of the base and an alcohol.

a. $CH_3-CH_2-\overset{O}{\overset{\|}{C}}-O^-Na^+ + CH_3-OH$

b. $CH_3-\overset{CH_3}{\overset{|}{C}H}-\overset{O}{\overset{\|}{C}}-O^-Na^+ + CH_3-CH_2-CH_2-OH$

c. $CH_3-CH_2-CH_2-\overset{O}{\overset{\|}{C}}-O^-Na^+ + C_6H_5-OH$

d. $C_6H_5-\overset{O}{\overset{\|}{C}}-O^-Na^+ + CH_3-OH$

16.129 Hydrolysis of an ester with a strong acid yields an alcohol and a carboxylic acid.
a. methanol and butanoic acid
b. methanol and ethanoic acid
c. 2-propanol and propanoic acid
d. 2-propanol and ethanoic acid

16.131 Hydrolysis of an ester with a strong acid yields an alcohol and a carboxylic acid.
a. ethanol and butanoic acid
b. 1-propanol and pentanoic acid
c. methanol and 3-methylpentanoic acid
d. ethanol and propanoic acid

16.133 Hydrolysis of an ester with a strong acid yields a carboxylic acid and an alcohol; saponification of an ester with a strong base yields the carboxylate salt of the base and an alcohol.

a. $CH_3-\underset{CH_3}{\underset{|}{C}H}-\overset{O}{\overset{\|}{C}}-O-CH_2-CH_3 + H_2O \overset{H^+}{\rightleftharpoons} CH_3-\underset{CH_3}{\underset{|}{C}H}-\overset{O}{\overset{\|}{C}}-OH + CH_3-CH_2-OH$

b. $CH_3-\underset{CH_3}{\underset{|}{C}H}-\overset{O}{\overset{\|}{C}}-O-CH_2-CH_3 \xrightarrow{NaOH} CH_3-\underset{CH_3}{\underset{|}{C}H}-\overset{O}{\overset{\|}{C}}-O^-Na^+ + CH_3-CH_2-OH$

c. $H{-}\overset{\displaystyle O}{\overset{\|}{C}}{-}O{-}CH_2{-}CH_2{-}CH_2{-}CH_3 + H_2O \overset{H^+}{\rightleftharpoons} CH_3{-}CH_2{-}CH_2{-}CH_2{-}OH + H{-}\overset{\displaystyle O}{\overset{\|}{C}}{-}OH$

d. $CH_3{-}\overset{\displaystyle O}{\overset{\|}{C}}{-}O{-}CH_2{-}\underset{\displaystyle CH_3}{\underset{|}{CH}}{-}\underset{\displaystyle CH_3}{\underset{|}{CH}}{-}CH_3 \xrightarrow{NaOH} CH_3{-}\overset{\displaystyle O}{\overset{\|}{C}}{-}O^- Na^+ + HO{-}CH_2{-}\underset{\displaystyle CH_3}{\underset{|}{CH}}{-}\underset{\displaystyle CH_3}{\underset{|}{CH}}{-}CH_3$

16.135 Thiols react with carboxylic acids to form thioesters, sulfur-containing analogs of esters; in a thioester an –SR group replaces the –OR group.

a. $CH_3{-}\overset{\displaystyle O}{\overset{\|}{C}}{-}S{-}CH_2{-}CH_3$

b. $CH_3{-}(CH_2)_8{-}\overset{\displaystyle O}{\overset{\|}{C}}{-}S{-}CH_3$

c. 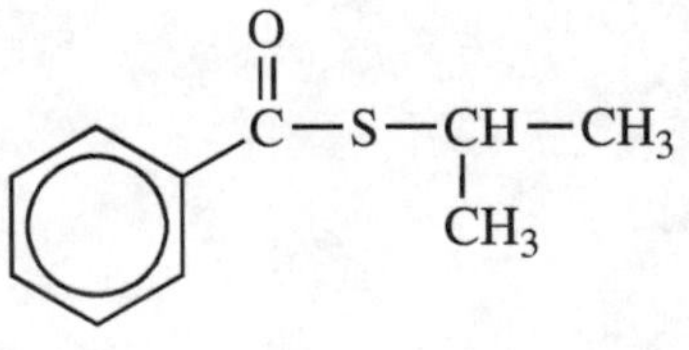 $C_6H_5{-}\overset{\displaystyle O}{\overset{\|}{C}}{-}S{-}\underset{\displaystyle CH_3}{\underset{|}{CH}}{-}CH_3$

d. $H{-}\overset{\displaystyle O}{\overset{\|}{C}}{-}S{-}CH_2{-}CH_2{-}CH_3$

16.137

a. $CH_3{-}\overset{\displaystyle O}{\overset{\|}{C}}{-}S{-}CH_3$

b. same as part a.

c. $H{-}\overset{\displaystyle O}{\overset{\|}{C}}{-}S{-}CH_2{-}CH_3$

d. same as part c.

16.139 A polyester is a condensation polymer in which the monomers are joined through ester linkages. Oxalic acid has carboxyl groups at either end, and 1,3-propanediol has hydroxyl groups at either end. The esterification reactions between these molecules forms a polymer; two repeating units of the polymer are shown below.

$$\left(\overset{\displaystyle O}{\overset{\|}{C}}{-}\overset{\displaystyle O}{\overset{\|}{C}}{-}O{-}(CH_2)_3{-}O{-}\overset{\displaystyle O}{\overset{\|}{C}}{-}\overset{\displaystyle O}{\overset{\|}{C}}{-}O{-}(CH_2)_3{-}O \right)$$

16.141 The given polyester is the product of esterification of a dicarboxylic acid with a diol. The diol has three carbon atoms (1,3-propanediol) and the dicarboxylic acid has four carbon atoms (succinic acid).

$$HO{-}\overset{\displaystyle O}{\overset{\|}{C}}{-}CH_2{-}CH_2{-}\overset{\displaystyle O}{\overset{\|}{C}}{-}OH, \quad HO{-}CH_2{-}CH_2{-}CH_2{-}OH$$

16.143 The monomers are ethylene glycol and terephthalic acid.

16.145 An acid chloride is a carboxylic acid derivative in which a portion of the carboxyl group has been replaced with a –Cl atom. Acid chlorides are named by replacing the *–ic acid* ending of an acid's common name with *–yl chloride*, or by replacing the *–oic acid* ending of the acid's IUPAC name *–oyl chloride*.

a. 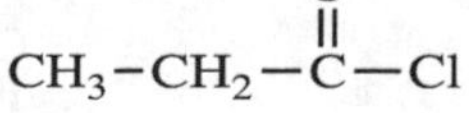 $CH_3-CH_2-\overset{O}{\overset{\|}{C}}-Cl$

b. 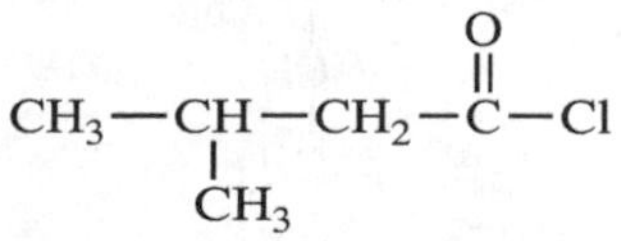 $CH_3-\underset{CH_3}{\underset{|}{CH}}-CH_2-\overset{O}{\overset{\|}{C}}-Cl$

An acid anhydride is a carboxylic acid derivative; it can be visualized as two carboxylic acid molecules bonded together after removal of a water molecule.

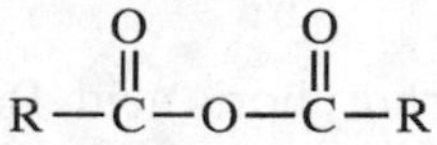

$$R-\overset{O}{\overset{\|}{C}}-O-\overset{O}{\overset{\|}{C}}-R$$

Symmetrical acid anhydrides (both R groups the same) are named by replacing the acid ending of the parent carboxylic acid name with the word *anhydride*. Mixed acid anhydrides (different R groups) are named by using the names of the individual parent carboxylic acids (in alphabetical order) followed by the word *anhydride*.

c. 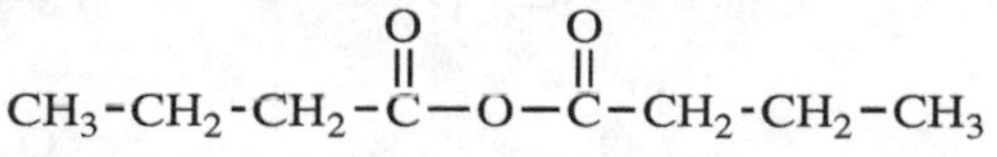 $CH_3-CH_2-CH_2-\overset{O}{\overset{\|}{C}}-O-\overset{O}{\overset{\|}{C}}-CH_2-CH_2-CH_3$

d. 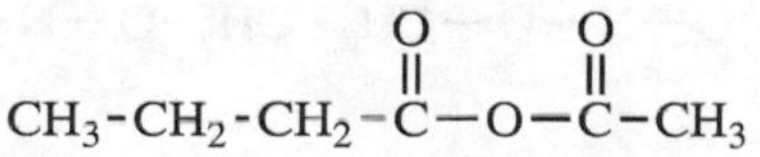 $CH_3-CH_2-CH_2-\overset{O}{\overset{\|}{C}}-O-\overset{O}{\overset{\|}{C}}-CH_3$

16.147 Compounds a. and d. are mixed acid anhydrides; they are named by using the names of the individual parent carboxylic acids (in alphabetical order) followed by the word *anhydride*. Compounds b. and c. are acid chlorides; an acid chloride is named by replacing the *–oic acid* ending of the IUPAC name of the acid from which it is derived with *–oyl chloride*.

a. ethanoic propanoic anhydride
b. pentanoyl chloride
c. 2,3-dimethylbutanoyl chloride
d. methanoic propanoic anhydride

16.149 a. 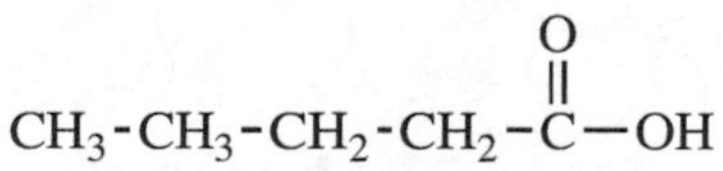 $CH_3-CH_3-CH_2-CH_2-\overset{O}{\overset{\|}{C}}-OH$

An acid chloride reacts with water, in a hydrolysis reaction, to regenerate the parent carboxylic acid. Pentanoyl chloride hydrolyzes to form pentanoic acid.

b. $CH_3-CH_3-CH_2-CH_2-\overset{O}{\overset{\|}{C}}-OH$

An acid anhydride undergoes hydrolysis to regenerate the parent carboxylic acids.

16.151 Reaction of an acid anhydride with an alcohol produces an ester and a carboxylic acid.

a. Acetic anhydride + ethyl alcohol → ethyl acetate + acetic acid

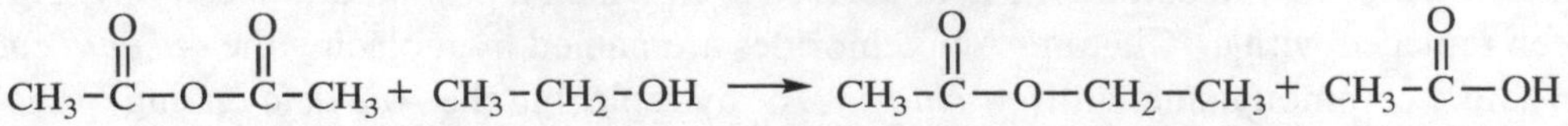

$$CH_3-\overset{\displaystyle O}{\overset{\|}{C}}-O-\overset{\displaystyle O}{\overset{\|}{C}}-CH_3 + CH_3-CH_2-OH \longrightarrow CH_3-\overset{\displaystyle O}{\overset{\|}{C}}-O-CH_2-CH_3 + CH_3-\overset{\displaystyle O}{\overset{\|}{C}}-OH$$

b. Acetic anhydride + 1-butanol → butyl acetate + acetic acid

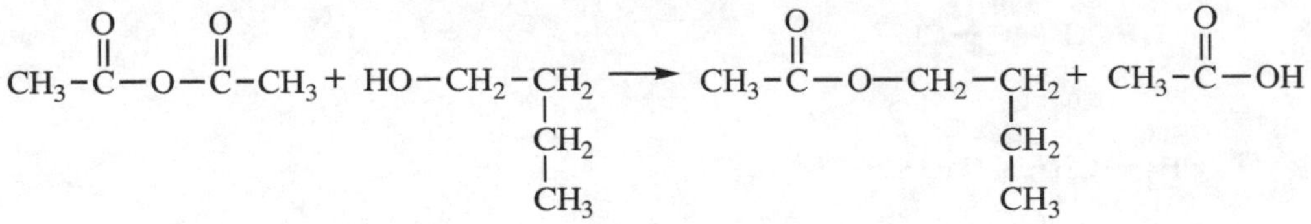

$$CH_3-\overset{\displaystyle O}{\overset{\|}{C}}-O-\overset{\displaystyle O}{\overset{\|}{C}}-CH_3 + HO-CH_2-CH_2-CH_2-CH_3 \longrightarrow CH_3-\overset{\displaystyle O}{\overset{\|}{C}}-O-CH_2-CH_2-CH_2-CH_3 + CH_3-\overset{\displaystyle O}{\overset{\|}{C}}-OH$$

16.153 a. propanoyl group b. butanoyl group c. butanoyl group d. ethanoyl group

16.155 The products are an ester and HCl.

16.157 A phosphate ester is formed by the reaction of an alcohol with phosphoric acid. Because phosphoric acid has three hydroxyl groups, it can form mono-, di-, and triesters. Nitric acid reacts with alcohols to form esters in a manner similar to that for carboxylic acids.

a. 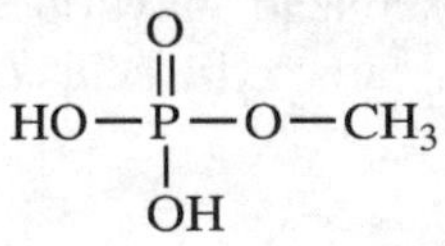 $HO-\overset{\displaystyle O}{\overset{\|}{P}}(OH)-O-CH_3$

b. 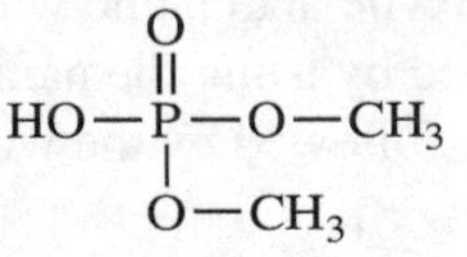 $HO-\overset{\displaystyle O}{\overset{\|}{P}}(O-CH_3)-O-CH_3$

c. 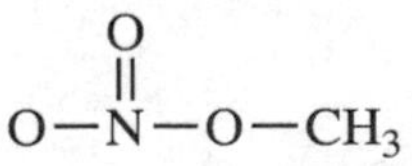 $O-\overset{\displaystyle O}{\overset{\|}{N}}-O-CH_3$

d. 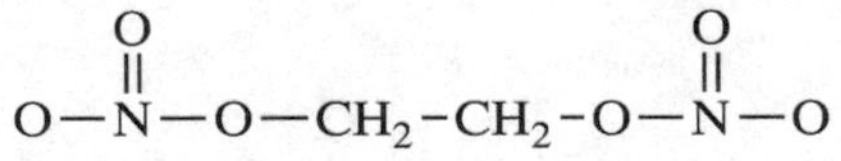 $O-\overset{\displaystyle O}{\overset{\|}{N}}-O-CH_2-CH_2-O-\overset{\displaystyle O}{\overset{\|}{N}}-O$

16.159 Phosphoric acid esters contain acidic hydrogen atoms, which are lost through ionization.

a. 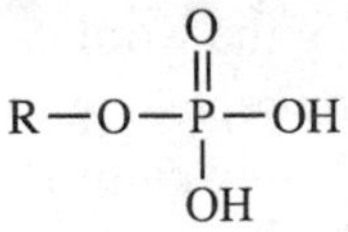 $R-O-\overset{\displaystyle O}{\overset{\|}{P}}(OH)-OH$

b. 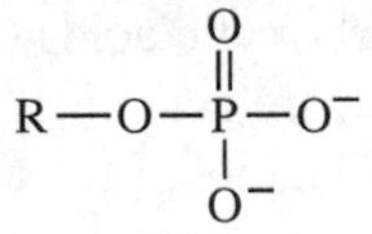 $R-O-\overset{\displaystyle O}{\overset{\|}{P}}(O^-)-O^-$

c. 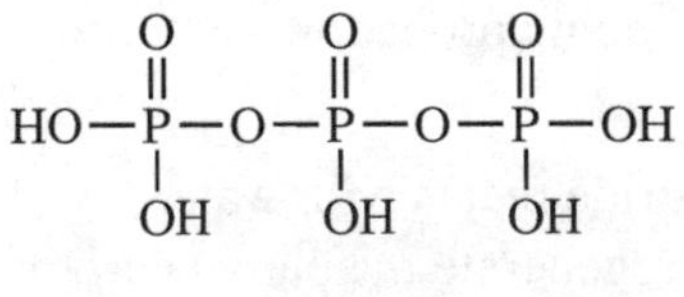 $HO-\overset{\displaystyle O}{\overset{\|}{P}}(OH)-O-\overset{\displaystyle O}{\overset{\|}{P}}(OH)-O-\overset{\displaystyle O}{\overset{\|}{P}}(OH)-OH$

d. 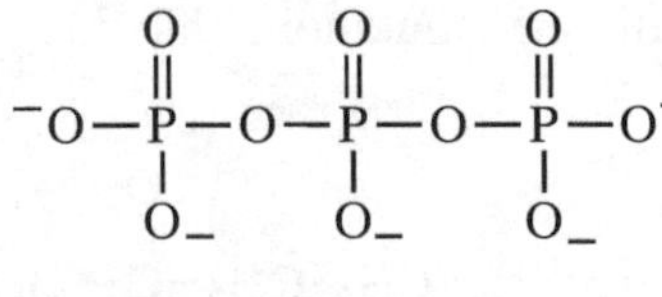 $^-O-\overset{\displaystyle O}{\overset{\|}{P}}(O^-)-O-\overset{\displaystyle O}{\overset{\|}{P}}(O^-)-O-\overset{\displaystyle O}{\overset{\|}{P}}(O^-)-O^-$

16.161 A phosphoryl group is a $-PO_3^{2-}$ group.

16.163 The structural formulas (ionic form) for the organic products produced from a phosphorylation reaction that involves the methyl ester of diphosphoric acid and ethyl alcohol are shown below. This reaction produces two monophosphate ester ions.

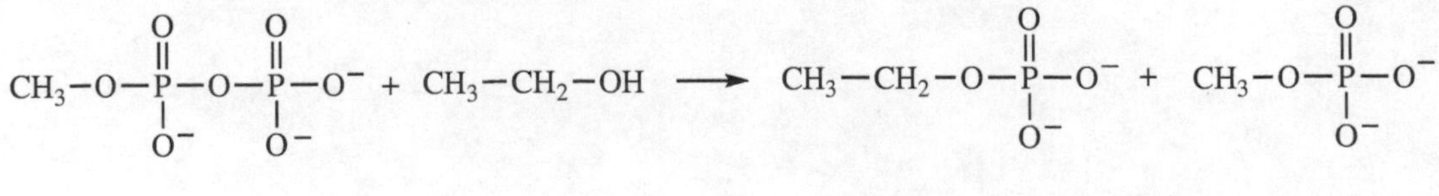

$$CH_3-O-\overset{\displaystyle O}{\overset{\|}{P}}(O^-)-O-\overset{\displaystyle O}{\overset{\|}{P}}(O^-)-O^- + CH_3-CH_2-OH \longrightarrow CH_3-CH_2-O-\overset{\displaystyle O}{\overset{\|}{P}}(O^-)-O^- + CH_3-O-\overset{\displaystyle O}{\overset{\|}{P}}(O^-)-O^-$$

Amines and Amides — Chapter 17

Solutions to Selected Problems

17.1 3 (N), 2(O), and 4 (C)

17.3 a. A primary amine has one R-group. b. A secondary amine has two R-groups.
c. A tertiary amine has three R-groups.

17.5 a. A primary amine has one nitrogen atom. b. A secondary amine has one nitrogen atom.
c. A tertiary amine has one nitrogen atom.

17.7 The amine functional group consists of a nitrogen atom with one or more alkyl, cycloalkyl, or aryl groups substituted for the hydrogen atoms in NH_3.
a. Yes, the compound contains an amine functional group; one alkyl group is attached to the nitrogen atom.
b. Yes, the compound contains an amine functional group; two alkyl groups are attached to the nitrogen atom.
c. No, the compound does not contain an amine functional group; the nitrogen atom is attached to a carbonyl functional group rather than an alkyl group. This is an amide.
d. Yes, the compound contains an amine functional group; three alkyl groups are attached to the nitrogen atom.

17.9 Of the compounds in Problem 17.7:
a. contains an amino group
b. contains a monosubstituted amino group
c. is not an amine
d. contains a disubstituted amino group

17.11 In a primary amine, the nitrogen atom is bonded to one hydrocarbon group and two hydrogen atoms, in a secondary amine the nitrogen atom is bonded to two hydrocarbon groups and one hydrogen atom, and in a tertiary amine the nitrogen atom is bonded to three hydrocarbon groups and no hydrogen atoms. The compounds in this problem are classified as:
a. a primary amine
b. a primary amine
c. a secondary amine
d. a tertiary amine

17.13 Primary, secondary, and tertiary amines contain a nitrogen atom bonded to (respectively) one, two, or three alkyl, cycloalkyl, or aryl groups. In a cyclic amine, the nitrogen is part of a ring. The compounds in this problem are classified as:
a. a secondary amine
b. a tertiary amine
c. a tertiary amine
d. a primary amine

17.15 The common name of an amine, like that of an aldehyde, is written as a single word. The alkyl groups attached to the nitrogen atom are named in alphabetical order, and the suffix *–amine* is added. Prefixes, *di-* and *tri-*, are added when identical groups are bonded the nitrogen atom.
a. ethylmethylamine
b. propylamine
c. diethylmethylamine
d. isopropylmethylamine

17.17 IUPAC rules for naming amines are similar to those for naming alcohols. Amines are named as *alkanamines*. Select as the parent chain the longest carbon chain to which the nitrogen atom is attached, number the chain from the end nearest the nitrogen atom; the location of the nitrogen atom on the chain is placed in front of the parent chain name. Identify and locate substituents. Secondary and tertiary amines are named as *N*-substituted primary amines. The largest carbon group bonded to the nitrogen atom is used as the parent amine name.

a. 3-pentanamine
b. 2-methyl-3-pentanamine
c. *N*-methyl-3-pentanamine
d. 2,3-butanediamine

17.19 IUPAC nomenclature for primary amines is similar to that for alcohols; the suffix is *–amine* rather than *–ol*. An $–NH_2$ group, like an –OH group, has priority in numbering the parent carbon chain. Secondary and tertiary amines are named as *N*-substituted primary amines. The largest carbon group bonded to the nitrogen atom is used as the parent amine name.

a. 1-propanamine
b. *N*-ethyl-*N*-methylethanamine
c. *N*-methyl-1-propanamine
d. *N*-methyl-2-butanamine

17.21 The simplest aromatic amine is called aniline. Aromatic amines with additional groups attached to the nitrogen atom are named as *N*-substituted anilines.

a. 2-bromoaniline
b. *N*-isopropylaniline
c. *N*-ethyl-*N*-methylaniline
d. *N*-methyl-*N*-phenylaniline

17.23 The longest carbon chain attached to the nitrogen atom furnishes the name of the parent compound. Additional groups bonded to the nitrogen atom are named as *N*-substituted groups. Aniline is the simplest aromatic amine. *N*-substituted anilines are aromatic amines with additional groups attached to the nitrogen atom

a. 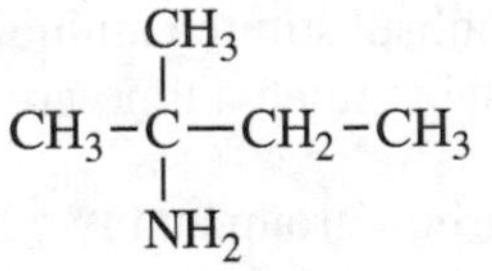
```
         CH3
          |
    CH3 - C - CH2 - CH3
          |
         NH2
```

b. $H_2N-CH_2-CH_2-CH_2-CH_2-CH_2-CH_2-NH_2$

c. 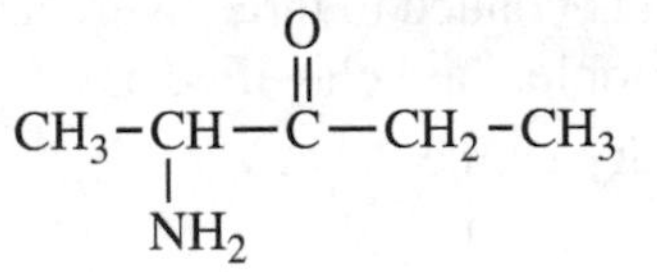
```
                O
                ||
    CH3 - CH - C - CH2 - CH3
          |
         NH2
```

d. 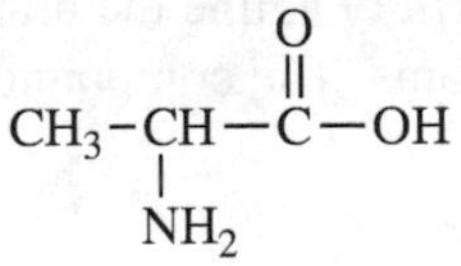
```
                O
                ||
    CH3 - CH - C - OH
          |
         NH2
```

17.25 a. Ethylpropylamine is a secondary amine. In a secondary amine, the nitrogen atom is bonded to two hydrocarbon groups and one hydrogen atom.

b. 2-Propanamine is a primary amine. In a primary amine, the nitrogen atom is bonded to one hydrocarbon group and two hydrogen atoms.

c. *N*-phenylaniline is a secondary amine. In a secondary amine, the nitrogen atom is bonded to two hydrocarbon groups and one hydrogen atom.

d. 2-Cyclohexylethanamine is a primary amine. In a primary amine, the nitrogen atom is bonded to one hydrocarbon group and two hydrogen atoms.

17.27 Constitutional isomerism in amines is due to either different carbon atom arrangements or different positioning of the nitrogen atom on the carbon chain. The two members of each of the following pairs of amines:

a. No, they are not constustional isomers.
b. Yes, they are constitutional isomers.
c. Yes, they are constitutional isomers.
d. No, they are not constustional isomers.

17.29 The number of saturated noncyclic amine constitutional isomers that exist for the given generalized formulations are:
a. four constitutional isomers
b. two constitutional isomers
c. two constitutional isomers
d. three constitutional isomers

17.31 There are eight isomeric primary amines that have the molecular formula $C_5H_{13}N$. Their IUPAC names are: 1-pentamine, 2-pentamine, 3-pentamine, 2-methyl-1-butanamine, 3-methyl-1-butanamine, 2-methyl-2-butanamine, 3-methyl-2-butanamine, 2,2-dimethyl-1-propanamine

17.33 The three methylamines (mono-, di-, and tri-) and ethylamine are gases at room temperature. Most other amines are liquids.
a. Butylamine is a liquid.
b. Dimethylamine is a gas.
c. Ethylamine is a gas.
d. Dibutylamine is a liquid.

17.35 The maximum number of hydrogen bonds that can form between
a. a methylamine molecule and other methylamine molecules is three (shown in Figure 17.4).
b. a methylamine molecule and water molecules is three (Each of the two hydrogen atoms bonded to the nitrogen atom forms one hydrogen bond to a water molecule. The nitrogen atom's non-bonding electron pair forms the third hydrogen bond with a water molecule, shown in Figure 17.6).

17.37 The boiling points of amines are higher than those of alkanes because hydrogen bonding is possible between amine molecules but not between alkane molecules.

17.39 a. $CH_3–CH_2–NH_2$ is more soluble in water because it has a shorter carbon chain (less nonpolar character than the longer carbon chain).
b. $H_2N–CH_2–CH_2–CH_2–NH_2$ is more soluble in water because it has two amine groups, both of which can form hydrogen bonds with water.

17.41 The result of the interaction of an amine with water is a basic solution containing substituted ammonium ions and hydroxyl ions.

a. $CH_3-CH_2-\overset{+}{N}H_3$

b. OH^-

c. $CH_3-\underset{\underset{CH_3}{|}}{CH}-NH-CH_3$

d. $CH_3-CH_2-\overset{+}{N}H_2-CH_2-CH_3 + OH^-$

17.43 To name a substituted ammonium ion, replace the word *amine* in the parent name with *ammonium ion*. The positive ion of aniline or a substituted aniline is named as an anilinium ion.
a. dimethylammonium ion
b. triethylammonium ion
c. *N,N*-diethylanilinium ion
d. *N*-isopropylanilinium ion

17.45 Ammonia and amines are weak bases; an amine molecule can accept a proton (H^+) from water to produce an ammonium ion. To determine the parent amine from the substituted ammonium or anilinium ion, simply remove H^+.

a. $CH_3-NH-CH_3$

b. $CH_3-CH_2-\underset{\underset{CH_2-CH_3}{|}}{N}-CH_2-CH_3$

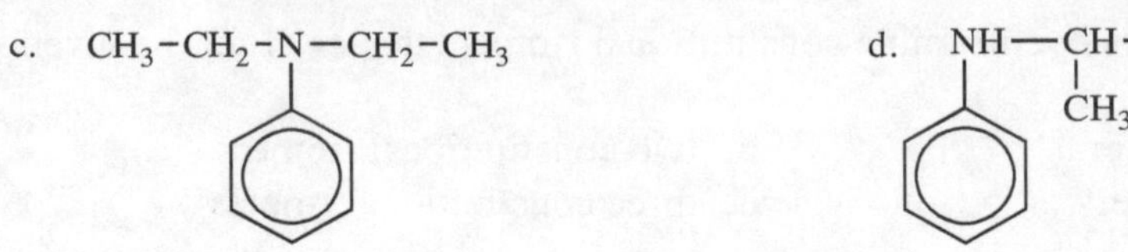

17.47 The skeletal structural formulas for the three isomeric disubstituted ammonium ions having the chemical formula $C_4H_{12}N^+$ are shown below.

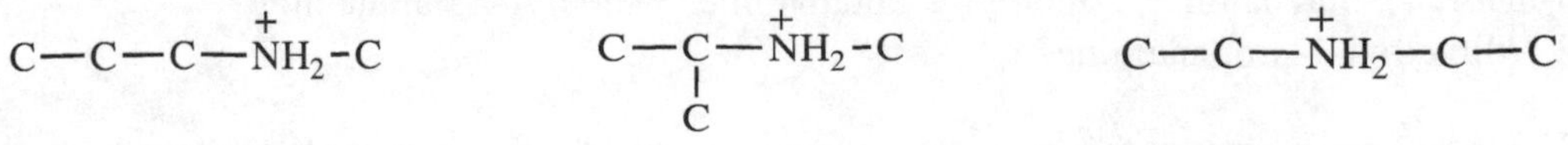

$\mathrm{C-C-C-\overset{+}{N}H_2-C}$ $\qquad$ $\mathrm{C-\underset{\displaystyle C}{\underset{|}{C}}-\overset{+}{N}H_2-C}$ $\qquad$ $\mathrm{C-C-\overset{+}{N}H_2-C-C}$

17.49 Amines react with strong acids to produce amine salts. The structures of the amine salts produced when the given amines react with HCl are:

a. $CH_3-CH_2-\overset{+}{N}H_3\ Cl^-$

b. 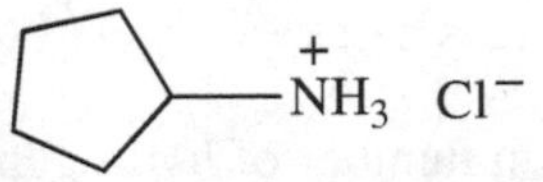 (cyclopentyl)$-\overset{+}{N}H_3\ Cl^-$

c. 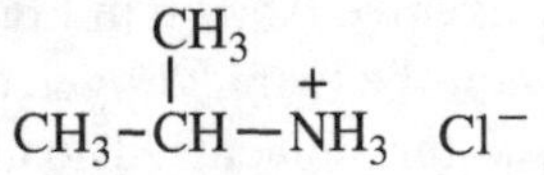 $CH_3-\overset{\displaystyle CH_3}{\overset{|}{C}H}-\overset{+}{N}H_3\ Cl^-$

d. $CH_3-CH_2-\overset{+}{N}H_2-CH_2-CH_3\ Cl^-$

17.51 Treating an amine salt with a strong base regenerates the parent amine. The structures of the amines produced by reaction of the given amine salts with NaOH are:

a. $CH_3-CH_2-NH_2$

b. $CH_3-NH-CH_2-CH_3$

c. 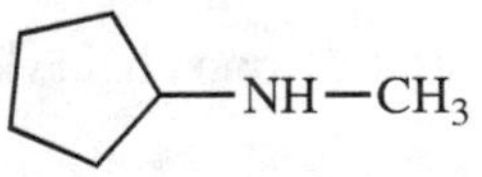 (cyclopentyl)$-NH-CH_3$

d. 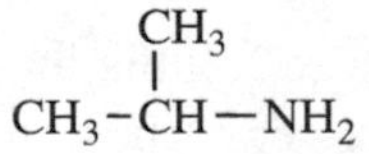 $CH_3-\overset{\displaystyle CH_3}{\overset{|}{C}H}-NH_2$

17.53 The structures of the missing substances in the given reactions of amines and amine salts:

a. 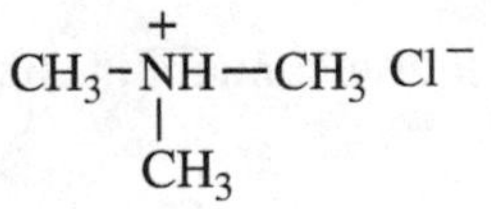 $CH_3-\underset{\displaystyle CH_3}{\underset{|}{\overset{+}{N}H}}-CH_3\ Cl^-$

b. HCl

c. 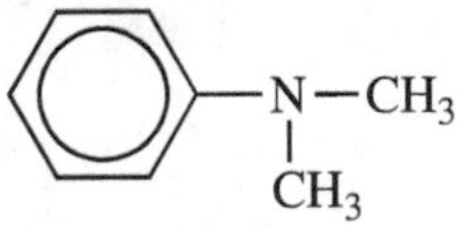 (phenyl)$-\underset{\displaystyle CH_3}{\underset{|}{N}}-CH_3$, NaBr

d. 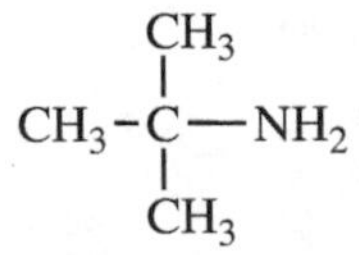 $CH_3-\overset{\displaystyle CH_3}{\overset{|}{\underset{\displaystyle CH_3}{\underset{|}{C}}}}-NH_2$

17.55 Amine salts are named in the same way as inorganic salts; the name of the positive ion, the substituted ammonium or anilinium ion, is given first and is followed by a separate word for the name of the negative ion.

a. propylammonium chloride
b. methylpropylammonium chloride
c. ethyldimethylammonium bromide
d. *N,N*-dimethylanilinium bromide

17.57 To find the common name of the parent amine from the name of the amine salt in Problem 17.53, remove "ammonium chloride" and add "amine." To determine the IUPAC name of the parent amine, use the longest carbon chain attached to the nitrogen atom as the parent compound. Additional groups bonded to the nitrogen atom are named as *N*-substituted groups.

a. The IUPAC name for propylamine is 1-propanamine.
b. The IUPAC name for methylpropylamine is *N*-methyl-1-propanamine.
c. The IUPAC name for ethyldimethylamine is *N,N*-dimethylethanamine.
d. The IUPAC name for *N,N*-dimethylaniline is *N,N*-dimethylphenylamine.

17.59 a. free amine, free base, deprotonated base b. free amine, free base, deprotonated base
c. protonated base d. protonated base

17.61 The reaction between an amine and an alkyl halide in the presence of a strong base results in the alkylation of the amine.
a. Yes, this is an alkylation reaction. b. Yes, this is an alkylation reaction.
c. Yes, this is an alkylation reaction. d. Yes, this is an alkylation reaction.

17.63 The classification of the organic product in each reaction in Problem 17.61:
a. a 1° amine b. a 3° amine c. a 2° amine d. a 1° amine

17.65 Alkylation under basic conditions to prepare an amine is a two-step process: in the first step an amine salt is produced; in the second step, the amine salt is converted by the base to the free amine. The other two products are the inorganic salt (from the base and the amine halide) and water.

a. $CH_3-CH_2-CH_2-NH_2$, NaCl, H_2O

b. 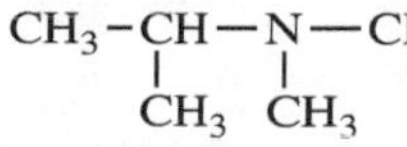 $\begin{array}{l} CH_3-\underset{\displaystyle CH_3}{\underset{|}{CH}}-\underset{\displaystyle CH_3}{\underset{|}{N}}-CH_3 \end{array}$, NaBr, H_2O

c. $CH_3-CH_2-NH-CH_2-CH_3$, NaCl, H_2O

d. 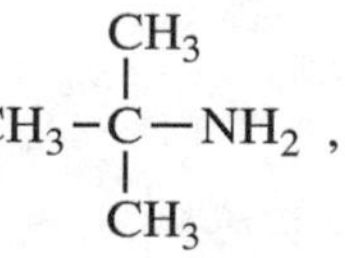 $CH_3-\overset{\displaystyle CH_3}{\overset{|}{\underset{\displaystyle CH_3}{\underset{|}{C}}}}-NH_2$, NaBr, H_2O

17.67 Since the tertiary amine (ethylmethylpropylamine) has three different alkyl groups, the secondary amine used to prepare it would have two of these alkyl groups and the alkyl halide would contain the third. The three possible combinations are: ethylmethylamine and propyl chloride, ethylpropylamine and methyl chloride, and methylpropylamine and ethyl chloride.

17.69 The reaction between an amine and an alkyl halide in the presence of a strong base results in the alkylation of the amine. Alkylation of a tertiary amine produces a quaternary ammonium salt (four alkyl groups attached to the nitrogen atom and the resulting positive charge balanced by the negative ion of the salt).

a. 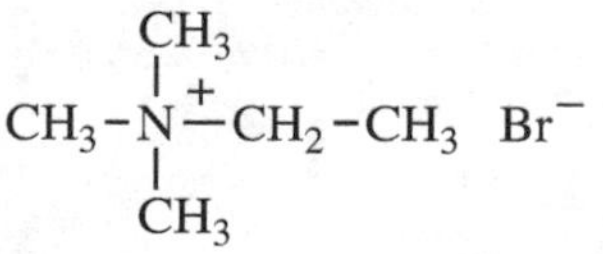 $CH_3-\overset{\displaystyle CH_3}{\overset{|\,+}{\underset{\displaystyle CH_3}{\underset{|}{N}}}}-CH_2-CH_3\ Br^-$

b. 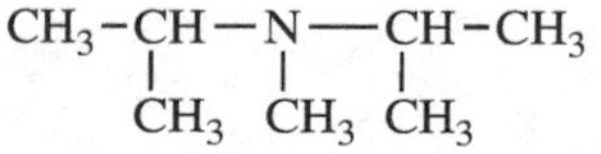 $CH_3-\underset{\displaystyle CH_3}{\underset{|}{CH}}-\underset{\displaystyle CH_3}{\underset{|}{N}}-\underset{\displaystyle CH_3}{\underset{|}{CH}}-CH_3$

c. 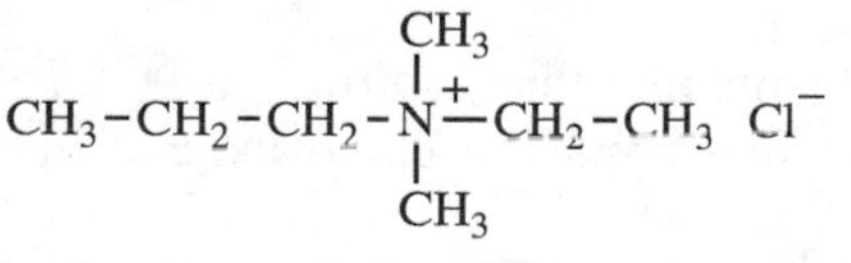 $CH_3-CH_2-CH_2-\overset{\displaystyle CH_3}{\overset{|\,+}{\underset{\displaystyle CH_3}{\underset{|}{N}}}}-CH_2-CH_3\ Cl^-$

d. $CH_3-CH_2-NH-CH_2-CH_3$

17.71 In amine salts, the nitrogen atom has an extra H^+ bonded to it; in a quaternary ammonium salt, four alkyl groups are bonded to the nitrogen atom. In each case, the resulting positive charge is balanced by the negative ion of an acid. The four salts are classified as follows:
a. amine salt b. quaternary ammonium salt
c. amine salt d. quaternary ammonium salt

17.73 Quaternary ammonium salts differ from amine salts in that the addition of a strong base does not convert quaternary ammonium salts back to their parent amines. There is no hydrogen atom on the nitrogen with which the OH^- can react.

a. Yes, this amine salt can be converted to the "parent" amine with a strong base.
b. No, a quaternary ammonium salt is not converted to the "parent" amine with a strong base.
c. Yes, this amine salt can be converted to the "parent" amine with a strong base.
d. No, a quaternary ammonium salt is not converted to the "parent" amine with a strong base.

17.75 To name amine salts and quaternary ammonium salts, first name the positive ion (the substituted ammonium ion); then with a separate word, give the name of the negative ion.

a. trimethylammonium bromide
b. tetramethylammonium chloride
c. ethylmethylammonium bromide
d. diethyldimethylammonium chloride

17.77 a. yes b. yes c. no d. yes

17.79 A fused-ring aromatic compound contains two or more rings fused together. Fig. 17.8 in your textbook gives structural formulas for some heterocyclic amines.

a. Pyrrolidine is a saturated heterocyclic amine.
b. Imidazole is an unsaturated heterocyclic amine.
c. Pyridine is an unsaturated heterocyclic amine.
d. Quinoline is an unsaturated, fused-ring heterocyclic amine.

17.81 There is no difference between a porphyrin ring and a heme unit in terms of number of pyrrole rings present.

17.83 a. True. Serotonin is a regulator of lactation in women.
b. True. Structurally, dopamine and norepinephrine differ by an –OH group.
c. False. L-Dopa and dopamine are not the same compound; L-dopa contains a carboxyl group and can pass through the blood-brain barrier.
d. False. Prozac is a drug that helps maintain serotonin levels in the brain.

17.85 a. Dopamine has a phenethylamine core.
b. Epinephrine has a phenethylamine core.
c. Amphetamine has a phenethylamine core.
d. Serotonin has a tryptamine core.

17.87 a. Dopamine is a neurotransmitter.
b. Methamphetamine is a central nervous system stimulant.
c. Pseudoephedrine is a decongestant.
d. Desoxyn is a central nervous system stimulant.

17.89 a. yes b. yes c. yes d. yes

17.91 a. True. Medicinally, atropine is used to dilate the pupil of the eye.
b. True. Quinine is a powerful antipyretic.
c. False. Morphine is a 100 times more powerful painkiller than aspirin.
d. False. The opium of the oriental poppy plant contains both morphine and codeine.

17.93 Codeine, morphine, hydrocodone, and oxycodone have a common central core, but they differ in the attached functional groups.

a. Codeine has a hydroxy and a methoxy group attached to the central core.
b. Morphine has two hydroxy groups attached to the central core.
c. Hydrocodone has a methoxy group and a keto group attached to the central core.
d. Oxycodone has a methoxy, a keto, and a hydroxy group attached to the central core.

17.95 a. A primary amide contains one R–group.
b. A tertiary amide contains three R–groups.
c. A monosubstitued amide contains two R–groups; monosubstituted is another name for a secondary amide.

17.97 Amides contain one nitrogen atom.
a. one nitrogen atom b. one nitrogen atom c. one nitrogen atom

17.99 An amide is a carboxylic acid derivative in which the carboxy –OH group has been replaced with an amino or a substituted amino group.
a. Yes. The –OH group of propanoic acid has been replaced by an amino group.
b. Yes. The –OH group of benzoic acid has been replaced by ethylmethylamine.
c. No. The amino group is not attached to the carbonyl carbon.
d. Yes. The nitrogen atom in the ring is attached to the carbonyl carbon; this is a lactam (a cyclic amide).

17.101 An unsubstituted amide has two hydrogen atoms bonded to its nitrogen atom, a monosubstituted amide has one hydrogen atom and one alkyl group bonded to the nitrogen atom, and a disubstituted amide has two alkyl bonded to the nitrogen atom. The amides are classified as follows:
a. monosubstituted b. disubstituted
c. unsubstituted d. monosubstituted

17.103 A primary amide is an unsubstituted amide, a secondary amide is a monsubstituted amide, and a tertiary amide is a disubstituted amide. The amides in Problem 17.101 are classified as
a. a secondary amide. b. a tertiary amide.
c. a primary amide. d. a secondary amide.

17.105 In an amine one or more alkyl, cycloalkyl, or aryl groups are attached to the nitrogen atom. An amide is a carboxylic acid derivative in which the carboxyl -OH group has been replaced with an amino or a substituted amino group.
a. The compound is an amide. b. The compound is an amine.
c. The compound is an amine d. The compound is an amine and an amide.

17.107 In assigning IUPAC names to amides, amides are considered to be derivatives of carboxylic acids, with the suffix changed from *–oic acid* to *–amide*. The names of groups attached to the nitrogen atom are placed in front of the base name, using an *N*- prefix as a locator.
a. *N*-ethylethanamide b. *N,N*-dimethylpropanamide
c. butanamide d. 2-chloropropanamide

17.109 The common names of amides are based on those of the corresponding carboxylic acids, with the suffix changed from *–ic acid* to *–amide*. The names of groups attached to the nitrogen atom are placed in front of the base name, using an *N*- prefix as a locator.
a. *N*-ethylacetamide b. *N,N*-dimethylpropionamide
c. butyramide d. α-chloropropionamide

17.111 IUPAC names of amides are based on the carboxylic acids from which the amides are derived. The suffix is changed from *–oic acid* to *–amide*. The names of groups attached to the nitrogen atom placed in front of the base name, using an *N*- prefix as a locator.
a. propanamide b. *N*-methylpropanamide
c. 3,5-dimethylhexanamide d. *N,N*-dimethylbutanamide

17.113 The names of groups attached to the nitrogen atom are placed in front of the base name, using an *N*- prefix as a locator. In part c. of this problem, the 3- refers to a methyl group on the four-carbon chain attached to the carbonyl group.

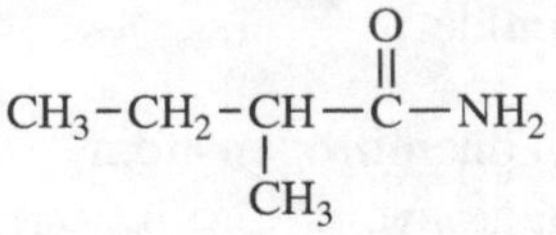

a. 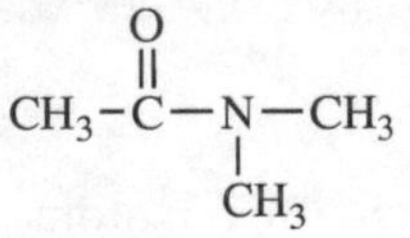 $CH_3-\overset{\overset{\displaystyle O}{\|}}{C}-\underset{\underset{\displaystyle CH_3}{|}}{N}-CH_3$

b. $CH_3-CH_2-\underset{\underset{\displaystyle CH_3}{|}}{CH}-\overset{\overset{\displaystyle O}{\|}}{C}-NH_2$

c. $CH_3-\underset{\underset{\displaystyle CH_3}{|}}{CH}-CH_2-\overset{\overset{\displaystyle O}{\|}}{C}-NH-CH_3$

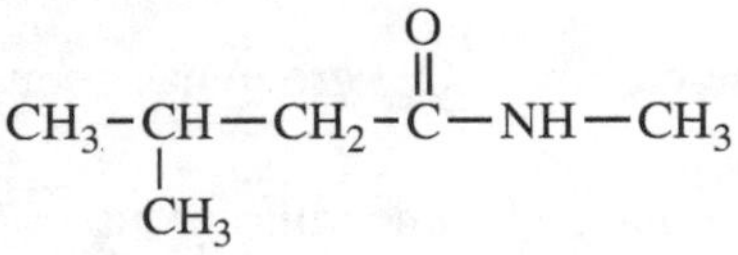

d. 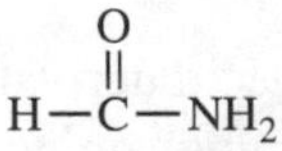 $H-\overset{\overset{\displaystyle O}{\|}}{C}-NH_2$

17.115 IUPAC nomenclature for primary amines is similar to that for alcohols, with the suffix *–amine* rather than *–ol*. An $–NH_2$ group has priority in numbering the parent carbon chain. Secondary and tertiary amines are named as *N*-substituted primary amines. The largest carbon group bonded to the nitrogen atom is used as the parent amine name.

IUPAC names of amides are based on the corresponding carboxylic acids, with the suffix changed from *–oic acid* to *–amide*. The names of groups attached to the nitrogen atom placed in front of the base name, using an *N*- prefix as a locator. The IUPAC names are

a. hexanamide.
b. 1-hexanamine.
c. 5-aminopentanamide.
d. 2,5-dimethyl-1,6-hexanediamine.

17.117 Urea is a naturally occurring amide whose formula is $(H_2N)_2CO$.

a. True.
b. False. In the pure state urea is a white solid.
c. True.
d. True.

17.119 The mode of action for the insect repellent known as DEET is inhibition of insect olfactory receptors.

17.121 A carbonyl group on a nitrogen atom that is bonded directly to the carbonyl carbon atom draws the nitrogen's lone pair of electrons closer to the nitrogen atom.

17.123 Amines are weak bases; nitrogen's nonbonding pair of electrons accepts a proton from water. Amides do not exhibit basic properties in solution; the nonbonding pair of electrons on the N atom do not bond to a H^+ ion because of the carbonyl group's electronegativity.

a. No, propanamide (an amide) does not exhibit basic behavior in aqueous solution.
b. Yes, 1-propanamine (an amine) does exhibit basic behavior in aqueous solution.
c. No, the given compound (an amide) does not exhibit basic behavior in aqueous solution.
d. Yes, the given compound does exhibit basic behavior in aqueous solution; it is both an amine and an amide.

17.125 The physical state, at room temperature, for these unbranched primary amides:

a. Methanamide is a liquid.
b. Ethanamide is a solid.
c. Butanamide is a solid.
d. Hexanamide is a solid.

17.127 a. The maximum number of hydrogen bonds that can form between an acetamide molecule and other acetamide molecules is four (two hydrogen bonds with the hydrogen atoms attached to nitrogen, and one with each pair of nonbonding electrons on the carbonyl oxygen atom).

b. The maximum number of hydrogen bonds that can form between an acetamide molecule and water molecules is four (the same type of hydrogen bonds as are listed above).

17.129 An amidification reaction is the reaction of a carboxylic acid with an amine (or ammonia) to produce an amide. All of the reactions shown are amidification reactions.

a. Yes b. Yes c. Yes d. Yes

17.131 A primary amide is an unsubstituted amide, a secondary amide is a monosubstituted amide, and a tertiary amide is a disubstituted amide. The amides in Problem 17.129 are classified as:

a. a 1° amide. b. a 2° amide. c. a 3° amide. d. a 3° amide.

17.133 The structure of the carboxylic acid used to produce each of the amides is:

a. $CH_3-\overset{O}{\overset{\|}{C}}-OH$ b. $H-\overset{O}{\overset{\|}{C}}-OH$

c. $CH_3-CH_2-\overset{O}{\overset{\|}{C}}-OH$ d. $CH_3-CH_2-\overset{O}{\overset{\|}{C}}-OH$

17.135 The structure of the nitrogen-containing compound used to produce each of the amides is:

a. NH_3 b. CH_3-NH_2

c. $CH_3-NH-CH_3$ d. CH_3-NH_2

17.137 The reaction of a carboxylic acid with ammonia, or a primary or a secondary amine, at a high temperature (greater than 100°C) produces an amide. The missing substances in this problem's amide preparation reactions are shown below.

a. CH_3-NH_2 b. 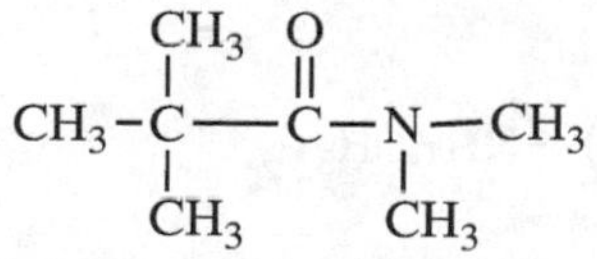 $CH_3-\underset{CH_3}{\underset{|}{\overset{CH_3}{\overset{|}{C}}}}-\overset{O}{\overset{\|}{C}}-\underset{CH_3}{\underset{|}{N}}-CH_3$ c. NH_3 d. 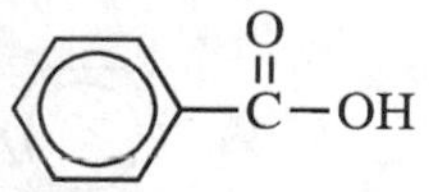 $C_6H_5-\overset{O}{\overset{\|}{C}}-OH$

17.139 The structure of the carbonyl-containing compound produced when each of the amides is hydrolyzed with HCl (acidic conditions) is:

a. $CH_3-\overset{O}{\overset{\|}{C}}-OH$ b. $H-\overset{O}{\overset{\|}{C}}-OH$

c. $CH_3-CH_2-\overset{O}{\overset{\|}{C}}-OH$ d. $CH_3-CH_2-\overset{O}{\overset{\|}{C}}-OH$

17.141 The structure of the carbonyl-containing compound produced when each of the amides in Problem 17.139 is hydrolyzed with NaOH (basic conditions) is:

a. $CH_3-\overset{O}{\overset{\|}{C}}-O^-\ Na^+$ b. $H-\overset{O}{\overset{\|}{C}}-O^-\ Na^+$

c. $CH_3-CH_2-\overset{O}{\overset{\|}{C}}-O^-\ Na^+$ d. $CH_3-CH_2-\overset{O}{\overset{\|}{C}}-O^-\ Na^+$

17.143 The structure of the nitrogen-containing compound produced when each of the amides in Problem 17.139 is hydrolyzed with HCl (acidic conditions) is:

a. $NH_4^+ Cl^-$ b. $CH_3{-}\overset{+}{N}H_3\ Cl^-$

c. $CH_3{-}\overset{+}{N}H_2{-}CH_3\ Cl^-$ d. $CH_3{-}\overset{+}{N}H_3\ Cl^-$

17.145 The structure of the nitrogen-containing compound produced when each of the amides in Problem 17.139 is hydrolyzed with NaOH (basic conditions) is:

a. NH_3 b. $CH_3{-}NH_2$

c. $CH_3{-}NH{-}CH_3$ d. $CH_3{-}NH_2$

17.147 In amide hydrolysis, the bond between the carbonyl carbon atom and the nitrogen is broken, and free acid and free amine are produced. Amide hydrolysis is catalyzed by acids, bases, and certain enzymes; sustained heating is also often required.

a. $CH_3{-}CH_2{-}CH_2{-}\overset{O}{\overset{\|}{C}}{-}OH$, $CH_3{-}NH_2$

b. $CH_3{-}CH_2{-}CH_2{-}\overset{O}{\overset{\|}{C}}{-}OH$, $CH_3{-}\overset{+}{N}H_3\ Cl^-$

c. $CH_3{-}CH_2{-}CH_2{-}\overset{O}{\overset{\|}{C}}{-}O^-\ Na^+$, $CH_3{-}NH_2$

d. 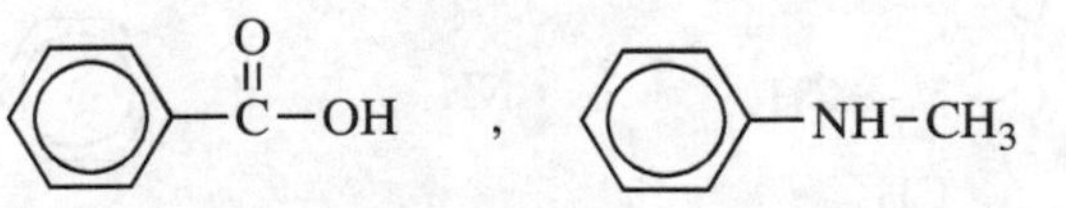 $C_6H_5{-}\overset{O}{\overset{\|}{C}}{-}OH$, $C_6H_5{-}NH{-}CH_3$

17.149 A polyamide is a condensation polymer in which the monomers are joined through amide linkages. The monomers are diacids and diamines. One acid group of the diacid reacts with one amine group of the diamine, leaving an acid group and an amine group on the two ends to react further; the process continues, generating a long polymeric molecule.

17.151 A structural representation of the polyamide formed from succinic acid and 1,4-butanediamine is shown below.

$$\left(\overset{O}{\overset{\|}{C}}{-}CH_2{-}CH_2{-}\overset{O}{\overset{\|}{C}}{-}\overset{H}{\overset{|}{N}}{-}(CH_2)_4{-}\overset{H}{\overset{|}{N}}\right)_n$$

17.153 $R{-}\overset{H}{\overset{|}{N}}{-}\overset{O}{\overset{\|}{C}}{-}O{-}R'$

Carbohydrates **Chapter 18**

Solutions to Selected Problems

18.1 a. False. Bioinorganic substances are about 75% of the mass composition of the human body.
b. True. Bioinorganic substances are much more abundant than bioorganic substances.

18.3 Use Figure 18.1 to answer this question.
a. Water is more abundant than protein.
b. Lipids are more abundant than carbohydrates.
c. Inorganic salts are more abundant than nucleic acids.
d. Proteins are more abundant than carbohydrates.

18.5 a. True.
b. True.
c. False. Plants have two main uses for carbohydrates. Carbohydrates serve as structural elements and provide energy reserves in the form of starch.
d. True.

18.7 a. A trisaccharide contains three monosaccharide units.
b. Oligosaccharides contain three to ten monosaccharide units.

18.9 a. When a tetrasaccharide undergoes hydrolysis the product produced is a monosaccharide.
b. When a polysaccharide undergoes hydrolysis the product produced is a monosaccharide.

18.11 a. False. In a monosaccharide, either an aldehyde group or a ketone group is present.
b. False. A monosaccharide is a carbohydrate that contains a single polyhydroxy aldehyde or polyhydroxy ketone unit.

18.13 Superimposable objects have parts that coincide exactly at all points when the objects are laid upon each other.

18.15 A chiral center in a molecule is a carbon atom that is bonded to four different groups.
a. No, the circled carbon atom is not a chiral center because two of the groups bonded to it are the same (hydrogen atoms).
b. No, the circled carbon atom is not a chiral center because two of the groups bonded to it are the same (methyl groups).
c. Yes, the circled atom is a chiral center because there are four different groups bonded to it.
d. Yes, the circled atom is a chiral center because there are four different groups bonded to it.

18.17 Chiral centers in organic molecules have four different groups bonded to a carbon atom. The chiral centers in the molecules below are marked with asterisks. Note that molecules may have more than one chiral center.

a. No chiral centers

b.
```
         Cl   Cl
        *|   *|
  CH2 ─ C ─ CH
   |     |    |
   Br   Br   Br
```

c.
```
                              O
                              ||
  CH2 ─*CH ─*CH ─*CH ─ C ─ H
   |      |      |      |
   OH    OH    OH    OH
```

d.
```
  CH2 ─*CH ─*CH ─*CH ─*CH ─ CH2
   |      |      |      |      |      |
   OH    OH    OH    OH    OH    OH
```

18.19 Chiral centers in organic molecules have four different groups bonded to a carbon atom; molecules may have more than one chiral center.

a. This symmetrical molecule has no chiral centers.
b. This molecule has two chiral centers; there are four different groups bonded to each of the carbon atoms with chlorine substituents.
c. This molecule has no chiral centers; the ring bonded to the carbon atom with the –OH substituent is the same in either direction.
d. This molecule has no chiral centers; the carbon atom with the –OH substituent is bonded to a total of three other atoms.

18.21 Chiral centers in organic molecules have four different groups bonded to a carbon atom.

a. does not have a chiral center
b. has a chiral center (C_2)
c. has a chiral center (C_2)
d. does not have a chiral center

18.23 An achiral molecule is a molecule whose mirror images are superimposable. Classifying the molecules in Problem 18.21:

a. Yes, it is an achiral molecule.
b. No, it is not an achiral molecule.
c. No, it is not an achiral molecule.
d. Yes, it is an achiral molecule.

18.25 Chiral organic molecules have left- or right-handed forms. For the molecules in Prob. 18.21:

a. No, it is not chiral and does not exist in left-handed or right-handed forms.
b. Yes, it is chiral and it exists in left-handed or right-handed forms.
c. Yes, it is chiral and it exists in left-handed or right-handed forms.
d. No, it is not chiral and does not exist in left-handed or right-handed forms.

18.27 Chiral molecules have nonsuperimposable mirror images; achiral molecules have superimposable mirror images. Indicating the nonsuperimposability of mirror image for the molecules in Problem 18.21:

a. No. Achiral molecules have superimposable mirror images.
b. Yes. The mirror images for chiral molecules are nonsuperimposable.
c. Yes. The mirror images for chiral molecules are nonsuperimposable.
d. No. Achiral molecules have superimposable mirror images.

18.29 Atoms in constitutional isomers are connected to each other in different ways; molecules that are stereoisomers have the same structural formulas but differ in orientation of atoms in space.

18.31 Two structural features that can generate stereoisomerism are the presence of a chiral center and the presence of "structural rigidity."

18.33 a. True. Stereoisomers always have the same molecular formula.
b. True. Stereoisomers always have the same structural formula.
c. False. Enantiomers are stereoisomers whose molecules are nonsuperimposable mirror images. Diastereomers are stereisomers whose molecules are not mirror images.
d. False. Enantiomers possess handedness; diasteromers do not.

18.35 In a Fischer projection, a chiral center is represented as the intersection of vertical and horizontal lines; the atom at the chiral center (usually carbon) is not explicitly shown. Vertical lines represent bonds from the chiral center directed into the printed page; horizontal lines represent bonds from the chiral center directed out of the printed page.

a. H (top); Br—+—Cl; CH_3 (bottom)

b. CH_3 (top); Br—+—Cl; H (bottom)

c. CH_3 (top); Br—+—H; Cl (bottom)

d. CH_3 (top); H—+—Br; Cl (bottom)

18.37 Enantiomers are stereoisomers whose molecules are nonsuperimposable mirror images. To draw the enantiomer of a monosaccharide whose Fischer projection is given, switch the –H and –OH groups attached to the left side of chiral centers to the right side, and those on the right side to the left side.

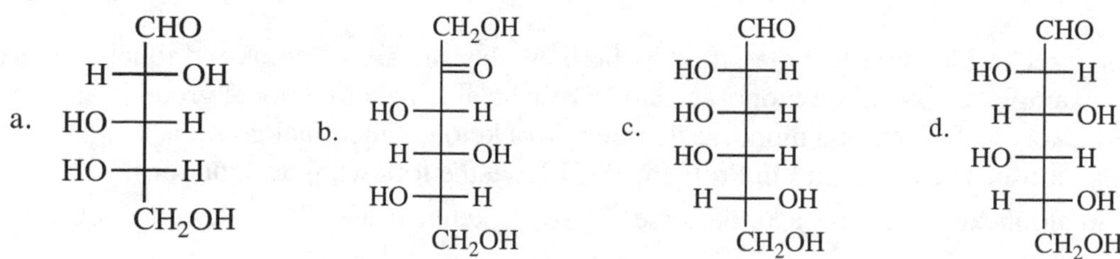

18.39 In the D,L system designating handedness of an enantiomer, the carbon chain of the monosaccharide is numbered from the carbonyl group end of the molecule; the highest-numbered chiral center determines D or L configuration. The enantiomer with the –OH group to the right is called right-handed and designated D, and the enantiomer with the –OH group to the left is left-handed and designated L. Using this system, the molecules in Prob. 18.37 have the designations:

a. D-enantiomer b. D-enantiomer c. L-enantiomer d. L-enantiomer

18.41 Enantiomers are stereoisomers whose molecules are nonsuperimposable mirror images. Diastereomers are stereoisomers whose molecules are not mirror images. Looking at the Fischer projections, imagine a mirror held up to one; if one projection is a "reflection" of the other then they are mirror images.

a. The molecules are diastereomers; they are not mirror images of one another.
b. The molecules are neither enantiomers nor diastereomers; they are not isomers (they have different molecular formulas).
c. The molecules are enantiomers; they are mirror images of one another, and their mirror images are not superimposable.
d. The molecules are diastereomers; they are not mirror images of one another.

18.43 Epimers are diastereomers whose molecules differ in the configuration of one chiral center.

a. Yes, the molecules are epimers.
b. No, the molecules are not epimers; they have different molecular formulas.
c. No, they are not epimers; they are enantiomers and differ at more than one chiral center.
d. Yes, the molecules are epimers.

18.45 a. There are three chiral centers in each of the two molecules.
b. There are two chiral centers in each of the two molecules.
c. There are four chiral centers in each of the two molecules.
d. There are two chiral centers in each of the two molecules.

18.47 D-glucose and L-glucose are enantiomers; nearly all their properties are the same, except their interactions with plane-polarized light and properties involving other chiral molecules (such as solubility in a chiral solvent). Properties would be:

a. different b. the same c. the same d. the same

18.49 (+)-Lactic acid and (–)-lactic acid are enantiomers. Nearly all their properties are the same, the exceptions being interaction with plane-polarized light and properties involving other chiral molecules (such as solubility in a chiral solvent). (+)-Lactic acid and (–)-lactic acid

a. have the same boiling point.
b. differ in their optical activity.
c. have the same solubility in ethanol.
d. differ in their reactions with (–)-2,3-butanediol.

18.51 An aldose is a monosaccharide that contains an aldehyde functional group (a polyhydroxy aldehyde); a ketose is a monosaccharide that contains a ketone functional group (a polyhydroxy ketone). The molecules in this problem are:

a. an aldose b. a ketose c. a ketose d. a ketose

18.53 Monosaccharides are often classified by both number of carbon atoms and functional group. For example: a six-carbon monosaccharide with an aldehyde functional group is an aldohexose; a five-carbon monosaccharide with a ketone functional group is a ketopentose. The molecules in Problem 18.51 have the following designations:

a. an aldohexose b. a ketohexose c. a ketotriose d. a ketotetrose

18.55 Using information from Figures 18.14 and 18.15, which show Fischer projection formulas and common names for D-aldoses and D-hexoses with three, four, five, and six carbons, we can name the monosaccharides in Problem 18.51.

a. D-galactose b. D-psicose c. dihydroxyacetone d. L-erythrulose

18.57 A chiral center is an atom in a molecule that has four different groups tetrahedrally bonded to it. The number of chiral centers in these monosaccharides are

a. four chiral centers. b. three chiral centers.
c. zero chiral centers. d. one chiral center.

18.59 Use structural diagrams from Section 18.8 to determine at which carbon atom(s) the structures of the monosaccharides in each pair differ.

a. D-Glucose and D-galactose differ at carbon 4.
b. D-Glucose and D-fructose differ at carbons 1 and 2; D-glucose is an aldose, and D-fructose is a ketose.
c. D-Glyceraldehyde and dihydroxyacetone differ at carbons 1 and 2; D-glyceraldehyde is an aldose and dihydroxyacetone is a ketone.
d. D-Ribose and 2-deoxy-D-ribose differ at carbon 2. As might be expected from its name, 2-deoxy-D-ribose does not have a –OH group on carbon 2, but instead has two hydrogen atoms.

18.61 a. D-Glucose and D-galactose are aldoses and hexoses; they are both aldohexoses.
b. D-Glucose and D-fructose are hexoses. D-Glucose is an aldose; D-fructose is a ketose.
c. D-Galactose and D-fructose are hexoses. D-Galactose is an aldose; D-fructose is a ketose.
d. D-Ribose and D-glyceraldehyde are both aldoses; neither is a hexose (D-ribose is a pentose, and D-glyceraldehyde is a triose).

18.63 Use Figures 18.14 and 18.15 to determine the Fischer projections for these monosaccharides

a.
CHO
H—OH
HO—H
H—OH
H—OH
CH_2OH
D-glucose

b.
CHO
H—OH
CH_2OH
D-glyceraldehyde

c.
CH_2OH
=O
HO—H
H—OH
H—OH
CH_2OH
D-fructose

d.
CHO
HO—H
H—OH
H—OH
HO—H
CH_2OH
L-galactose

18.65 a. D-fructose is also known as levulose and fruit sugar.
b. D-glucose is known as grape sugar; two other names are dextrose and blood sugar.
c. D-galactose is known as brain sugar.

18.67 The cyclic forms of monosaccharides result from the ability of their carbonyl groups to react intramolecularly with a hydroxyl group, forming a cyclic hemiacetal.
a. An aldohexose, upon intramolecular cyclization, forms a six-membered ring.
b. A ketohexose, upon intramolecular cyclization, forms a five-membered ring.
c. An aldopentose, upon intramolecular cyclization, forms a five-membered ring.
d. A ketopentose, upon intramolecular cyclization, forms a four-membered ring.

18.69 In a cyclic aldose, there is one carbon atom outside the ring; in a cyclic ketose, there are two carbon atoms outside the ring.
a. D-glucose becomes a cyclic aldose, with one carbon atom outside the ring.
b. D-galactose becomes a cyclic aldose, with one carbon atom outside the ring.
c. D-fructose becomes a cyclic ketose, with two carbon atoms outside the ring.
d. D-ribose becomes a cyclic aldose, with one carbon atom outside the ring.

18.71 A hemiacetal is a compound in which a carbon atom (the hemiacetal carbon) is bonded to both a hydroxyl group and an alkoxy group. An anomeric carbon atom is the hemiacetal carbon atom present in a cyclic monosaccharide structure.
a. In the cyclic form of D-glucose, there is one anomeric carbon atom present.
b. In the cyclic form of D-glucose, there is one hemiacetal carbon atom present.

18.73 The structure for the anomer of thc monosaccharide in Problem 18.71 is shown below.

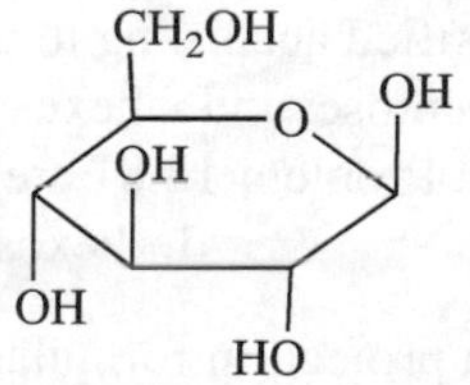

CH_2OH OH O OH OH HO

18.75 The α-stereoisomer in the cyclic form of a monosaccharide has the –OH on the opposite side of the ring from the $-CH_2OH$ group. The stereoisomer in Problem 18.71 is the α-anomer.

18.77 The structure of D-glucose is sometimes written in an open-chain form and sometimes in a cyclic form because the cyclic and noncyclic forms interconvert; an equilibrium exists between the forms.

18.79 Structural representations of cyclic forms of monosaccharides are called Haworth projections. Conventions for drawing Hayworth projection formulas: 1) –OH groups on the right in the Fischer projection are below the ring; –OH groups to the left are above the ring. 2) The –OH group formed from the carbonyl group may be abovc or below the plain of the ring, depending on how ring closure occurs.

a. 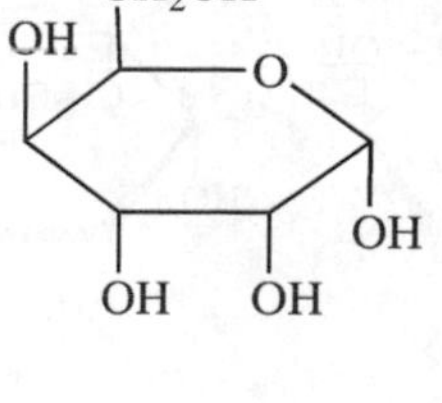 CH_2OH OH O OH OH OH

b. 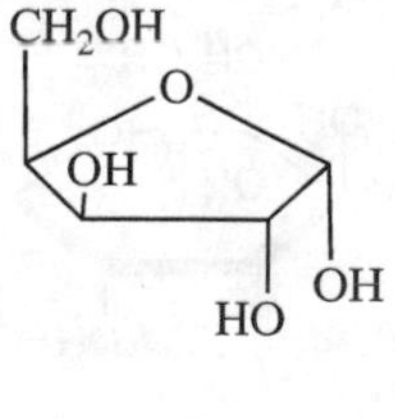 CH_2OH O OH HO OH

18.81 The α-anomer in the cyclic form of a monosaccharide has the –OH on the opposite side of the ring from the $-CH_2OH$ group. The β-anomer has the –OH group on the same side of the ring as the $-CH_2OH$ group. The Haworth projection formula shows

a. an α-D-monosaccharide. b. an α-D-monosaccharide.
c. a β-D-monosaccharide. d. an α-D-monosaccharide.

18.83 Any –OH group at a chiral center that is to the right in a Fischer projection formula points down in the Haworth projection formula. Any group to the left in a Fischer projection formula points up in the Haworth projection formula.

a.
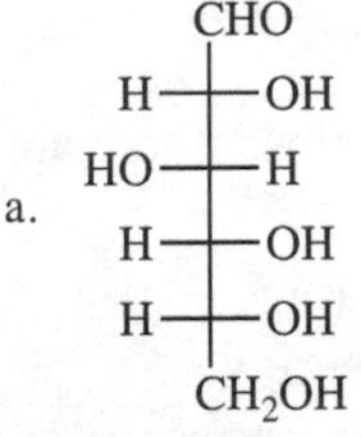

b.
CHO
H–OH
HO–H
HO–H
H–OH
CH_2OH

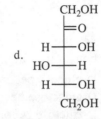

c.
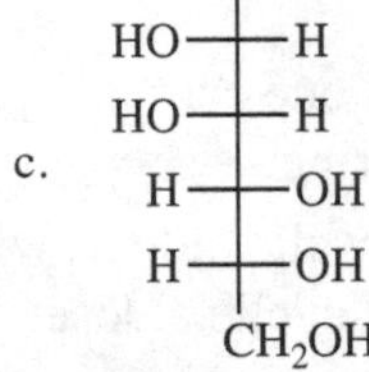

d.
CH_2OH
=O
H–OH
HO–H
H–OH
CH_2OH

18.85 The cyclic forms of monosaccharides result from the ability of their carbonyl groups to react intramolecularly with a hydroxyl group, forming a cyclic hemiacetal. Compare the Fischer projection formulas (open chain) with the Haworth projection formulas (cyclic). The numbers assigned to the left-most carbon atom of the monosaccharides in Problem 18.81 are:

a. C-4 b. C-4 c. C-4 d. C-5

18.87 The anomeric carbon atom is the hemiacetal carbon atom present in a cyclic monosaccharide structure. In the monosaccharides in Problem 18.81 the anomeric carbon atom is:

a. C-1 b. C-1 c. C-1 d. C-2

18.89 Monosaccharides with three to seven carbon atoms are classified according to the number of carbon atoms in the monosaccharide. A triose, a tetrose, a pentose, and a hexose contain 3, 4, 5, and 6 carbon atoms respectively. The monosaccharides in Problem 18.81 are classified as:

a. hexose b. hexose c. hexose d. hexose

18.91 There are three rules that will help you to interpret Haworth projection formulas. 1) –OH groups at chiral centers that point to the right in a Fischer projection formula will point down in the Haworth projection formula. Any group to the left in a Fischer projection formula will point up in the Haworth projection formula. 2) In writing the D form, the terminal CH_2OH is positioned above the ring; in the L form, it is positioned below the ring. 3) α or β configuration is determined by the position of the –OH group on carbon 1 (relative to the CH_2OH group that determines D or L forms): in a β configuration, both of these groups point in the same direction; in an α configuration, the groups point in opposite directions. The monosaccharides in Problem 18.81 are named:

a. α-D-glucose b. α-D-galactose c. β-D-mannose d. α-D-sorbose

18.93 First, draw the correct Fischer projection formulas for D-galactose and L-galactose. Hint: The D- and L- forms of galactose are mirror images. Use the three guidelines given in the answer above (Problem 18.91) to help you in drawing these Haworth projection formulas.

a.
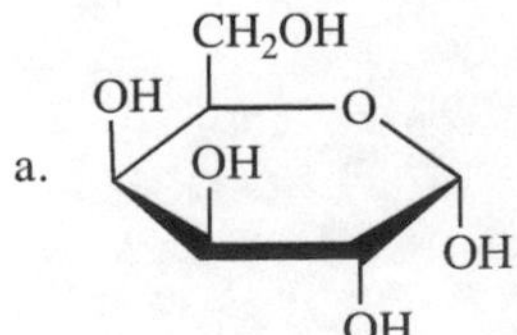

b.
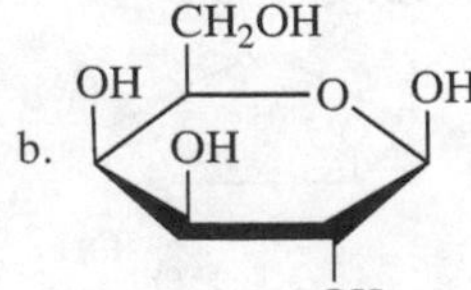

c.
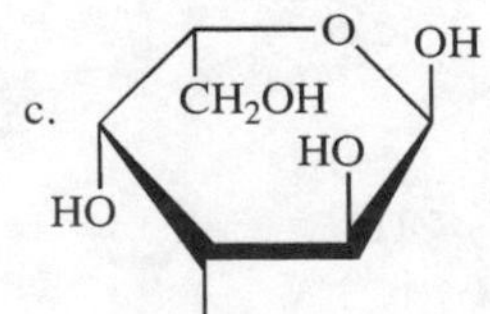

d.
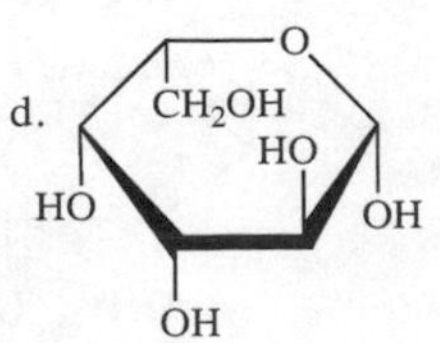

18.95 The Fischer projection formulas, with the given changes, are shown below.

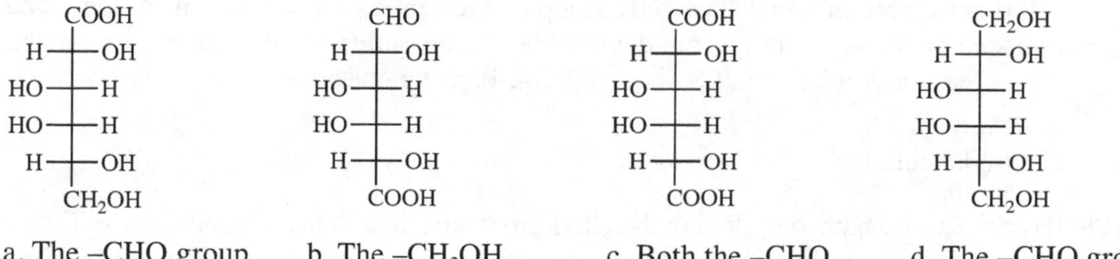

a. The –CHO group is oxidized.

b. The $-CH_2OH$ group is oxidized.

c. Both the –CHO group and $-CH_2OH$ group are oxidized.

d. The –CHO group is reduced.

18.97 The galactose derivatives with a carboxyl group are acidic sugars; those in which the –CHO group has been reduced to $-CH_2OH$ are sugar alcohols. The galactose derivatives in Problem 18.89 are

a. an acidic sugar
b. an acidic sugar
c. an acidic sugar
d. a sugar alcohol

18.99 The galactose derivatives in Problem 18.95 are named as follows.

a. When the –CHO end of an aldose is oxidized to give an aldonic acid, the aldonic acid is named by dropping *–ose* and adding *–onic acid*. The compound is named galactonic acid.
b. When the $-CH_2OH$ end of an aldose is oxidized to give an alduronic acid, the alduronic acid is named by dropping *–ose* and adding *–uronic acid*. The compound is galacturonic acid.
c. When the –CHO group and the $-CH_2OH$ group of an aldose are oxidized to produce a dicarboxylic acid (aldaric acid), the aldaric acid is named by dropping the *–ose* and adding *–aric acid*. The compound is named galactaric acid.
d. When the –CHO group of an aldose is reduced to a $-CH_2OH$ group to give a sugar alcohol, the sugar alcohol is named by dropping *–ose* and adding *–itol*. The compound is named galactitol.

18.101 Weak oxidizing agents, such as Tollens and Benedict's solutions, oxidize the aldehyde end of an aldose; under the basic conditions of these solutions, ketoses are also oxidized. Since the aldoses and ketoses act as reducing agents in such reactions, they are called reducing sugars. All monosaccharides are reducing sugars. All four of the monosaccharides named in this problem (D-glucose, D-galactose, D-fructose, and D-ribose) are monosaccharides and thus reducing sugars.

a. reducing sugar b. reducing sugar c. reducing sugar d. reducing sugar

18.103 A glycoside is an acetal formed from a cyclic monosaccharide by replacement of the hemiacetal carbon –OH group with an –OR group. All of the four structures in this problem have an –OR group on the hemiacetal carbon, so all four are glycosides.

a. yes b. yes c. yes d. yes

18.105 Glycosides, like the hemiacetals from which they are formed, can exist in both α and β forms: in a β configuration, both the terminal $-CH_2OH$ and –OR groups point in the same direction; in an α configuration, the groups point in opposite directions. For the acetals in Problem 18.103, the configuration at the acetal carbon (carbon 1 in the pyranose rings and carbon 2 in the furanose ring) is:

a. alpha b. beta c. alpha d. beta

18.107 Since the cyclic form of a monosaccharide is a hemiacetal, it can react with an alcohol in acid solution to form an acetal. The –OH group on the hemiacetal carbon (on carbon 1 or 2) is replaced by the –OR group from the alcohol. To determine which alcohol was needed to form each acetal in Problem 18.108, look at the –OR group on carbon 1 (or on carbon 2 in part c).

a. methyl alcohol
b. ethyl alcohol
c. ethyl alcohol
d. methyl alcohol

18.109 Glycosides are named by listing the alkyl group attached to the oxygen, followed by the name of the monosaccharide involved, with the suffix *–ide* appended to it. The names of the compounds in Problem 18.97 are as follows:

a. methyl-α-D-alloside
b. ethyl-β-D-altroside
c. ethyl-α-D-fructoside
d. methyl-β-D-glucoside

18.111 Structures for a. α-D-galactose-1-phosphate and b. β-D-galactose-1-phosphate are shown.

a. 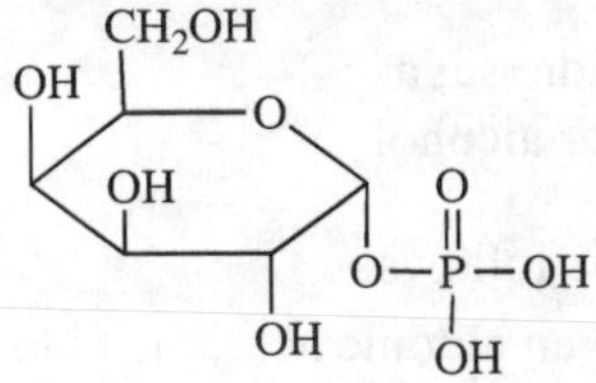 CH_2OH OH O OH O O–P–OH OH OH

b.

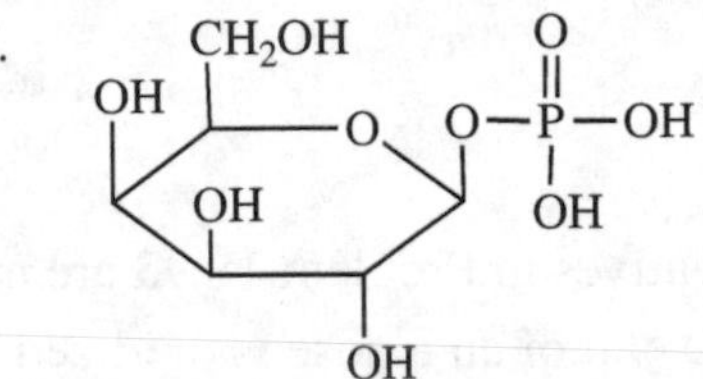

18.113 Structure for a. α-D-gulosamine and b. *N*-acetyl-α-D-gulosamine are shown below.

a. 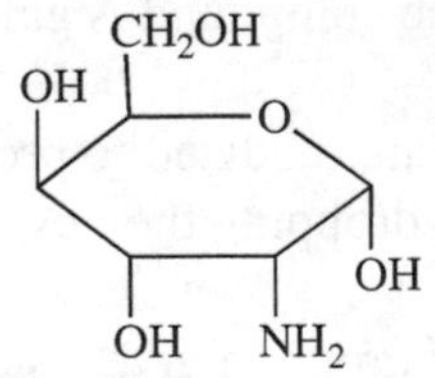 CH_2OH OH O OH OH NH_2

b. 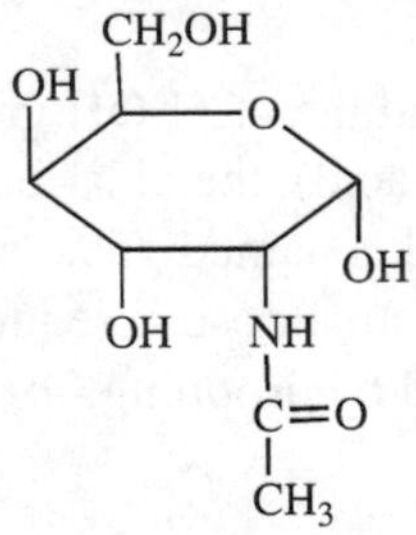 CH_2OH OH O OH OH NH C=O CH_3

18.115 At least one glucose "monosaccharide building block" is present in the dissacharide.

a. yes
b. yes
c. yes
d. yes

18.117 A glycosidic linkage is formed in the reaction between the hemiacetal carbon atom –OH group of one monosaccharide and the –OH group on the other monosaccharide. It is always a carbon-oxygen-carbon bond in a disaccharide. The notation $\alpha(1 \rightarrow 4)$ indicates that the linkage is between carbon 1 and carbon 4; the bond at carbon 1 is pointing down (opposite to the CH_2OH group on that unit), an α configuration.

a. Yes, in maltose there is an $\alpha(1 \rightarrow 4)$ glycosidic linkage.
b. No, in cellobiose the glycosidic linkage is not $\alpha(1 \rightarrow 4)$.
c. No, in lactose the glycosidic linkage is not $\alpha(1 \rightarrow 4)$.
d. No, in sucrose the glycosidic linkage is not $\alpha(1 \rightarrow 4)$.

18.119 Anomers are the α and β forms of cyclic monosaccharides. Anomeric forms exist for three of these disaccharides but not for sucrose because sucrose forms an acetal, which is not easily hydrolyzed.

a. yes
b. yes
c. yes
d. no

18.121 A reducing sugar is a carbohydrate that gives a positive test (is oxidized) with Tollens and Benedict's solutions. To be a reducing sugar, a disaccharide must contain a hemiacetal group that is in equilibrium with an open-chain form with a –CHO group.

a. Sucrose is not a reducing sugar; both of its carbonyl groups are tied up in the glycosidic linkage.
b. Maltose is a reducing sugar; the open-chain form of one of the glucose units has a –CHO group that can be oxidized.
c. Lactose is a reducing sugar; the open-chain form of the glucose unit has a –CHO group that can be oxidized.
d. Cellobiose is a reducing sugar; the open-chain form of one of the glucose units has a –CHO group that can be oxidized.

18.123 A hemiacetal is a compound in which a carbon atom (the hemiacetal carbon) is bonded to both a hydroxyl (–OH) group and an alkoxy (–OR) group. An acetal is a compound in which a carbon atom is bonded to two alkoxy (–OR) groups. The given disaccharides contain:

a. one acetal and one hemiacetal
b. one acetal and one hemiacetal
c. one acetal and one hemiacetal
d. one acetal and one hemiacetal

18.125 In this problem we are looking at the configuration of the hemiacetal part of the disaccharide (that is, the configuration at carbon 1 of the monosaccharide containing the hemiacetal).

a. The bond at carbon 1 is pointing down (opposite to the CH_2OH group), an α configuration.
b. The bond at carbon 1 is pointing up (in the same direction as the CH_2OH group on that unit), a β configuration.
c. The bond at carbon 1 is pointing down (opposite to the CH_2OH group), an α configuration.
d. The bond at carbon 1 is pointing up (in the same direction as the CH_2OH group on that unit), a β configuration.

18.127 A disaccharide is a reducing sugar if it has a hemiacetal center that opens to yield an aldehyde (which can be oxidized by a weak oxidizing agent such as Tollens or Benedict's solution). All of the disaccharides in Problem 18.113 are reducing sugars.

18.129 The cyclic structures of the monosaccharides produced when each of the disaccharides in Problem 18.113 undergoes hydrolysis are shown below.

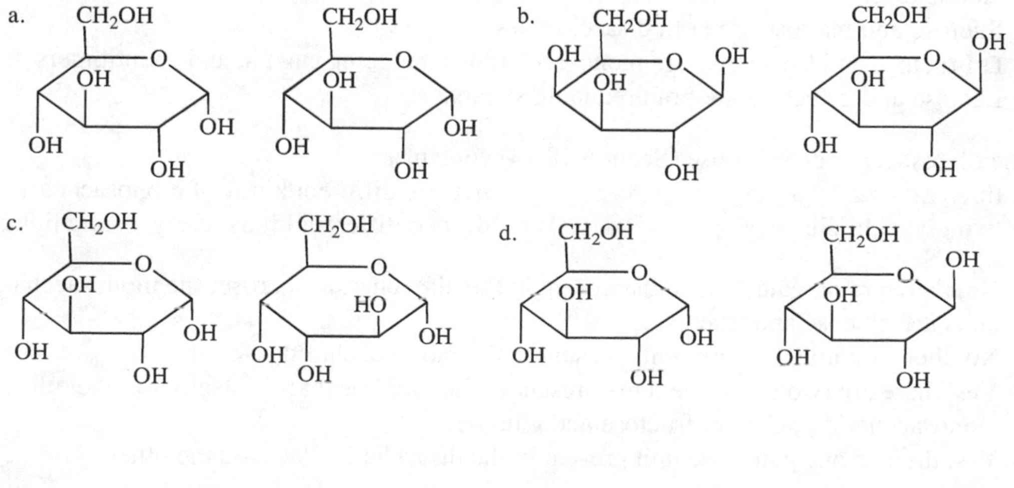

a. CH_2OH, O, OH, OH, OH, OH — CH_2OH, O, OH, OH, OH, OH

b. CH_2OH, OH, O, OH, OH, OH — CH_2OH, OH, O, OH, OH, OH

c. CH_2OH, O, OH, OH, OH, OH — CH_2OH, O, HO, OH, OH, OH

d. CH_2OH, O, OH, OH, OH, OH — CH_2OH, OH, O, OH, OH, OH

18.131 The glycosidic linkage between the two units in a disaccharide is between the –OH group on carbon 1 of the first monosaccharide and one of the –OH groups (usually carbon 4 or carbon 6) on the second monosaccharide. The position of the –OH group on the carbon in the glycosidic linkage that is numbered 1 determines whether the glycosidic linkage is α or β.

a. The linkage is between carbon 1 and carbon 6; the bond at carbon 1 is pointing down (opposite to the CH_2OH group on that unit), an α configuration. Notation is $\alpha(1 \rightarrow 6)$.
b. The linkage is between carbon 1 and carbon 4; the bond at carbon 1 is pointing up (in the same direction as the CH_2OH group on that unit), a β configuration. Notation is $\beta(1 \rightarrow 4)$.
c. The linkage is between carbon 1 and carbon 4; the bond at carbon 1 is pointing down (opposite to the CH_2OH group), an α configuration. Notation is $\alpha(1 \rightarrow 4)$.
d. The linkage is between carbon 1 and carbon 4; the bond at carbon 1 is pointing down (opposite to the CH_2OH group on that unit), an α configuration. Notation is $\alpha(1 \rightarrow 4)$.

18.133 The structure of the disaccharide (sophorose) that contains an α-D-glucose unit, a β-D-glucose unit, and a $\beta(1\rightarrow2)$ glycosidic linkage is shown below.

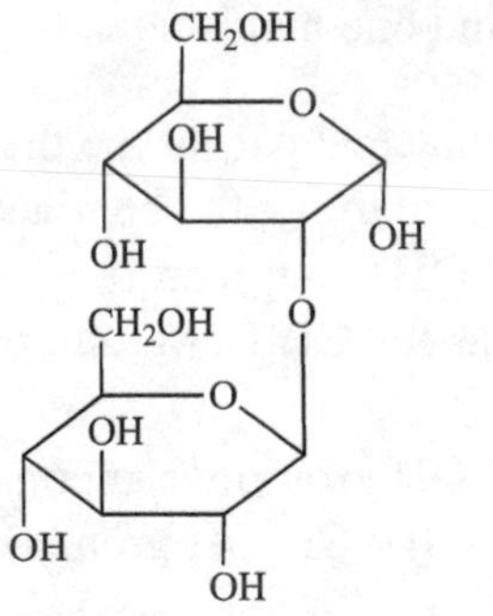

18.135 Anomers are the α and β forms of cyclic monosaccharides. Enantiomers are stereoisomers whose molecules are nonsuperimposable mirror images of each other. An aldohexose is a C_6 sugar with –CHO in the C1 position. All monosaccharides are reducing sugars; a dicaccharide with a "free" (hemiacetal rather than acetal) carbonyl group is a reducing sugar. Disaccharides contain two monosaccharide units.

a. α-D-Glucose and β-D-glucose are both monosaccharides, reducing sugars, anomers, and aldohexoses.
b. Sucrose and maltose are both disaccharides.
c. D-Fructose and L-fructose are monosaccharides, reducing sugars, and enantiomers.
d. Lactose and galactose are both reducing sugars.

18.137 The oligosaccharide raffinose (Section 18.14) contains:

a. three monosaccharides
b. two glyosidic linkages
b. three different kinds of monosaccharides
d. two different kinds of glycosidic linkages

18.139 a. No, there are no galactose units present in the disaccharide sucrose; the monosaccharide units are glucose and fructose.
b. No, there are no galactose units present in the monosaccharide ribose.
c. Yes, there are two galactose units present in the oligosaccharide stachyose; the other monosaccharide units are fructose and glucose.
d. Yes, there is one galactose unit present in the disaccharide lactose; the other monosaccharide is glucose.

18.141 a. The glycosidic linkage for maltose is $\alpha(1 \rightarrow 4)$; the two –OH groups that form the linkage are attached, respectively, to carbon 1 of the first glucose unit (in an α configuration), and to carbon 4 of the second.
b. Galactose has no glycosidic linkages; it is a monosaccharide.
c. Stachyose has two glycosidic linkages. The first is $\alpha(1 \rightarrow 6)$; the two –OH groups that form the linkage are attached, respectively, to carbon 1 of the galactose unit (in an α configuration), and to carbon 6 of glucose. The second glycosidic linkage is $\alpha,\beta(1 \rightarrow 2)$; the two –OH groups that form the linkage are attached, respectively, to carbon 1 of the glucose unit (in an α configuration), and to carbon 2 of β-D-fructose.
d. Fructose has no glycosidic linkages; it is a monosaccharide.

18.143 They are two names for the same thing.

18.145 The range is from less than 100 monomer units up to a million monomer units.

18.147 a. correct b. incorrect c. incorrect d. correct

18.149 Comparing amylopectin and glycogen:
a. No difference in glycosidic linkage: amylopectin and glycogen both have $\alpha(1 \rightarrow 4)$ glycosidic linkages.
b. No difference in monosaccharide monomers: amylopectin and glycogen both consist of glucose monomers.
c. Glycogen has more branching.
d. Glycogen has more monomer units (up to 1,000,000); amylopectin has up to 100,000 glucose units.

18.151 a. correct b. incorrect c. incorrect d. correct

18.153 a. to neither b. to cellulose only c. to neither d. to both

18.155 a. Amylopectin, amylose, and glycogen all have $\alpha(1 \rightarrow 4)$ glycosidic linkages; amylopectin and glycogen also have $\alpha(1 \rightarrow 6)$ linkages.
b. In Amylose, cellulose, and chitin, the glycosidic linkages for each polymer type are all the same. For amylose, the linkages are all $\alpha(1 \rightarrow 4)$; for cellulose and chitin, the linkages are all $\beta(1 \rightarrow 4)$.
c. The polymer chain in amylose, cellulose, and chitin is unbranched.
d. In chitin, the monsaccharide repeating unit is not glucose; it is an N-acetyl amino derivative of D-glucose.

18.157 Hyaluronic acid is a polysaccharide.
a. True. One of its monosaccharide building blocks is NAG.
b. False. One of its monosaccharide building blocks (glucuronate) has a –1 charge.
c. True. Two types of glycosidic linkages are present, $\beta(1 \rightarrow 3)$ and $\beta(1 \rightarrow 4)$.
d. True. One of its biochemical functions is as a lubricant in joints.

18.159 Homopolysaccarides contain one type of monosaccharide monomer; heteropolysaccharides contain more than one type of monosaccharide monomer. Unbranched polysaccharides contain straight chains of monomer units with the same linkage; branched polysaccharides are a mixture of straight chains and side chains with different glycosidic linkages.

a. Starch and cellulose are homopolysaccharides;
b. Glycogen and pectin are homopolysaccharides and branched polysaccharides.
c. Amylose and chitin are homopolysaccharides and unbranched polysaccharides.
d. Heparin and hyaluronic acid are heteropolysaccharides and unbranched polysaccharides.

18.161 The four compounds are classified by function and by structure.

a. Amylose is a storage polysaccharide.
b. Stachyose is not a polysaccharide; it is an oligosaccharide.
c. Hyaluronic acid is an acidic polysaccharide.
d. Cellulose is a structural polysaccharide.

18.163 a. All glycosidic linkages present in amylose, cellulose, chitin, and heparin are the same for each of the individual polysaccharides.

b. For amylopectin and glycogen, some but not all glycosidic linkages are $\alpha(1 \rightarrow 4)$.
c. Hyaluronic acid has both $\beta(1 \rightarrow 3)$ and $\beta(1 \rightarrow 4)$ glycosidic linkages.
d. All of the glycosidic linkages in amylose and heparin are $\alpha(1 \rightarrow 4)$.

18.165 A simple carbohydrate is a dietary monosaccharide or a dietary disaccharide.

a. Yes, glucose is a simple carbohydrate.
b. Yes, sucrose is a simple carbohydrate.
c. No, starch is not a simple carbohydrate.
d. No, cellulose is not a simple carbohydrate.

18.167 A refined sugar has been separated from its plant source whereas a natural sugar has not.

18.169 A glycolipid is a lipid molecule that has a carbohydrate molecule covalently bonded to it.

18.171 In glycoproteins associated with cell membrane structure, the protein part of glycoprotein is incorporated into the protein part of the cell membrane structure and the carbohydrate (oligosaccharide) part functions as a marker on the outer cell membrane surface. Cell recognition generally involves the interaction between the carbohydrate marker of one cell and the protein embedded in the cell membrane of another cell.

Lipids — Chapter 19

Solutions to Selected Problems

19.1 a. False. All lipids are insoluble or only sparingly soluble in water. b. True.

19.3 Lipids are insoluble in water but soluble in nonpolar solvents.
- a. Lipids are insoluble in water because it is a polar solvent.
- b. Lipids are soluble in diethyl ether because it is a nonpolar solvent.
- c. Lipids are insoluble in methanol because it is a polar solvent.
- d. Lipids are soluble in pentane because it is a nonpolar solvent.

19.5 In terms of biochemical function, the five major categories of lipids are: energy-storage lipids, membrane lipids, emulsification lipids, messenger lipids, and protective-coating lipids.
- a. triacylglycerols – energy storage
- b. bile acids – emulsification
- c. cholesterol – membrane
- d. steroid hormones – messenger

19.7 In terms of carbon chain length, fatty acids are characterized as long-chain fatty acids (C_{12} to C_{26}), medium-chain fatty acids (C_8 and C_{10}), or short-chain fatty acids (C_4 and C_6).
- a. Myristic acid (14:0) is a long-chain fatty acid.
- b. Caproic acid (6:0) is a short-chain fatty acid.
- c. Arachidic acid (20:0) is a long-chain fatty acid.
- d. Capric acid (10:0) is a medium-chain fatty acid.

19.9 In a saturated fatty acid molecule all carbon-carbon bonds are single bonds. In a monounsaturated fatty acid, one carbon-carbon double bond is present in the carbon chain; in a polyunsaturated fatty acid, two or more carbon-carbon double bonds are present in the carbon chain. The notation in parentheses after the fatty acid name gives the number of carbon atoms followed by the number of carbon-carbon double bonds in the carbon chain.
- a. Stearic acid (18:0) has no double bonds in its carbon chain, and so is a saturated fatty acid.
- b. Linolenic acid (18:3) has three double bonds; it is a polyunsaturated fatty acid.
- c. Docosahexaenoic acid (22:6) has six double bonds; it is a polyunsaturated fatty acid.
- d. Oleic acid (18:1) has one double bond; it is a monounsaturated fatty acid.

19.11 MUFA stands for monounsaturated fatty acid.

19.13 Use Table 19.1. The given fatty acids are
- a. saturated
- b. monounsaturated
- c. monounsaturated
- d. polyunsaturated

19.15 Use Table 19.1 to classify the fatty acids in Problem 19.13.
- a. The omega classification system does not apply; the fatty acid is saturated.
- b. omega-7
- c. omega-9
- d. omega-6

19.17 The numerical shorthand designation 18:2 ($\Delta^{9,12}$) means that the fatty acid chain has 18 carbon atoms and two carbon-carbon double bonds, and that the locations of the double bonds are between carbon 9 and carbon 10 and between carbon 12 and carbon 13 (numbering from the carboxyl group).

$$CH_3-(CH_2)_4-CH{=}CH-CH_2-CH{=}CH-(CH_2)_7-COOH$$

19.19 The IUPAC name of a fatty acid gives the length of the carbon chain and the degree of unsaturation of the fatty acid. IUPAC names can be determined from Table 19.1.

a. Myristic acid is a C_{14} saturated fatty acid; its IUPAC system name is tetradecanoic acid.

b. Palmitoleic acid is a C_{16} acid with one *cis* double bond between carbon 9 and carbon 10; its IUPAC system name is *cis*-9-hexadecenoic acid.

19.21 As carbon chain length increases, melting point increases.

19.23 The introduction of a *cis*-double bond into a fatty acid molecule is associated with a bend in the carbon chain.

19.25 Melting points for fatty acids are influenced by both carbon chain length and degree of unsaturation (number of double bonds present). Melting point increases with increasing chain length. Melting point decreases as the degree of unsaturation increases.

a. The 18:1 acid has the lower melting point because it has one double bond, a higher degree of unsaturation than the 18:0 acid has.

b. The 18:3 acid has a lower melting point because it has a higher degree of unsaturation than the 18:2 acid does.

c. The 14:0 acid has a lower melting point because it has a shorter carbon chain than the 16:0 acid does.

d. The 18:1 acid has lower a melting point because it has both a higher degree of unsaturation and a shorter carbon chain than the 20:0 acid does.

19.27 The four structural subunits that contribute to the structure of a triacylglycerol are a glycerol molecule and three fatty acid molecules.

19.29 a. There are three functional groups in a triacyglycerol molecule with saturated fatty acid residues: the three ester functional groups.

b. There is one kind of functional group (ester).

c. The name for the functional group is ester.

d. There are three subunit linkages (ester linkages).

19.31 Palmitic acid is a saturated fatty acid containing 16 carbon atoms. Three molecules of palmitic acid are esterified with glycerol in the structure below.

$$H_2C-O-\overset{\overset{\displaystyle O}{\|}}{C}-(CH_2)_{14}-CH_3$$
$$HC-O-\overset{\overset{\displaystyle O}{\|}}{C}-(CH_2)_{14}-CH_3$$
$$H_2C-O-\overset{\overset{\displaystyle O}{\|}}{C}-(CH_2)_{14}-CH_3$$

19.33 A block diagram of a triacylglycerol molecule shows the four subunits present in the structure: glycerol and three fatty acids. In the diagrams below, the fatty acids are stearic acid (S) and linolenic acid (L); they are shown in all possible combinations.

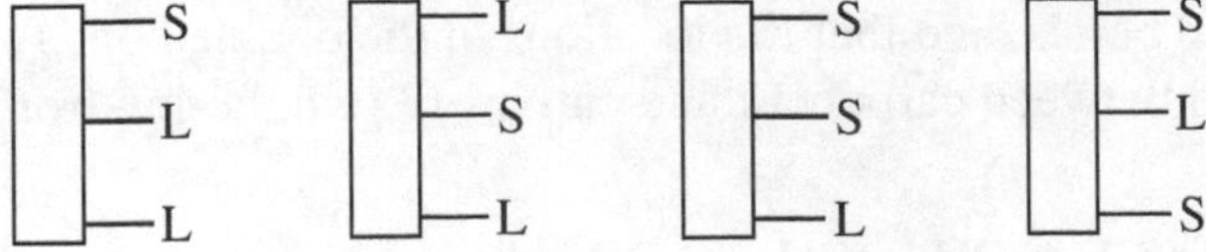

19.35 Table 19.1 gives the names and structures for selected fatty acids.
a. This triacylglycerol molecule contains palmitic acid, myristic acid, and oleic acid.
b. This triacylglycerol molecule contains oleic acid, palmitic acid, and palmitoleic acid.

19.37 a. On the basis of its melting point, this mixture would be classified as a fat.
b. The substance is a solid at room temperature.

19.39 a. The triacylglycerol molecule in part a of Problem 19.35 has three acyl groups, zero 18:2 fatty acid residues, two saturated fatty acid residues, and zero linolenic acid residues.
b. The triacylglycerol molecule in part b of Problem 19.35 has three acyl groups, zero 18:2 fatty acid residues, one saturated fatty acid residue, and zero linolenic acid residues.

19.41 a. Pairing "Saturated fat" and "good fat" is not correct; saturated fat in the diet can increase heart disease risk, so it is a "bad fat."
b. Pairing "Polyunsaturated fat" and "bad fat" is not correct; polyunsaturated fat in the diet can reduce the risk of heart disease but increase the risk of certain kinds of cancer, so it is a "good and bad fat."

19.43 a. Pairing "Cold-water fish" and "high in omega-3 fatty acids" is correct.
b. Pairing "Fatty fish" and "low in omega-3 fatty acids" is not correct; cold-water fish, also called fatty fish, contain more omega-3 acids than leaner, warm-water fish.

19.45 An essential fatty acid is necessary to the human body but cannot be synthesized by the human body; it must be obtained in the diet. There are two essential fatty acids: linoleic acid and linolenic acid.
a. Lauric acid (12:0) is a nonessential fatty acid.
b. Linoleic acid (18:2) is an essential fatty acid.
c. Myristic acid (14:0) a nonessential fatty acid.
d. Palmitoleic acid (16:1) is a nonessential fatty acid.

19.47 a. The triacylglycerol molecule in part a, Prob. 19.35, has zero omega-3 fatty acid residues, zero omega-6 fatty acid residues, one "good" fatty acid residue, and one Δ^9 fatty acid residue.
b. The triacylglycerol molecule in part b has zero omega-3 fatty acid residues, zero omega-6 fatty acid residues, two "good" fatty acid residues, and three Δ^9 fatty acid residue.

19.49 Acidic hydrolysis of a triacylglycerol molecule gives glycerol and fatty acid as products.
a. Adding 3 molecules of water produces 1 glycerol molecule and 3 fatty acid molecules.
b. If 2 molecules of water are added, 1 monoacylglycerol molecule and 2 fatty acid molecules are produced.
c. If 1 molecule of water is added, 1 diacylglycerol molecule and 1 fatty acid molecule are produced.

19.51 In Problem 19.49 the reactions are classified as:
a. complete hydrolysis b. partial hydrolysis c. partial hydrolysis

19.53 Complete hydrolysis of a triacylglycerol molecule gives one glycerol molecule and three fatty acid molecules as products.

$$\underset{\text{OH}}{\underset{|}{CH_2}}-\underset{\text{OH}}{\underset{|}{CH}}-\underset{\text{OH}}{\underset{|}{CH_2}}$$

$$CH_3-(CH_2)_{14}-COOH$$

$$CH_3-(CH_2)_{12}-COOH$$

$$CH_3-(CH_2)_7-CH=CH-(CH_2)_7-COOH$$

19.55 Saponification of a triacylglycerol molecule with NaOH gives one glycerol molecule and the sodium salts of three fatty acids molecules.

$$\underset{\text{OH}}{\underset{|}{CH_2}}-\underset{\text{OH}}{\underset{|}{CH}}-\underset{\text{OH}}{\underset{|}{CH_2}} \qquad CH_3-(CH_2)_{14}-COO^-\ Na^+$$

$$CH_3-(CH_2)_{12}-COO^-\ Na^+ \qquad CH_3-(CH_2)_7-CH{=}CH-(CH_2)_7-COO^-\ Na^+$$

19.57 Hydrogenation involves hydrogen addition across carbon-carbon multiple bonds, which increases the degree of saturation. Carbon chains that have no double bonds are already saturated.

19.59 One molecule of H_2 will react with each double bond in the triacylglycerol molecule. Since there are six double bonds in the molecule, six molecules of H_2 will react with one triacylglycerol molecule.

19.61 Partial hydrogenation of a triacylglycerol molecule with two molecules of H_2 will result in the addition of hydrogen to two of the double bonds. If there are three double bonds in the molecule, one will remain after the partial hydrogenation. The three possible products are shown below.

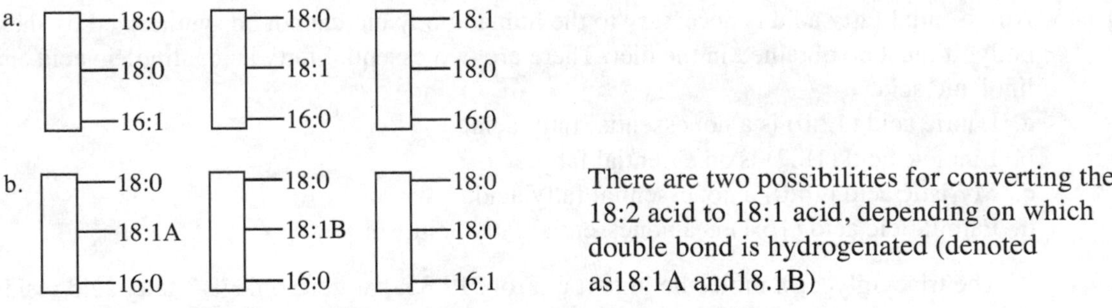

There are two possibilities for converting the 18:2 acid to 18:1 acid, depending on which double bond is hydrogenated (denoted as18:1A and18.1B)

19.63 Rancidity results from the hydrolysis of ester linkages and the oxidation of carbon-carbon double bonds, which produce aldehyde and carboxylic acid products that often have objectionable odors.

19.65 a. Carbon-carbon double bonds are broken in both oxidation and hydrogenation reactions involving an alkene.
b. Fatty acids are among the products of hydrolysis and oxidation of a triacylglycerols.
c. During hydrogenation of *cis*-double bonds, some *cis*-double bonds are converted to *trans*-double bonds.
d. Water is a reactant in hydrolysis and saponification reactions.

19.67 a. When a simple triacylglycerol undergoes hydrolysis two different types of organic molecules are obtained: glycerol and three identical fatty acids.
b. When a mixed triacylglycerol undergoes hydrolysis, three or four different types of organic molecules are obtained: glycerol and two or three different fatty acids.
c. When a simple triacylglycerol undergoes saponification, two different types of organic molecules are obtained: glycerol and the fatty acid salt of the three identical fatty acids.
d. When a mixed triacylglycerol undergoes saponification, three or four different types of organic molecules are obtained: glycerol and two or three different fatty acid salts.

19.69 In the block diagram for a glycerophospholipid, the building blocks are labeled with letters and the linkages between building blocks are labeled with numbers.
a. The building blocks labeled B and C are fatty acid residues.
b. The building block labeled E is an alcohol residue.
c. The linkages labeled 1, 2, 3, and 4 are ester linkages.
d. The linkages labeled 3 and 4 involve a phosphate residue.

19.71 A phosphatidyl group is made up of fatty acid, glycerol, and phosphate subunits.

19.73 a. False. Choline has one –OH group. b. True.
c. True. d. True.

19.75 Based on the "head and two tails" model for the structure of a glycerophospholipid
a. a fatty acid is a part of the "tail" structure.
b. an amino alcohol is part of the "head" structure.
c. a phosphate group is part of the "head" structure.

19.77 a. True. Four functional groups are present in the sphingosine molecule.
b. True. The carbon chain of the sphingosine molecule is unsaturated.
c. False. The sphingosine molecule has only one tail.
d. True. An amino functional group is present in the sphingosine molecule.

19.79 In the block diagram of a sphingophospholipid the building blocks are labeled with letters and the linkages between the building blocks are labeled with numbers.
a. The building block labeled B is a fatty acid residue.
b. The building block labeled C is a phosphate residue.
c. The linkage labeled 1 is an amide linkage.
d. The linkages labeled 1 and 2 involve a sphingosine residue.

19.81 Sphingomyelin contains an amino alcohol called choline.

19.83 a. The "head" of a sphingophospholipid contains a phosphate and an amino alcohol.
b. The fatty acid chain and the carbon chain of sphingosine have hydrophobic properties.

19.85 a. Triacylglycerols and sphingophospholipids contain four building blocks.
b. The linkages in triacyglycerols and glycerophospholipids are all ester linkages.
c. An alcohol building block is present in glycerophospholipids and sphingophospholipids.
d. Triacylglycerol is an energy-storage lipid.

19.87 a. None of the three compounds must contain ethanolamine.
b. Glycerophospholipids contain phosphatidylcholine.
c. Sphingophospholipids must contain an amide linkage.
d. None of the three types of compounds are nonsaponifiable (all are saponifiable).

19.89 In the block diagram of a sphingoglycolipid the building blocks are labeled with letters and the linkages between the building blocks are labeled with numbers.
a. The building block labeled B is a fatty acid residue.
b. The building block labeled C is a carbohydrate residue.
c. The linkage labeled 1 is an amide linkage.
d. The linkages labeled 2 involves a monosaccharide.

19.91 a. A monosaccharide is a component of (1) a cerebroside but not a ganglioside.
b. An oligosaccharide is a component of (2) a ganglioside but not a cerebroside.
c. A fatty acid is a component of (3) both a cerebroside and a ganglioside.
d. Sphingosine is a component of (3) both a cerebroside and a ganglioside.

19.93 a. All three types of compounds (glycerophospholipids, sphingophospholipids, and sphingoglycolipids) are membrane lipids.
b. All three types of compounds (glycerophospholipids, sphingophospholipids, and sphingoglycolipids) are saponifiable lipids.
c. Sphingophospholipids and sphingoglycolipids have an amide linkage present.
d. All three types of compounds (glycerophospholipids, sphingophospholipids, and sphingoglycolipids) have a "head-and-two-tail" structure.

19.95 a. True. Rings A, B, and C are identical.
b. False. Rings A and C are part of a fused-ring system, with B between A and C.
c. True. Rings A and B share a common side.
d. False. There are 17 carbon atoms present in a steroid nucleus.

19.97 a. There are three six-membered rings present in a cholesterol molecule.
b. There are zero amide linkages in a cholesterol molecule.
c. There is one hydroxyl substituent in a cholesterol molecule.
d. There are two functional groups in a cholesterol molecule, a hydroxyl group and a double bond.

19.99 Use the information in Table 19.4 to answer this question. The four foods, ranked in order of increasing amounts of cholesterol present, are: Swiss cheese, fish filet, chicken, liver.

19.101 The cholesterol associated with LDLs contributes to increased blood cholesterol levels, and so is often called "bad cholesterol." The cholesterol associated with HDLs contributes to reduced blood cholesterol levels, and is often called "good cholesterol."

19.103 The general structural characteristic associated with lipids present in a lipid bilayer is a "head-and-two-tail" structure.

19.105 a. False. The outside surface positions in a lipid bilayer are occupied by hydrophilic entities.
b. True. The interaction between the fluid inside a cell and the surface of a lipid bilayer is primarily an interaction between polar entities.
c. False. The outside surface positions in a lipid bilayer are occupied by the polar heads of phospholipids and glycolipids.
d. False. The interactions between adjacent lipids in a lipid bilayer usually involve intermolecular interactions, not covalent bonding.

19.107 In the lipid bilayer, the presence of unsaturated acids, with the kinks in their carbon chains, prevents tight packing of fatty acids chains. The open packing creates "open" areas in the lipid bilayer through which biochemicals can pass into and out of the cell.

19.109 It is a membrane protein that penetrates the interior of the lipid bilayer (cell membrane).

19.111 Passive transport means that a substance moves across a cell membrane by diffusion from an area of high concentration to one of lower concentration without the expenditure of any cellular energy. In facilitated transport a substance moves across a cell membrane with the aid of membrane proteins, from a region of higher concentration to a region of lower concentration, without the expenditure of cellular energy.

19.113 a. Active transport is the movement across a membrane against a concentration gradient.
b. Facilitated transport is a process in which proteins serve as "gates."
c. Active transport is a process in which expenditure of cellular energy is required.
d. Passive transport and facilitated transport are both processes in which movement across the membrane is from a high to a low concentration.

19.115 An emulsifier is a substance that can disperse and stabilize water-insoluble substances as colloidal particles in an aqueous solution.

19.117 a. False. Cholic acid is a C_{24} molecule, but cholesterol is a C_{27} molecule.
b. True.
c. True.
d. False. Cholesterol does not contain a carboxyl group.

19.119 a. False. Glycocholic acid is a complexed bile acid, but 7-deoxycholic acid is not.
b. False. Taurocholic acid contains the element sulfur, but glycocholic does not.
c. False. Glycocholic acid contains a side-chain amide linkage, but cholic acid does not.
d. True.

19.121 A molecule of glycocholic acid has
a. six attachments to the steroid nucleus.
b. three different functional groups.
c. a total of five functional groups.
d. four different elements.

19.123 The part of the taurocholic acid molecule that is strongly hydrophilic is the amino-acid-bearing carbon chain.

19.125 The medium through which bile acids are supplied to the small intestine is bile, an emulsifying agent secreted by the liver.

19.127 Bile acids are stored in the gall bladder and released into the small intestine during digestion.

19.129 a. Taurocholic acid is an emulsification lipid.
b. Sphingoglycolipids are membrane lipids.
c. Triacylglycerols are energy-storage lipids.
d. Cholesterol is a membrane lipid.

19.131 a. True.
b. False. Progesterone is a female sex hormone; testosterone is a male sex hormone.
c. False. Cortisone is a synthetic derivative of cortisol (an adrenocorticoid hormone); aldosterone is an adrenocorticoid hormone (a mineralcorticoid).
d. False. Cortisol is a glucocorticoid; aldosterone is a mineralcorticoid.

19.133 a. Aldosterone is a naturally occurring steroid hormone.
b. Cortisone is a synthetic steroid hormone.
c. Estradiol is a naturally occurring steroid hormone.
d. Norethynodrel is a synthetic steroid hormone.

19.135 a. adrenocorticoid hormone
b. sex hormone
c. adrenocorticoid hormone
d. sex hormone

19.137 a. control Na^+/K^+ ion balance in cells and body fluids
b. responsible for secondary male characteristics
c. controls glucose metabolism and is an anti-inflammatory agent
d. responsible for secondary female characteristics

19.139 The anabolic steroid with the given characteristics is:

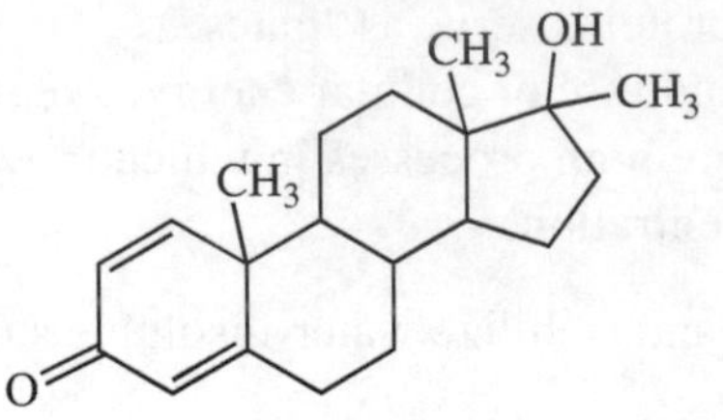

19.141 The prostaglandin structure is based on a straight-chain 20-carbon fatty acid that is converted into a prostaglandin structure when the eighth and twelfth carbon acids of the fatty acid become connected to form a five-membered ring, a cyclopentane ring.

19.143 A prostaglandin is similar to a leukotriene but has a cyclopentane ring formed by a bond between carbon 8 and carbon 12.

19.145 a. Bile acids are steroid-nucleus-based lipids. b. Oils are glycerol-based lipids.
c. Prostaglandins are fatty-acid-based lipids. d. Thromboxanes are fatty-acid-based lipids.

19.147 a. Bile acids are emulsification lipids. b. Cholesterol is a membrane lipid.
c. Eicosanoids are messenger lipids. d. Sphingophospholipids are membrane lipids.

19.149 A biological wax is a lipid that is a monoester of a long-chain fatty acid and a long-chain alcohol.

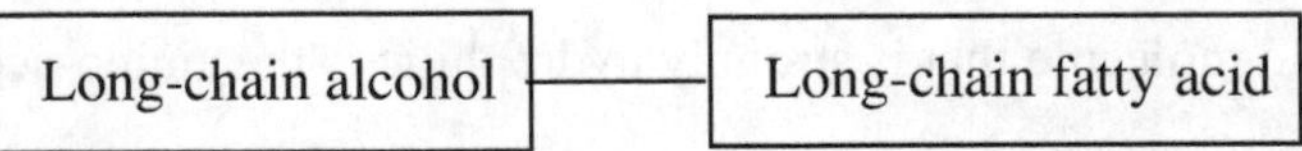

19.151 A biological wax is a lipid that is a monoester of a long-chain fatty acid and a long-chain alcohol; a mineral wax is a mixture of long-chain alkanes obtained from the processing of petroleum.

19.153 a. The structures of bile acids are based on a steroid nucleus.
b. Biological waxes and sphingophospholipids are saponifiable lipids.
c. Biological waxes and sphingophospholipids contain at least one fatty acid building block.
d. Sphingophospholipids contain at least one amide linkage.

19.155 a. Yes, a biological wax has a "head-and-two-tails" structure.
b. No, cholesterol does not have a "head-and-two-tails" structure.
c. Yes, sphingoglycolipids have a "head-and-two-tails" structure.
d. Yes, fats have a "head-and-two-tails" structure.

19.157 a. A biological wax has an ester linkage.
b. A triacylglycerol has an ester linage.
c. None apply. A bile acid does not have ester, amide, or glycosidic linkages.
d. None apply. Sphingosine does not have ester, amide, or glycosidic linkages.

19.159 a. Yes, sphingoglycolipids and sphingophospholipids are both saponifiable.
b. No. Biological waxes are saponifiable; mineral waxes are not.
c. No. Triacylglycerols are saponifable; steroid hormones are not.
d. No. Eicosanoids are saponifiable; cholesterol is not.

19.161 a. Cholesterol has zero saponifiable linkages.
b. Triacylglycerols have three saponifiable linkages.
c. Bile acids have zero saponifiable linkages.
d. Biological waxes have one saponifiable linkage.

Proteins — Chapter 20

Solutions to Selected Problems

20.1 The monomers are called amino acids.

20.3 The percent protein in a cell is 15% by mass.

20.5 In an α-amino acid both the amino group and the carboxyl group are attached to the α-carbon atom.

20.7 a. Yes, this is an α-amino acid.
b. No. This is an amino acid, but not an α-amino acid.

20.9 The R group present in an α-amino acid is called the amino acid side chain. The number of carbon atoms present in the R group of each of the given standard amino acids is:
a. 1 b. 4 c. 3 d. 0

20.11 Use Table 20.1 to determine the name and three-letter abbreviation of each of the standard amino acids in Problem 20.9.
a. cysteine, Cys
b. lysine, Lys
c. valine, Val
d. glycine, Gly

20.13 Use Table 20.1 to classify the amino acids in Problem 20.9
a. polar neutral b. polar basic c. nonpolar d. nonpolar

20.15 Use Table 20.1 to classify the amino acids in Problem 20.9 as hydrophobic or hydrophilic.
a. hydrophilic b. hydrophilic c. hydrophobic d. hydrophobic

20.17 Use Table 20.1 to identify the amino acid side chains with these characteristics.
a. contains only C and H: alanine, valine, leucine, isoleucine, proline, phenylalanine
b. contains a carboxyl group: aspartic acid, glutamic acid
c. contains an amide group: asparagine, glutamine
d. contains the element O: serine, threonine, asparagine, glutamine, tyrosine, aspartic acid, glutamic acid

20.19 The standard amino acids having three-letter abbreviations that are not the first three letters of their names are: isoleucine (Ile), tryptophan (Trp), asparagine (Asn), glutamine (Gln)

20.21 The number of amino groups and the number of carboxyl groups present in
a. polar neutral – one amino group and one carboxyl group
b. polar acidic – one amino group and two carboxyl groups

20.23 In the amino acid proline, the side chain covalently bonds to the amino acid's amino group.

20.25 An essential amino acid is an amino acid needed by the human body that must be obtained from dietary sources.
a. Yes, valine is an essential amino acid.
b. Yes, isoleucine is an essential amino acid.
c. Yes, tryptophan is an essential amino acid.
d. No. Proline is not an essential amino acid.

20.27 A complete dietary protein contains all the essential amino acids needed by the human body.

a. Yes, an egg does contain complete protein.
b. No, oats do not contain complete protein.
c. No, corn does not contain complete protein.
d. Yes, soy does contain complete protein.

20.29 A limiting amino acid is an essential amino acid that is missing or present in inadequate amounts in an incomplete dietary protein.

a. No, an egg does not contain a limiting amino acid; eggs are a complete protein.
b. Yes, oats do contain a limiting amino acid; oats are not a complete protein.
c. Yes, corn does contain a limiting amino acid; corn is not a complete protein.
d. No, soy does not contain a limiting amino acid; soy is a complete protein.

20.31 Complementary proteins are two or more incomplete dietary proteins that when combined provide an adequate amount of all essential amino acids.

a. No, soy and rice are not complementary proteins. Soy is a complete protein; rice is not.
b. No, egg and milk are not complementary proteins; egg and milk are both complete proteins.
c. No, beef and oats are not complementary proteins. Soy is a complete protein; rice is not.
d. No, rice and corn are not complementary proteins; both corn and rice lack adequate lysine.

20.33 A limiting amino acid is an essential amino acid that is missing or present in inadequate amounts in an incomplete dietary protein.

a. The limiting amino acid in wheat is lysine.
b. The limiting amino acids in beans are methionine and tryptophan.
c. Soy has no limiting amino acids.
d. The limiting amino acid in peas is methionine.

20.35 With few exceptions, amino acids found in nature and in proteins are from the L-family of isomers.

20.37 Fischer projection formulas for amino acids are drawn with the –COOH at the top, the –R group at the bottom (positioned vertically), and the $-NH_2$ group to the left of the α carbon for the L isomer and to the right of the α carbon for the D isomer.

a.
$$\begin{array}{c} COOH \\ H_2N-\!\!\!+\!\!\!-H \\ CH_2 \\ | \\ OH \end{array}$$
L-serine

b.
$$\begin{array}{c} COOH \\ H-\!\!\!+\!\!\!-NH_2 \\ CH_2 \\ | \\ OH \end{array}$$
D-serine

c.
$$\begin{array}{c} COOH \\ H-\!\!\!+\!\!\!-NH_2 \\ CH_3 \end{array}$$
D-alanine

d.
$$\begin{array}{c} COOH \\ H_2N-\!\!\!+\!\!\!-H \\ CH_2 \\ | \\ CH-CH_3 \\ | \\ CH_3 \end{array}$$
L-leucine

20.39 The structural formula of the amino acid is a Fischer projection formula

a. This is a D-amino acid; in a Fischer projection formula the D form has the $-NH_2$ group on the right.
b. This is a nonpolar amino acid; it has a nonpolar side chain.
c. This is an essential amino acid; its name is valine.
d. This is a standard amino acid, one of 20 amino acids normally found in proteins.

20.41 α-Amino acids are white crystalline solids with high decomposition points, a characteristic of ionic compounds. An amino acid exists as a charged species called a zwitterion; the amino group is protonated and thus has a positive charge, and the carboxyl group has lost a proton and has a negative charge. The strong intermolecular forces between these positive and negative charges are the cause of the high melting point of amino acids.

20.43 A zwitterion is a molecule that has a positive charge on one atom and a negative charge on another one. In the zwitterion form of an α-amino acid, the amino group is protonated and has a positive charge, and the carboxyl group has lost a proton and has a negative charge.

a. $\begin{array}{c} COO^- \\ H_3\overset{+}{N}-C-H \\ CH_2 \\ CH-CH_3 \\ CH_3 \end{array}$ leucine

b. $\begin{array}{c} COO^- \\ H_3\overset{+}{N}-C-H \\ CH-CH_3 \\ CH_2 \\ CH_3 \end{array}$ isoleucine

c. $\begin{array}{c} COO^- \\ H_3\overset{+}{N}-C-H \\ CH_2 \\ SH \end{array}$ cysteine

d. $\begin{array}{c} COO^- \\ H_3\overset{+}{N}-C-H \\ H \end{array}$ glycine

20.45 In solution, three different amino acid forms can exist (zwitterion, negative ion, and positive ion). The zwitterion predominates in neutral solution. In acidic solution (low pH) the positively charged species (protonated amino group) predominates; in basic solution (high pH) the negatively charged species (carboxylate ion) predominates.

a. $\begin{array}{c} COO^- \\ H_3\overset{+}{N}-C-H \\ CH_2 \\ OH \end{array}$ serine at pH 5.68

b. $\begin{array}{c} COOH \\ H_3\overset{+}{N}-C-H \\ CH_2 \\ OH \end{array}$ serine at pH 1.0

c. $\begin{array}{c} COO^- \\ H_2N-C-H \\ CH_2 \\ OH \end{array}$ serine at pH 12.0

d. $\begin{array}{c} COOH \\ H_3\overset{+}{N}-C-H \\ CH_2 \\ OH \end{array}$ serine at pH 3.0

20.47 An isoelectric point is the pH at which an amino acid has no net charge because an equal number of positive and negative charges are present. At the isoelectric point, zwitterion concentration in a solution is maximized.

20.49 The two –COOH groups in glutamic acid have different acidities; they deprotonate at different pH values. Side chain carboxyl groups are weaker acids than α-carbon carboxyl groups.

20.51 In a low pH aqueous solution (an acidic solution) all acid and amino groups are protonated. The net charge on the amino acid will be positive, and the magnitude of the charge will depend on the number of amino groups present in the amino acid.

a. Valine has one acid group and one amino group; valine's net charge in a low pH aqueous solution is +1.
b. Lysine has one acid group and two amino groups; lysine's net charge in a low pH aqueous solution is +2.
c. Aspartic acid has two acid groups and one amino group; aspartic acid's net charge in a low pH aqueous solution is +1.
d. Serine has one acid group and one amino group; serine's net charge in a low pH aqueous solution is +1.

20.53 Cysteine has a side chain that contains a sulfhydryl group (–SH). In the presence of mild oxidizing agents, cysteine dimerizes to form a cystine molecule; cystine contains two cysteine residues linked by a disulfide bond.

20.55 A peptide bond is a covalent bond between the carboxyl group of one amino acid and the amino group of another amino acid.

20.57 For the tripeptide Gly-Ala-Cys

a. Gly is located at the peptide's N-terminal end.
b. Cys is located at the peptide's C-terminal end.
c. two peptide bonds are present.
d. two amide linkages are present.

20.59 The condensed structural representation of a tripeptide is:

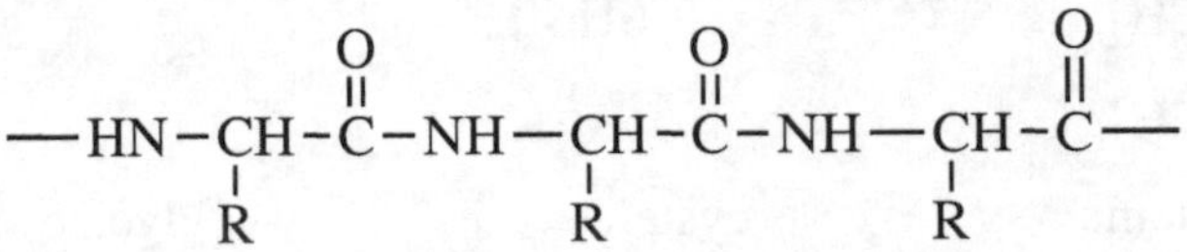

$$\text{—HN–}\underset{\displaystyle R}{\underset{|}{\text{CH}}}\text{–}\overset{\displaystyle O}{\overset{\|}{\text{C}}}\text{–NH–}\underset{\displaystyle R}{\underset{|}{\text{CH}}}\text{–}\overset{\displaystyle O}{\overset{\|}{\text{C}}}\text{–NH–}\underset{\displaystyle R}{\underset{|}{\text{CH}}}\text{–}\overset{\displaystyle O}{\overset{\|}{\text{C}}}\text{—}$$

20.61 The condensed structural representation of the tripeptide Val-Ser-Cys is:

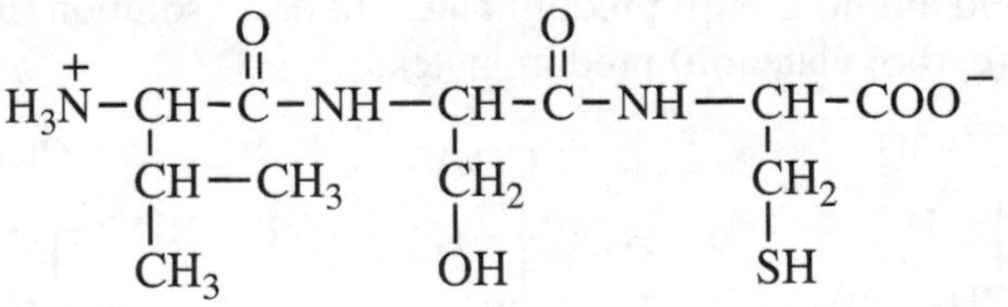

$$H_3\overset{+}{N}\text{–}\underset{\begin{array}{c}|\\ CH\text{—}CH_3\\ |\\ CH_3\end{array}}{CH}\text{–}\overset{\displaystyle O}{\overset{\|}{C}}\text{–NH—}\underset{\begin{array}{c}|\\ CH_2\\ |\\ OH\end{array}}{CH}\text{–}\overset{\displaystyle O}{\overset{\|}{C}}\text{–NH—}\underset{\begin{array}{c}|\\ CH_2\\ |\\ SH\end{array}}{CH}\text{–}COO^-$$

20.63 Table 20.1 gives the names and structures of the 20 standard amino acids. Abbreviated names for the two tripeptides are:

a. Ser–Ala–Cys b. Asp–Thr–Asn

20.65 In naming a peptide according to the IUPAC system, the rules are: 1) the C-terminal amino acid residue (on the right) keeps its full amino acid name, 2) all other amino acid residue names end in *–yl* (replacing the *–ine* or *–ic acid* ending of the amino acid name, with the exceptions of tryptophyl, cysteinyl, glutaminyl, and asparaginyl), and 3) the amino acid naming sequence begins with the N-terminal amino acid residue.

a. Ser–Cys is serylcysteine.
b. Gly–Ala–Val is glycylalanylvaline.
c. Tyr–Asp–Gln is tyrosylaspartylglutamine.
d. Leu–Lys–Trp–Met is leucyllysyltryptophylmethionine.

20.67 The condensed structural formulas for a. glycylalanine and b. cysteinylalanylglycine are:

a. 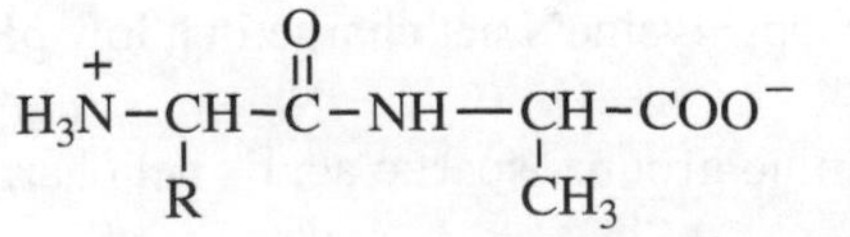 $$H_3\overset{+}{N}\text{–}\underset{\begin{array}{c}|\\ R\end{array}}{CH}\text{–}\overset{\displaystyle O}{\overset{\|}{C}}\text{–NH—}\underset{\begin{array}{c}|\\ CH_3\end{array}}{CH}\text{–}COO^-$$

b. 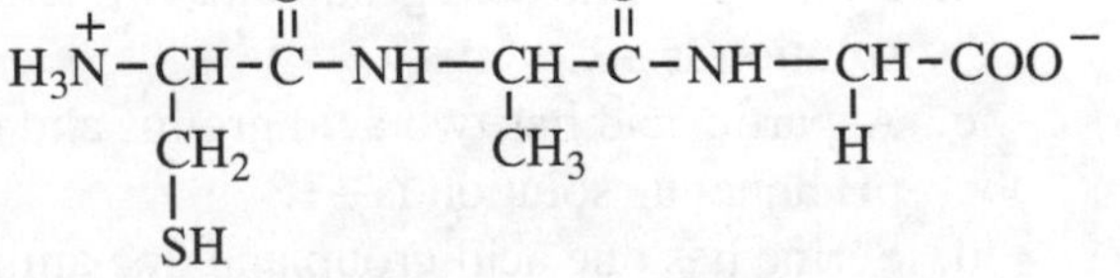 $$H_3\overset{+}{N}\text{–}\underset{\begin{array}{c}|\\ CH_2\\ |\\ SH\end{array}}{CH}\text{–}\overset{\displaystyle O}{\overset{\|}{C}}\text{–NH—}\underset{\begin{array}{c}|\\ CH_3\end{array}}{CH}\text{–}\overset{\displaystyle O}{\overset{\|}{C}}\text{–NH—}\underset{\begin{array}{c}|\\ H\end{array}}{CH}\text{–}COO^-$$

20.69 For the tripeptide Ala-Val-Gly

a. none of the amino acid residues are hydrophilic.
b. the amino acid residues of Ala, Val, and Gly are hydrophobic.
c. the amino acid residues of Ala, Val, and Gly possess nonpolar R groups.
d. Val participates in two amide linkages.

20.71 The tripeptide is tyrosylleucylisoleucine. In naming peptides, the C-terminal amino acid residue (located at the far right of the structure) keeps its full amino acid name. All of the other amino acid residues have a name that end in a *–yl* suffix that replaces the *–ine* or *–ic* acid ending of the amino acid name.

a. The structure of the tripeptide using three-letter symbols for the amino acids is Tyr-Leu-Ile.
b. There are two peptide bonds present within the tripeptide.
c. Tyr is the amino acid residue with the largest R group.
d. None of the amino acid residue has an acidic side chain.

20.73 Peptides that contain the same amino acids but in different order are different molecules. The sequence of amino acids in a peptide is written with the N-terminal amino acid on the left and the C-terminal on the right. Ser is the N-terminal amino acid in Ser–Cys, and Cys is the N-terminal amino acid in Cys–Ser; the two dipeptides are structural isomers.

20.75 Peptides that contain the same amino acids but in different order are different molecules (structural isomers). Six different tripeptides can be formed from one molecule each of serine, valine, and glycine: Ser–Val–Gly, Val–Ser–Gly, Gly–Val–Ser, Ser–Gly–Val, Val–Gly–Ser, and Gly–Ser–Val.

20.77 The two best-known peptide hormones, both produced by the pituitary gland, are oxytocin and vasopressin.

a. Both are nonapeptides with six residues held in a loop by a disulfide bond.
b. They differ in the identity of the amino acids in positions 3 and 8 of the peptide chain.

20.79 Enkephalins are neurotransmitters produced by the brain; they bind at receptor sites in the brain to reduce pain. The action of the prescription painkillers morphine and codeine is based on their binding at the same receptor sites as the naturally occurring enkephalins.

20.81 The glutathione structure is unusual in that the amino acid Glu is bonded to Cys through the side-chain carboxyl group rather than through its α-carbon carboxyl group.

20.83 In a monomeric protein, only one peptide chain is present; in a multimeric protein, more than one peptide chain is present.

20.85 a. True. By definition, a multimeric protein contains more than one peptide chain.

b. False. A simple protein contains only amino acid residues (with no restriction on the types of amino acids present).
c. True. A conjugated protein contains one or more peptide chains and at least one non-amino acid component.
d. True. Glycoproteins contain carbohydrate groups as their prosthetic groups.

20.87 The primary structure of a protein is the order in which the amino acids are bonded to each other.

20.89 The number of different primary structures possible for a four-amino acid segment of a protein consisting of
a. two glycine units and two alanine units is six.
b. two glycine units, one alanine unit, and one valine unit is twelve.

20.91 The number of different primary structures possible for a four-amino acid segment of a protein if there are
a. no restrictions on the amino acids that can be present is 160,000 ($20 \times 20 \times 20 \times 20$).
b. two each of two different amino acids present is 2280; $6(20 \times 1 \times 19 \times 1)$

20.93 This segment of a protein "backbone" is long enough to show the positions where three R groups (amino acid side chains) can be attached.

```
        O              O
        ||             ||
—HC—C—NH—CH—C—NH—CH—
  |             |              |
```

20.95 In an alpha-helix secondary structure for a protein
a. the general shape of the protein backbone is a helix (coiled spring).
b. the amino acid R groups extend outward from the coil.

20.97 a. False. Hydrogen bonds present in an α-helix structure always involve a carbonyl oxygen atom of a peptide linkage and the hydrogen atom of an amino group of another peptide linkage further along the protein backbone.
b. True. Both α-helix and β-pleated sheet structures can be present in the same protein.
c. False. In a β-pleated sheet structure the protein backbone segments involved in hydrogen bonding can be two segments from different backbones or two segments of the same backbone that has folded back upon itself.
d. False. In an α-helix all of the amino acid R groups extend outward from the spiral.

20.99 The portions of the secondary structure of a protein that have an arrangement other than α-helix or β-pleated sheet structure are called "unstructured" segments.

20.101 The four types of attractive forces are disulfide bonds, electrostatic interactions, hydrogen bonds, and hydrophobic interactions.

20.103 a. nonpolar
b. polar neutral R groups
c. –SH groups
d. acidic and basic R groups

20.105 The four types of attractive interactions that contribute to the tertiary structure of a protein (hydrophobic, electrostatic, hydrogen bonding, and disulfide bonds) are all interactions between amino acid R groups.
a. The interactions between the nonpolar side chains of phenylalanine and leucine are hydrophobic.
b. The interactions between the charged side chains of arginine and glutamic acid are electrostatic (sometimes called salt bridges).
c. The interaction between the sulfur atoms in two cysteine molecules is a disulfide bond.
d. The interaction between the polar side chains (containing –OH groups) of serine and tyrosine is hydrogen bonding.

20.107 a. Salt bridges between amino acids with acidic and basic side chains are a part of tertiary protein structure.
b. Hydrogen bonds are part of secondary and tertiary protein structure.
c. A segment of a protein chain that folds back upon itself is part of that protein's secondary protein structure.
d. The C-terminal amino acid in the specified protein chain is part of that protein's primary structure.

20.109 The quaternary structure of a protein is the organization among the various peptide chains in a multimeric protein.

20.111 They are the same.

20.113 When the tripeptide Ala-Ala-Val undergoes complete hydrolysis, the possible products are: Ala and Val.

20.115 When the tripeptide Ala-Ala-Val undergoes partial hydrolysis, the possible products are: Ala, Val, Ala-Ala, and Ala-Val.

20.117 We know that the amino acids in the di- and tripeptides produced by hydrolysis must be present in the same order as they were in the original tetrapeptide, Ala-Gly-Ser-Tyr. There are five possible di- and tripeptides: Ala–Gly–Ser, Gly–Ser–Tyr, Ala–Gly, Gly–Ser, Ser–Tyr

20.119 Peptides can undergo partial hydrolysis of their peptide bonds, yielding a mixture of smaller peptides. We know that the amino acids in the smaller peptides must be present in the same order as they were in the hexapeptide. By overlapping the smaller peptide segments, we can determine the amino acid sequence in the hexapeptide: Ala–Gly–Met–His–Val–Arg

20.121 Shown below are structural formulas for the products obtained from the complete hydrolysis of the tripeptide Ala-Gly-Ser under a. acidic conditions and b. basic conditions.

a. $H_3\overset{+}{N}-\underset{\substack{|\\CH_3}}{CH}-COOH$, $H_3\overset{+}{N}-\underset{\substack{|\\H}}{CH}-COOH$, $H_3\overset{+}{N}-\underset{\substack{|\\CH_2\\|\\OH}}{CH}-COOH$

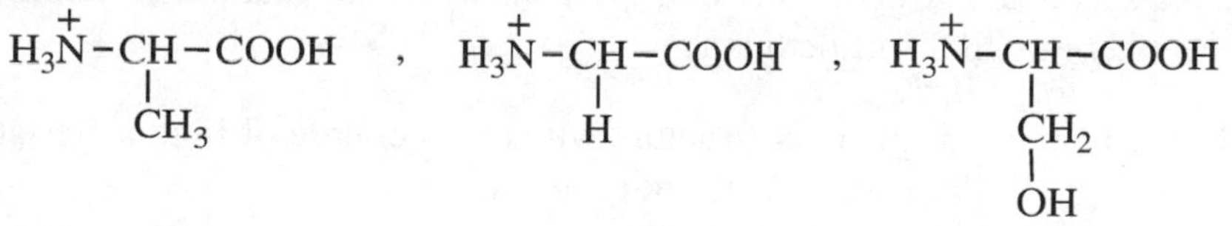

b. $H_2N-\underset{\substack{|\\CH_3}}{CH}-COO^-$, $H_2N-\underset{\substack{|\\H}}{CH}-COO^-$, $H_2N-\underset{\substack{|\\CH_2\\|\\OH}}{CH}-COO^-$

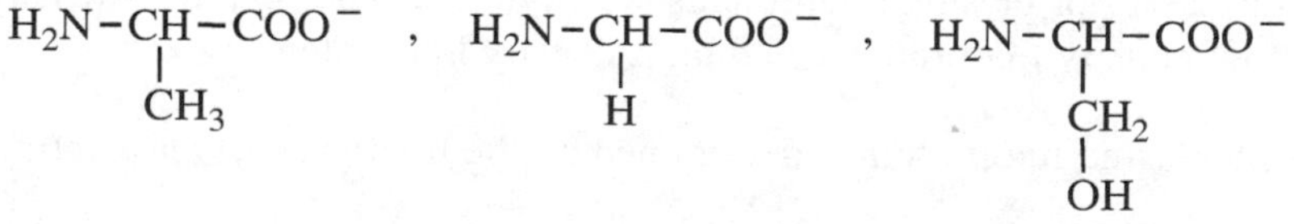

20.123 Protein denaturation is the partial or complete disorganization of a protein's characteristic three-dimensional shape as a result of disruption of its secondary, tertiary, and quaternary structural interactions.

20.125 The primary structure of the protein in the cooked egg remains the same, but the secondary, tertiary, and quaternary structures of protein structure are disrupted.

20.127 a. Yes, microwave radiation is a denaturing agent.
b. Yes, detergent is a denaturing agent.
c. No, water is not a denaturing agent.
d. Yes, strong acid is a denaturing agent.

20.129 a. Fibrous proteins are generally water-insoluble; globular proteins are generally water-soluble, enabling them to travel through the blood and other body fluids.
b. Fibrous proteins generally have structural functions that provide support and external protection; globular proteins are involved in metabolic reactions, performing functions such as catalysis, transport, and regulation.

20.131 a. α-keratin is a fibrous protein found in protective coatings for organisms (feathers, hair, wool, etc.)
b. Collagen is a fibrous protein found in tendons, bone, and other connective tissue.
c. Hemoglobin is a globular protein involved in oxygen transport in blood.
d. Myoglobin is a globular protein involved in oxygen storage in muscle.

20.133 a. Actin is a contractile protein.
b. Myoglobin is a storage protein.
c. Transferrin is a transport protein.
d. Insulin is a messenger protein.

20.135 The functional classification for proteins known as enzymes is catalytic protein.

20.137 The two non-standard amino acids present in collagen are 4-hydroxyproline and 5-hydroxylysine, derivatives of the standard amino acids proline and lysine.

20.139 The function of the carbohydrate groups in collagen is related to cross-linking; they direct the assembly of collagen triple helices into more complex aggregations called collagen fibrils.

20.141 An antigen is a substance foreign to the human body (such as a bacterium or virus) that invades the human body; an antibody is a biochemical molecule that counteracts a specific antigen.

20.143 The basic structural features of a typical immunoglobulin molecule are: four polypeptide chains (two identical long heavy chains and two identical short light chains) that have constant and variable amino acid regions, carbohydrate content varying from 1% to 12% by mass, and a secondary structure involving a Y-shaped conformation with long and short chains connected through disulfide linkages.

20.145 A plasma lipoprotein has a spherical structure with an inner core of lipid material surrounded by a shell of phospholipids, cholesterol, and proteins.

20.147 The four major classes of plasma lipoproteins are chylomicrons, very-low-density lipoproteins, low-density lipoproteins, and high-density lipoproteins.

20.149 The density of a plasma lipoprotein is determined by the lipid/protein mass ratio.

20.151 a. Chylomicrons transport dietary triacylglycerols from the intestine to various locations.
b. Low-density lipoproteins transport cholesterol from the liver to cells throughout the body.

Enzymes and Vitamins — Chapter 21

Solutions to Selected Problems

21.1 The general role of enzymes in the human body is to act as catalysts for biochemical reactions.

21.3 Enzymes differ from inorganic laboratory catalysts in two ways: they are larger in size, and their activity is regulated by other substances.

21.5 A simple enzyme is composed only of protein (amino acid chains); a conjugated enzyme has a nonprotein part in addition to a protein part.

a. An enzyme that has both protein and nonprotein parts is a conjugated enzyme.
b. An enzyme that requires Mg^{2+} for enzyme activity is a conjugated enzyme.
c. An enzyme in which only amino acids are present is a simple enzyme.
d. An enzyme in which a cofactor is present is a conjugated enzyme.

21.7 A metal ion can function as a cofactor but not as a coenzyme because a cofactor can be inorganic or organic, but a coenzyme must be organic.

21.9 a. True.
b. False. An apoenzyme is the protein part of a conjugated enzyme.
c. True.
d. True

21.11 Based on its name (*–ase* ending for an enzyme)

a. sucrose is not an enzyme.
b. lactase is an enzyme.
c. fructose oxidase is an enzyme.
d. creatine kinase is an enzyme.

21.13 An enzyme name may indicate the type of reaction catalyzed by the enzyme and/or the substrate upon which the enzyme acts. Table 21.1 gives the main classes and subclasses of enzymes and the types of reactions they catalyze.

a. The function of pyruvate carboxylase is to add a carboxylate group to pyruvate.
b. The function of alcohol dehydrogenase is to remove H_2 from an alcohol.
c. The function of L-amino acid reductase is to reduce an L-amino acid.
d. The function of maltase is to hydrolyze maltose.

21.15 a. The substrate for pyruvate carboxylase is pyruvate.
b. The substrate for alcohol dehydrogenase is an alcohol.
c. The substrate for L-amino acid reductase is an L-amino acid.
d. The substrate for maltase is maltose.

21.17 a. Sucrase (or sucrose hydrolase) would be a possible name for an enzyme that catalyzes the hydrolysis of sucrose.
b. Pyruvate decarboxylase would be a possible name for an enzyme that catalyzes the decarboxylation of pyruvate.
c. Glucose isomerase would be a possible name for an enzyme that catalyzes the isomerization of glucose.
d. Lactate dehydrogenase would be a possible name for an enzyme that catalyzes the removal of hydrogen from lactate.

21.19 For the given enzymes, the correctness of the pairing of enzyme with function:
a. Yes, it is correct.
b. Yes, it is correct,
c. No, it is not correct.
d. Yes, it is correct.

21.21 Table 21.1 gives the six main enzyme classes and some of the reactions they catalyze.
a. An enzyme that converts a *cis* double bond to a *trans* double bond is an isomerase.
b. An enzyme that dehydrates an alcohol to form a compound with a double bond is a lyase.
c. An enzyme that transfers an amino group from one substrate to another is a transferase.
d. An enzyme that hydrolyses an ester linkage is a hydrolase.

21.23 a. CO_2 is removed in the reaction so the enzyme is a decarboxylase.
b. A triacylglycerol (a lipid) is hydrolyzed; the enzyme is a lipase.
c. A phosphate-ester bond is hydrolyzed; the enzyme is a phosphatase.
d. H_2 is removed; the enzyme is a dehydrogenase.

21.25 a. False. Substrate molecules are attached.
b. False. It sometimes has a fixed, rigid geometry.
c. False. Its geometric shape can be flexible.
d. False. Amount of accommodation varies with the enzyme and the substrate.

21.27 The statements in Problem 21.25 refer to the two different models of enzyme action.
a. The statement refers to neither model.
b. This statement refers to the lock-and-key model.
c. This statement refers to the lock-and-key model.
d. This statement refers to the induced-fit model.

21.29 The forces that hold a substrate at an enzyme active site are electrostatic forces, hydrogen bonds, and hydrophobic interactions with amino acid R groups.

21.31 In the following equation: $E + S \rightleftharpoons ES \rightarrow EP \rightarrow E + P$
a. ES represents the enzyme-substrate complex.
b. P represents the product of the enzyme reaction.

21.33 a. False. According to the lock-and-key model for enzyme action, the active site of an enzyme has a fixed geometric shape.
b. True. In an enzyme-catalyzed reaction, the compound that undergoes a chemical change is called the substrate.
c. False. The nonprotein portion of a conjugated enzyme is the enzyme's cofactor.
d. False. Simple enzymes are composed only of protein; conjugated enzymes have nonprotein cofactors.

21.35 a. Absolute specificity means than an enzyme will catalyze a particular reaction for only one substrate.
b. Linkage specificity means that an enzyme will catalyze a reaction that involves a particular type of bond.

21.37 a. An enzyme that exhibits absolute specificity (only one substrate) is more limited than one that exhibits group specificity (only one type of functional group).
b. An enzyme that exhibits stereochemical specificity (only one stereoisomer or one of a pair of enantiomers) is more limited than an enzyme that exhibits linkage specificity (one type of bond).

21.39 Enzyme specificity is the extent to which an enzyme's activity is restricted to a specific substrate or type of chemical reaction.
a. Sucrase is an enzyme with absolute specificity; it catalyzes only one reaction.
b. Lipase is an enzyme with linkage specificity; it catalyzes the reaction of a particular type of bond.
c. A decarboxylase is an enzyme with group specificity; it catalyzes reactions of a particular functional group (carboxyl).
d. L-Glutamate oxidase is an enzyme with stereochemical specificity; it catalyzes the reaction of only the L-stereoisomer of glutamate.

21.41 Based on the graph showing enzyme activity and pH and temperature:
a. The optimum pH (highest activity) for enzyme A is 7.0.
b. The optimum temperature for enzyme B is 38°C.
c. At a pH of 7.4, enzyme B has the greater activity.
d. At a temperature of 37.8°C, enzyme B has the greater activity.

21.43 Based on the graph in Problem 21.41 showing enzyme activity and pH and temperature:
a. When the pH decreases from 7.2 to 7.0, the activity of enzyme A increases.
b. When the pH increases from 7.2 to 7.4, the activity of enzyme A decreases.
c. When temperature decreases from 36.8°C to 36.6°C, the activity of enzyme A decreases.
d. When temperature increases from 36.8°C to 37.8°C, the activity of enzyme A decreases.

21.45 For human enzymes, the optimum temperature is around 37°C; the optimum pH is 7.0 – 7.5. The activity of a typical non-digestive human enzyme
a. decreases as the temperature is decreased from 35°C to 34°C.
b. decreases as the pH decreases from 7.1 to 6.8.

21.47 At constant temperature, pH, and enzyme concentration, the rate of a reaction increases as substrate concentration increases. However, at some point, enzyme capabilities are being used to their maximum extent (active sites are saturated) and no further reaction rate increase is possible. This activity pattern, shown below, is called a saturation curve.

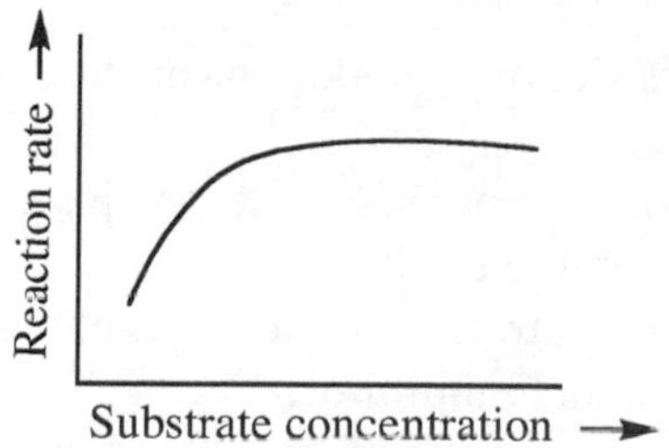

21.49 If each enzyme molecule is working to full capacity (saturated), a further increase in substrate concentration will have no effect on the rate of the reaction; the rate will remain constant.

21.51 In the biochemical reaction that involves the substrate arginine and the enzyme arginase:

a. Decreasing the substrate (arginine) concentration decreases the rate of reaction.
b. Increasing the temperature from its optimum value decreases the rate of reaction; the optimum value is the temperature at which the enzyme exhibits maximum activity.
c. Increasing the amount of enzyme (arginase) increases the rate of reaction; higher enzyme concentration allows more substrate molecules to be accommodated.
d. Decreasing pH from its optimum value decreases the rate of reaction; the optimum pH is the value at which the enzyme exhibits maximum activity.

21.53 An extremophile is a microorganism that thrives in extreme environments of temperature, pressure, pH, etc.

21.55 The correct pairing of extremophile subtype and environmental survival conditions:

a. No. It should be halophile and high salinity.
b. Yes.
c. Yes.
d. No. It should be alkaliphile and high solution pH.

21.57 In the production of commercial laundry detergent formulations, extremophiles that can withstand a hot-water environment and/or cold-water environment are used.

21.59 No, a competitive inhibitor and a substrate cannot bind to an enzyme at the same time. A competitive enzyme inhibitor is a molecule that can compete with the substrate for occupancy of the enzyme's active site; this slows enzyme activity because only one molecule may occupy the active site at a given time.

21.61 The statements concern different types of enzyme inhibitors.

a. False. The block on an enzyme's active site by a reversible competitive inhibitor can be reversed.
b. False. A noncompetitive inhibitor binds to a site other than the active site.
c. True.
d. False. Pb^{2+} is an irreversible enzyme inhibitor.

21.63 A reversible competitive inhibitor resembles a substrate enough that it can compete with the substrate for occupancy of the enzyme's active site; a reversible noncompetitive inhibitor decreases enzyme activity by binding to a site on an enzyme other than the active site; an irreversible inhibitor inactivates enzymes by forming a strong covalent bond to an amino acid side-chain group at the enzyme's active site.

a. If both the inhibitor and the substrate bind at the active site on a random basis, the inhibitor is called a reversible competitive inhibitor.
b. If the inhibitor effect cannot be reversed by the addition of more substrate, the inhibitor is either a reversible noncompetitive inhibitor or an irreversible inhibitor.
c. If the inhibitor structure does not have to resemble the substrate structure, the inhibitor is either a reversible noncompetitive inhibitor or an irreversible inhibitor.
d. If the inhibitor can bind to the enzyme at the same time as the substrate, it is a reversible noncompetitive inhibitor.

21.65 a. True.
b. True.
c. False. Allosteric enzymes can be positive regulators or negative regulators.
d. True.

21.67 The correct pairing of concepts related to regulation of enzyme activity:
a. Yes, the pairing is correct.
b. No. Feedback control should be paired with regulator in the final reactant in the final reaction sequence.
c. Yes, the pairing is correct.
d. No. Protein kinases effect the addition of phosphate groups; phosphatases catalyze the removal of the phosphate groups.

21.69 a. During the covalent modification of an enzyme using a phosphorylation reaction, a phosphate group is added.
b. During the covalent modification of an enzyme using a phosphatase enzyme, a phosphate group is removed.

21.71 The phosphorylated version of an enzyme can be either the "turned-on" or "turned-off" form of the enzyme.

21.73 a. An apoenzyme is the protein portion of a conjugated enzyme; a proenzyme is the inactive precursor of an enzyme.
b. A simple enzyme contains only protein; an allosteric enzyme contains two or more protein chains and two binding sites.

21.75 a. Angiotensinogen is a decapeptide; angiotensin in an octapeptide.
b. Angiotensinogen is the inactive form of the hormone; angiotensin is the active form.

21.77 Sulfa drugs act by the competitive inhibition of the enzyme necessary in bacteria for the synthesis of folic acid from PABA. Humans do not have this enzyme and acquire folic acid from the diet.

21.79 TPA stands for tissue plasminogen activator, which activates an enzyme that dissolves blood clots.

21.81 a. LDH is the acronym for lactate dehydrogenase.
b. AST is the acronym for aspartate transaminase.

21.83 a. The enzyme CPK in the bloodstream is an indicator of possible heart disease.
b. The enzyme ALT in the bloodstream is an indicator of possible heart disease, liver disease, and muscle damage.

21.85 There are nine water-soluble vitamins (vitamin C and eight B vitamins) and four fat-soluble vitamins (A, D, E, and K).
a. Vitamin K is a fat-soluble vitamin.
b. Vitamin B_{12} is a water-soluble vitamin.
c. Vitamin C is a water-soluble vitamin.
d. Thiamin is a water-soluble vitamin.

21.87 Because water-soluble vitamins are rapidly eliminated in the urine, they are unlikely to be toxic except when taken in unusually large doses. Fat-soluble vitamins are stored in fat tissues and are more likely to be toxic when consumed in excess of need.

a. Vitamin K would be likely to be toxic when consumed in excess.
b. Vitamin B_{12} would be unlikely to be toxic when consumed in excess.
c. Vitamin C would be unlikely to be toxic when consumed in excess.
d. Thiamin would be unlikely to be toxic when consumed in excess.

21.89 The B vitamins usually have a cofactor function in the human body.

a. No, vitamin K would not be likely to have a cofactor function.
b. Yes, vitamin B_{12} would have a cofactor function.
c. No, vitamin C would not have a cofactor function.
d. Yes, thiamin would have a cofactor function.

21.91 Ascorbic acid has two hydroxyl groups while dehydroascorbic acid has two ketone groups. Ascorbic acid's ring contains a double bond while in dehydroascorbic acid's ring all bonds are single.

21.93 a. Vitamin C is a cosubstrate in collagen formation.
b. Vitamin C reactivates spent vitamin E.
c. Vitamin C reactivates enzymes needed for dopamine synthesis.
d. Vitamin C is reactivated by niacin (NADH).

21.95 In the biosynthesis of vitamin C in two steps, the reactants and products are as follows:
Step 1 — L-gulonic acid $\rightarrow$ γ-L-gulonolactone
Step 2 — γ-L-gulonolactone $\rightarrow$ L-ascorbic acid

21.97 The most characterized role for the B vitamins in the human body is as a precursor for enzyme cofactors.

21.99 a. Vitamin B_1 is the alternative name for the vitamin; thiamin is its preferred name.
b. Vitamin B_6 is the preferred name for the vitamin.
c. Cobalamin is the alternative name for the vitamin; vitamin B_{12} is its preferred name.
d. Niacin is the preferred name for the vitamin.

21.101 a. Thiamin and biotin are the two B vitamins that contain the element sulfur.
b. Folate and biotin each contains a fused two-ring component.
c. Niacin, vitamin B_6, and folate each exists in two or more different structural forms.
d. Riboflavin contains a monosaccharide component.

21.103 One form of niacin contains a carboxyl substituent (nicotinic acid); the other form contains an amide constituent (nicotinamide).

21.105 a. Vitamin B_6 is the precursor of PLP (pyridoxal-5-phosphate).
b. Thiamin is the precursor of TPP (thiamin pyrophosphate).
c. Pantothenic acid is the precursor of coenzyme A.
d. Niacin is the precursor of $NADP^+$ (nicotinamide adenine dinucleotide phosphate).

21.107 a. The coenzyme form of thiamin is involved in the transfer of carbon dioxide (carbonyl group).
b. The coenzyme form of folate is involved in the transfer of a one-carbon group other than CO_2.
c. The coenzyme form of pantothenic acid is involved in the transfer of an acyl group.
d. The coenzyme form of vitamin B_{12} is involved in the transfer of a methyl group and hydrogen atoms.

21.109 There are three preformed vitamin A forms called retinoids. They differ in the functional group attached to the terpene structure. The functional group for retinol is CH_2–OH; for retinal it is CHO; and for retinoic acid it is COOH.

21.111 Cell differentiation is the process whereby immature cells change in structure and function to become specialized cells. In the cell differentiation process, Vitamin A binds to protein receptors, and these vitamin A–protein receptor complexes bind to regulatory regions of DNA molecules.

21.113 Preformed vitamin A forms are called retinoids. The retinoids include retinal, retinol, and retinoic acid.
a. Yes, retinal can be converted to retinol in the human body.
b. No, retinoic acid cannot be converted to retinal in the human body.

21.115 Vitamin D_2 (ergocalciferol) differs from vitamin D_3 (cholecalciferol) only in the side-chain structure; the vitamin D_2 side chain contains a double bond and a methyl substituent not present in the vitamin D_3 side chain.

21.117 The principal function of vitamin D in humans is to maintain normal blood levels of calcium ions and phosphate ions so that bones can absorb these minerals.

21.119 In the biosynthetic pathway that produces active vitamin D_3 from cholesterol
a. 7-dehydrocholesterol is encountered before cholecalciferol.
b. calcidiol is encountered before calcitriol.

21.121 The alpha form of vitamin E exhibits the greatest biochemical activity.

21.123 The principal function of vitamin E in the human body is as an antioxidant – a compound that protects other compounds from oxidation by being oxidized itself.

21.125 Vitamin K_1 and vitamin K_2 differ structurally in the length and degree of unsaturation of a side chain.

21.127 Menaquinones and phylloquinones are forms of vitamin K. Menaquinones (vitamin K_2) are found in fish oil and meats and are synthesized by bacteria, including those in the human intestinal tract. Phylloquinones (vitamin K_1) are found in plants.

21.129 Pairing of structure name and vitamin name:
a. No, it is not correct. Retinal is a form of vitamin A.
b. Yes, it is correct.
c. No, it is not correct. Tocopherol should be paired with vitamin E.
d. No, it is not correct. Naphthoquinone should be paired with vitamin K.

21.131 a. Vitamin C is a water-soluble antioxidant.
b. Vitamin E is a fat-soluble antioxidant.
c. Vitamin A is involved in the process of vision.
d. Vitamin C is involved in the formation of collagen.

21.133 a. The molecules of vitamins A, D, E, and K do not contain nitrogen atoms.
b. The molecule of vitamin B_{12} contains a metal atom (cobalt).
c. The molecular structures of vitamins A, D, E, and K have a saturated or an unsaturated carbon chain.
d. The molecules of niacin, pantothenic acid, folate, and biotin each contain one or more carboxyl (–COOH) functional groups.

Nucleic Acids Chapter 22

Solutions to Selected Problems

22.1 a. True.
b. False. Both are unbranched polymers.
c. True.
d. True

22.3 a. No, ribose is not present in DNA molecules.
b. Yes, deoxyribose is present in DNA molecules.
c. Yes, this pentose (deoxyribose) is present in DNA molecules.
d. No, this pentose (ribose) is not present in DNA molecules.

22.5 Use Figure 22.2 to identify the nitrogen-containing bases.
a. Cytosine is a pyrimidine.
b. Adenine is a purine.
c. This base (guanine) is a purine.
d. This base (uracil) is a pyrimidine.

22.7 The one-letter abbreviation for the four bases in Problem 22.5:
a. Cytosine is C. b. Adenine is A. c. Guanine is G. d. Uracil is U.

22.9 a. Cytosine is found in both DNA and RNA.
b. Adenine is found in both DNA and RNA.
c. Guanine is found in both DNA and RNA.
d. Uracil is found in RNA, but not in DNA.

22.11 a. DNA contains only 2-deoxyribose as its pentose sugar subunit.
b. There are four choices for the nitrogen-containing base subunits in RNA nucleotides: adenine, guanine, cytosine, and uracil.
c. There is only one type of phosphate subunit (a –2-charged phosphate ion) in DNA nucleotides.

22.13 a. False. RNA nucleosides contain ribose.
b. False. Nucleosides do not contain phosphate groups.
c. True.
d. False. The two subunits found in a nucleoside are a pentose sugar and a heterocyclic base.

22.15 The name of the nucleoside that contains
a. ribose and adenine is adenosine.
b. uracil and ribose is uridine.
c. deoxyribose and cytosine is deoxycytidine.
d. thymine and deoxyribose is deoxythymidine.

22.17 a. False. RNA nucleosides contain ribose; DNA nucleosides do not.
b. False. All nucleotides contain a phosphate group.
c. False. Some nucleotides contain a purine base.
d. False. One of the three subunits in a nucleotide is a nitrogen-containing base.

22.19 a. A nucleotide that contains ribose, uracil, and phosphate is a RNA nucleotide.
b. A nucleotide that contains deoxyribose, adenine, and phosphate is a DNA nucleotide.
c. dTMP is a DNA nucleotide.
d. CMP is a RNA nucleotide.

22.21 The base T is not found in RNA (ribose) nucleotides.

22.23 a. AMP is the abbreviation for adenosine 5′-monophosphate; it contains adenine and ribose.
b. dGMP is the abbreviation for deoxyguanosine 5′-monophosphate; it contains guanine and deoxyribose.
c. dTMP is the abbreviation for deoxythymidine 5′-monophosphate; it contains thymine and deoxyribose.
d. UMP is the abbreviation for uridine 5′-monophosphate; it contains uracil and ribose.

22.25 a. AMP is the abbreviation for adenosine 5′-monophosphate.
b. dGMP is the abbreviation for deoxyguanosine 5′-monophosphate.
c. dTMP is the abbreviation for deoxythymidine 5′-monophosphate.
d. UMP is the abbreviation for uridine 5′-monophosphate.

22.27 a. The name of the nucleotide is deoxythymidine 5′-monophosphate.
b. This nucleotide would be found in DNA only.
c. The name for the type of bond that connects the phosphate and sugar subunits is phosphoester bond.
d. The name for the type of bond that connects the sugar and base subunits is *β-N*-glycosidic bond.

22.29 A nucleoside consists of a nitrogen-containing base and a pentose. A nucleotide consists of a pentose sugar, a nitrogen-containing base, and a phosphate group.
a. Adenosine is a nucleoside consisting of adenine and ribose.
b. Adenine is neither a nucleoside nor a nucleotide; it is a nitrogen-containing base.
c. dAMP is a nucleotide consisting of adenine, deoxyribose, and a phosphate group.
d. Adenosine 5′-monophosphate is a nucleotide consisting of adenine, ribose, and a phosphate group.

22.31 For the trinucleotide 5′ G–C–A 3′
a. There are 6 subunits (sugar and phosphate subunits) in its "backbone."
b. There are 3 "nonbackbone" subunits (nitrogen-containing bases) present.
c. There are 2 phosphodiester linkages.
d. The overall charge carried by the trinucleotide is –4.

22.33 The trinucleotide in Problem 22.31 is found in both DNA and RNA because G, C, and A are bases found in both DNA and RNA.

22.35 In the lengthening of a polynucleotide chain, the 3′ end of the chain would have a phosphate subunit bonded to it.

22.37 The structure of the RNA dinucleotide 5′ U–G 3′ is:

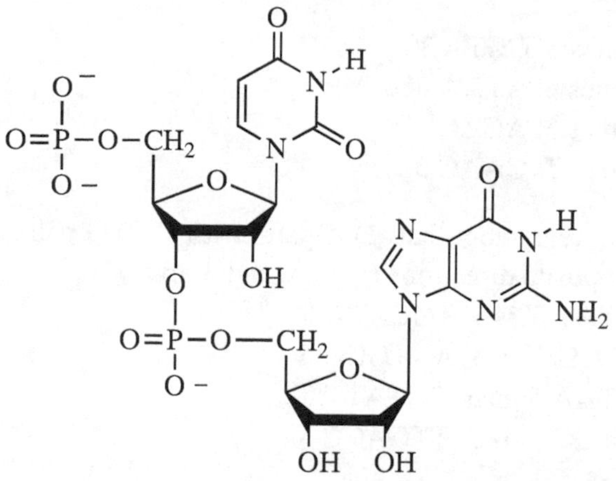

22.39 For the trinucleotide 5′ T–G–A 3′

a. there are three *β-N*-glycosidic linkages.
b. there is one phosphoester linkage.
c. there are two phosphodiester linkages.
d. there were two byproduct water molecules produced during the formation of the trinucleotide.

22.41 a. True.
b. False. Bases extend inward to the interior of the helix.
c. False. Hydrogen bonding occurs between the two polynucleotide strands.
d. False. The two polynucleotide strands run in opposite directions.

22.43 In DNA molecules, A–T and G–C pairings are most favorable for hydrogen bonding and are said to be complementary.

a. no b. no c. yes d. no

22.45 Because of hydrogen bonding that exists between certain base pairs in DNA, the two paired bases are present in equal amounts: %A = %T; %C = %G

a. Since %T = 36%, %A = 36%.
b. Since %T + %A = 72%, %C + %G = (100-72)% = 28%; %C = %G = 14%
c. %G = %C = 14%

22.47 The bases A and G do not form a complementary base pair because two double-ring bases are too large to fit in the helix interior.

22.49 The base composition for the other strand of the DNA double helix is the same (19% A, 34% C, 28% G, and 19% T).

22.51 In reversing the order of a DNA base sequence 3′ to 5′ to the order 5′ to 3′, simply reverse the order of the bases.

a. When 3′ ATCG 5′ is reversed, it becomes 5′ GCTA 3′.
b. When 3′ AATA 5′ is reversed, it becomes 5′ ATAA 3′.
c. When 3′ CACA 5′ is reversed, it becomes 5′ ACAC 3′.
d. When 3′ CAAC 5′ is reversed, it becomes 5′ CAAC 3′.

22.53 In writing the complementary DNA strand, remember that: 1) A pairs with T, 2) G pairs with C, and 3) complementary strands are in opposite directions (5′ to 3′ and 3′ to 5′).

a. The complementary strand of 5′ ACGTAT 3′ is 3′ TGCATA 5′.
b. The complementary strand of 5′ TTACCG 3′ is 3′ AATGGC 5′.
c. The complementary strand of 3′ GCATAA 5′ is 5′ CGTATT 3′.
d. The complementary strand of 5′ AACTGG 3′ is 3′ TTGACC 5′.

22.55 For the DNA segment 5′ TTGCAC 3′ there are

a. six nucleotides
b. two purine bases (A and G)
c. six phosphate groups
d. zero ribose subunits

22.57 In writing the complementary DNA strand, remember that: 1) A pairs with T, 2) G pairs with C, and 3) complementary strands are in opposite directions (5′ to 3′ and 3′ to 5′).

a. When the template base sequence is 3′ AATGC 5′, the newly formed base sequence is 5′ TTACG 3′.
b. When the template base sequence is 5′ AATGC 3′, the newly formed base sequence is 3′ TTACG 5′ or, on reversal, 5′ GCATT 3′.
c. When the template base sequence is 3′ GCAGC 5′, the newly formed base sequence is 5′ CGTCG 3′.
d. When the template base sequence is 5′ GCAGC 3′, the newly formed base sequence is 3′ CGTCG 5′ or, on reversal, 5′ GCTGC 3′.

22.59 In writing a complementary DNA strand, remember that: 1) A pairs with T, 2) G pairs with C, and 3) complementary strands are in opposite directions (5′ to 3′ and 3′ to 5′). The two daughter strands Q and R are complementary. Parent and daughter strands are complementary.

a. The complementary strand of 5′ ACTTAG 3′ (Q) is 3′ TGAATC 5′ (R).
b. The parent strand of Q (5′ ACTTAG 3′) is 3′ TGAATC 5′.
c. The parent strand of R (3′ TGAATC 5′) is 5′ ACTTAG 3′.

22.61 As the DNA helix unwinds, the two template strands are in opposite directions as they emerge from the replication fork. The synthesis of a daughter DNA strand growing toward a replication fork in the 5′ to 3′ direction is continuous; the synthesis of the other daughter DNA strand in the 5′ to 3′ direction is away from the replication fork, and therefore must take place in segments.

22.63 a. False. The lagging strand grows in a direction opposite to that of the replication fork.
b. False. Growth of the lagging strand involves the production of Okazaki fragments.
c. False. Both lagging and leading strands involve daughter DNA segments.
d. False. DNA helicase effects the unwinding of a DNA double helix.

22.65 a. True.
b. False. Chromosomes occur in matched (homologous) pairs.
c. True.
d. False. Most of the mass of a chromosome comes from protein.

22.67 The nucleotides in a given DNA molecule are 28% thymine-containing. Since T and A are complementary and C and G are complementary, the DNA molecule is 28% A, 22% C, and 22% G. The base mixture available contains 22% A, 22% C, 28% G, and 28% T, so the adenine (A) will be depleted first in the replication process.

22.69 a. True.
b. True
c. False. During the translation phase mRNA molecules are deciphered and protein is synthesized.
d. True.

22.71 a. False. Base pairing occurs in both RNA and DNA.
b. False. DNA is double-stranded; RNA is not.
c. False. DNA molecules are larger.
d. False. Thymine in DNA is replaced by uracil in RNA.

22.73 In the designation of various types of RNA the small letter stands for
a. transfer
b. messenger
c. ribosomal
d. heterogeneous nuclear.

22.75 The predominant location in which each of these various types of RNA carries out its biochemical function:
a. mRNA acts in the nucleus and cytoplasm.
b. tRNA acts in the cytoplasm.
c. Ribosomal RNA acts in the cytoplasm.
d. Small nuclear RNA acts in the cell nucleus.

22.77 a. DNA transcription occurs in the nuclear region.
b. hnRNA is formed from DNA in the nuclear region.
c. tRNA is needed in the cytoplasmic region.
d. snRNA is needed in the nuclear region.

22.79 a. True.
b. False. They align themselves along the DNA's template strand.
c. True.
d. False. The initially-produced RNA strand is hnRNA.

22.81 The base sequence of the RNA initially synthesized from the given DNA templates is:
a. 5′ AUAU 3′
b. 5′ GGGC 3′
c. 5′ CGAU 3′
d. 5′ UUUU 3′

22.83 DNA contains the bases adenine, thymine, guanine and cytosine; in RNA the base uracil replaces thymine. Base pairing between the DNA template strand and the DNA informational strand is complementary, and the following base pairings occur: T–A, A–T, G–C, C–G. Base pairing between the DNA template strand and the newly formed hnRNA strand is complementary, and the following base pairings occur: T–A, A–U, G–C, C–G.

a. Using the template DNA strand 3′ TACGGC 5′, the DNA informational strand is 5′ ATGCCG 3′ and the hnRNA strand is 5′ AUGCCG 3′.
b. Using the template DNA strand 3′ CCATTA 5′, the DNA informational strand is 5′ GGTAAT 3′ and the hnRNA strand is 5′ GGUAAU 3′.
c. Using the template DNA strand 3′ ACATGG 5′, the DNA informational strand is 5′ TGTACC 3′ and the hnRNA strand is 5′ UGUACC 3′.
d. Using the template DNA strand 3′ ACGTAC 5′, the DNA informational strand is 5′ TGCATG 3′ and the hnRNA strand is 5′ UGCAUG 3′.

22.85 Base pairing between the DNA template strand and the newly formed hnRNA strand is complementary, and the following base pairings occur: T–A, A–U, G–C, C–G.

a. The hnRNA strand 5′ CCUUAA 3′ was formed from the template DNA strand 5′ TTAAGG 3′.
b. The hnRNA strand 5′ ACGUAC 3′ was formed from the template DNA strand 5′ GTACGT 3′.
c. The hnRNA strand 5′ ACGACG 3′ was formed from the template DNA strand 5′ CGTCGT 3′.
d. The hnRNA strand 5′ UACCAU 3′ was formed from the template DNA strand 5′ ATGGTA 3′.

22.87 A gene is segmented; it has portions called exons that convey genetic information and portions called introns that do not convey genetic information.

22.89 A gene is segmented; it has portions called exons that convey genetic information and portions called introns that do not convey genetic information. During post-transcription processing, hnRNA is edited to remove the introns and join together the remaining exons to form a shortened mRNA strand. Therefore, the mRNA strand would have the following base sequence: 5′ UACGCAUU 3′

22.91 Both exons and introns are transcribed during the formation of hnRNA. The hnRNA is then edited (under enzyme direction) to remove the introns and join together the remaining exons to form a shortened RNA strand. The DNA segment in this problem has one intron (5′ TAGC 3′), which is removed, and two exons (5′ TCAG 3′ and 5′ TTCA 3′), which are joined together to form 5′ TCAGTTCA 3′. The complementary mRNA strand formed from the hnRNA is 5′ UGAACUGA 3′.

22.93 Splicing is the process of removing introns from an hnRNA molecule and joining the remaining exons together to form an mRNA molecule.

a. hnRNA undergoes the splicing.
b. snRNA is present in the spliceosomes (a large assembly of snRNA molecules and proteins involved in the conversion of hnRNA molecules to mRNA molecules).

22.95 Two different mRNA molecules (exons 1, 2, 3 and exons 1, 3) can be produced from a hnRNA molecule containing three exons, with the middle one being an "alternative" exon.

22.97 The characterization for each base-pairing situation is
a. involves a DNA strand and an RNA strand
b. involves two DNA strands
c. involves a DNA strand and an RNA strand
d. could involve either two DNA strands or a DNA strand and an RNA strand

22.99 A codon is a three-nucleotide mRNA sequence that codes for a specific amino acid.

22.101 Table 22.2 gives the universal genetic code, composed of 64 three-nucleotide sequences (codons) and the amino acids that the sequences code for.
a. CUU codes for the amino acid leucine (Leu).
b. AAU codes for the amino acid asparagine (Asn).
c. AGU codes for the amino acid serine (Ser).
d. GGG codes for the amino acid glycine (Gly).

22.103 The possible codons for
a. tyrosine are UAA and UAC.
b. alanine are GCU, GCC, GCA, and GCG.
c. Leu are UUA, UUG, CUU, CUC, CUA, and CUG
d. Cys are UGU, and UGC

22.105 The base sequence ATC could not be a codon, because a codon is a segment of RNA, and the base T cannot be present in RNA.

22.107 Use Table 22.2 to find the amino acids coded for by the given codons: AUG codes for Met, AAA codes for Lys, GAA codes for Glu, GAC codes for Asp, and CUA codes for Leu. The amino acid sequence is: Met–Lys–Glu–Asp–Leu.

22.109 Table 22.2 gives the universal genetic code, composed of 64 three-nucleotide sequences (codons) and the amino acids for which the sequences code. The mRNA sequence 3′ AUG-AAA-GAA-GAC-CUA 5′ must be reversed to the mRNA sequence 5′ AUC-CAG-AAG-AAA-GUA 3′, which when broken down to codons, codes for the amino acid sequence: Ile-Gln-Lys-Lys-Val

22.111 The hnRNA nucleotide sequence is 5′ UCCG-CCAU-UAACA 3′.
a. Base pairing between the DNA template strand and the hnRNA strand is complementary, and the following base pairings occur: T–A, A–U, G–C, C–G. Therefore, the base sequence of the DNA template strand is 3′ AGGC-GGTA-ATTGT 5′.
b. The informational strand is the complementary strand of the DNA template strand. 5′ TCCG-CCAT-TAACA 3′
c. The intron (5′ CCAU 3′) is removed produced from the hnRNA and the remaining base sequence is the mRNA : 5′ UCCG-UAACA 3′
d. The mRNA (broken into 3-base sequences) is 5′ UCC-GUA-ACA 3′, which codes for the amino acid sequence Ser-Val-Thr.

22.113 a. False. Three hairpin loops are present.
b. True.
c. False. The amino acid is attached to the 3′ end.
d. False. An amino acid is covalently bonded to its attachment site through an ester linkage.

22.115 a. True.
b. False. Anticodons are found on tRNA molecules.
c. False. The base U can be present in both codons and anticodons.
d. True.

22.117 The anticodons, written in the 5′-3′ direction, are:
a. UCU b. ACG c. AAA d. UUG

22.119 a. Gly b. Leu c. Ala d. Leu

22.121 a. 5′ ACG 3′ encodes for the amino acid Thr (threonine).
b. 3′ ACG 5′ must be reversed to 5′ GCA 3′, which encodes for Ala (alanine).
c. First, reverse the anticodon, 5′ ACG 3′, to the correct direction, 3′ GCA 5′. Then find the complementary base pairing, 5′ CGU 3′. This is the codon; the amino acid for which it encodes is Arg (arginine).
d. The anticodon 3′ ACG 5′ is written in the correct direction. The codon is the complementary base pairing 5′ UGC 3′, which encodes for the amino acid Cys (cysteine).

22.123 a. The base sequence for the DNA template strand is 5′ TCC GCA TTA ACA 3′. Using base pairing, we obtain the hnRNA sequence: 3′ AGG CGU AAU UGU 5′.
b. Since the hnRNA contains no introns, the hnRNA base sequence and the base sequence of mRNA are the same: 3′ AGG CGU AAU UGU 5′.
c. Reverse the mRNA base sequence, 3′ AGG CGU AAU UGU 5′, to the correct direction, 5′ UGU UAA UGC GGA 3′. Break this base sequence into codons: UGU, UAA, UGC, and GGA.
d. To find the tRNA molecule anticodons, use base pairing with the codons above. The anticodons are: ACA, AUU, ACG, and CCU

22.125 a. False. Ribosomes have two subunits, one of which is much larger than the other.
b. True. The active site of a ribosome is predominantly RNA rather than protein.
c. True. The mRNA involved in protein synthesis binds to the small subunit of a ribosome.
d. True. A ribosome functions as an enzyme in protein synthesis.

22.127 a. False. Elongation occurs after activation and initiation.
b. True.
c. False. The second site in an mRNA-ribosome complex is called the A site.
d. False. A polyribosome involves one mRNA molecule and several ribosomes.

22.129 The amino acids in the pentapeptide with their mRNA codon synonyms are:
Gly: GGU, GGC, GGA or GGG; Ala: GCU, GCC, GCA or GCG Cys: UGU or UGC
Val: GUU, GUC, GUA or GUG; Tyr: UAU or UAC
Choosing one codon for each amino acid in the sequence, we can obtain one of the <u>many</u> possible base sequences coding for this peptide: 5′ GGUGCUUGUGUUUAU 3′

22.131 The following mRNA base sequence is used during protein synthesis 5′ CAA-CGA-AAG 3′.

a. The codons are 5′ CAA 3′, 5′ CGA 3′, and 5′ AAG 3′.
b. The tRNA anticodons are the complements of the codons: 3′ GUU 5′, 3′ GCU 5′, and 3′ UUC 5′.
c. The amino acids specified for by the codons are 5′ CAA 3′ codes for the amino acid Gln, 5′ CGA 3′ codes for the amino acid Arg, and 5′ AAG 3′ codes for the amino acid Lys.

22.133 a. A codon is a three-nucleotide sequence in a mRNA molecule.

b. An intron is a segment of hnRNA that does not convey genetic information.
c. A tRNA molecule is a specific amino acid carrier.
d. mRNA interacts with the ribosome.

22.135 The mRNA base sequence is 5′ CUU CAG 3′.

a. The dipeptide coded for by 5′ CUU CAG 3′ is Leu-Gln.
b. The dipeptide formed if a mutation converts CUU to CUC is Leu-Gln because CUU and CUC both code for leucine.
c. The dipeptide formed if a mutation converts CAG to AAG is Leu-Lys.
d. The dipeptide formed if a mutation converts CUU to CUC and CAG to AAG is Leu-Lys.

22.137 The DNA sequence is 3′ TTA ATA 5′.

a. Transcription of the DNA sequence by base pairing gives the hnRNA segment 5′ AAU UAU 3′, which gives two codons, 5′ AAU 3′ and 5′ UAU 3′. Look up the amino acids corresponding to these two codons. The dipeptide formed is Asn-Tyr.
b. If a DNA mutation converts ATA to ATG, the new codon formed by base pairing is 5′ UAC 3′, which corresponds to Tyr. The dipeptide does not change (Asn-Tyr).
c. If a DNA mutation converts ATA to AGA, the new codon formed by base pairing is 5′ UCU 3′, which corresponds to Ser. The new dipeptide is Asn-Ser.
d. If a DNA mutation converts TTA to TTT, the new codon formed by base pairing is 5′ AAA 3′, which corresponds to Lys. The new dipeptide is Lys-Tyr.

22.139 A frameshift mutation is a mutation that inserts or deletes a base in a DNA molecule sequence. The amino acid sequence produced by the frameshift mutation in this problem is Lys-Leu-Ala-(2 bases).

22.141 a. False. A virus contains DNA or RNA but not both.

b. True.
c. False. An RNA-containing virus is called a retrovirus.
d. False. Vaccines contain an inactive or weakened form of the virus.

22.143 Recombinant DNA is DNA that contains genetic material from two different organisms.

22.145 Plasmids (small, circular, double-stranded molecules that carry only a few genes) are transferred relatively easily from one cell to another. A desired foreign gene can be inserted into the plasmid to form the recombinant DNA.

22.147 Transformation is the process of incorporating recombinant DNA into a host cell. The transformed cells then reproduce, resulting in a large number of identical cells called clones.

22.149 The individual strands of DNA are cut at different points, giving a "staircase" cut (both cuts are between A and A). These ends with unpaired bases are called "sticky ends" because they are ready to stick to (pair up with) a complementary section of DNA if they find one.

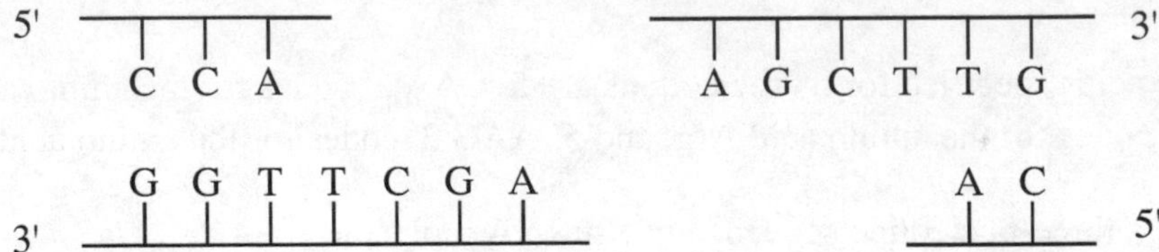

22.151 a. During replication of DNA a complete unwinding of a DNA molecule occurs.
b. During the transcription phase of protein synthesis, partial unwinding of a DNA molecule occurs.
c. During the translation phase of protein synthesis, an mRNA-ribosome complex is formed.
d. During the formation of recombinant DNA, the process of transformation occurs.

22.153 a. True.
b. True.
c. False. The primer is bound to a complementary strand of DNA that functions as a template.
d. False. After four cycles of the PCR, sixteen DNA molecules have been produced.

Biochemical Energy Production — Chapter 23

Solutions to Selected Problems

23.1 During anabolism small, biochemical molecules are joined together to form larger ones (a synthetic process). During catabolism, large biochemical molecules are broken down to smaller ones (a process of degradation).

23.3 A metabolic pathway is a series of consecutive biochemical reactions used to convert a starting material into an end product.

23.5 During anabolism, synthesis occurs; during catabolism, degradation occurs and energy is produced.
a. Synthesis of a protein from amino acids is anabolic.
b. Formation of a triacylglycerol from gycerol and fatty acids is anabolic.
c. Hydrolysis of a polysaccharide to produce monsaccharides is catabolic.
d. Formation of a polynucleotide from nucleotides is anabolic.

23.7 Prokaryotic cells have no nucleus and are found only in bacteria; their DNA is usually a single circular molecule found near the center of the cell in a region called the nucleoid. Eukaryotic cells, which are found in all higher organisms, have their DNA in a membrane-enclosed nucleus.

23.9 a. True.
b. True.
c. False. The cytosol is the water-based fluid part of cytoplasm.
d. True.

23.11 a. False. The inner membrane of a mitochondrion is highly impermeable to most substances.
b. True.
c. True.
d. True.

23.13 ATP stands for *a*denosine *t*ri*p*hosphate.

23.15 The structural subunits of an ATP molecule are: adenine (1), ribose (1), and phosphate (3).

23.17 An ADP molecule has one phosphoanhydride bond and one phosphoester bond.

23.19 a. ATP has three phosphate groups; AMP has one phosphate group.
b. ADP contains adenine; GDP contains guanine.

23.21 The designation P_i stands for inorganic phosphate.

23.23 The generalized chemical equation for the hydrolysis of ATP to ADP is:
$ATP + H_2O \rightarrow ADP + P_i + H^+ + \text{energy}$

23.25 FAD stands for *f*lavin *a*denine *d*inucleotide.

23.27 Block diagram for FAD:

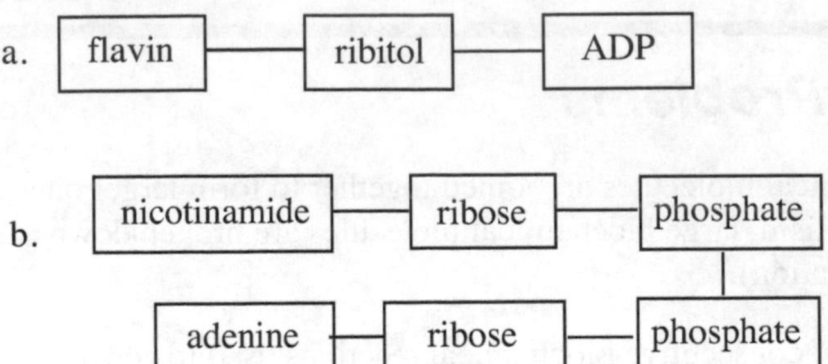

23.29 The B vitamin present in
a. NAD^+ is nicotinamide.
b. $FADH_2$ is riboflavin.

23.31 a. Yes, the B-vitamin portion of FAD is the "active" subunit in redox reactions.
b. Yes, the B-vitamin portion of NADH is the "active" subunit in redox reactions.

23.33 a. FAD denotes the oxidized form of the molecule.
b. $FADH_2$ denotes the reduced form of the molecule.

23.35 a. NAD^+ and NADH both contain two ribose units, adenine and two phosphate groups.
b. NAD^+ and $FADH_2$ both contain two phosphate groups, ribose, and adenine.

23.37 The three-subunit block diagram for CoA:

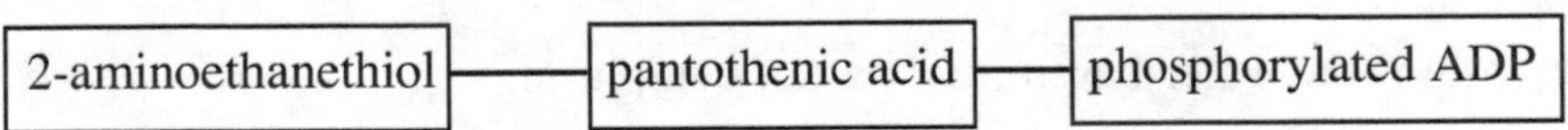

23.39 a. The B vitamin is pantothenic acid.
b. The "active" subunit is 2-aminoethanethiol.

23.41 a. The biochemical function of ATP involves phosphate groups.
b. The biochemical function of $FADH_2$ involves electrons.
c. The biochemical function of NAD^+ involves electrons.
d. The biochemical function of CoA-SH involves acetyl groups.

23.43 a. $FADH_2$ is a reducing agent.
b. ADP is neither an oxidizing agent nor a reducing agent.
c. Coenzyme A is neither an oxidizing agent nor a reducing agent.
d. NAD^+ is an oxidizing agent.

23.45 a. NAD^+ contains two ribose subunits.
b. CoA-SH, FAD, and NAD^+ contain two phosphate subunits.
c. ATP, CoA-SH, FAD, and NAD^+ contain one adenine subunit.
d. CoA-SH, FAD, and NAD^+ contain four different kinds of subunits.

23.47 Structural formulas for the four substances.

a. oxaloacetic acid: $HOOC-\overset{\overset{\displaystyle O}{\|}}{C}-CH_2-COOH$

b. oxaloacetate ion: ${}^{-}OOC-\overset{\overset{\displaystyle O}{\|}}{C}-CH_2-COO^{-}$

c. citric acid: $HOOC-CH_2-\underset{\underset{\displaystyle COOH}{|}}{\overset{\overset{\displaystyle OH}{|}}{C}}-CH_2-COOH$

d. citrate ion: ${}^{-}OOC-CH_2-\underset{\underset{\displaystyle COO^{-}}{|}}{\overset{\overset{\displaystyle OH}{|}}{C}}-CH_2-COO^{-}$

23.49 a. Malate, oxaloacetate, and fumarate contain four carbon atoms.
b. Oxaloacetate and α–ketoglutarate contain a keto functional group.
c. Malate, oxaloacetate, fumarate, and α–ketoglutarate are ions with a charge of –2.
d. Malate, oxaloacetate, and fumarate are succinic acid derivatives.

23.51 A high energy compound has a greater free energy of hydrolysis than that of a typical compound.

23.53 The notation P_i is used as a general designation for any free monophosphate species present in cellular fluid.

23.55 Table 23.1 gives the free energies of hydrolysis of phosphate-containing metabolic molecules.
a. Phosphoenolpyruvate releases more free energy on hydrolysis than ATP.
b. Creatine phosphate releases more free energy on hydrolysis than ADP.
c. 1,3-Bisphosphoglycerate releases more free energy on hydrolysis than glucose 1-phosphate.
d. AMP releases more free energy on hydrolysis than glycerol 3-phosphate.

23.57 In the conversion of ADP to AMP and P_i
a. one phosphorus-oxygen bond is broken.
b. one new phosphorus-oxygen bond is formed.

23.59 The four general stages of the process by which biochemical energy is obtained from food are: 1) digestion, 2) acetyl group formation, 3) citric acid cycle, and 4) electron transport chain and oxidative phosphorylation

23.61 Two other names for the citric acid cycle are the tricarboxylic acid cycle, in reference to the three carboxylate groups present in the citric acid cycle, and the Krebs cycle, after its discoverer Hans Adolf Krebs.

23.63 Oxaloacetate and acetyl CoA are needed to start a turn of the citric acid cycle.

23.65 Seven out of the eight reactions of the citric acid cycle take place in the mitochondrial matrix.

23.67 In the citric acid cycle
a. malate is a reactant in an oxidation reaction.
b. malate is a product in a hydration reaction.
c. isocitrate is a reactant in an oxidation/decarboxylation reaction.
d. isocitrate is a product in an isomerization reaction.

23.69 a. Two molecules of CO_2 are formed in the citric acid cycle, in steps 3 and 4.
b. One molecule of $FADH_2$ is formed the citric acid cycle, in step 6.
c. A secondary alcohol is oxidized two times in the citric acid cycle, in steps 3 and 8.
d. Water adds to a carbon-carbon double bond two times in the cycle, in steps 2 and 7.

23.71 a. The steps in the citric acid cycle that involve oxidation are steps 3, 4, 6, and 8.
b. Isomerization takes place in the citric acid cycle in step 2.
c. Hydration takes place in step 7 of the cycle; the loss and gain of water in step 2 is isomerization rather than hydration.

23.73 The four dicarboxylic acid species in the citric acid cycle are shown below.

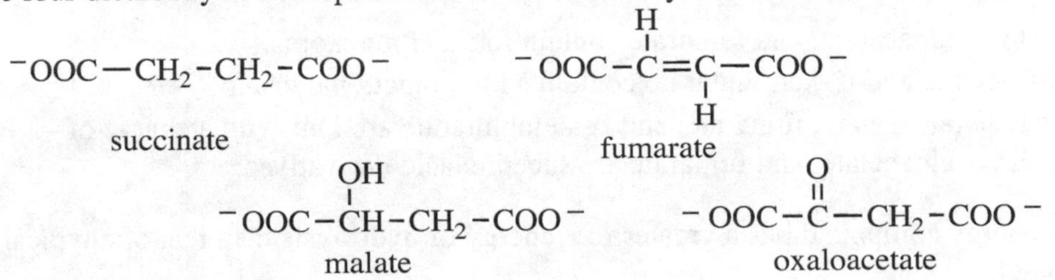

$^-OOC-CH_2-CH_2-COO^-$
succinate

$^-OOC-CH=CH-COO^-$ (each double-bond carbon bearing one H, drawn above and below)
fumarate

$^-OOC-CH(OH)-CH_2-COO^-$
malate

$^-OOC-C(=O)-CH_2-COO^-$
oxaloacetate

23.75 a. Citrate to isocitrate is a C_6 to C_6 change.
b. Succinate to fumarate is a C_4 to C_4 change.
c. Malate to oxaloacetate is a C_4 to C_4 change.
d. Isocitrate to α-ketoglutarate is C_6 to C_5 change.

23.77 a. In step 3 of the citric acid cycle, isocitrate is oxidized and then decarboxylated to form α-ketoglutarate. During the oxidation step, NAD^+ is reduced; one hydrogen and two electrons are transferred to NAD^+ to form NADH.
b. In step 6 of the citric acid cycle, succinate is oxidized to fumarate; FAD is reduced to $FADH_2$ in the process.

23.79 a. In step 3, the reaction catalyzed by isocitrate dehydrogenase, the reactant isocitrate is oxidized to form oxalosuccinate and decarboxylated to form α-ketoglutarate.
b. In step 7, catalyzed by fumarase, the reactant fumarate is hydrated to form L-malate.
c. In step 8, catalyzed by malate dehydrogenase, the reactant L-malate is oxidized to form oxaloacetate.
d. In step 2, catalyzed by aconitase, the reactant citrate is isomerized (by dehydration and hydration) to form isocitrate.

23.81 NADH and $FADH_2$ provide the electrons needed for operation of the electron transport chain.

23.83 O_2 is the final electron acceptor in the electron transport chain.

23.85 The given order for electron carriers as they are first encountered in the electron transport chain:
a. Yes, it is correct.
b. No, it is not correct.
c. No, it is not correct.
d. No, it is not correct.

23.87 a. Fe(III)SP is in its oxidized form; the iron atom has a +3 oxidation state.
b. Cyt b (Fe^{3+}) is in its oxidized form; it contains a +3 iron atom.
c. NADH is in its reduced form; NAD^+ has gained a –H and two electrons.
d. FAD is in its oxidized form; the reduced form of FAD is $FADH_2$.

23.89 a. The reaction, $CoQH_2 \rightarrow CoQ$, is an oxidation because hydrogen is lost.
b. The reaction, $NAD^+ \rightarrow NADH$, is a reduction because hydrogen is gained.
c. The reaction, cyt c (Fe^{2+}) $\rightarrow$ cyt c (Fe^{3+}) is an oxidation because electrons are lost.
d. The reaction, cyt b (Fe^{3+}) $\rightarrow$ cyt b (Fe^{+2}), is a reduction because electrons are gained.

23.91 a. NADH is associated with protein complex I; the net result of electron movement through protein complex I is the transfer of electrons from NADH to CoQ.
b. CoQ is associated with protein complex I (electrons transferred from NADH to CoQ), protein complex II (electrons transferred from $FADH_2$ to CoQ), and protein complex III ($CoQH_2$, carrying electrons from complex I and complex II, transfers electrons to FeSP).
c. Cyt b is associated with protein complex III; cyt b receives electrons from FeSP and transfers them to another FeSP.
d. Cyt a is associated with protein complex IV; cyt c (carrying electrons from complex III) transfers electrons to cyt a, which in turn transfers electrons to cyt a_3.

23.93 $CoQH_2$ carries electrons from both complexes I and II to complex III.

23.95 $CoQH_2$ carries two electrons per trip from complex II to complex III.

23.97 a. The missing substances are $FADH_2$, 2Fe(II)SP, and $CoQH_2$ as shown in the electron transport chain reaction sequence below.

FAD 2Fe(III)SP $CoQH_2$

$FADH_2$ 2Fe(II)SP CoQ

b. The missing substances are $FMNH_2$, $2Fe^{2+}$, and $CoQH_2$, as shown in the electron transport chain reaction sequence below.

$FMNH_2$ $2Fe^{3+}$ $CoQH_2$

FMN $2Fe^{2+}$ CoQ

23.99 a. NADH is a reactant in the ETC.
b. O_2 is a reactant in the ETC.
c. Fumarate is a reactant in the CAC.
d. Cytochrome a is a reactant in the ETC.

23.101 The order in which the four substances are first encountered in the common metabolic pathway is CO_2, succinate, $FADH_2$, FeSP.

23.103 In the common metabolic pathway:
a. Oxaloacetate is neither oxidized nor reduced.
b. NAD^+ is reduced.
c. $FADH_2$ is oxidized.
d. FeSP is both oxidized and reduced.

23.105 Oxidative phosphorylation is the biochemical process by which ATP is synthesized from ADP and P_i using energy from the electron transport chain.

23.107 The protons that are involved in the proton gradient are found in the mitochondrial matrix and the intermembrane space.

23.109 When a NADH molecule is oxidized, 10 protons cross the inner mitochondrial membrane.

23.111 When two electrons from a NADH molecule are passed through the electron transport chain, the number of protons that cross the inner mitochondrial membrane is
a. complex I – four electrons
b. complex II – zero electrons
c. complex III – four electrons
d. complex IV – two electrons

23.113 ATP synthases are the coupling factors that link the processes of oxidative phosphorylation and the electron transport chain; ATP synthases "power" the synthesis of ATP.

23.115 In oxidative phosphorylation, coenzymes are oxidized and ADP is phosphorylated.

23.117 An ATP molecule produced in oxidative phosphorylation must be moved from the matrix back into the intermembrane space before it becomes available for cellular us.

23.119 a. 2.5 molecules of ATP are formed from each NADH molecule that enters the ETC.
b. 1.5 molecules of ATP are formed from each $FADH_2$ molecule that enters the ETC.

23.121 a. In steps 3, 4, and 8 of the CAC a molecule of NADH, which generates 2.5 units of ATP, is produced.
b. In step 6 of the CAC a molecule of $FADH_2$, which generates 1.5 units of ATP, is produced.
c. In step 5 of the CAC a molecule of GDP, which generates 1.0 unit of ATP, is produced.

23.123 If four moles of NADH and three moles of $FADH_2$ are oxidized within a mitochondrion, 14.5 moles of ATP become available for cellular use.

23.125 ROS is the designation for several highly reactive oxygen species. Among them are hydrogen peroxide (H_2O_2), superoxide ion (O_2^-), and hydroxyl radical (OH).

23.127 a. Superoxide ions (free radicals) are generated within cells by the reaction of O_2 with a phosphorylated version of the coenzyme NADH.

$$2O_2 + NADPH \rightarrow 2O_2^- + NADP^+ + H^+$$

b. Superoxide ions are converted to hydrogen peroxide by reaction with hydrogen ions:

$$2O_2^- + 2H^+ \rightarrow H_2O_2 + O_2$$

23.129 a. Yes, thiamin is important in the functioning of the CAC; in step 4 of the citric acid cycle thiamin is in the form TPP.
b. No, biotin is not involved in the CAC.
c. Yes, niacin is important in the functioning of the CAC; niacin is a part of NAD^+, which is reduced to NADH in steps 3, 4, and 8.
d. No, vitamin B_6 is not involved in the CAC.

23.131 a. No, thiamin is not involved in the ETC.
b. Yes, riboflavin is needed for the proper functioning of the ETC; it is a part of FAD, $FADH_2$, and FMN.
c. No, folate is not involved in the ETC.
d. No, vitamin B_{12} is not involved in the ETC.

Carbohydrate Metabolism Chapter 24

Solutions to Selected Problems

24.1 Starch digestion begins in the mouth; the enzyme salivary α-amylase catalyzes the hydrolysis of α-glycosidic linkages in starch (from plants) and glycogen (from meats) to produce smaller polysaccharides and the disaccharide maltose.

24.3 The primary site for carbohydrate digestion is the small intestine. The pancreas secretes α-amylase into the small intestine; pancreatic α-amylase breaks down polysaccharide chains into shorter segments until maltose and glucose are the dominant species.

24.5 The digestion of sucrose (and other disaccharides) occurs on the outer membranes of intestinal mucosal cells; the conversion of sucrose to glucose and fructose is a hydrolysis reaction.

24.7 The three major monosaccharides produced by the digestion of carbohydrates are glucose, galactose, and fructose.

24.9 Glycolysis intermediates:
a. Fructose 6-phosphate – C_6 molecule.
b. Phopoenolpyruvate – C_3 molecule.
c. 3-Phospoglycerae – C_3 molecule.
d. Dihydroxyacetone phosphate – C_3 molecule

24.11 Phosphate groups in molecules in Problem 24.9
a. Fructose 6-phosphate – 1 phosphate
b. Phopoenolpyruvate – 1 phosphate
c. 3-Phosphoglycerate – 1 phosphate
d. Dihydroxyacetone phosphate – 1 phosphate

24.13 The step in glycolysis that each molecule in Problem 24.9 is encountered:
a. Fructose 6-phosphate – step 3
b. Phopoenolpyruvate – step 10
c. 3-Phosphoglycerate – step 8
d. Dihydroxyacetone phosphate – step 5

24.15 Type of reaction that occurs in the given steps of glycolysis:
a. step 1 – phosphorylation using ATP
b. step 3 – phosphorylation using ATP
c. step 5 – isomerization
d. step 7 – phosphorylation of ADP

24.17 Of the ten steps in glycolysis
a. two produce ATP.
b. five involve phosphorylation.
c. one involves NAD^+ as a reactant.
d. two involve a compound with a high energy bond as a reactant.

24.19 a. Glucose + ATP $\xrightarrow{\text{Hexokinase}}$ glucose 6-phosphate + ADP

b. 2-Phosphoglycerate $\xrightarrow{\text{Enolase}}$ phosphoenolpyruvate + water

c. 3-Phosphoglycerate $\xrightarrow{\text{Phosphoglyceromutase}}$ 2-phosphoglycerate

d. 1,3-Bisphosphoglycerate + ADP $\xrightarrow{\text{Phosphoglycerokinase}}$ 3-phosphoglycerate + ATP

24.21 a. A second substrate-level phosphorylation reaction takes place in step 10; phosphoenolpyruvate transfers its high-energy phosphate group to an ADP molecule to produce ATP and pyruvate.
b. The first ATP-consuming reaction is step 1; the phosphorylation of glucose to yield glucose 6-phosphate uses a phosphate group from an ATP molecule.
c. The third isomerization reaction is step 8; the phosphate group of 3-phosphoglycerate is moved from carbon 3 to carbon 2.
d. NAD^+ is used as an oxidizing agent in step 6; a phosphate group is added to glyceraldehyde 3-phosphate to produce 1,3-bisphosphoglycerate, and the hydrogen of the aldehyde group becomes a part of NADH.

24.23 Galactose enters the glycolysis metabolic pathway in the form of glucose 6-phosphate.

24.25 The processing of one fructose molecule and one glucose molecule through the glycolysis metabolic pathway produce the same number of net ATPs.

24.27 The first step is the conversion of glucose to glucose 6-phosphate. Glucose 6-phosphate has no membrane transporter, so glucose is retained inside the cell.

24.29 Glycolysis occurs within the cytosol.

24.31 a. Hexokinase is associated with the glycolysis metabolic pathway; it is an enzyme in step 1 at the C_6 stage.
b. Lactase is associated with carbohydrate digestion; it is an enzyme for the hydrolysis of disaccharides.
c. Hydrolysis is associated with carbohydrate digestion; polysaccharides and disaccharides are broken down in hydrolysis reactions.
d. Dehydration is associated with the glycolysis metabolic pathway; dehydration takes place in step 9 of glycolysis at the C_3 stage.

24.33 Under aerobic conditions, pyruvate is converted to acetyl CoA in the human body.

24.35 NADH is needed to convert pyruvate to lactate in the human body.

24.37 Pyruvate is converted to acetyl CoA in the mitochondrial matrix.

24.39 The chemical purpose for lactate fermentation is the oxidation of NADH to NAD^+.

24.41 In ethanol fermentation a C_3 pyruvate molecule is converted to a C_2 ethanol molecule. A CO_2 molecule is formed from the third pyruvate carbon atom.

24.43 The overall reaction equation for the conversion of pyruvate to acetyl CoA is:
pyruvate + CoA–SH + NAD^+ $\rightarrow$ acetyl CoA + NADH + CO_2

24.45 a. CO_2 is produced in the formation of acetyl CoA and ethanol from pyruvate.
b. NADH is a reactant in the production of lactate and ethanol from pyruvate.
c. NAD^+ is a reactant in the production of acetyl CoA from pyruvate.
d. Under anaerobic conditions, pyruvate is converted to the C_3 molecule lactate.

24.47 a. CO_2 is associated with (2) pyruvate oxidation and (4) ethanol fermentation.
b. Acetyl CoA is associated with (2) pyruvate oxidation.
c. ATP is associated with (1) glycolysis.
d. NADH is associated with (1) glycolysis, (2) pyruvate oxidation, (3) lactate fermentation, and (4) ethanol fermentation.

24.49 NADH produced in the cytosol cannot directly participate in the ETC because mitochondria are impermeable to NADH. A transport system shuttle (dihydroxyacetone phosphate-glycerol 3-phosphate shuttle) transports the electrons across the outer mitochondrial membrane.

24.51 The electrons from NADH are the "cargo" for glycerol 3-phosphate in the dihydroxyacetone phosphate-glycerol 3-phosphate shuttle.

24.53 For every glucose molecule converted into two pyruvates, there is a net gain of 2ATP molecules. The net gain for every glucose molecule converted to CO_2 and H_2O, through the CAC and the ETC, is 30 ATP.

24.55 Of the 30 ATP molecules produced from the complete oxidation of one glucose molecule
a. 2 are produced in glycolysis.
b. 0 are produced by the oxidation of pyruvate.

24.57 The initial reactant or reactants in
a. glycogenesis is glucose 6-phosphate.
b. glycogenolysis is glycogen.

24.59 a. Glucose 6-phosphate is involved in both glycogenesis and glycogenolysis.
b. UDP is involved in glycogenesis, but not glycogenolysis.
c. Glycogen is involved in both glycogenesis and glycogenolysis.
d. P_i is involved in both glycogenesis and glycogenolysis.

24.61 a. True.
b. False. An isomerization reaction changes glucose 6-phosphate to glucose 1-phosphate.
c. False. A UTP molecule is used to activate a glucose 1-phosphate molecule.
d. False. The equivalent of two ATP molecules are consumed to add a glucose unit to glycogen.

24.63 In glycogenolysis, the type of reaction that occurs
a. in step 1 is phosphorolysis.
b. in step 2 is isomerization.

24.65 The first two steps of glycogenolysis are the same in liver cells and in muscle cells. However, muscle cells cannot form free glucose from glucose 6-phosphate because they lack the enzyme glucose 6-phosphatase; in liver cells, which have this enzyme, the product is glucose.

24.67 Complete glycogenolysis takes place mainly in the liver; it produces glucose. Glycogenolysis in muscle and brain cells produces glucose 6-phosphate, which can enter the glycolytic pathway as the first intermediate in that pathway. Since brain and muscle cells do not produce glucose, these cells can use glycogen for energy production only.

24.69 During glycogenesis, glycogen is synthesized from glucose 6-phosphate. During glycogenolysis, glucose 6-phosphate is produced from glycogen. During glycolysis, glucose is converted into two molecules of pyruvate.

a. Glucose 6-phosphate is associated with (1) glycolysis, (2) glycogenesis, and (3) glycogenolysis.
b. UTP is associated with (2) glycogenesis; formation of UDP-glucose activates glucose 1-phosphate so it can be added to a growing glycogen chain.
c. Phosphoglucomutase is associated with glycogenesis; this enzyme effects the change from glucose 6-phosphate to glucose 1-phosphate.
d. Pyruvate is associated with glycolysis; in glycolysis, glucose is converted into two molecules of pyruvate.

24.71 The starting reactants for the process of gluconeogenesis are pyruvate, CO_2, and ATP.

24.73 In the first three steps of gluconeogenesis, the reactants are pyruvate, oxaloacetate, and phophoenolpyruvate; in the last three steps of glycolysis, the reactants are 2-phosphoglycerate, phosphoenolpyruvate, and pyruvate.

24.75 In gluconeogenesis and glycolysis, respectively, there are

a. 0 and 2 triphosphate nucleotides produced.
b. 3 and 2 triphosphate nucleotides consumed.

24.77 In gluconeogenesis

a. a C_3 molecule is converted to a C_4 molecule in step 1.
b. CO_2 is a product in step 2.

24.79 a. True.

b. False. Lactate travels from the muscle cell into the blood and is transported to the liver.
c. False. Lactate is produced in the muscle tissue.
d. True.

24.81 Equation for the conversion of lactate to pyruvate in the Cori cycle:

$$\begin{array}{c}COO^-\\ |\\ H-C-OH\\ |\\ CH_3\end{array} + NAD^+ \longrightarrow \begin{array}{c}COO^-\\ |\\ C=O\\ |\\ CH_3\end{array} + NADH + H^+$$

24.83 During glycolysis, glucose is converted into two molecules of pyruvate. Gluconeogenesis is the process by which glucose is synthesized from noncarbohydrate materials.

a. Hexokinase is involved in glycolysis but not in gluconeogenesis.
b. Phosphofructokinase is involved in glycolysis but not in gluconeogenesis.
c. Pyruvate carboxylase is involved in gluconeogenesis but not in glycolysis.
d. Phosphoglyceromutase is involved in both glycolysis and gluconeogenesis.

24.85 During glycolysis, glucose is converted into two molecules of pyruvate. Gluconeogenesis is the process by which glucose is synthesized from noncarbohydrate materials.
a. 2-Phosphoglycerate is involved in both glycolysis and gluconeogenesis.
b. Phosphoenolpyruvate is involved in both glycolysis and gluconeogenesis.
c. Fructose 1,6-bisphosphate is involved in both glycolysis and gluconeogenesis.
d. Glucose 6-phosphate is involved in both glycolysis and gluconeogenesis.

24.87 a. There are eight different C_3 molecules involved in the Cori cycle; they are dihydroxyacetone phosphate, glyceraldehyde 3-phosphate, 1,3-bisphosphoglycerate, 3-phosphoglycerate, 2-phosphoglycerate, phosphoenolpyruvate, pyruvate, and oxaloacetate.
b. There are seven different C_3 molecules involved in gluconeogenesis; gluconeogenesis is the reverse of glycolysis with one added step.
c. There are seven different C_3 molecules involved in glycolysis; they are dihydroxyacetone phosphate, glyceraldehyde 3-phosphate, 1,3-bisphosphoglycerate, 3-phosphoglycerate, 2-phosphoglycerate, phosphoenolpyruvate, and pyruvate.
d. There are two different C_3 molecules involved in lactate fermentation; they are lactate and pyruvate.

24.89 a. False. Glucose is the initial reactant for glycolysis.
b. False. Pyruvate is the initial reactant for gluconeogenesis.
c. True.
d. True.

24.91 a. all four processes
b. glycolysis and gluconeogenesis
c. gluconeogenesis
d. glycogenesis

24.93 a. glycolysis
b. glycolysis
c. glycogenesis
d. glycolysis

24.95 a. In gluconeogenesis, there is a loss of six triphosphates, four ATP molecules and two GTP molecules.
b. In glycogenesis there is a loss of two triphosphates; one ATP molecule is used in the formation of glucose 6-phosphate and one in the regeneration of UTP.
c. In the transfer of a glycogen glucose unit to pyruvate, there is a gain of three triphosphates (rather than the usual two from glycolysis). This is because a glucose 6-phosphate produced by glycogenolysis enters the glycolytic pathway at step 2; it does not have to be activated by the addition of a phosphate group.
d. In the Cori cycle there is a loss of four triphosphates; two ATP molecules are obtained from glycolysis and six ATP molecules are used in gluconeogenesis.

24.97 a. True.
b. False. The initial reactant for the PPP is glucose 6-phosphate.
c. True.
d. True.

24.99 The number of phosphate group present in the structure of
a. NAD^+ is 2.
b. $NADP^+$ is three.

24.101 There are two stages in the pentose phosphate pathway, an oxidative stage and a nonoxidative stage. The oxidative stage (involving three steps) occurs first; glucose 6-phosphate is converted to ribulose 5-phosphate and CO_2. The overall reaction is:
glucose 6-phosphate + $2NADP^+$ + $H_2O \rightarrow$ ribulose 5-phosphate + CO_2 + 2NADPH + $2H^+$

24.103 When glucose 6-phosphate is converted to ribulose 5-phosphate in the oxidative stage of the pentose phosphate pathway, the carbon lost from glucose is converted to CO_2.

24.105 a. Lactate is involved in the Cori cycle and lactate fermentation; lactate from lactate fermentation enters the Cori cycle as pyruvate.
b. NAD^+ is involved in the Cori cycle, lactate fermentation, and glycolysis.
c. Glucose 6-phosphate is involved in the pentose phosphate pathway (it is converted to ribulose 5-phospate), the Cori cycle, and glycolysis; it is the activated form of glucose in these processes.
d. Ribose 5-phospate is produced in the pentose phosphate pathway; its production is necessary for further nucleic acid and coenzyme production.

24.107 Insulin is a hormone that promotes the uptake and utilization of glucose by cells; its function is to lower blood glucose levels, which it does by increasing the rate of glycogen synthesis.

24.109 Glucagon is a hormone released when blood-glucose levels are low; its function is to increase blood-glucose concentrations by speeding up the conversion of glycogen to glucose in the liver.

24.111 Insulin is produced by the beta cells of the pancreas.

24.113 Epinephrine binds to a receptor site on the outside of the cell membrane, stimulating the production of a second messenger, cyclic AMP from ATP; cAMP, released in the cell's interior, activates glycogen phophorylase, the enzyme that initiates glycogenolysis.

24.115 The function of glucagon is to speed up the conversion of glycogen to glucose in liver cells. The function of epinephrine is similar to that of glucagon (stimulation of glycogenolysis), but its primary target is muscle cells.

24.117 a. Thiamin is not involved in any of these four processes; as TPP, thiamin is involved in the formation of acetyl CoA from pyruvate.
b. Riboflavin is not involved in any of these four processes; as FAD, riboflavin is involved in the citric acid cycle.
c. Pantothenic acid is not involved in any of these four processes; pantothenic acid (as CoA) is involved in the conversion of pyruvate to acetyl CoA.
d. Vitamin B_6, in the form of PLP, is involved in glycogenolysis.

Lipid Metabolism — Chapter 25

Solutions to Selected Problems

25.1 In triacylglycerol digestion, the location where
- a. chyme is produced is the stomach.
- b. gastric lipases are active is the stomach.
- c. initial production of monoacylglycerols occurs is the stomach.
- d. fatty acid micelles are formed is the small intestine.

25.3
- a. True.
- b. False. Bile is a fluid containing emulsifying agents.
- c. True.
- d. True.

25.5
- a. True.
- b. False. Gastric lipases present in the stomach activate the hydrolysis of TAGs.
- c. False. Churning action in the stomach breaks up TAG materials into small globules.
- d. True.

25.7 A fatty acid micelle is a spherical droplet containing fatty acids, monoacylglycerols, and bile.

25.9 The products of the complete hydrolysis of a triacylglycerol molecule are three fatty acid molecules and a glycerol molecule.

25.11 Free fatty acids and monoacylglycerols are "repackaged" into triacylglycerols for the first time in the intestinal cells.

25.13 Adipocytes have a large storage capacity for triacylglycerols; they are among the largest cells in the body and differ from other cells in that most cytoplasm has been replaced with a large triacylglycerol droplet.

25.15 Triacylglycerol mobilization is the hydrolysis of triacylglycerols stored in adipose tissue, followed by release of the hydrolysis products (fatty acids and glycerol) into the bloodstream.

25.17 Several hormones, including epinephrine and glucagon, interact with adipose cell membranes to stimulate the production of cAMP; cAMP activates hormone-sensitive lipase, the enzyme needed for triacylglycerol hydrolysis.

25.19 a. Step 1 b. Step 1 c. Step 2 d. Step 1

25.21 Dihydroxyacetone phosphate has an O–P group on carbon 3 rather than an O–H, as in glycerol.

25.23 a. true b. false c. true d. false

25.25 a. intermembrane space b. matrix
c. intermembrane space d. matrix

25.27 The oxidizing agent needed in step 1 of a turn of the β-oxidation pathway is FAD.

25.29 In a turn of the β-oxidation pathway, the functional group change
a. in step 1 is alkane to alkene.
b. in step 2 is alkene to 2° alcohol.
c. in step 3 is 2° alcohol to ketone.

25.31 The unsaturated enoyl CoA formed by dehydrogenation in Step 1 of the β–oxidation pathway is a *trans* isomer.

25.33 The compounds represented by the structural formulas below are associated with the fatty acid β-oxidation pathway.

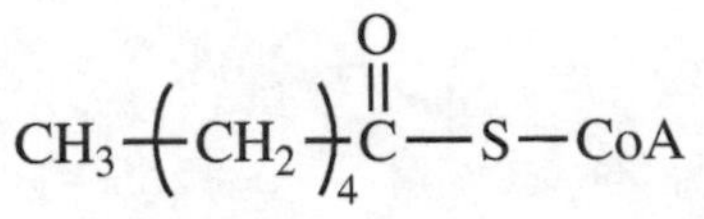

$$CH_3\!+\!CH_2\!+_4\!\overset{\overset{\large O}{\|}}{C}-S-CoA$$

a. C_6 acyl CoA

$$CH_3-\overset{\overset{\large OH}{|}}{CH}-CH_2-\overset{\overset{\large O}{\|}}{C}-S-CoA$$

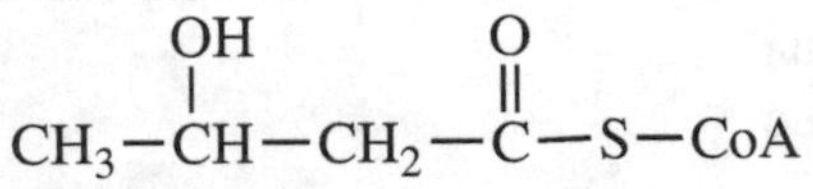

b. C_4 hydroxyacyl CoA

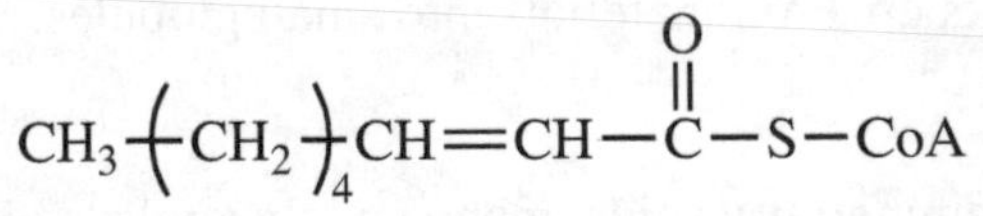

$$CH_3\!+\!CH_2\!+_4\!CH=CH-\overset{\overset{\large O}{\|}}{C}-S-CoA$$

c. C_8 enoyl CoA

$$CH_3\!+\!CH_2\!+_6\!\overset{\overset{\large O}{\|}}{C}-CH_2-\overset{\overset{\large O}{\|}}{C}-S-CoA$$

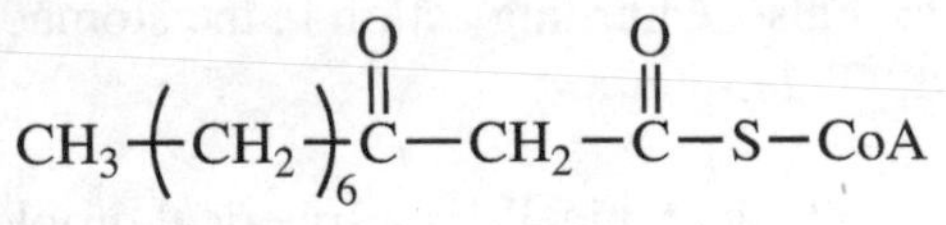

d. C_{10} ketoacyl CoA

25.35 a. The reactant in step 1 of the β-oxidation pathway is acyl CoA.
b. The product in step 2 of the β-oxidation pathway is hydroxyacyl CoA.
c. The product in step 3 of the β-oxidation pathway is ketoacyl CoA.
d. The reactant in step 4 of the β-oxidation pathway is ketoacyl CoA.

25.37 a. AMP is involved with fatty acid activation; AMP is produced from ATP when the free fatty acid is activated.
b. FAD is involved in the β-oxidation pathway; in step 1 (the dehydrogenation of acyl CoA), FAD acts as an oxidizing agent. It is a hydrogen acceptor, forming $FADH_2$.
c. Acetyl CoA is involved in (2) fatty acid transport, (1) fatty acid activation, and (3) the β-oxidation pathway.
d. H_2O is involved in the β-oxidation pathway in step 2, the hydration of enoyl CoA to hydroxyacyl CoA.

25.39 a. The compound is a reactant in Step 3 of turn 2 of the β–oxidation pathway (it still has eight carbon atoms).
b. The compound is a reactant in Step 2 of turn 3 (it has lost two carbon atoms in the first turn).
c. The compound is a reactant in Step 4 of turn 3 (it lost two carbons in the first turn).
d. The compound is a reactant in Step 1 of turn 2 (it still has eight carbon atoms).

25.41 Compound a in Problem 25.39 undergoes a dehydrogenation reaction during Step 3 of the β–oxidation pathway; compound d undergoes a dehydrogenation reaction in Step 1 of the β–oxidation pathway.

25.43 Each turn of the β–oxidation pathway produces one acetyl CoA molecule (one C_2 unit); the number of acetyl CoA molecules (C_2 units) is equal to half the number of carbon atoms in the fatty acid, but the number of turns of the β–oxidation pathway is always one less than the number of C_2 units because in the last turn a C_4 unit splits into two C_2 units.

a. A C_{16} fatty acid requires: 16/2 – 1 = 7 turns

b. A C_{12} fatty acid requires: 12/2 – 1 = 5 turns

25.45 One step; the isomerase gives a *trans* 2,3-enoyl product which is hydrated normally to a L-hydroxy product.

25.47 a. NAD^+ is involved in both glycerol metabolism and fatty acid metabolism; in both cases it is an oxidizing agent.

b. ADP is produced from ATP in glycerol metabolism to dihydroxyacetone phosphate.

c. Kinase is an enzyme in glycerol metabolism to dihydroxyacetone phosphate.

d. Ketoacyl CoA is the product in step 3 of fatty acid metabolism to acetyl CoA.

25.49 a. In an active state, the major fuel for skeletal muscle is glucose (from glycogen).

b. In a resting state, the major fuel for skeletal muscle is fatty acids.

25.51 a. A C_{10} acid requires: 10/2 – 1 = 4 turns of the β–oxidation pathway.

b. A C_{10} fatty acid yields 5 acetyl CoA molecules (C_2 units).

c. A C_{10} fatty acid, in 4 turns of the β–oxidation pathway, yields 4 NADH molecules.

d. A C_{10} fatty acid, in 4 turns of the β–oxidation pathway, yields 4 $FADH_2$ molecules.

25.53 Further processing through the common metabolic pathway of the products from the first turn of the β–oxidation pathway results in the production of 14 ATP molecules.

25.55 The net ATP production for the complete oxidation of the C_{10} fatty acid in Problem 25.51:

$$5 \text{ acetyl CoA} \times \frac{10 \text{ ATP}}{1 \text{ acetyl CoA}} = 50 \text{ ATP}$$

$$4 \text{ FADH}_2 \times \frac{1.5 \text{ ATP}}{1 \text{ FADH}_2} = 6 \text{ ATP}$$

$$4 \text{ NADH} \times \frac{2.5 \text{ ATP}}{1 \text{ NADH}} = 10 \text{ ATP}$$

Activation of fatty acid = –2 ATP

Net ATP production = 64 ATP

25.57 Less $FADH_2$ is produced from an unsaturated fatty acid, since $FADH_2$ is not generated in producing a carbon-carbon double bond (it is already there).

25.59 One gram of carbohydrate yields 4 kcal of energy; one gram of fat yields 9 kcal of energy.

25.61 Under certain conditions an excess of acetyl CoA is converted to ketone bodies; these conditions are: (1) dietary intakes high in fat and low in carbohydrates, (2) diabetic conditions where the body cannot adequately process glucose even though it is present, and (3) prolonged fasting.

25.63 When oxaloacetate supplies are too low for all acetyl CoA present to be processed through the citric acid cycle, ketone bodies are formed from the excess acetyl CoA.

25.65 The structures of the three ketone bodies (acetoacetate, β-hydroxybutyrate, and acetone) are:

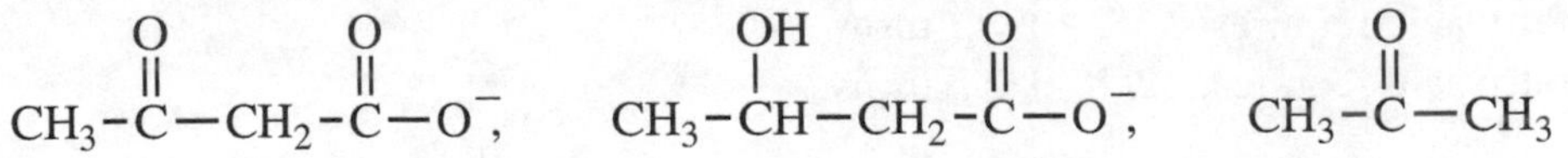

$$CH_3-\overset{\overset{O}{\|}}{C}-CH_2-\overset{\overset{O}{\|}}{C}-O^-, \quad CH_3-\overset{\overset{OH}{|}}{CH}-CH_2-\overset{\overset{O}{\|}}{C}-O^-, \quad CH_3-\overset{\overset{O}{\|}}{C}-CH_3$$

25.67 a. acetoacetate, acetone
b. β-hydroxybutyrate
c. acetoacetate
d. acetone

25.69 a. In step 1 of ketogenesis, two acetyl CoA molecules ($C_2 + C_2$) combine to form acetoacetyl CoA (C_4).
b. In step 4 of ketogenesis, acetoacetate (C_4) is reduced to β-hydroxybutyrate (C_4).
c. In step 3 of ketogenesis, chain cleavage of HMG-CoA (C_6) produces acetyl CoA (C_2) and acetoacetate (C_4).
d. In step 2 of ketogenesis, a condensation reaction occurs between acetoacetyl CoA (C_4) and acetyl CoA (C_2) to produce HMG-CoA (C_6).

25.71 a. Acetoacetyl CoA is a C_4 species involved in ketogenesis.
b. 3-Hydroxy-3-methylglutaryl CoA is a C_6 species involved in ketogenesis.
c. Acetoacetate is a C_4 species involved in ketogenesis.
d. β-Hydroxybutyrate is a C_4 species involved in ketogenesis.

25.73 Acetoacetate and succinyl CoA are reactants; acetoacetyl CoA and succinate are products.

25.75 Certain abnormal metabolic conditions (diabetic conditions, fasting, or a diet high in fat and low in carbohydrates) lead to accumulation of ketone bodies in blood and urine, a condition called ketosis.

25.77 a. Acyl CoA is a reactant in step 1 and a product in step 4 in the β-oxidation pathway.
b. Enoyl CoA is the product of step 1 in the β-oxidation pathway.
c. Acety CoA is associated with both the β-oxidation pathway and ketogenesis.
d. β-Hydroxybutyrate is associated with ketogenesis; it is the product of the reduction of acetoacetate in step 4.

25.79 a. In ketogenesis two different condensation reactions occur (in steps 1 and 2), but not in the β-oxidation pathway.
b. Both the β-oxidation pathway and ketogenesis have four distinct steps.
c. Thiolysis occurs in the β-oxidation pathway at step 4.
d. A hydrogenation reaction occurs in step 4 of ketogenesis, but does not occur in the β-oxidation pathway.

25.81 a. False. Lipogenesis takes place in the cell cytosol; the β-oxidation pathway occurs in the mitochondrial matrix.
b. False. Different enzymes are involved in the two processes.
c. False. The carrier for β-oxidation pathway intermediates is CoA; lipogenesis intermediates are bonded to ACP (acyl carrier protein).
d. False. Lipogenesis is dependent on the reducing agent NADPH; the β-oxidation pathway is dependent on the oxidizing agents FAD and NAD^+.

25.83 a. False. Both ACP and CoA contain a 2-ethanethiol structural subunit.
b. False. Both ACP and CoA contain a pantothenic acid structural subunit.
c. True.
d. False. ACP is a much larger molecule than CoA.

25.85 a. matrix b. cytosol c. cytosol d. cytosol

25.87 a. malonyl CoA b. acetyl ACP c. malonyl CoA d. acetyl ACP

25.89 a. Step 2 b. Step 3 c. Step 1 d. Step 4

25.91 a. acetoacetyl ACP b. crotonyl ACP c. crotonyl ACP d. acetoacetyl ACP

25.93 a. condensation b. hydrogenation c. dehydration d. hydrogenation

25.95 Production of unsaturated fatty acids uses an oxidation step in which hydrogen is removed from a fatty acid and combined with O_2 to form water.

25.97 The biosynthesis of a C_{14} saturated fatty acid requires:
a. 6 turns of the biosynthesis cycle (each turn adds C_2; the first turn produces a C_4 compound).
b. 6 malonyl ACP molecules (malonyl ACP is the source of the C_2 group added in each turn).
c. 6 ATP bonds (each malonyl CoA molecule requires an ATP for formation).
d. 12 NADPH (two hydrogenation steps in the elongation process, Step 2 and Step 4, require one NADPH molecule each).

25.99 a. The β-oxidation pathway, ketogenesis, and the chain elongation phase of lipogenesis all have four distinct reaction steps.
b. Two different hydrogenation reactions occur (steps 2 and 4) in the chain elongation phase of lipogenesis.
c. Two different dehydrogenation reactions occur (steps 1 and 3) in the β-oxidation pathway.
d. A thiolysis reaction occurs (step 4) in the β-oxidation pathway.

25.101 a. The carniting shuttle system is used in the β-oxidation pathway to transport fatty acids across the inner mitochondrial membrane.
b. Malonyl ACP is a reactant in lipogenesis; in step 1 malonyl ACP condenses with acetyl ACP to form acetoacetyl ACP.
c. CO_2 is a product lipogenesis; it is given off in the step 1 condensation reaction.
d. Molecular O_2 is sometimes needed in lipogenesis.

25.103 a. Enoyl CoA is the product in step 1 of the β-oxidation pathway.
b. FAD is the oxidizing agent in step 1 of the β-oxidation pathway.
c. β-Hydroxybutyrate is the product in step 4 of ketogcnesis.
d. The first step of glycerol metabolism involves phosphorylation of a primary hydroxyl group of glycerol to give glycerol 3-phosphate.

25.105 a. succinate b. oxaloacetate c. malate d. fumarate

25.107 a. unsaturated acid b. keto acid c. keto acid d. hydroxy acid

25.109 a. Cholesterol can be produced from acetyl CoA in a multi-step process.
b. Acetoacetyl CoA is produced in one step from acetyl CoA in the step 1 condensation of ketogenesis.
c. Malonyl CoA can be produced in one step in the process of ACP complex formation.
d. Pyruvate cannot be produced from acetyl CoA.

25.111 a. Ketogenesis occurs within the mitochondria of a cell.
b. Glycolysis occurs in the cell's cytosol.
c. The citric acid cycle occurs within the mitochondria of a cell.
d. The β-oxidation pathway occurs within the mitochondria of a cell.

25.113 a. Niacin is involved in all four of the processes: the β-oxidation pathway, ketogenesis, lipogenesis, and conversion of ketone bodies to acetyl CoA.
b. Thiamin is not involved in any of these four processes.
c. Pantothenic acid is involved in all four of the processes: the β-oxidation pathway, ketogenesis, lipogenesis, and conversion of ketone bodies to acetyl CoA.
d. Folate is not involved in any of these four processes.

Protein Metabolism — Chapter 26

Solutions to Selected Problems

26.1 In protein digestion
a. protein is denatured in the stomach.
b. trypsin is active in the small intestine.
c. breaking of peptide bonds is completed in the small intestine.
d. hydrochloric acid is secreted in the stomach.

26.3 a. No, it is not correct. Gastrin causes hydrochloric acid secretion.
b. Yes, it is correct.
c. No, it is not correct. Secretin stimulates production of HCO_3^-.
d. Yes, it is correct.

26.5 a. False. Pepsin is the active form of the zymogen pepsinogen.
b. False. Approximately 10% of protein digestion occurs in the stomach.
c. True.
d. False. Secretin is a hormone that stimulates pancreatic production of HCO_3^-.

26.7 The partially digested protein mixture in the stomach is very acidic; the partially digested protein mixture in the small intestine is slightly basic (addition of HCO_3^-).

26.9 a. Gastrin is a hormone secreted by the cells in the stomach's mucosa.
b. Trypsin is an enzyme in the small intestine for the hydrolysis of peptide bonds.
c. HCl is found in the stomach; it denatures proteins and activates pepsinogen.
d. Carboxypeptidase is found in the small intestine; it is an enzyme for the hydrolysis of peptide bonds.

26.11 a. Gastrin is a hormone.
b. Trypsin is a digestive enzyme.
c. HCl is neither a hormone nor a digestive enzyme.
d. Carboxypeptidase is a digestive enzyme.

26.13 a. True.
b. False. The amino acid pool is not in a specific location in the body; free amino acids are present throughout the body.
c. True.
d. False. The amino acid pool is the total supply of free amino acids in the body, both essential and nonessential.

26.15 a. A negative nitrogen balance is produced when protein degradation exceeds protein synthesis.
b. A positive nitrogen balance is produced during pregnancy.
c. A positive nitrogen balance is produced when during convalescence from an emaciating illness.
d. A negative nitrogen balance is produced when there is a protein-poor diet.

26.17 The three major sources of amino acids for the amino acid pool are dietary protein, protein turnover, and biosynthesis of amino acids.

26.19 Use Table 26.1 to classify amino acids as essential or nonessential.
a. Lysine is essential.
b. Cysteine is nonessential.
c. Serine is nonessential.
d. Tryptophan is essential.

26.21 Essential amino acids are those amino acids that cannot be synthesized by the body, and so must be obtained in the diet. The amino acids in Problem 26.19 may be classified as follows:
a. Lysine cannot be synthesized in the body.
b. Cysteine can be synthesized in the body.
c. Serine can be synthesized in the body.
d. Tryptophan cannot be synthesized in the body.

26.23 An amino acid and a keto acid are the two types of reactants in a transamination reaction.

26.25 A keto acid contains both carbonyl and carboxyl functional groups.
a. No, glutamate is not a keto acid.
b. Yes, oxaloacetate is a keto acid
c. No, aspartate is not a keto acid.
d. No, glutarate is not a keto acid.

26.27 a. Glutamate is a C_5 species.
b. Oxaloacetate is not a C_5 species.
c. Asparate is not a C_5 species.
d. Glutarate is a C_5 species.

26.29 Transamination reactions are shown in the structural equations below.

a. Serine and oxaloacetate

$$\mathrm{HO{-}CH_2{-}\underset{}{CH}(\overset{+}{N}H_3){-}COO^-} + \mathrm{{}^-OOC{-}CH_2{-}C({=}O){-}COO^-} \longrightarrow \mathrm{HO{-}CH_2{-}C({=}O){-}COO^-} + \mathrm{{}^-OOC{-}CH_2{-}CH(\overset{+}{N}H_3){-}COO^-}$$

b. Alanine and oxaloacetate

$$\mathrm{CH_3{-}CH(\overset{+}{N}H_3){-}COO^-} + \mathrm{{}^-OOC{-}CH_2{-}C({=}O){-}COO^-} \longrightarrow \mathrm{CH_3{-}C({=}O){-}COO^-} + \mathrm{{}^-OOC{-}CH_2{-}CH(\overset{+}{N}H_3){-}COO^-}$$

c. Glycine and α-ketoglutarate

$$\mathrm{H{-}CH(\overset{+}{N}H_3){-}COO^-} + \mathrm{{}^-OOC{-}CH_2{-}CH_2{-}C({=}O){-}COO^-} \longrightarrow \mathrm{H{-}C({=}O){-}COO^-} + \mathrm{{}^-OOC{-}CH_2{-}CH_2{-}CH(\overset{+}{N}H_3){-}COO^-}$$

d. Threonine and α-ketoglutarate

$$\mathrm{CH_3{-}CH(CH_3){-}CH(\overset{+}{N}H_3){-}COO^-} + \mathrm{{}^-OOC{-}CH_2{-}CH_2{-}C({=}O){-}COO^-} \longrightarrow \mathrm{CH_3{-}CH(CH_3){-}C({=}O){-}COO^-} + \mathrm{{}^-OOC{-}CH_2{-}CH_2{-}CH(\overset{+}{N}H_3){-}COO^-}$$

26.31 A transamination reaction involves the interchange of the amino group of an α-amino acid with the keto group of an α-keto acid.

a. Yes, oxaloacetate can be converted to asparatate by transamination.
b. Yes, aspartate can be converted to oxaloacetate by transamination.
c. No, aspartate cannot be converted to glutamate by transamination.
d. No, oxaloacetate cannot be converted to α-ketoglutarate by transamination.

26.33 An amino acid and a keto acid are the two types of reactants in a transamination reaction.

a. No, oxaloacetate and α-ketoglutarate cannot be the reactants in a transamination reaction; both are keto acids.
b. Yes, glutamate and oxaloacetate can be the reactants in a transamination reaction.
c. No, glutarate and glutamate cannot be the reactants in a transamination reaction; there is no α-keto acid.
d. No, oxaloacetate and succinate cannot be the reactants in a transamination reaction; there is no amino acid.

26.35 The two α-keto acids that are usually reactants in transamination reactions are pyruvate and α-ketoglutarate.

26.37 The purpose of the transamination phase of protein catabolism is to collect amino groups from many amino acids into one amino acid.

26.39 Pyridoxal phosphate, a coenzyme produced from pyridoxine, is an integral part of the transamination process; the amino group of an amino acid is transferred first to the pyridoxal phosphate and then to an α-keto acid.

26.41 In the further processing of glutamate via oxidative deamination:

a. The other two reactants are NAD^+ and H_2O.
b. The four products are α-ketoglutarate, NH_4^+, NADH, and H^+.
c. The type of enzyme needed is dehydrogenase.
d. The nitrogen carrier for further reactions is NH_4^+.

26.43 The products of the oxidative deamination of the given reactants are the α-keto compounds below.

a. $^{-}OOC-CH_2-CH_2-\overset{O}{\overset{\|}{C}}-COO^-$

b. $HS-CH_2-\overset{O}{\overset{\|}{C}}-COO^-$

c. $CH_3-\overset{O}{\overset{\|}{C}}-COO^-$

d. 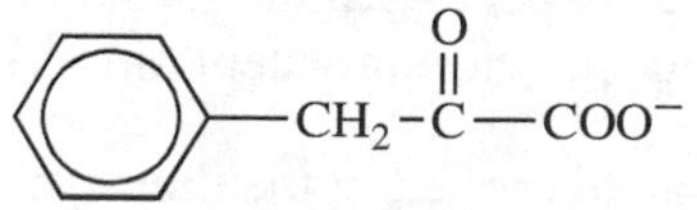 $C_6H_5-CH_2-\overset{O}{\overset{\|}{C}}-COO^-$

26.45 a. α-Ketoglutarate is a product in both oxidative deamination reactions and in the transamination reaction of glutamate.

b. Glutamate is a reactant in both transamination reactions and oxidative deamination to give α-ketoglutarate.
c. Glutamate dehydrogenase is the enzyme involved in the oxidative deamination reaction of glutamate to give α-ketoglutarate.
d. NH_4^+ is one of the products of the oxidative deamination of glutamate.

26.47 The two nitrogen-containing entities that are further processed in the urea cycle are aspartate and NH_4^+.

26.49 Of the two urea cycle fuels, aspartate enters the urea cycle directly and NH_4^+ enters indirectly.

26.51 The desired end product of urea cycle operation is urea.

26.53 The chemical structure of urea is: $H_2N-\overset{\overset{\displaystyle O}{\|}}{C}-NH_2$

26.55 A carbamoyl group is an amide group: $-\overset{\overset{\displaystyle O}{\|}}{C}-NH_2$

26.57 The chemical compounds needed to produce carbamoyl phosphate are NH_4^+, CO_2, H_2O, and ATP.

26.59 The two standard amino acids that participate in the urea cycle are aspartate and arginine.

26.61 Four nitrogen atoms participate in the urea cycle; two of them are removed as urea.
a. Ornithine is an N_2 compound.
b. Citrulline is an N_3 compound.
c. Aspartate is an N_1 species.
d. Arginosuccinate is an N_4 species.

26.63 a. Citrulline (Step 1) is encountered before arginine (Step 3).
b. Ornithine (Step 1) is encountered before aspartate (Step 2).
c. Argininosuccinate (Step 2) is encountered before fumarate (Step 3).
d. Carbamoyl phosphate, a reactant in Step 1, is encountered before citrulline, a product in Step 1.

26.65 a. Asparatate enters the urea cycle in step 2.
b. A condensation reaction occurs in step 2 of the urea cycle.
c. Ornithine is a product in step 4 of the urea cycle.
d. The reaction $N_4 \rightarrow N_2 + N_2$ occurs in step 4 of the urea cycle.

26.67 Step 1 of the urea cycle occurs in the mitochondrial matrix.

26.69 The energy expended in the formation of a molecule of urea from ammonium ion and oxaloacetate is the equivalent of 4 ATPs.

26.71 The fumarate formed in the urea cycle enters the citric acid cycle, where it is converted to malate and then to oxaloacetate, which is then converted to aspartate through transamination.

26.73 a. Oxaloacetate is associated with transamination; it reacts with an amino acid to produce aspartate.
b. Arginine is involved in the urea cycle; hydrolysis of arginine produces urea.
c. H_2O is involved in oxidative deamination and in the urea cycle.
d. ATP is involved in the urea cycle; ATP is used in the condensation reaction between citrulline and aspartate.

26.75 a. Carbamoyl phosphate is a N_1 species.
b. Glutamate is a N_1 species.
c. Urea is a N_2 species.
d. Citrulline is a N_3 species.

26.77 Each of the 20 amino acid carbon skeletons undergoes a different degradation sequence; there are only seven degradation products of these sequences. Four of the products are also intermediates in the citric acid cycle: α-ketoglutarate, succinyl CoA, fumarate, oxaloacetate.

26.79 Figure 26.9 shows the fates of the carbon skeletons of amino acids.
a. Leucine is metabolized to acetoacetyl CoA and acetyl CoA.
b. Isoleucine is metabolized to succinyl CoA and acetyl CoA.
c. Aspartate is metabolized to fumarate and oxaloacetate.
d. Arginine is metabolized to α-ketoglutarate.

26.81 Figure 26.9 shows the fates of the carbon skeletons of amino acids.
a. Leucine is ketogenic.
b. Isoleucine is both ketogenic and glucogenic.
c. Aspartate is glucogenic
d. Arginine is glucogenic.

26.83 A glucogenic amino acid has a carbon-containing degradation product that can be used to produce glucose via gluconeogenesis.

26.85 Figure 26.10 gives a summary of the starting materials for the biosysnthesis of the 10 nonessential amino acids. These five starting materials are pyruvate, α-ketoglutarate, 3-phosphoglycerate, oxaloacetate, and phenylalanine.

26.87 The starting materials for the biosynthesis of these nonessential amino acids are:
a. Pyruvate is the starting material for alanine.
b. 3-Phosphoglycerate is the starting material for serine.
c. 3-Phosphoglycerate is the starting material for cysteine.
d. α-Ketoglutarate is the starting material for proline.

26.89 The four nonessential amino acids that can be biosynthesized using glycolysis intermediates are serine, cysteine, glycine, and alanine.

26.91 Globin is the protein portion of the conjugated protein hemoglobin. During the breakdown of hemoglobin, the globin protein is hydrolyzed to amino acids, which become part of the amino acid pool.

26.93 Heme (the nonprotein portion of hemoglobin) contains four pyrrole groups joined together in a ring, with an iron atom at the center. Degradation of heme begins with a ring-opening reaction in which one carbon atom is lost, and the iron atom is released; the product of this reaction is biliverdin.

26.95 The order in which these four substances appear during the catabolism of heme is: biliverdin, bilirubin, bilirubin diglucuronide, urobilin.

26.97 The characteristic yellow color of urine is due to the bile pigment urobilin.

26.99 Excess bilirubin in the blood causes the skin and the white of the eyes to acquire a yellowish tint known as jaundice. Jaundice occurs when the balance between degradation of heme to form bilirubin and removal of bilirubin from the blood by the liver is upset.

26.101 a. CO is produced at the same time as biliverdin.
b. Bilirubin is associated with the condition called jaundice.
c. Molecular O_2 is a reactant in the production of biliverdin.
d. Stercobilin is a bile pigment that has a brownish color.

26.103 a. Citrulline is an intermediate in the urea cycle.
b. In oxidative deamination, glutamate reacts to form α-ketoglutarate.
c. In the second step of heme degradation, biliverdin is converted to bilirurubin.
d. Ammonium ion produced by oxidative deamination is then incorporated into another molecule (carbamoyl phosphate), which then enters the first step of the urea cycle.

26.105 a. False. Two of the twenty standard amino acids contain sulfur (cysteine and methionine).
b. True.
c. False. Pyruvate is the degradation product from the carbon skeleton; H_2S is the sulfur-containing product from the degradation of the amino acid.
d. False. An acetyl-CoA molecule is the activating agent in the first step of the conversion of serine to cysteine.

26.107 In the process called sulfate assimilation:
a. The two reactants in the first step of the process are SO_4^{2-} and ATP.
b. The sulfur-containing reactant in the second step of the process is APS.
c. The nonsulfur-containing product in the third step of the process is 3′-phosphoadenosine 5′-phosphate.
d. The sulfur-containing product in the fourth step of the process is S^{2-}.

26.109 The connecting link from the urea cycle to the citric acid cycle is fumarate.

26.111 The numerous metabolic pathways of carbohydrates, lipids, and proteins are linked by various compounds that participate in more than one pathway. During protein degradation, amino acid carbon skeletons are degraded to acetyl CoA or acetoacetyl CoA; these degradation products are converted by ketogenesis to ketone bodies.

26.113 When dietary proteins produce amino acids in amounts that exceed body needs, the excess amino acids are degraded and converted to body fat stores.

26.115 a. Niacin is involved as a cofactor in the process of oxidative deamination and in carbon skeleton degradation to non-CAC intermediates.
b. Folate is involved as a cofactor in carbon skeleton degradation to non-CAC intermediates.
c. Biotin is involved as a cofactor in carbon skeleton degradation to CAC intermediates.
d. Vitamin B_6 is involved as a cofactor in transamination, carbon skeleton degradation to CAC intermediates, and in carbon skeleton degradation to non-CAC intermediates.

CPSIA information can be obtained
at www.ICGtesting.com
Printed in the USA
FFHW02n0149310818
48166567-51892FF